Emulsion Polymers and Emulsion Polymerization

Emulsion Polymers and Emulsion Polymerization

David R. Bassett, Editor
Union Carbide Corporation

Alvin E. Hamielec, Editor
McMaster University

Based on a symposium cosponsored by the Divisions of Organic Coatings and Plastics Chemistry and Polymer Chemistry at the Second Chemical Congress of the North American Continent (180th ACS National Meeting), Las Vegas, Nevada, August 25–29, 1980.

ACS SYMPOSIUM SERIES 165

AMERICAN CHEMICAL SOCIETY
WASHINGTON, D. C. 1981

Library of Congress CIP Data

Emulsion polymers and emulsion polymerization.
(ACS symposium series, ISSN 0097-6156; 165)

"Symposium on Emulsion Polymerization"—Pref.
Includes bibliographies and index.

1. Addition polymerization—Congresses.
I. Bassett, David R., 1939- . II. Hamielec, Alvin E., 1935- . III. Symposium on Emulsion Polymerization (1980: Las Vegas, Nev.). IV. American Chemical Society. Division of Organic Coatings and Plastics Chemistry. V. American Chemical Society. Division of Polymer Chemistry. VI. Series.

QD281.P6E47 547'.28 81-10823
ISBN 0-8412-0642-2 AACR2
ASCMC 8 165 1–605 1981

PRINTED IN THE UNITED STATES OF AMERICA

ACS Symposium Series

M. Joan Comstock, *Series Editor*

FOREWORD

The ACS SYMPOSIUM SERIES was founded in 1974 to provide a medium for publishing symposia quickly in book form. The format of the Series parallels that of the continuing ADVANCES IN CHEMISTRY SERIES except that in order to save time the papers are not typeset but are reproduced as they are submitted by the authors in camera-ready form. Papers are reviewed under the supervision of the Editors with the assistance of the Series Advisory Board and are selected to maintain the integrity of the symposia; however, verbatim reproductions of previously published papers are not accepted. Both reviews and reports of research are acceptable since symposia may embrace both types of presentation.

CONTENTS

PREFACE

The symposium on Emulsion Polymerization held at the National Meeting of the American Chemical Society in Las Vegas followed a similar symposium held five years earlier at an ACS meeting in April, 1975. The proceedings of the 1975 symposium, organized by I. Piirma and J. L. Gardon, were subsequently published as Volume 24 in the ACS Symposium Series. The remarkable growth in emulsion polymerization technology was noted at that time. This growth has not only continued during the succeeding half-decade, but has accelerated. The present volume documents recent advances made by an international body of scientists working on a wide range of fundamental and applied problems in emulsion polymerization and emulsion polymers.

In planning the program, it was felt that tutorial lectures by recognized authorities would be an appropriate way of reviewing the state of the art and also an efficient means of introducing each of the diverse areas of emulsion polymerization in preparation for the contributed papers to follow. Invited lectures (Chapters 1–6) included a treatment of several important aspects of emulsion polymer particles: their nucleation, their growth and stabilization, and their characterization by light scattering. A discussion of the synthesis and study of model polymer colloids revealed their wide use in diverse applications. Also included were lectures on molecular weight development and on the design and operation of continuous latex reactors.

One area of intense current effort is the design and control of molecular and particle structure in emulsion polymerization. At least eight of the chapters touch on this subject. Recent advances include the use of novel reaction pathways to achieve nonuniform as well as uniform structures when desired. Light scattering, electron microscopy, and NMR techniques have been utilized to probe particle morphologies and molecular architecture. Reaction kinetics has always been a central concern in emulsion polymerization investigations. The effects of chain entanglement, monomer diffusion, and nonreactive components on kinetics have been treated. The engineering aspects of emulsion polymerization are closely aligned with the foregoing subjects since the reaction process has such a strong effect on the properties of emulsion polymers. Of particular interest are methods for control of reactors and for real-time monitoring of reactor dynamics.

This collection of papers indicates the rapid advancement in our understanding and use of the many facets of emulsion polymerization and emulsion polymers. It also suggests likely advances during the next five years: further development of analytical techniques for investigating the surface chemistry, internal structure, and physical properties of emulsion polymers; inverse emulsion polymerization of water-soluble monomers; development of a better understanding of coagulation—for particle size control as well as for the prevention of reactor fouling; a more complete understanding of high-conversion kinetics; and an increase in on-line instrumentation to permit the use of continuous reactors to produce high-quality emulsion polymers.

In addition to the support of the cosponsoring Divisions of Organic Coatings and Plastics Chemistry and Polymer Chemistry, we gratefully acknowledge financial contributions to the symposium from the following corporations: Air Products and Chemicals, Diamond Shamrock, Dow Chemical, Eastman Kodak, Nalco Chemical, PPG Industries, SOHIO, and Union Carbide. We must also acknowledge the referees who reviewed the papers and made many helpful suggestions to the authors.

D. R. Bassett
Technical Center
Union Carbide Corporation
South Charleston, West Virginia 25303

A. E. Hamielec
Department of Chemical Engineering
McMaster University
Hamilton, Ontario, Canada

April 17, 1981

1

Latex Particle Nucleation and Growth

ROBERT M. FITCH

Department of Chemistry and Institute of Materials Science, The University of Connecticut, Storrs, CT 06268

Emulsion polymerization was first reported in 1927 by Ray P. Dinsmore of the Goodyear Tire and Rubber Co. (1). He made aqueous emulsions of various methyl-butadienes using oleic acid salts and casein or egg albumin as emulsifiers, and allowed them to stand at 50°-70°C for six months to polymerize. Although not what we would consider a practical process today, it led over the next two decades to an entire industry. It was natural that the term "emulsion polymerization" should be applied: one started with a liquid emulsion and ended with a polymer emulsion. As it turned out, the appellation has been an unfortunate one in that, except in rare circumstances, the mechanism does not involve polymerization in emulsified monomer droplets. It was observed by both McBain and Harkins independently in 1932 that polymeric latex particles could be formed in the absence of emulsifying agents from monomers of low water-solubility (2, 3). Since the particles were much smaller than the droplets of monomer which may have been formed by agitation, it was concluded that homogeneous nucleation of the polymer particles had occurred. In 1937 Fikentscher showed that even in the presence of micellar emulsifier the "aqueous phase" was the principal locus of polymerization, not the emulsified monomer droplets (4). Heller and Klevens reported in 1943-1945 on their quantitative studies on the strong influence of emulsifier concentration on the number of polymer particles formed both below and above the critical micelle concentration (CMC) (5). Two years later Harkins published the results of a series of quantitative investigations on the polymerization of styrene and isoprene both in the absence and presence of monomer-swollen soap micelles (6). Harkins observed that the rates of polymerization were much greater when micelles were present and therefore proposed that these were the principal locus of particle formation. This led to the landmark work of Smith and Ewart, published the following year, which presented quantitative theories for the prediction of the absolute particle concentration, N, and for the absolute rate of polymerization (7). All of the above studies involved monomers with very low solubilities in water.

0097-6156/81/0165-0001$07.25/0

At about the same time Baxendale, Evans and coworkers published work on a more water-soluble monomer, methyl methacrylate (MMA), both in the absence and presence of a cationic surface-active agent, and concluded that nucleation of the polymer particles was by a homogeneous mechanism, in which soap micelles played no role (8). Thus, two contending schools of thought were established concerning the mechanism of particle formation, one embracing the theory of homogeneous nucleation, the other, the micellar, or heterogeneous, mechanism. The latter had much greater success for many years, probably because of its applicability to an industry which was well developed (synthetic rubber), and because it received experimental support in the work of several investigators (9). This was, however, limited almost entirely to styrene and a very few comonomers.

It is the purpose of this paper to present the case for the homogeneous nucleation school, which will be seen to apply to a large number of monomers more water-soluble than styrene, as well as to resolve the differences between the two schools. The nucleation period during an emulsion polymerization will be taken as the time during which the number of particles is changing, either increasing or decreasing. This involves processes which are not strictly nucleation, but which are important to the prediction of the final particle size and size distribution. The growth of particles will be discussed only to the extent that it affects the mechanism or kinetics of particle formation.

General Nucleation Theory

Colloidal particles are formed from a homogeneous medium by the clustering of smaller units to form "embryos" of various sizes. In the case of polymers in solution the aggregates may be of repeat units of the same or different molecules. The specific surface area of such embryos is very great, and its creation requires the expenditure of an amount of energy equal to the area, A, times the interfacial free energy, γ:

$$\Delta G_s = A \cdot \gamma \text{ joules.} \tag{1}$$

This energy is supplied by the condensation of hydrophobic units from the aqueous medium and by the energy of polymerization, ΔG_v and ΔG_p, both expressed per unit volume of polymer. For isotropic, amorphous polymer particles of radius r, the free energy of formation of an embryo will be

$$\Delta G_f = -\tfrac{4}{3}\pi r^3 \; (\Delta G_v + \Delta G_p) + 4\pi r^2 \gamma . \tag{2}$$

When the size of the particles is very small, the last term dominates, ΔG_f is positive and the embryos are unstable, i.e. they will disaggregate. When r reaches a critical size, r*, the slope of $d\Delta G_f/dr$ becomes negative, so that further growth is favored. Thus chains, initiated by free radicals generated in the continuous phase, must grow until they reach a critical size such that when they separate as a new phase, they will form primary particles of a size greater than r*. This is represented schematically in Figure 1, in which the free energy of formation, ΔG_f, is plotted against embryo size, r (10). The maximum in the curve, ΔG_f^*, may be taken as an energy of activation for the formation of primary colloidal particles.

If one provides the system with preformed nuclei, for example in the form of "seed" latex particles, monomer droplets (in sufficient number), dust, or monomer-swollen soap micelles, the ΔG_s will have been provided and the system progresses on a downward slope (dashed curve in Figure 1) from the beginning. This constitutes heterogeneous nucleation. Since it is energetically favored, it will tend to occur whenever such conditions obtain.

The rates at which these processes occur will not only depend upon ΔG_f^* and the rate of generation of free radicals, R_i, but also upon the concentration of monomer in the aqueous phase. Thus monomers such as styrene, isoprene and octyl acrylate, which have low solubilities in water, nucleate and grow very slowly during the early stages of an emulsion polymerization, whereas more soluble monomers such as methyl acrylate, vinyl acetate and acrylonitrile give rapid rates. In the presence of surfactant micelles the less soluble monomers are solubilized (11), so that they nucleate and grow more rapidly.

As mentioned earlier, this paper will stress homogeneous nucleation not only because of its applicability to a broad range of monomers and to systems both aqueous and organic, but also because several features are applicable to systems undergoing heterogeneous nucleation and which were not considered in the original Smith-Ewart theory. To begin with, a brief historical review is in order.

Historical

Baxendale, Evans and coworkers reported in 1946 that the polymerization of methyl methacrylate (MMA) in aqueous solution was characterized by homogeneous solution kinetics, i.e. where mutual termination of free radicals occurred, in spite of the fact that the polymer precipitated as a separate phase. Increases in the rates of polymerization upon the addition of the surfactant cetyl trimethyl ammonium bromide (CTAB) were attributed to the retardation of the rate of coagulation of particles, which was manifested in a reduction in the effective rate constant for mutual termination, k_t.

In 1952 W. J. Priest, in an important paper, laid out all of the basic qualitative features of the theory of homogeneous nucleation in emulsion polymerization as it is known today (12). This was based upon his studies of particle size distributions in vinyl acetate polymerization initiated by potassium persulfate ($K_2S_2O_8$) in the presence of varying amounts of different stabilizers and inhibitors at several temperatures. Priest proposed that (1) "polymerization in solution is the initial process";

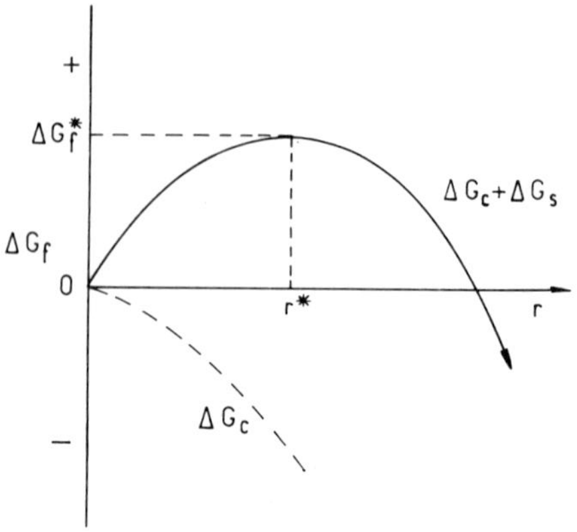

Figure 1. Free energy of nucleus formation as a function of size

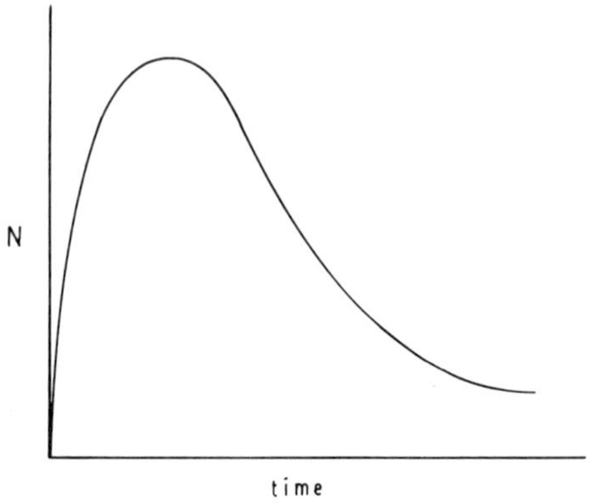

Figure 2. Particle formation kinetics according to Priest (12)

(2) "with the growth of a given polymer chain, a point is reached at which the polymer is no longer soluble" and nucleates to a primary particle; (3) "the number of primary particles would be sensibly equivalent to the number of chains initiated in solution"; (4) the number of particles which would otherwise be formed may be reduced "by combination of incompletely developed chains with polymer droplets before the former are adequately stabilized"; (5) the number of particles may be further reduced by coagulation, depending upon "the quantity and efficiency of the material employed as an emulsion stabilizer"; (6) "the ultimate particle size---is presumably a function of the relative number of sulfate groups per particle", introduced by the initiator and stabilizing the particles against coagulation because of their charge; (7) "the primary particle size would be of the order of 10^{-20} cm^3" in volume, or 1.3 nm in radius; and finally (8), that "almost all of the polymerization (99.9%) occurs within the swollen polymer droplets, and only a maximum of 0.1% may be identified with primary particles". On the basis of these points Priest drew a figure for the kinetics of particle formation in a typical vinyl acetate polymerization. This is shown in Figure 2.

Ten years later Napper and Alexander, in studying the kinetics of vinyl acetate polymerization in the presence of anionic, cationic and nonionic emulsifiers arrived at the same conclusions as Priest's, although they did not cite his work (13). They also observed an acceleratory effect of added emulsifier like that found by Baxendale et al. (8), but they seemed unaware of that work as well. They showed that when the charge on primary particles (due to initiator fragments) was opposite to that of the emulsifier, the rate of polymerization was slower than that in the absence of emulsifier. This was presumably due to the greater instability of the colloid formed and the consequent production of fewer polymerizing centers.

Peggion, Testa and Talamini proposed a mechanism identical to that of Priest for the formation of particles in vinyl chloride emulsion polymerization (14). This was based on an extensive study of the kinetics of polymerization as a function of concentrations of various emulsifiers above and below the CMC.

In 1965 Dunn and Taylor confirmed the theory for vinyl acetate polymerization (15), and proposed, in the light of the presumed importance of rapid coagulation during the earliest stages of reaction, that the "DLVO" theory for colloid stability (16) be applied. Fitch proposed a kinetic basis for a quantitative theory and observed that for observation of particle formation kinetics, "fast" reaction techniques must be used because "particle formation occurs in a matter of seconds or even less (17)".

That growing oligomeric radicals in solution might be captured (Priest's point 4 above) with different efficiencies by different types of particles was first put forward by Parts, Moore and Watterson (18). They proposed that polymer particles must be more efficient than micelles, when both coexisted, in order to make their theory agree with experiments. Although they were concerned with heterogeneous nucleation, this concept of the efficiency of radical capture turns out to be of great importance in many cases, as we shall see later.

The polymerization of MMA in aqueous media was re-investigated by Fitch, Prenosil and Sprick in 1969 (19). They reaffirmed the findings of Baxendale and coworkers (8) that homogeneous kinetics obtained in the early stages of polymerization, but they found heterogeneous kinetics in the later stages, presumably when particle coagulation stopped. They also proposed that there was "thermodynamic control" of particle nucleation, implying a critical degree of polymerization of radicals in solution prior to nucleation, determined by the free energy considerations discussed in the preceeding section.

In 1970 Dunn and Chong carefully measured surface electrical charge densities on a series of polyvinyl acetate latexes in order to calculate rates of coagulation by application of the DLVO theory (20). They had to assume that all of the sulfate groups analyzed by solution methods would be found on the surface of the particles, thereby obtaining probably unrealistically high surface charge densities. From these were calculated surface potentials after applying a Stern correction based on a fixed layer thickness equal to the diameter of a water molecule. Also, by assuming 100% efficiencies of placement of sulfate groups from initiator fragments (supported by experiments) and from knowledge of both initiator decomposition and polymerization kinetics, they could estimate surface electrical potentials at various percent conversions. Values of the Fuchs stability factor*, W, were then estimated in order to obtain relative coagulation rates. Dunn and Chong concluded that, in the absence of emulsifier, primary particles coagulated at approximately the Smoluchowski fastest rate*, and that as they accumulated charge by coagulation and capture of charged oligomeric radicals, they would become more stable. At about 5% conversion coagulation would stop. With added emulsifier, of course, stability would be obtained earlier in the reaction. They also made the important observation that if new primary particles were nuclested during later stages of the polymerization (beyond ∿5% conversion), they would be unstable and tend to coagulate with the larger particles already present. Thus was raised the possibility of continuous nucleation of particles throughout the reaction even when the total number, N, remains constant. Although the authors did not point this out, their theory also explains how a new generation of smaller particles may be obtained by the addition of sufficient emulsifier at some later stage to stabilize the new nuclei as they are formed.

*These terms are discussed in greater detail in the section on coagulation.

Quantitative Prediction of the Particle Number

The first attempt to formulate a homogeneous nucleation theory to predict the absolute number concentration of particles, N, was made by Fitch and Tsai in 1970 (21). This was supported by a large number of experiments on the polymerization of MMA. It was further developed the following year (22), and was based primarily on the scheme of Priest (12) with an idea from Gardon (9d). The latter suggested that the rate of capture of oligomeric radicals in solution by pre-existing particles, R_c, should be proportional to the collision cross-section, or the square of the radius of the particles, r^2. This has been called the "collision theory" of radical capture. In 1975 Fitch and Shih measured capture rates in MMA seeded polymerizations and came to the conclusion that R_c was proportional to the first power of the radius, as would be predicted by Fick's theory of diffusion (23). In his book, K. J. Barrett also pointed out that diffusion must govern the motions of these species in condensed media (10).

The theory says that the number of particles increases at a rate which is equal to the rate of generation of free radicals, R_i, (Priest's point 3), reduced by the rate at which the radicals are captured by particles, R_c, (Priest's point 4). Coagulation will further reduce the number of particles at a rate R_f (Priest's point 5). If the mutual termination of radicals in solution is neglected, then

$$\frac{dN}{dt} = R_i - R_c - R_f. \tag{3}$$

This assumes that only individual radicals nucleate ("self-nucleation") primary particles. If aggregation of radicals is involved, then the first term on the right of Equation 3 should be modified to bR_i, where b is the reciprocal of the aggregation number in a primary particle. The absolute number of particles is obtained by integration:

$$N = \int_0^t (R_i - R_c - R_f)\,dt. \tag{4}$$

The individual terms must be evaluated in order to obtain numerical values for N. In the simplest case the rate of coagulation, in the presence of adequate stabilizer, will be equal to zero. There is some question as to whether this can be accomplished in practice because of the extreme instability of primary particles which are 1 to 2 nm in size (16, 20). The rate of initiation may be obtained from literature values or from independent measurements. The initiator efficiency, f, must also be obtained. For example, in the thermal decomposition of persulfate ion the rate of initiation is given by

$$R_i \equiv \frac{dM_o^{\bullet}}{dt} = -2k_d^{*} fI \tag{5}$$

where $M_o^{\bullet}$ and I are concentrations of free radicals and initiator, respectively, and k_d^* is the specific rate constant.

The rate of capture is (23)

$$R_c = \frac{4\pi D_{op} C_s r r_h N}{(r_h - r)} , \qquad (6)$$

where D_{op} is the average mutual diffusion coefficient of oligoradical and polymer particle, C_s is the steady state concentration of radicals in solution, r and r_h are the radii of the particle and the diffusion barrier surrounding it. It is usually assumed that r_h>>r, a reasonably safe assumption when particles are very small and/or not too numerous. This gives the capture rate proportional to the product of the particle number and the radius:

$$R_c \simeq 4\pi D_{op} C_s N r. \qquad (7)$$

Experimentally one can obtain absolute values of the capture rates by conducting a series of seeded polymerizations in which N and r are varied. If conditions are set so that no coagulation takes place (R_f = 0), then Equations 3 and 7 combined give

$$\frac{dN}{dt} = R_i - 4\pi D_{op} C_s N r. \qquad (8)$$

At low seed concentrations, $R_i > R_c$; dN/dt> 0, and new particles will be formed. As the number of seed particles is increased at a given size r, a condition will be reached at which no new particles are formed, i.e. dN/dt = 0. At this point $R_i = R_c$. If R_i is independently known, R_c is immediately obtained. In seeded polymerizations with MMA, Fitch and Shih (23), and with vinyl chloride, Gatta and coworkers (24) found the rate of capture proportional to N•r, in support of Equation 7. A problem with Equation 7 arises from the fact that the concentration of free radicals in solution, C_s, cannot be determined, so that absolute values of R_c cannot be predicted. There is the further complication that C_s^c and R_c may be interdependent, so that strict proportionality of R_c to N•r may not hold in all cases. A way around this difficulty was given by Ugelstad and will be discussed later.

In unseeded polymerizations, the number of particles increases and they grow in size by polymerization and imibition of monomer, so that both N and r are functions of time. The value of N is computed by iterative numerical integration of Equation 4, whilst the value of r must be calculated from the rate of polymerization: Since the particles are spheres,

$$r = \left(\frac{3v}{4\pi}\right)^{1/3} , \qquad (9)$$

where v is the volume of a particle. The volume of polymer in a particle is given by

$$v_p = V_p/N = \frac{1}{N\rho} \int_0^t Rpdt, \qquad (10)$$

where V_p is the total volume of polymer in a dm^3 of latex, ρ is the polymer density and Rp is the overall rate of polymerization. The particles are swollen with monomer at a rate which is fast compared to the rate of consumption of monomer by polymerization, especially when the particles are very small (25). Thus the problem is treated in terms of swelling equilibrium (26):

$$\ln\phi_m = (\frac{1}{j} - 1)\phi_p - \chi\phi_p^2 - \frac{2\overline{V}_m\gamma}{RTr} \quad , \qquad (11)$$

in which the volume fractions of monomer and polymer in the particle are given by ϕ_m and ϕ_p, j is the degree of polymerization of the polymer (when j is large, $(1/j)\ll 1$), χ is the Flory-Huggins interaction parameter between monomer and polymer, $\overline{V}_m$ is the partial molar volume of monomer, and γ is the interfacial tension between the particle and the surrounding medium. Gardon has provided solutions to Equation 11 in the form of curves for the cases where $j = \infty$ (27), and Ugelstad and coworkers have done the same for small values of j (28). The volume of the monomer-swollen particle is thus

$$V = \frac{V_p}{1 - \phi_m} \quad , \qquad (12)$$

from which the particle radius is obtained upon substitution into Equation 9.

In order to apply these equations, the kinetics of polymerization must be known (Rp in Equation 10). Fitch and Tsai assumed homogeneous kinetics (22) for their MMA polymerizations, based on the earlier results of Fitch, Prenosil and Sprick (19) and of Baxendale et al. (8). Such an assumption is probably not valid for other monomers, especially those less water-soluble than MMA. In such cases the non-steady-state modification of the Smith-Ewart kinetics (9d) may be applied, or a direct numerical calculation of the distribution of radicals among particles by means of Equation 7 may be made.

In all of these considerations concerning the rate of radical capture and its effect on the rate of particle formation we have been hampered by the lack of a mathematical solution for the concentration of oligoradicals in the aqueous phase, Cs. This problem was finally resolved by Ugelstad and Hansen in 1976 (29), and further elucidated in a series of important papers published in 1978-1979 (30, 31, 32).

Absolute Capture Rates

Prior to nucleation the free radicals generated in the continuous phase propagate by reaction with dissolved monomer. Propagation continues stepwise until the radicals have reached the critical chain length for nucleation, jcr, at which point phase separation occurs and primary particles are formed. The rate of primary particle formation may therefore be expressed in terms of the polymerization process (29):

$$\frac{dN_1}{dt} = k_p M_w M^{\bullet}_{jcr-1} \qquad (13)$$

where k_p is the propagation rate constant, M_w is the monomer concentration in the water phase and $M^{\bullet}_{jcr-1}$ is the concentration of free radicals of a size one repeat unit less than the critical value for self-nucleation. During their growth the oligoradicals may mutually terminate in the aqueous phase or be captured by particles already present. Radicals of any given chain-length j may also disappear by addition of another monomer unit. The primary radicals are formed at a rate equal to that of initiation, R_i. Thus there are three aspects to the kinetics involved:

a) Radical appearance:

$$\frac{dM^{\bullet}_o}{dt} = R_i \qquad (14)$$

b) Propagation, termination and capture in the water (or continuous) phase:

$$\frac{dM^{\bullet}_j}{dt} = R_{pj-1} - R_{pj} - R_{twj} - R_{cj} \qquad (15)$$

c) Particle nucleation:

$$\frac{dN_1}{dt} = R_{p(jcr-1)} \qquad (13)$$

Here the symbol R represents a rate, as before, while the subscripts i, p, tw and c represent initiation, propagation, water-phase termination, and capture respectively. Reactions of the primary radicals, $M^{\bullet}_o$, other than propagation, are assumed to be negligible. The subscript 1 on N_1 signifies that only primary particles are being discussed. No consideration of subsequent coagulation is given yet; this is treated more fully in a subsequent section. Equation 13 is solved upon addition of the series of Equations 14, 15, 13, expressing them in terms of appropriate constants and concentrations, and for all values of j from 1 to (jcr-1). An exact solution for the number of particles as a function ot time, t, involves numerical integration of these equations. However the authors show that by assuming a steady state in $M^{\bullet}_j$, the concentration of oligomers, C_s, is obtained. Further, by assuming that C_s is governed solely by termination in the aqueous phase, and by taking an average value for the rate constant for radical capture by particles, $\bar{k}_c$, an analytical solution is obtained (30) for the particle number:

$$N_{1(t)} = \frac{1}{k_1}\left[\{k_1 R_i t + (k_2+1)^{jcr}\}^{1/jcr} - (k_2+1)\right] \qquad (16)$$

$$k_1 \equiv \frac{\bar{k}_c}{k_p M_w} \quad (dm^3)$$

$$k_2 \equiv \frac{(k_{tw} R_i)^{1/2}}{k_p M_w} \quad (-)$$

$$\bar{k}_c = \sum_{j=1}^{jcr-1} \frac{k_{cj} M_j^\bullet}{C_s} \quad (dm^3 s^{-1}) \qquad (17)$$

Equation 16 tends to underestimate the number of particles except during the earliest few seconds of reaction, but serves as an extremely useful predictor for assessing the effect of experimental variables on the number of primary particles formed as a function of time. In Figure 3 are shown some calculations for styrene polymerization in which results from this approximative equation (curves A) are compared to those for the full numerical solution (curves C) at two values for jcr (30). It can be seen that when the oligoradical solubility is reduced (jcr = 10 → 5), the rate of nucleation and final number of particles are greatly increased. This is, of course, in the absence of change in any other variable.

In the discussion thus far it has been assumed that the rate of capture of radicals by particles is equal to the collision frequency calculated on the basis of Fickian diffusion, i.e. that all collisions are irreversible. Fitch and Shih have pointed out that there is an electrostatic repulsion between charged radicals and particles which may retard the rate of capture (23). By application of the DLVO theory they calculated that this effect should be negligible with primary particles, but perhaps important as the particle size and surface charge density increased (e.g. in the presence of adsorbed ionic emulsifier). More recently Hansen and Ugelstad have shown that even upon capture, the oligomeric radical may desorb again, leading to the important concept of reversible capture (29). In order to assess the consequences of these effects, these authors recalculated the primary particle nucleation kinetics assuming a 100-fold reduction in the rate of capture. The results are shown in Figure 4, and should be compared to those in Figure 3. The effect is enormous: an increase of an order of magnitude in the number of primary particles and a nucleation period of over twelve minutes (instead of 1 min) are observed.

To see the effects of monomer solubility and other variables on nucleation rate, calculations using Equation 16 were made for a series of monomers with reactivity similar to that for methyl methacrylate. These are shown in Figure 5. Curve 1 represents

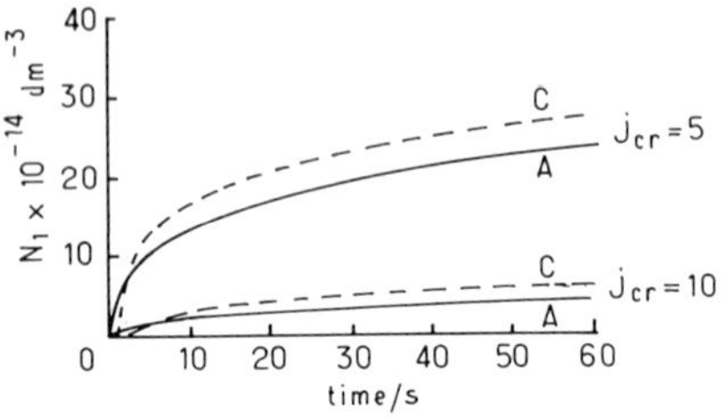

Figure 3. Kinetics of primary particle appearance calculated from full numerical solution of Equations 13–15 (Curve C), and from approximative Equation 16 (Curve A); irreversible capture.

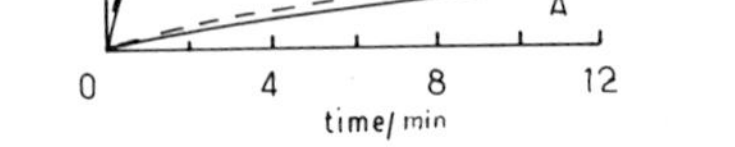

Figure 4. Kinetics of primary particle appearance calculated from full numerical solution of Equations 13–15 (Curve C), and from approximative Equation 16 (Curve A); reversible capture, where $R_{c(rev.)} = R_{c(irrev.)} \times 10^{-2}$.

the results for methyl methacrylate at a concentration in water close to saturation, i.e. 0.10 molar. The other parameters, given in the caption to Figure 5, are taken from the literature. To assess the effect of monomer solubility, the parameters jcr and M_w were varied (the values relative to those in Curve 1 are shown on the right side of the figure). In Curve 3 the rate of initiation was also changed. Curves 4, 5 and 6 show the results for a fictitious higher alkyl homolog, less water-soluble than the methyl ester. When jcr is reduced slightly, more particles are nucleated, as in Figures 4 and 5. When the monomer concentration is sufficiently reduced, however, mutual termination dominates over propagation and fewer particles are formed, as seen in Curves 2 and 6 of Figure 5 (the calculations assume that particles are not produced upon water-phase termination of two oligoradicals, each with j<jcr). This is further illustrated by a small increase in N found when R_i is reduced to one half (curve not shown): even more particles are produced because of the reduction in mutual termination in the water-phase. A still further reduction in R_i, however, leads to no change in the particle number (Curve 3). Thus no variable has a simple, direct effect on nucleation rate. There are optimal levels depending, however, upon the values of the other parameters. These computations, based on Equation 16, have assumed a constant, average value for the specific rate constant for capture:

$$\bar{k}_c = \sum_{j=1}^{jcr-1} \frac{k_{cj} M_j^{\bullet}}{C_s} \qquad (17)$$

As mentioned earlier, k_c may be reduced by an electrostatic interaction, whose magnitude may be given the Fuchs symbol W', but also because of reversibility (30). Because the radical, upon collision with a particle, may add monomer units and/or terminate with a radical already present in the particle, Hansen and Ugelstad applied the theory of Dankwerts for diffusion with reaction to determine the overall capture rate. Under these conditions the diffusion equation of Fick must be modified to read:

$$\frac{dc}{dt} = D \frac{\partial^2 c}{\partial R^2} - kc \qquad (18)$$

where c is the concentration of diffusing species, D is the diffusion coefficient and k is the first order rate constant for the chemical reaction(s) involved:

$$k = k_p M_p + \frac{k^*_{tp} n}{v} \qquad (18')$$

where n is the number of free radicals already present in the particle (n=0 or 1), k_p and k^*_{tp} are the propagation and termination (* represents molecular units) rate constants, M_p is the monomer concentration in the particles and v is the volume of the monomer-swollen polymer particle. An overall effectivity factor for the capture rate can be obtained by coupling Equation 18 with the equation for diffusion flux to a spherical particle under the influence of an interaction potential. The specific rate constant for capture of a radical of size j then becomes:

$$k_{cj} = 4\pi D_{wj} r F_j \qquad (19)$$

where F_j is the rate-reduction factor which takes into account both the degree of reversibility of adsorption and the electrostatic retardation. When the particles and the oligoradicals are both small, the efficiency of capture, F, is very low, on the order of 10^{-6} to 10^{-4}. That is, a radical in solution could collide with 10,000 to 1,000,000 particles before it was irreversibly captured. If the particle contains a radical already, the capture of another from the water phase is much more probable (in Equation 18' $(k^*_{tp}/v) >> k_p M_p$ ordinarily) and nearly all radicals of j-values greater than a few units will be irreversibly captured by the particle. At the outset of a polymerization all primary particles contain radicals, but this rapidly changes until in a few seconds or minutes the average number of radicals per particle, $\bar{n}$, has declined to approximately 1/2. At any instant there will be a mixture of oligoradicals of all values of j. The largest of these, with $j \rightarrow jcr$, will be captured irreversibly regardless of the nature of the particle.

Many experimental parameters contribute to the magnitude of the capture efficiency, F. A detailed discussion is beyond the scope of this paper, but a summary is given in Table I and some further elucidation, below: Hansen and Ugelstad distinguish two cases, A and B, each with two sub-cases, under which approximate solutions for F_i are obtained. Case A obtains for very small particles, produced in the earliest stages of particle formation, while B obtains for larger sizes. Subcase 2 in either case involves large oligoradicals and irreversible capture. The factors given in Table I were calculated for common monomers such as styrene, acrylate and methacrylate esters. Differences in monomer solubility would be accounted for by the choice of values for a_j. Under Case A, the particles are so small that they will not be greatly swollen by monomer (small r in Equation 11 leads to small ϕ_m), and as a result, the value of D_{pj} will be low and k_{tp} will be diffusion-controlled. This leads to a lower value for F than that for a more fully swollen particle, because of less likelihood for termination of the adsorbed oligoradical.

As particles grow, two things occur with opposite effects: (1) they become more highly swollen with monomer (M_p increases), with the result that radicals in temporary contact with particles will increase in j and approach irreversible capture faster; simultaneously the capture rate constant is increasing with r, the radius; and (2) the surface electrical potential is likely to

Table I

The Radical Capture Efficiency: Approximate Solutions

$$k_{cj} = 4\pi D_{wj} r F_j$$

Case	Conditions	Reduction Factor, F_j
A	X_j<<1: very small particles or micelles. Electrostatic effects negligible.	
A-1	Small, soluble oligoradicals (a_j small)	$F_j = \frac{a_j r^2 k}{3D_{wj} E} \simeq \begin{cases} 10^{-6} a_j & \text{for } n=0 \\ 8 \times 10^{-5} a_j & \text{for } n=1 \end{cases}$
A-2	Large oligoradicals	$F_j = 1$
B	X_j>>1:(a) Very large particles containing no radicals (r>1.0μm) (b) Secondary particles containing a radical (r> ∿ 50nm) (c) Small, monomer-starved particles	5
B-1	Small, very soluble oligoradicals	$F_j = \frac{a_j r (kD_{pj})^{1/2}}{ED_{wj}} \gtrsim 0.05 a_j$
B-2	Large oligoradicals	$F_j = 1$

Notes:

For case B, particles with high electrical potential, ψ_o, may have lower values of F_j than those shown.

$X_j \equiv \left(\frac{r^2 k}{D_{pj}}\right)^{1/2}$ a_j is relative solubility of j-mer in organic phase compared to that in water

$E \equiv \exp\left(\frac{e\psi_o}{k'T}\right)$

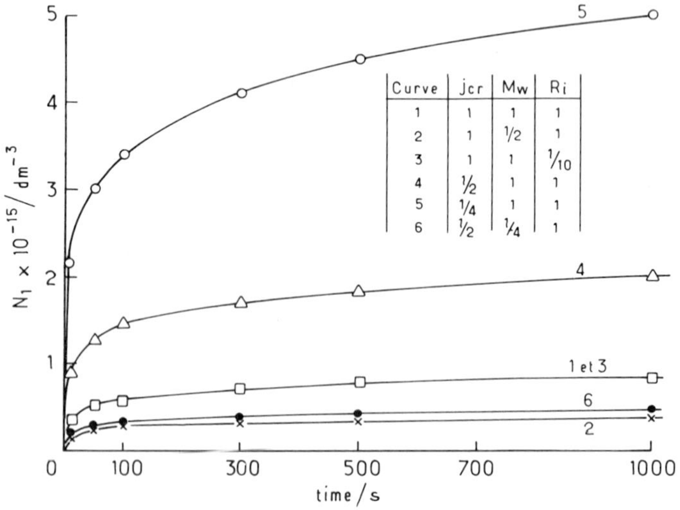

Figure 5. Kinetics of primary particle nucleation calculated from Equation 16 for methacrylate-like monomers. Parameters for Curve 1 are $k_p = 350\ dm^3mol^{-1}s^{-1}$; $D_w = 5 \times 10^{-10}\ m^2s^{-1}$; $r = 2 \times 10^{-9} m$; $\bar{k}_c = 6.3 \times 10^{-18}\ m^3s^{-1}$; $k^*_{tw} = 10^{-17}\ dm^3s^{-1}$; $M_w = 0.10\ mol\ dm^{-3}$; $R_i = 10^{20}\ m^{-3}s^{-1}$; $jcr = 60$. *Parameters for Curves 2–6 relative to those for Curve 1 shown at upper right.*

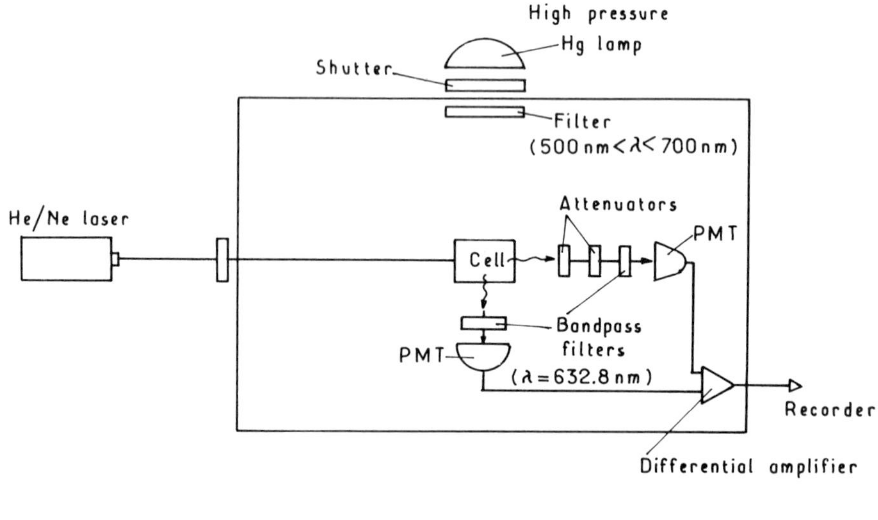

Figure 6. Light scattering apparatus for nucleation kinetics

increase as more charged radicals are captured. This gives higher values of the parameters E and W_i' which lead to reduced capture efficiency. This electrostatic effect will be much more pronounced in the presence of added surface-active agent or copolymerized ionogenic monomer. Higher rates of nucleation, of course, result from lower rates of radical capture. The former effect, which leads to reduced nucleation, dominates during the early stages when the electrostatic effect is negligible. These are the conditions given in Table I, where W'=1 and E ≃ 1.6 - 2.0. As the particles become several tens of nanometers in size, electrostatic repulsions will become significant. Even so, the real value of W' may be lower than that calculated because of the "tunneling effect", first proposed by Fitch and Shih (23), in which the hydrophobic part of an oligomeric free radical in solution may "tunnel under" the electrostatic barrier, which affects only the ionic head of the radical, as it approaches a charged particle. Reduction in surface charge may occur by burial of ionic groups during polymerization and particle coagulation.

Experiments on Nucleation Kinetics

Recently attempts have been made to determine experimentally the kinetics of nucleation during the first few tens of seconds of reaction (33). Since the nuclei will scatter light as they are formed, it is possible to study the reaction by following the Rayleigh scattering intensity as a function of time. Since the total volume fraction of polymer produced in this time is on the order of 10^{-9} to 10^{-8} dm^3 polymer/dm^3 aqueous phase, one is pressed to the very limits of detection of this technique. The method employed involves continuous photo-initiation at a short wavelength of light, and scattering at another, longer wavelength. The apparatus is shown schematically in Figure 6. By having two photomultiplier tube (PMT) detectors situated at 0° and 90° to the incident laser beam, and by putting their output through a differential amplifier, high sensitivity of detection is obtained (34). The initiator used was butanedione-2, 3, which undergoes homolytic cleavage to produce two acetyl radicals upon radiation with violet-blue light from the mercury lamp. The red laser light only is detected by the PMT's because of the bandpass filters in front of each. The reaction cell was filled with a solution of methyl methacrylate (MMA) monomer and initiator in water. The chemical system had to be scrupulously clean, devoid of dust particles and reactive impurities at concentrations above a few parts per billion. Typical results are shown in Figure 7. The conditions for these experiments are given below:

Exp. No.	Code	[MMA]	[SDS]	[butanedione]
1	723	0.10M	0	0.34 M
2	715	0.10M	6.0×10^{-4}M	0.16 M

Also shown in Figure 7 is the theoretical curve based upon Equation 16, and taking into account particle swelling by monomer and growth by polymerization. In contrast to it, both experimental curves rise much more steeply. The experiment run in the absence of surfactant, sodium dodecyl sulfate (SDS), has the steeper slope. This pronounced difference due to surfactant concentration, even at levels as low as the 6 X 10^{-4} molar used here, has always been found in the many experiments which have been run (33).

It is instructive to make a quantitative comparison between the experimental and theoretical curves. The light scattering intensity, expressed as the Rayleigh ratio, is a function of the size and number of particles. If all three reactions shown in Figure 7 polymerize at the same rate, Rp, then at a given time, say 10 seconds, the total volume of polymer formed, V_T, will be the same in all three. Under these conditions, the scattering intensity becomes a direct measure of the relative number of particles:

$$R_{90} \propto \frac{1}{N}$$

Comparison of the two experimental curves with the theoretical one at two times, 10s and 70s, is given in Table II. The approximations made in calculating these ratios of particle

Table II

Relative Particle Numbers from Light Scattering

Curve*	time/s	R_{90}X 10^9/cm^{-1}	$N_{theor.}/N_{exp.}$
theor	10	0.06	--
	70	1.06	--
Exp. 1	10	6.3	105
	70	80.0	76
Exp. 2	10	5.3	88
	70	22.3	21

*Ref. to Figure 7.

numbers were chosen so as to minimize their values. The results force us to the conclusion that coagulation must be an important process from the outset, and that it is retarded by the presence of surface active agent. In the first ten seconds the number of particles is reduced to something like 1/100 the value expected if only nucleation and capture were involved. The rate of coagulation appears to be reduced as the reaction progresses, and more so in the presence of surfactant: the average degree of aggregation, $N_{theor.}/N_{exp.}$, was reduced to 76 and 21 in Exp. 1

and Exp. 2, respectively. Further discussion of the role of coagulation in particle formation is given in the following section.

Coagulation

Throughout the previous discussion mention has been made of the probable role of coagulation in the mechanism of particle formation. Experiments by Fitch and Watson have quantified its role (35). Using a light scattering apparatus similar to that of Fitch and Palmgren shown in Figure 6, but with a single photomultiplier detector, they followed the increase in R_{90} with time as a function of surfactant (SDS) concentration in a series of MMA polymerizations. The experiments differed from those described in the previous section in that photo-initiation was confined to a single burst of light of about 1 ms duration. Under these conditions initiation, growth and termination all occur within a few seconds, so that by the time the first measurement is made at $t \simeq 10s$, essentially all polymerization has stopped ($V_{T(SW)}$ is a constant), and any change observed in R_{90} is due to a change in N (35). Typical results are shown in Figure 8 in which the variable is SDS concentration. The steepest slope, corresponding to the most rapid coagulation rate, R_f, occurs at $[SDS] = 0$. But even at SDS concentrations on the order of 10^{-4} molar, the surfactant has an enormous effect on R_f. To obtain a quantitative measure, one must apply the coagulation theory of Smoluchowski and Fuchs which states that

$$\frac{1}{N_{(t)}} = \frac{1}{N_o}\left(1 + \frac{t}{\tau}\right) \qquad (20)$$

where τ is the half-life, or the time required for the number of particles, $N_{(t)}$, to decrease to one-half of its original value, N_o. When this is combined with the light scattering equation of Rayleigh, and allowance made for a distribution in the particle size because of the stochastic nature of the process, one obtains (37)

$$R_{90(t)} = R_{90(0)}\left(1 + \frac{2t}{\tau}\right) \qquad (21)$$

This is an equation for a straight line, the slope of which is $2R_{90}/\tau$, and describes the plots in Figure 8. From the experimental slopes one may calculate the half-life, τ, and the Fuchs stability factor, W. These values are given in Table III. They confirm that coagulation is an important process in determining particle size and number from the outset. With no surfactant present the particles apparently have a net attraction for each other ($W < 1$), and disappear with a half-life on the order of four milliseconds. As surfactant concentration is increased it

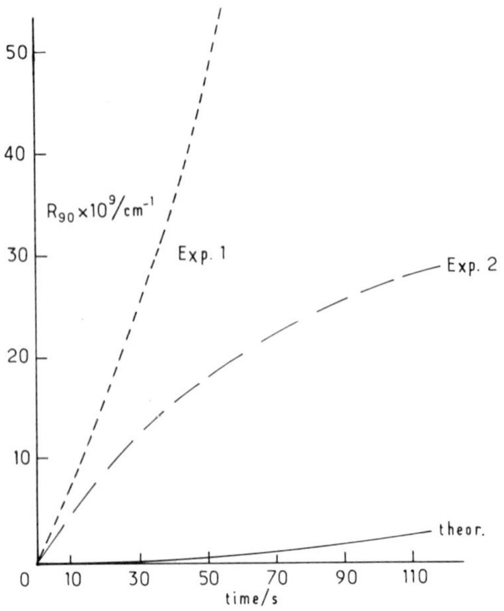

Figure 7. Particle nucleation and growth kinetics determined by laser light scattering: Rayleigh ratio R_{90} vs. time. Exp. 1: no surfactant present; Exp. 2: (SDS) = 6×10^{-4} mol dm^{-3}. Theor: Calculated from Curve 4, Figure 5, considering particle growth by polymerization.

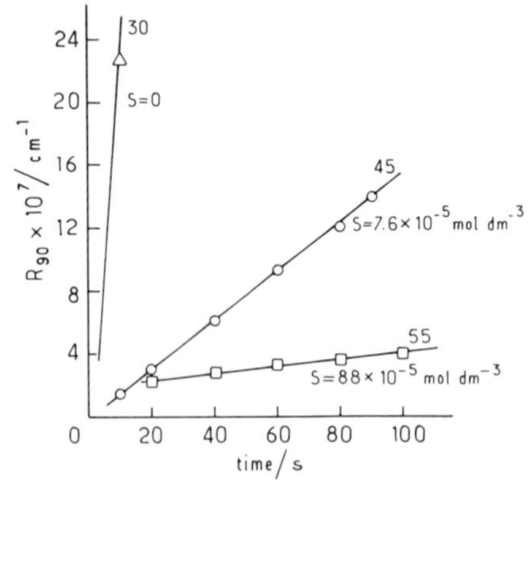

Figure 8. Particle growth by coagulation, flash initiation. Rayleigh ratio R_{90} as a function of time (S represents concentration of SDS). Exp. 30: (SDS) = 0; Exp. 45: (SDS) = 7.6×10^{-5} mol dm^{-3}; Exp. 55: (SDS) = 88×10^{-5} mol dm^{-3}.

rapidly will adsorb onto the particles, building up their surface electrical potential, and causing them increasingly to repel each other.

Table III
Half-lives of Coagulating Primary Particles as a Function of Surfactant Concentration

Exp. No.	(SDS)/mol dm^{-3}	slope/ cm^{-1}s^{-1} X 10^{8}	τ/s	W
30	0	94	0.0043	0.36
45	7.6 X 10^{-5}	1.5	0.30	25
55	88.0 X 10^{-5}	0.21	2.2	180

The so-called "Smoluchowski fast coagulation rate" may be characterized by a half-life independent of particle size (36):

$$\tau = \frac{3\eta}{4Nk'T} \qquad (22)$$

where η is the viscosity of the continuous medium. If the particle concentration, N, is 10^{16} dm^{-3}, the half-life will be 12 milliseconds. Thus, according to the results in Table III, the primary particles coagulate almost 3 times as fast as this in the absence of surfactant (due to van der Waals attractions). In the presence of SDS at 7.6 X 10^{-5} molar and at 88 X 10^{-5} molar the half-life is increased 25 and 180 times, respectively. Dunn and Chong estimated lower values of W ≃ 5 and 100 at higher SDS concentrations of 5 X 10^{-4} and 2.2 X 10^{-3} mol dm^{-3}, respectively, in vinyl acetate emulsion polymerizations (20). The difference in stability or in efficacy of sodium dodecyl sulfate as a stabilizer for the two polymers may be attributed to the difference in the adsorption isotherms of SDS on polyvinyl acetate (PVAc) and PMMA (37). The area occupied per adsorbed SDS molecule is 1.1nm^2 and 0.79nm^2 on PVAc and on PMMA, respectively (37). The surface electrical charge densities would be proportional to the reciprocal of these areas. This, in turn, determines the surface potential, ψ_o (16). The relationship between W and ψ_o at various ionic strengths is dealt with by the DLVO theory (16).

In later stages of the polymerization it is possible to have a situation in which the larger particles bear enough charges to be colloidally stable, but where N•r is not large enough to prevent nucleation. A steady state may soon be reached in which dN/dt = 0. Then, according to Equation 3,

$$R_i = R_c + R_f$$

even though R_f may be positive. Thus, under such conditions, coagulation may occur throughout the polymerization. Under these circumstances the primary particles are continuously generated, but quickly coagulate onto larger particles formed earlier in the reaction, as predicted by Dunn and Chong (20). Thus at very low surfactant concentrations, the particle size distribution may be completely determined by coagulation rather than nucleation, in which case surface electrical potential, particle size and ionic strength become critical factors.

In the presence of increasing amounts of emulsifier, then, the mechanism of particle formation becomes increasingly determined by particle nucleation rather than coagulation. Similar electrostatic effects may be brought about by copolymerization of ionic monomers in the absence of conventional emulsifiers (38, 39), although it has been shown that these may result in the production of considerable water-soluble polyelectrolyte (40, 41). Not considered above is the likely interaction of growing oligomeric radicals with emulsifier molecules (not micelles) prior to nucleation (42). More work in this area is indicated.

Effects of Monomer Solubility

Few systematic studies have been made on the effects of monomer solubility on particle size, in which all other experimental variables have been kept constant. An exception is the work of Sütterlin (43) and coworkers who studied homologous series of acrylate and methacrylate monomers. For example, in the acrylate series the value of k_p/k_t is relatively independent of the number of carbon atoms in the alkyl group (44), whereas jcr and monomer (and oligomer) solubility vary greatly. They studied the effect of SDS concentration on N at constant R_i. All experiments were conducted at a monomer/water ratio of 1/4, so that the aqueous phase was saturated and an external monomer phase existed at the beginning. Typical results are shown in Figure 9, in which the logarithm of the particle number, N, has been plotted against the logarithm of the SDS concentration, over approximately four orders of magnitude, for the emulsion polymerization of ethyl acrylate (EA) and 2-ethylhexyl acrylate (EHA). The latter monomer has a very low solubility in water compared to the former. The differences in the two curves are representative of the entire series as well as many other monomers differing in water-solubility. The soluble monomer, EA, shows an almost linear rise in log N until the SDS concentration exceeds the critical micelle concentration, CMC, when the slope diminishes. In contrast, the poorly soluble monomer, EHA, exhibits relatively low N values until the CMC is passed, after which it has more particles than the EA system. Furthermore, N increases more rapidly with [SDS] above the CMC in the EHA system. The more hydrophobic monomer will be solubilized in surfactant micelles above the CMC to a much greater extent than will the

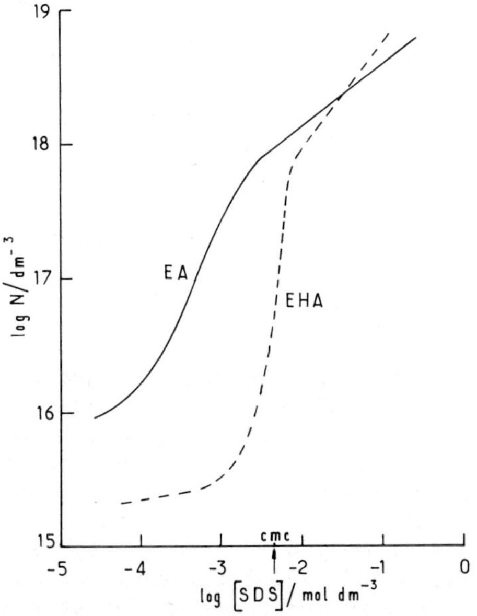

Figure 9. Dependence of particle number on concentration of sodium dodecyl sulfate for EA and EHA (44)

more hydrophilic monomer, because of more favorable interactions with the hydrocarbon core of the micelles (11). A plot of the amount of monomer dissolved or solubilized in the aqueous phase would closely resemble the N-vs.-[SDS] plot for EHA in Figure 9. On the other hand the solubility of EA is relatively independent of surfactant concentration. For this monomer N increases with [SDS] primarily because of progressive reduction in R_f, as observed for MMA in Figures 7 and 8. Below the CMC, the more soluble monomer produces more particles primarily because of its greater propagation rate in the aqueous phase (k_pM_w), as predicted by Curves 1 and 6 in Figure 5. Above the CMC, there is apparently more of the hydrophobic monomer available for nucleation in the micelles because of its greater degree of solubilization. The micelles are, as stated at the outset of this paper, energetically favored sites for nucleation where they exist. Thus EHA produces more particles above the CMC than does EA. This effect may be enhanced by the fact that SDS is adsorbed more strongly on the more hydrophobic polymer (37), giving it a higher ψ_o above the CMC. Apparently below CMC this effect is insufficient to overcome the vast differences in M_w between the two monomers (EA is 340 times more soluble than EHA).

Prospects for the Future

The theory for particle formation and growth as it has been developed indicates the need for certain kinds of information on a variety of systems. The important parameters not sufficiently available in the literature are jcr and the interfacial tension, γ, between monomer-swollen polymer and surfactant solutions along with the corresponding adsorption isotherms. The values of jcr have been determined in only a few cases: by Priest for vinyl acetate (12), by Fitch and Tsai for methyl methacrylate (45) and, indirectly, by Hansen and Ugelstad for styrene (29, 30, 31, 32). These values need to be confirmed by independent methods, and values obtained for other systems as well. The effect of monomer and surfactant concentrations on the value of jcr should also be investigated, as suggested by the results of Fitch and Palmgren (33) and of Chen and Piirma (42). Study of surfactant adsorption is required because it affects, through γ, the swelling of particles by monomer, and, through ψ_o, the rates of capture and coagulation. A general theory for surfactant adsorption based on unifying principles is required for predicting the behavior of any chosen system. Considerable progress has been made by Vijayendran (37) in this direction to improve upon the archaic and overly qualitative "HLB" system.

The interactions of charged oligomeric radicals with charged particles needs to be studied in greater detail, since these affect the rate of capture, R_c. What is the magnitude of the "tunneling effect" in overcoming the electrostatic barrier? Can experiments be devised to measure the Dankwerts diffusion-with-reaction kinetics directly? The answers to these questions not only will bear on the kinetics of nucleation, but also on the kinetics of polymerization throughout the remainder of the reaction after particles have been formed.

A largely unexplored area is that involving ionic emulsion polymerization in nonaqueous media. This has been shown to be practicable for methacrylates by Barrett and coworkers (46) and for styrene by Dawkins and Taylor (47). These systems combine the advantages of ionic polymerizations, which provide stereoregular and monodisperse polymers, with those of emulsion polymerization, which provides high rates of heat transfer and low viscosity in the presence of high molecular weight polymer at high concentrations. The mechanisms of particle nucleation and growth are largely unexplored in ionic emulsion polymerization systems.

In conclusion, it is clear that much has been accomplished in this field which combines challenging scientific problems with practical applications. But there are more problems awaiting those who would be inspired to solve them. Their solution will involve increasingly sophisticated experimental techniques involving relatively fast processes.

List of Symbols

A - interfacial area m^2

a_j - partition coefficient for j-mers between organic and aqueous phases --

b - reciprocal aggregation number --

c - concentration dm^{-3}

C_j^*, C_{ja} - concentration of j-mers inside and outside particle/water interface dm^{-3}

C_s - steady state concentration of radicals dm^{-3}

CMC - critical micelle concentration mol dm^{-3}

$\overline{D}_{op}$ - average mutual diffusion coefficient, for radical and particle $dm^2 s^{-1}$

D_{pj} - diffusion coefficient of j-mers in a particle $dm^2 s^{-1}$

e - charge on the electron C

E - surface electrical potential function --

f - initiator efficiency --

F - reduction coefficient for radical capture --

ΔG_f - free energy of formation of a particle J

ΔG_p - free energy of polymerization per unit volume J dm^{-3}

ΔG_s - interfacial free energy J dm^{-2}

ΔG_v - free energy of condensation per unit volume J dm^{-3}

I - concentration of initiator molecules dm^{-3}
j - ratio of molar volumes of polymer and monomer (approx. degree of polymerization of polymer) --
k - Dankwerts reaction rate coefficient dm^3s^{-1}
k' - Boltzmann constant JK^{-1}
k_{cj} - specific rate constant for capture of j-mers dm^3s^{-1}
k^*_d - specific rate constant for initiator decomposition s^{-1}
k_p - propagation rate constant $dm^3mol^{-1}s^{-1}$
k_t - specific rate constant for termination $dm^3mol^{-1}s^{-1}$
k^*_{tw}, k^*_{tp} - specific rate constants for termination in water and in particles (molecular units) dm^3s^{-1}
m - relative refractive index --
M_j - concentration of free radicals of chain length j in water phase dm^{-3}
$M\cdot_o$ - concentration of initiator radicals dm^{-3}
M_w - monomer concentration in water phase $mol\ dm^{-3}$
n - number of free radicals in a monomer/polymer particle --
N - particle number concentration dm^{-3}
N_1 - number concentration of primary particles dm^{-3}
n_o - refractive index of solvent --
r - particle radius m
R - Gas constant $JK^{-1}mol^{-1}$
$R\cdot$ - concentration of initiator radicals dm^{-3}
R_{90} - Rayleigh ratio cm^{-1}
R_c - rate of capture of growing radicals by monomer-polymer particles $dm^{-3}s^{-1}$
R_{cj} - rate of capture of a j-mer by any particle $dm^{-3}s^{-1}$
R_f - rate of particle coagulation $dm^{-3}s^{-1}$
r_h - radius of diffusion barrier surrounding a particle m
R_i - rate of generation of free radicals $dm^{-3}s^{-1}$
R_{ij} - radial distance between two particles, i and j m
R_{pj} - rate of propagation of an oligomer j units long in continuous phase $dm^{-3}s^{-1}$
R_{twj} - rate of termination of a j-mer in water phase with any other radical $dm^{-3}s^{-1}$
t - time s
T - temperature K
v - volume of a monomer-swollen particle dm^3
$\overline{V}_m$ - partial molar volume of monomer in particle dm^3mol^{-1}
v_p - volume of polymer in a particle dm^3

$V_{T(sw)}$ - volume fraction of monomer-swollen particles dm^3dm^{-3}
W', W - Fuchs stability factors between radical and particle, and particle and particle --
X_j - reactivity parameter for j-mers --
γ - interfacial tension Jm^{-2}
λ_o - vacuum wavelength of light cm
ϕ_m, ϕ_p - volume fractions of monomer and polymer --
τ - half-life for coagulation s
ψ_o - surface electrical potential V
χ - Flory-Huggins interaction parameter --

Literature Cited

1. Dinsmore, R. P., U. S. Pat. 1,732,795 (1929).
2. McBain, M. E. S., private communication to W. D. Harkins (1932).
3. Harkins, W. D., "private communication" to himself (1932).
4. Fikentscher, L. Z. angew. Chem. 1938, 51, 433.
5. Heller, W.; Klevens, H. B. CR (Copolymer Research Reports to Office of Rubber Reserve, War Production Board) 1943, 124, 237; 1944, 241; 1945, 563, 670.
6. Harkins, W. D. J. Amer. Chem. Soc. 1947, 69, 1428.
7. Smith, W. V.; Ewart, R. H. J. Chem. Phys. 1948, 16, 592.
8. Baxendale, J. H.; Evans, M.G.; Kilham, J.K. Trans. Faraday Soc. 1946, 42, 668; Baxendale, J. H.; Bywater, S.; Evans, M. G., ibid. 1946, 42, 675.
9. (a) Smith, W. V. J. Amer. Chem. Soc. 1948, 70, 3695.
   (b) Gerrens, H. Fortschr. Hochpolym.-Forsch. 1959, Bd. 1, 234-328.
   (c) Van der Hoff, B. M. E. J. Polym. Sci. 1958, 33, 487.
   (d) Gardon, J. L. J. Polym. Sci., A-1 1968, 6, 643, 665, 687, 2853, 2859.
10. Barrett, K. E. J., Ed. "Dispersion Polymerization in Organic Media"; J. Wiley: New York, 1975.
11. Elworthy, P. H.; Florence, A. T.; Macfarlane, C. B. "Solubilization by Surface-active Agents"; Chapman and Hall: London, 1968.
12. Priest, W. J. J. Phys. Chem. 1952, 56, 1077.
13. Napper, D. H.; Alexander, A. E. J. Polym. Sci. 1962, 61, 127.
14. Peggion, E.; Testa, F.; Talamini, G. P. Makromol. Chem. 1964, 71, 173.
15. Dunn, A. S.; Taylor, P. A. Makromol. Chem. 1965, 83, 207.

16. (a) Deryagin, B. V.; Landau, L. D. Acta phys. - chim. USSR 1941, 14, 633.
(b) Verwey, E. J. W.; Overbeek, J. Th. G., "Theory of the Stability of Lyophobic Colloids"; Elsevier: Amsterdam, 1948.
17. Fitch, R. M. Offic. Dig. J. Paint Technol. Eng. 1965, 37, (489), Pt. 2, 32.
18. Parts, A. G.; Moore, D. E.; Watterson, J. G. Makromol. Chem. 1965, 89, 156.
19. Fitch, R. M.; Prenosil, M. B.; Sprick, K. J. J. Polym. Sci., Part C 1969, 27, 95.
20. Dunn, A. S.; Chong, L. C. H. Brit. Polym. J. 1970, 2, 49.
21. Fitch, R. M.; Tsai, C. H. Polymer Letters 1970, 8, 703.
22. Fitch, R. M.; Tsai, C. H., in "Polymer Colloids", Fitch, R.M. Ed.; Plenum: New York, 1971; p. 73.
23. Fitch, R. M.; Shih, L. B. Prog. Colloid Polym. Sci. 1975, 56, 1.
24. Gatta, G.; Benetta, G.; Talamini, G.; Vianello, G. Adv. in Chem. Ser. 1969, 91, 158.
25. Liang, S. J.; Fitch, R. M., results to be published.
26. Morton, M.; Kaizerman, S.; Altier, M.W. J. Colloid Sci. 1968, 6, 2859.
27. Gardon, J. L. J. Polym. Sci. A-1 1968, 6, 2859.
28. (a) Ugelstad, J. Makromol. Chem. 1978, 179, 815.
(b) Ugelstad, J.; Kaggerud, K. Herder; Fitch, R. M., in "Polymer Colloids II", Fitch, R. M. Ed.; Plenum: New York, 1980.
29. Ugelstad, J.; Hansen, F. K. Rubber Chem. Technol. 1976, 49 (3), 536.
30. Hansen, F. K.; Ugelstad, J. J. Polym. Sci., Polym. Chem. Ed. 1978, 16, 1953.
31. Hansen, F. K.; Ugelstad, J., ibid. 1979, 17, 3033.
32. Hansen, F. K.; Ugelstad, J., ibid. 1979, 17, 3047.
33. Fitch, R. M.; Palmgren, T. H., unpublished results.
34. Kegeles, G., private communication.
35. Fitch, R. M.; Watson, R. C. J. Colloid Interface Sci. 1979, 68 (1), 14.
36. Overbeek, J. Th. G., in "Colloid Science", Kruyt, H. R., Ed.; Vol. I; Elsevier: New York/Amsterdam, 1952; pp. 278-298.
37. Vijayendran, B. R., in "Polymer Colloids II", Fitch, R. M. Ed.; Plenum: New York, 1980; p. 209.
38. Wright, H. J.; Bremmer, J. F.; Bhimani, N.; Fitch, R. M., U. S. Pat. 3,501,432.
39. Krieger, I. M.; Juang, M. S. J. Polym. Sci., Polym. Chem. Ed. 1976, 14, 2089.
40. Schild, R. L.; El-Aasser, M. S.; Poehlein, G. W.; Vanderhoff, J. W., in "Emulsions, Latices and Dispersions", Becher, P.; Yudenfreund, M. Y., Eds.; Dekker: New York, 1978; p. 99.

41. Chonde, Y.; Krieger, I. M. J. Colloid Interface Sci. 1980, 77 (1), 138.
42. Chen, C. Y.; Piirma, I. J. Polym. Sci., Polym. Chem. Ed. 1980, 18, 1979.
43. Sütterlin, N. in "Polymer Colloids II", Fitch R. M., Ed.; Plenum: New York, 1980; p. 583.
44. Brandrup, J.; Immergut, E. H., Eds.; "Polymer Handbook"; Wiley: New York, 1975.
45. Fitch, R. M.; Tsai, C. H. in "Polymer Colloids"; Fitch, R.M., Ed.; Plenum: New York, 1971; p. 103.
46. Barrett, K. E. J., "Dispersion Polymerization in Organic Media"; Wiley: New York, 1975.
47. Dawkins, J. V.; Taylor, G. in "Polymer Colloids II", Fitch, R. M., Ed.; Plenum: New York, 1980; p. 447.

RECEIVED April 6, 1981.

2

Latex Particle Stabilization

R. H. OTTEWILL

School of Chemistry, University of Bristol, England

In the early work of Schulze (1), Linder and Picton (2) and Hardy (3) the sensitivity of colloidal dispersions to the addition of electrolytes was clearly demonstrated. Then in 1900 Hardy (4) showed that the stability of sols was connected with the electrophoretic mobility of the particles and he demonstrated, i) that the valency of the ion opposite in charge to that of the sol particles determined the ability of an electrolyte to coagulate a sol and that, ii) the effectiveness of the electrolyte increased rapidly with increase in valency of the counter-ion. These observations formed the basis of the so-called Schulze-Hardy rule.

During the 1930's a clearer idea of the role of the electrical double layer in stabilising colloidal particles began to emerge, particularly in the work of Verwey (5), Kruyt (6) and Derjaguin (7). In 1938 in a classic paper Langmuir (8) showed that when an overlap of double layers occurred, with two flat plates whose surfaces were at the same electrostatic potential, then a repulsion pressure was developed between them.

Also, in the 1930's London (9) indicated the quantum mechanical origin of dispersion forces between apolar molecules and in subsequent work extended these ideas to interaction between particles (10). It was shown that whereas the force between molecules varied inversely as the seventh power of the separation distance, that between "thick" flat plates varied inversely as the third power of the distance of surface separation. These ideas lead directly to the concept of a "long range" van der Waals attractive force. A similar relationship was found for interaction between spheres (10).

A considerable advance was made in the 1940's when in the theories of Derjaguin and Landau (11) and Verwey and Overbeek (12) a theory of the stability of lyophobic colloids was obtained by assuming the pairwise additivity of the potential energy of electrostatic repulsion, V_R, and the van der Waals attraction, V_A. The sum of the electrostatic repulsive energy and the van der Waals attraction, V_T, ($= V_R + V_A$), with the addition of a

0097-6156/81/0165-0031$07.25/0

short range Born repulsion energy gave a curve of the total potential energy of repulsion against distance of surface separation of the form shown in Figure 1. This type of curve exhibits a number of characteristic features. At short distances, a deep potential-energy minimum occurs, the position of which decides the distance of closest approach, h_o and is hence termed the primary minimum. At intermediate distances the electrostatic repulsion makes the largest contribution and hence a maximum occurs of magnitude, V_m; this is usually termed the primary maximum . At larger distances, the exponential decay of the electrical double layer term causes it to fall off more rapidly than the power law of the attractive term and another minimum occurs in the curve, of depth V_{SM}; this is termed the secondary minimum. The latter is shallow for small particles but can become appreciable for particles greater than 1 micron at high electrolyte concentrations

The ideas underlying the form of the curve shown in Figure 1 form the basis of our present-day approach to discussing the stability and instability of colloidal dispersions in aqueous media in the presence of electrolytes.

Polymer Latices

Before applying the ideas summarised in the first section to polymer latices it is appropriate to consider the nature of polymer latex particles. We know, for example, that each particle is composed of a large number of polymer chains, with the chains having molecular weights in the range of about 10^5 to 10^7. Moreover, the particles themselves can be amorphous, crystalline, rubbery, glassy or monomer swollen, either extensively or minutely. It follows, therefore, that the properties of the system on drying depends directly on the physical state of the particles, for example, if the particles are soft, coalescence can occur to form a continuous film, whereas with hard particles their individuality is retained. The nature of the particle obtained is directly related to the preparative method employed and the surface properties are often determined by :-

i) groupings arising from the initiator used,
ii) adsorbed or grafted surface active agents,
iii) adsorbed or grafted polymers soluble in the dispersion medium,
iv) the basic property of the polymer forming the particle.

Arising from the initiator used we can expect to find surface groupings (13) such as:-

weak acid, $-COOH^-$
strong acid, $-O-SO_3^-$
non-ionic $-OH$

basic

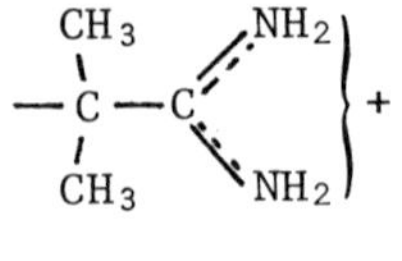

The hydrophobic portions of the surface accompanying these groups will be determined by the polymer and could be, for example, polystyrene or polytetrafluoroethylene. The possibility therefore exists of obtaining a wide variety of particles and Figure 2 attempts to illustrate this in a schematic fashion. Clearly, particle shape should be added as an additional variable. In the situations where molecular extension occurs into the solvent medium, as illustrated in Figures 2b, 2c and 2d an additional component, allowing for the properties of this region, has to be added to considerations of the potential energy of interaction (see later).

In the practical utilisation of latices, either dilute or concentrated, the word stability can be used in many different ways, and stability may be required:-

a) to electrolyte addition,
b) to shear or mechanical work,
c) to freezing,
d) to heat, i.e. sterilization
e) to long term storage
f) to drying conditions such that instantaneous redispersion is obtained on re-wetting.

In the current article category a) will primarily be considered.

In terms of usage of the word instability it is often required to destabilize particles in order to obtain:-

i) a coarse aggregation of particles with open pores for easy filtration,
ii) coarse compact particles for rapid settling or the fabrication of materials.

For situations i) and ii) the coagulated state, i.e. with the particles in intimate contact, is desirable. For other purposes, the flocculated state is required, i.e. with the particles still essentially individual and separated by a thin layer of liquid, thus giving control of the rheological properties of the sytem. Frequently, the secondary minimum plays a significant role in flocculated systems.

The Basis of Stability

In an ionizing medium such as water the latex particles will in general be electrically charged either by the ionisation of surface groupings of the type illustrated above or by the adsorption of ionic materials, e.g. surface active agents. Moreover, the solvency of water is good for small ions, which are well-solvated and stay in the medium, and poor for the latex particle in that most polymers used for latices are totally insoluble. In this situation the condition of electroneutrality is maintained by balancing the charge on the latex surface by the charges on the ions of opposite sign in the solution phase to give the so-called electrical double layer. As a consequence of its surface charge, the latex particle surface has an electro-

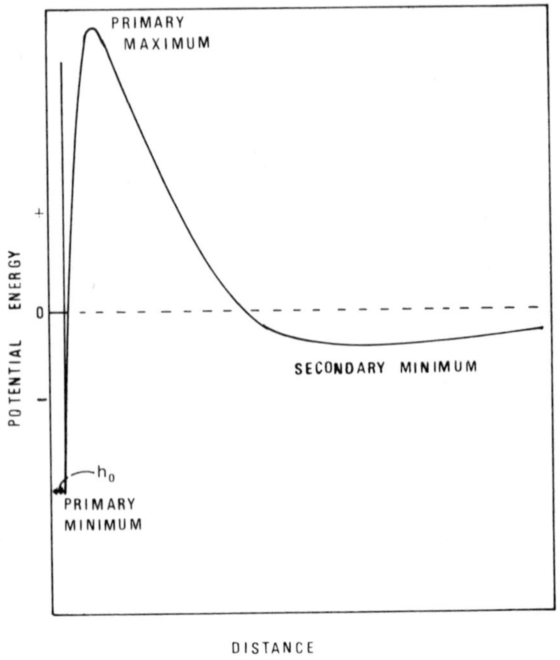

Figure 1. Schematic illustrating the form of the curve of potential energy vs. distance for the interaction between two spheres. Total interaction involving electrostatic repulsion and van der Waals attraction.

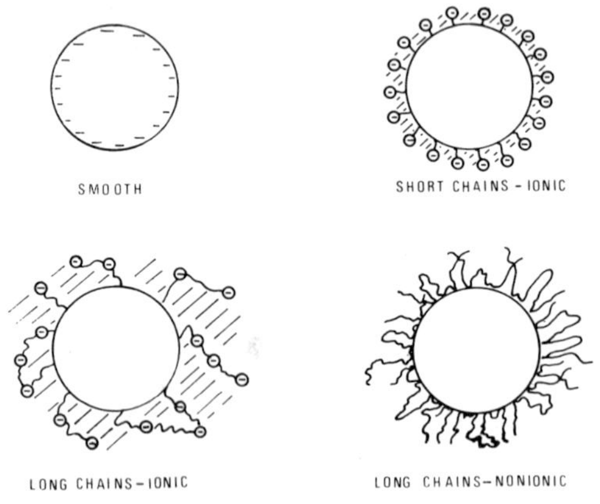

Figure 2. Schematic of the nature of the surface on various types of latex particles

static potential, ψ_s, which can be either positive or negative, depending on the surface groupings, relative to earth.

In more quantitative terms ψ_s can now be taken in conjunction with the properties of the medium to obtain an expression for the electrostatic repulsive potential energy, V_R. As an example, Reerink and Overbeek (14) found for two spheres of radius a, with their surfaces separated by a distance h, the expression:-

$$V_R = 3.469 \times 10^{19}\ \varepsilon(kT)^2\ a\gamma^2\ \exp(-\kappa h)/v^2 \qquad (1)$$

with κ = the reciprocal Debye-Huckel double layer thickness,
ε = the dielectric constant of the medium,
k = Boltzmann's constant,
T = absolute temperature,
v = the magnitude of the valency of the counter-ion,
$\gamma = [\exp(ve\ \psi_s/2kT) - 1]/[\exp(ve\ \psi_s/2kT + 1]$
e = fundamental electronic charge.

For the attractive part of the interaction energy it has been shown that, V_A, can be given for two spheres of the same radius by the expression,

$$V_A = -\frac{A}{12}\left\{\frac{1}{x^2 + 2x} + \frac{1}{x^2 + 2x + 1} + 2\ \ln\frac{x^2 + 2x}{x^2 + 2x + 1}\right\} \qquad (2)$$

where A = the composite Hamaker Constant for the particles in the medium as given by,

$$A = (\sqrt{A_{11}} - \sqrt{A_{22}})^2 \qquad (3)$$

with A_{11} = the Hamaker Constant of the particles and A_{22} that of the medium. The value of x is given by h/2a, and as a useful approximation when x << 1 we obtain,

$$V_A = -\frac{A.a}{12h} \qquad (4)$$

Provided that $|V_R| > |V_A|$ for most values of h then the form of curve shown in Figure 1 is obtained. When the magnitude of V_m is substantial, say >> 10 kT, a stable dispersion is obtained. The form of the potential energy curve obtained by this approach shows immediately that the stability of a dispersion to electrolyte is kinetic in origin rather than thermodynamic, that is, the lowest free energy state is in the primary minimum and entry into this is prevented by the presence of the large activation energy represented by ΔV_f. A more sophisticated and detailed representation of these ideas can be found elsewhere (12,15,16).

The Onset of Instability - The Critical Coagulation Concentration. Provided that the magnitude of the primary maximum is substantial, then the probability of the transition of the approaching particle into the primary minimum is small. However, as V_m becomes small or tends to zero (Figure 3) then the transition becomes facile and the system becomes unstable because the particles associate. Theoretically convenient definitions for the onset of instability then can be taken as,

$$V_T = 0 \quad \text{and} \quad \frac{d\,V_T}{dh} = 0$$

Adopting these definitions and combining equations (1) and (4) we find,

$$\kappa_{crit} = \frac{2.04 \times 10^{-5}\,\gamma^2}{A\,v^2}\ \text{cm}^{-1} \tag{5}$$

and since for a symmetrical electrolyte κ can be related directly to the concentration of electrolyte, C, expressed in mol dm^{-3} by

$$\kappa^2 = \frac{8\pi\,v^2 e^2 N_A\,C}{\varepsilon kT\,10^3}\ \text{cm}^{-2} \tag{6}$$

with N_A = Avogadro's number, we find from (5) and (6)

$$C_{crit} = \frac{3.856 \times 10^{-25}\,\gamma^4}{A^2\,v^6}\ \text{mol dm}^{-3} \tag{7}$$

The concentration C_{crit} is usually termed the critical coagulation concentration (c.c.c.) and we note the inverse dependence on the 6th power of the valency of the counter-ion.

For low surface potentials, ψ_s < 25 mV, the expression can be simplified to give,

$$C_{crit} = \frac{8.821 \times 10^{-19}\,\psi_s^4}{A^2\,v^2}\ \text{mol dm}^{-3} \tag{8}$$

and we now note the dependence of C_{crit} on the second power of the valency.

C_{crit} is clearly an important quantity for a latex dispersion since it essentially represents the electrolyte concentration at which complete loss of stability occurs. It may be obtained experimentally by a variety of different methods (14,17, 18,19). It should be noted, however, that since coagulation is a kinetic phenomena time enters as a variable and consequently the various methods may yield somewhat different numerical results. This effect is illustrated by results obtained for the coagulation of polytetrafluoroethylene (PTFE) latices with sodium chloride as a function of pH (19). From Figure 4 it can be seen that different results are obtained according to whether the system was examined after 2 h or 24 h. As expected the results indicate that the state of aggregation is more advanced after 24 h and consequently systems at a lower electrolyte concentration have coagulated. Care must therefore be taken when comparing values

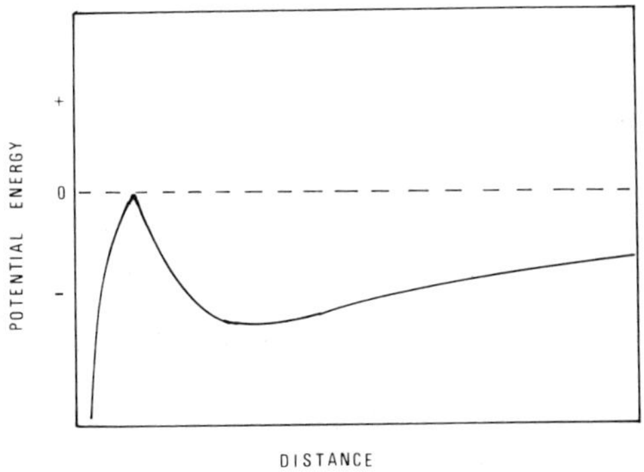

Figure 3. Diagram of a potential energy–distance curve to illustrate the condition for defining the critical coagulation concentration

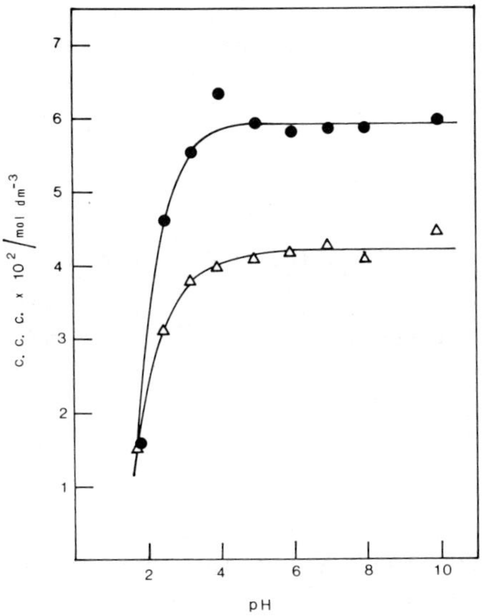

Figure 4. Plot of c.c.c. vs. pH for a PTFE latex in HCl solutions (19) ((–●–) 2 h after mixing; (–△–) 24 h after mixing)

of the c.c.c. obtained by different methods or different workers. As can be seen from Figure 4, however, the value of the c.c.c. obtained at a particular time remains constant above ca. pH 4 but drops rapidly below this pH.

These results were obtained on a latex (19) which had been extensively dialysed to remove as much as possible of the surfactant used in the original preparation. The results suggest the presence of surface groups with a pKa of the order of 3 from the rapid decrease of the c.c.c. with pH between 2 and 4. It seems likely that these arose from the initiator and were either succinic or propionic acid attached to fluorocarbon chains or fluorocarbon carboxylic acids. In this type of system where the individual PTFE chains have a high molecular weight, > 10^6, a low surface charge density is expected; this was confirmed by conductometric titrations which gave a value of 0.38 $\mu C\ cm^{-2}$ equivalent to one charged group per 42 nm^2. The c.c.c. values obtained with PTFE latices using several electrolytes (19) are given in Table 1, wherein it can also be seen that addition of 2.5×10^{-2} mol dm^{-3} ammonium perfluoro-octanoate to the latex increased its stability to the addition of sodium chloride and aluminium nitrate. The perfluoro-octanoate ions are adsorbed onto the surface of the PTFE particles via the perfluorocarbon chains thus increasing the surface charge density, and hence the surface potential, by exposing an array of ionised carboxyl groups to the solution phase. In the case of addition of barium nitrate, however, insoluble barium perfluoro-octanoate was formed which apparently stripped the surfactant from the particle surface so that consequently no increase in stabilisation was observed

Table I

c.c.c. Values for PTFE Latices at pH 3.3

Electrolyte used	c.c.c./mol dm^{-3} Dialysed Latex	c.c.c./mol dm^{-3}. Latex with 2.5×10^{-2} mol dm^{-2} ammonium perfluoro-octanoate
NaCl	4.7×10^{-2}	0.22
$Ba(NO_3)_2$	7.4×10^{-3}	7.4×10^{-3}
$Al(NO_3)_3$	1.6×10^{-4}	5.4×10^{-3}

The results obtained for PTFE latices show clearly that the surface charge of the latex particle plays an important role in determining its stability to electrolyte addition and also that the stability can be affected by the presence of surfactant.

One of the prime reasons for studying PTFE systems was that the PTFE surface has a very low surface energy (18.5 mN m^{-1}) and also it has a Hamaker Constant (4.40×10^{-20} J) close to that of water (3.72×10^{-20} J), which should aid the overall stability of such systems (19). In practice, however, for dialysed systems it appears that the behaviour is dominated by the very low surface charge density of the particles and by the possible uneven distribution of the charged groups because of the high degree of crystallinity of the particles.

The stability of a number of polymer latices has been examined and some typical values are reported in Table II. The trends observed are qualitatively in agreement with the trends predicted by the theoretical approach outlined in this section for particles with smooth surfaces, with a homogeneous distribution of surface charges (ψ_s everywhere the same) using simple electrolytes, i.e. those which do not interact chemically with water to form new ionic species.

Table II

c.c.c. values for various polymer latices

Latex	Counter-ion	c.c.c./mmol dm^{-3}	Ref.
Polystyrene (Carboxyl Surface)	H^+	1.3	(20)
	Na^+	160	(21)
	Ba^{2+}	14.3	(17)
	La^{3+} (pH 4.6)	0.3	(20)
Polystyrene (Amidine Surface)	Cl^-	150	(22)
	Br^-	90	(22)
	I^-	43	(22)
Divinyl-styrene	Na^+	160-560	(23)
Styrene-butadiene	Na^+	200	(18)
	K^+	320	(18)
	Mg^{2+}	6	(18)
	Ba^{2+}	6	(18)
	La^{3+} (pH 3)	0.5	(18)
Polyvinylchloride	Na^+	50-200	(24)
	Mg^{2+}	2-10	(24)

The values listed in Table 2 should only be used for qualitative guidance since these can be variations of the c.c.c. with particle size, type and density of surface groupings, time of measurement after adding electrolyte and the presence or absence of stabilising materials such as surfactants. In practice it is advisable to determine the actual value for a particular latex system.

The Effects of Ions which Interact with Water. In the last section the premise was made that the counter-ions used for coagulation did not interact with water. However, it is well-known that many multivalent ions do react with water to form hydrolysed species which can sometimes be polymeric in nature. For example, in the case of aluminium, the Al^{3+} ion exists at pH values below ca. pH 3.3 as the hexa-aquo ion, with six water molecules, in the octahedral co-ordinate positions. As the pH is slowly increased reaction occurs with water to form a sequence of species. The chemistry involved in these reactions is somewhat complex and has not been fully resolved but a plausible reaction scheme can be proposed, for the present purpose, as,

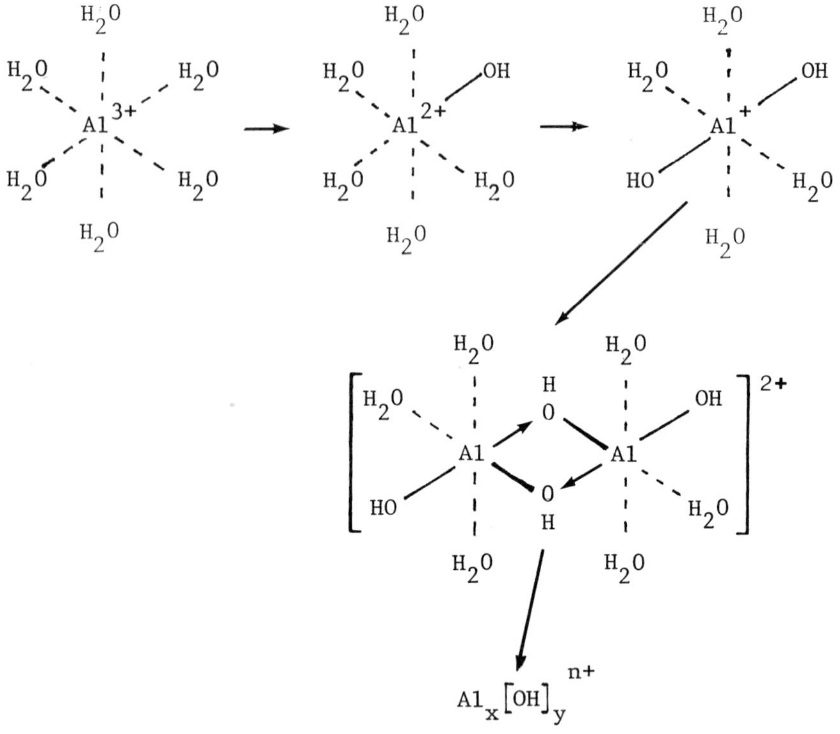

where $Al_x[OH]_y^{n+}$ represents an inorganic polymer soluble in water. Polymeric species of this type can adsorb strongly onto negatively charged particles and reduce the effective surface charge on the particle to zero. As anticipated from equation (7) this situation leads immediately to coagulation. At higher concentrations of the aluminium species, super-equivalent adsorption can take place, thus conferring a positive charge on the particle and leading to restabilisation of the dispersion as one containing positively charged rather than the original negatively charged particles. In addition, it is also possible for the positive polymeric species to "bridge" two negatively charged particles.

The exact nature of the polymeric species in solutions of aluminium salts at the pH conditions for charge reversal is not known with certainty. It is possible that several species co-exist, depending on their stability constants, and that these also change with time with the ultimate product of hydrolysis being aluminium hydroxide particles. A number of species have been proposed in the literature and Matijević, Janauer and Kerker (25) have provided evidence for the existence of $Al_8(OH)_{20}^{4+}$ from coagulation studies. The higher valency of this type of species again reduces the concentration of ions required to produce coagulation. The combined effects of high valency and reduction of the surface charge to zero makes aluminium salts very effective coagulants in the pH range of ca. 4 to 5.5 Coagulation can be achieved at very low salt concentrations and since most of the aluminium is adsorbed by the particles there is little salt left in the filtrate after removal of the coagula. This factor is exploited in the use of aluminium salts for the treatment of potable waters.

The basic pattern of the coagulation of polymer latices with aluminium salts has been clearly demonstrated by the work of Matijević and his collaborators using styrene-butadiene (26) polyvinyl chloride (27) and PTFE (28) latices. The results obtained by Matijević and Force (26) for the coagulation of styrene-butadiene latices using aluminium nitrate are shown in Figure 5. From these it can be seen that up to a pH of ca. 3.4 the c.c.c. remains constant at 5×10^{-4} mol dm^{-3} and then decreases between pH 3.4 and 4.8 to reach a constant value of ca. 2.5×10^{-6} mol dm^{-3} between pH 4.8 and 6.0. The region of restabilization as positively charged particles can also be seen on this "domain" plot.

In the case of PTFE latices the behaviour seems to be strongly dependent on the amount of stabiliser present in the system. Kratohvil and Matijević (28) found evidence for both charge reversal and restabilization. On the other hand, Ottewill and Rance (29) using well-dialysed PTFE latices obtained the results shown in Figure 6. The change in c.c.c. with pH is clearly defined but the c.c.c. above pH 5 is at a fairly high concentration of electrolyte and insufficient positive charge is built up to stabilise the dispersion. It is also possible with this system that the very hydrophobic polytetrafluoroethylene parts of the surface do not adsorb the hydrolysed species and in the well-dialysed system adsorption can only occur on the sparse charged sites.

Secondary Minimum Effects. One of the distinctive features on the potential energy diagram shown in Figure 1 is the secondary minimum. This as shown in Figure 7 is a feature which although not very pronounced for small diameter particles becomes enhanced as particle size increases, particularly at the higher electrolyte concentrations. As can be seen the secondary minimum becomes deeper with increase in salt concentration and although the magnitude of the primary maximum is reduced it is nevertheless maintained. The form of the curve indicates the possibility that

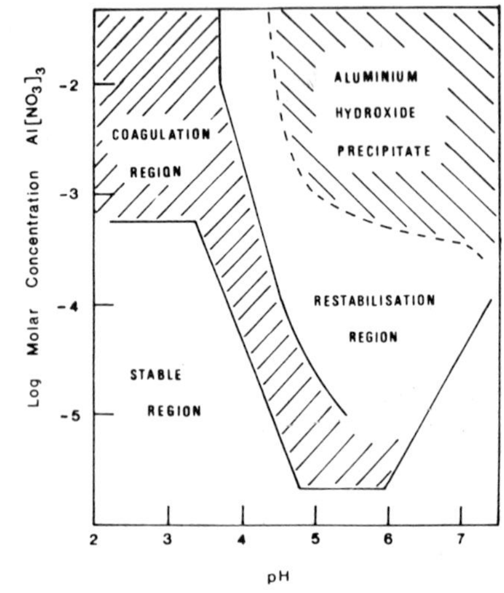

Figure 5. Log (molar concentration of $Al(NO_3)_3$) vs. pH showing the positions of the coagulation domains for a styrene–butadiene latex. Drawn from the data of Matijević and Force (26).

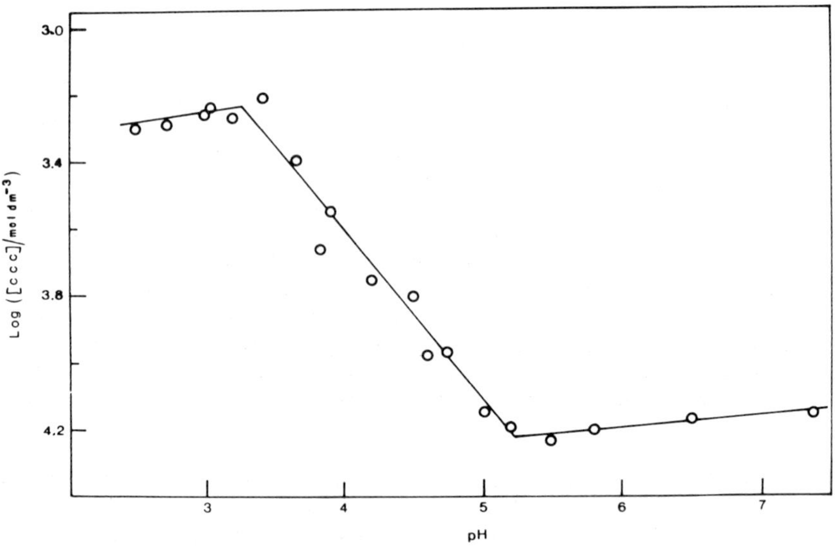

Figure 6. Log (c.c.c.) vs. pH for a PTFE latex using aluminum nitrate as the coagulating electrolyte (19)

over this range of electrolyte concentration, once a particle enters a secondary minimum, it will have a long residence time there and remain separated from the second spherical particle by distances of the order of 6 to 10 nm. However, there still remains a substantial primary maximum in potential energy to be overcome before the particles can come into contact or enter the primary minimum. Association in the latter state clearly corresponds to a condition where the particles come into close physical contact, providing the possibility with subsequent thermal diffusion, that they will fuse together. Under these conditions, therefore, one would expect the units formed to be hard, compact and essentially nonreversible and there is compelling logic to term this state COAGULATION. On the other hand, when association occurs in a secondary minimum, the particles remain separated by a liquid film, which renders thermal diffusion between the solid particles unlikely, and leaves the possibility that by decreasing the salt concentration that the particles can redisperse. The logical term for this state is FLOCCULATION and there are strong reasons for distinguishing between the coagulated and flocculated states when simple electrolytes are used to produce them. It is clear from Figure 7 that with continued addition of electrolyte, a transition from the flocculated to the coagulated state should also occur.

The spherical nature of the particles and the high degree of monodispersity in polystyrene latices makes them ideal systems for testing such a hypothesis and exploring its practical implications. In some preliminary experiments (30) we have determined the c.c.c. values of a series of latices of different particle sizes and then over a range of salt concentrations at and above the c.c.c. examined the effect on the associated state of using dialysis to remove the salt. The results obtained are summarized in Table III.

These data require extension but in a tentative manner the conclusions can be summarized in Figure 8 where the domains of coagulation and flocculation are represented. Moreover, these ideas have only been applied to sodium chloride. With higher valency electrolytes more specific effects may occur which could dominate the phenomena.

On a kinetic basis the presence of a pronounced secondary minimum should lead to a steady state condition in which the rate of particles entering the secondary minimum to form associated units should be balanced by their rate of egress back to single particles. Direct evidence for this situation has been obtained recently using optical microscope observations on particles of diameter 2 μm (31). In addition, by coupling the microscope with a high speed camera, observations were made on particles over a period of time. This gave values for the life-time of doublets in the associated state and also revealed the fact that the particles in an associated unit could be quite

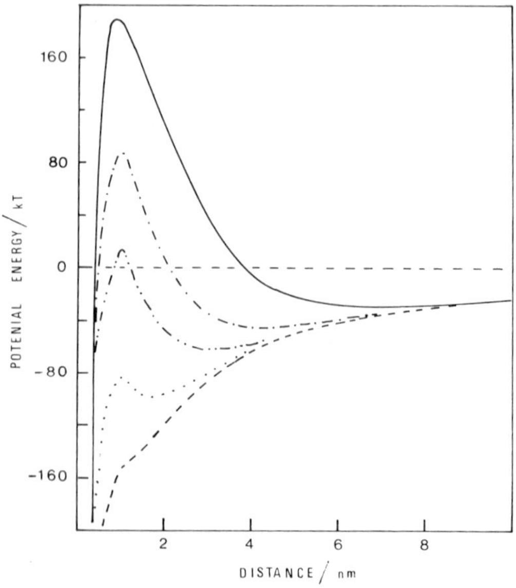

Figure 7. Potential energy vs. distance curves at different electrolyte concentrations for a latex with a particle diameter of 3.24 μm, a surface potential of 26 mV, and a Hamaker Constant A = 8 × 10^{-21} J ((——) 0.05; (– · –) 0.16; (– · · –) 0.16; (· · ·) 0.30; (– – –) 0.40 mol dm^{-3}).

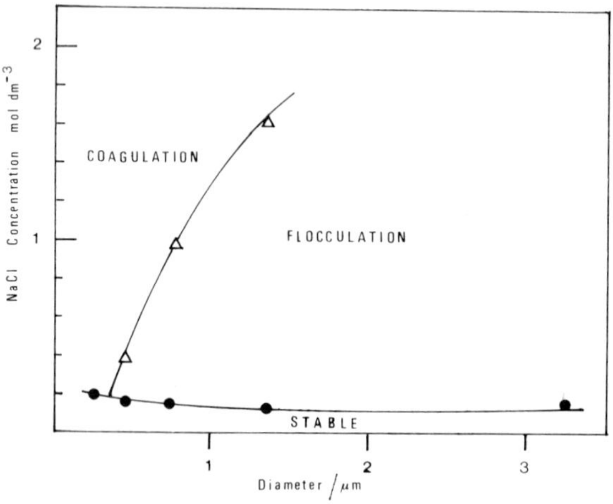

Figure 8. Diagram illustrating the domains of stability, coagulation, and flocculation. Data obtained using sodium chloride as the electrolyte, and polystyrene latices of various sizes.

Table III

Reversibility of Aggregates Formed on Addition of Electrolyte

Particle Diameter/μm	c.c.c./mol dm^{-3} value for NaCl	Electrolyte Conc. NaCl mol dm^{-3}	Behaviour on Dialysis
0.21	0.208	0.208	large aggregates non-reversible
0.40	0.155	0.155	some redispersion
		0.25	some redispersion
		0.40	non-reversible
0.58	0.150	0.150	some redispersion
		0.60	some redispersion
		0.70	some redispersion
		> 1.00	non-reversible
1.34	0.108	0.108	redispersion
		0.400	redispersion
		1.500	redispersion
3.24	0.158	0.158	redispersion
		0.400	redispersion
		1.0	redispersion

mobile. It was observed that as well as some particles leaving the aggregated unit as single particles and returning to the disperse phase there was a continued rearrangement of the particles. This was also observed with floccules at salt concentrations well above the c.c.c. These observations clearly support the contention that association can occur in a secondary minimum and that in this situation a liquid film is maintained between the particles.

The Effects of Added Surfactants

In general, the types of surfactant added to a latex will be either anionic, cationic or anionic as classified by the nature of the head group (32). If it is assumed that the latex particle has a surface free of adsorbed materials and is negatively charged then the various possibilities of adsorption of the surfactant can be envisaged by the schematic diagram given in Figure 9. The discussion of the various phenomena observed can then be based on these models.

Ionic Head Group with a Charge of the Same Sign as the Particle. As mentioned in the previous section in discussion of the coagulation of PTFE latices the addition of perfluoro-octanoate ions to the latex increases its stability to electrolyte addition. This is a consequence of the adsorption of the

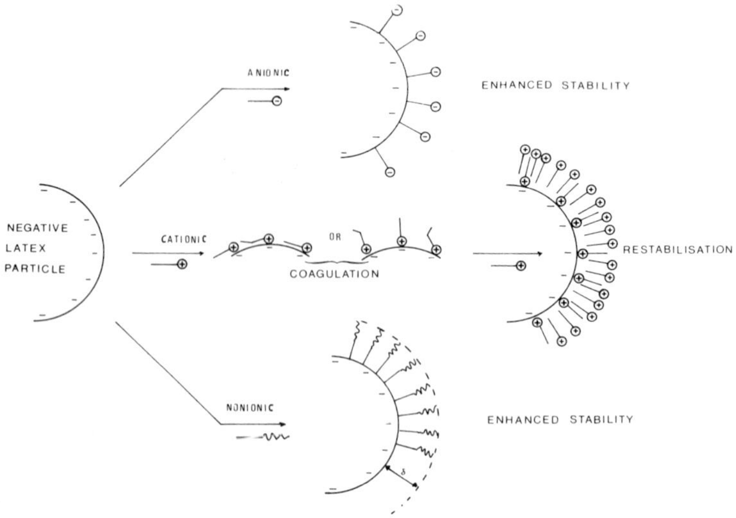

Figure 9. Diagram illustrating the effect of anionic, cationic, and nonionic surfactants on anionic polystyrene latex particles

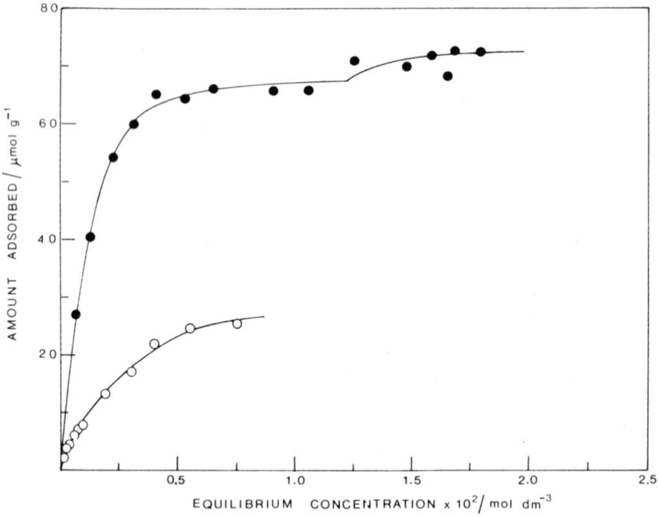

Figure 10. Adsorption isotherms for surfactants on a PTFE latex at 25°C: (–●–) ammonium perfluoro-octanoate at pH 6.0 in 10^{-3} mol dm^{-3} ammonium nitrate; (–○–) sodium dodecanoate at pH 9.0 in 10^{-3} mol dm^{-3} sodium nitrate.

surfactant on to the surface via the perfluorocarbon chain (Figure 9). An adsorption isotherm (33) is given in Figure 10 and it can be seen from this that using ammonium perfluoro-octanoate in the presence of 10^{-2} mol dm^{-3} ammonium chloride at pH 6 an adsorption plateau was reached at a concentration of ca.20% of the critical micelle concentration of the surface active agent; on the plateau the area per adsorbed perfluoro-octanoate ion was found to be 54 Å^2 which can be compared with 47 Å^2 obtained at the air-water interface. Thus the c.c.c. data essentially parallel the effects of adsorption.

However, two important points should be noted. Firstly, that when one of the ions of an added electrolyte can interact with the surfactant to form an insoluble salt this reaction can remove the adsorbed layer from the particle (see Table I). Secondly, the hydrophobic chain of the surfactant must be compatible with the particle surface. This point is illustrated in Figure 10 which shows the adsorption of dodecanoic acid on to PTFE particles. The adsorption of the $C_{11}H_{23}$ hydrocarbon chain to PTFE is clearly much less favourable than the adsorption of the, C_7F_{15}, fluorocarbon chain; in fact, a C_7H_{15} chain, in the form of octanoic acid showed no apparent adsorption on to a PTFE surface. Thus, although both acids have hydrophobic chains, there is clearly a remarkable difference between their affinities for the substrate.

On basically hydrocarbon-hydrophobic substrates such as polystyrene, it is well established that even on the negatively charged particles there is adsorption of surfactant anions via the hydrocarbon chains.

Ionic Head Group with a Charge of the Opposite Sign to the Particle. As illustrated in a very simple fashion in Figure 9 in the case of a negatively charged particle the first stage of adsorption of a cationic surfactant is via the positive head group to neutralise the charge on the particle, whence $\psi_s = 0$ and consequently $V_R = 0$ also. Some data obtained for the coagulation of polystyrene latices by a series of alkyl trimethylammonium halides are shown in Figure 11; for comparison results are also included for a simple 1:1 electrolyte, potassium bromide (34,35). It can be seen from these data that the range of concentrations over which coagulation occurs is very narrow and that the c.c.c. is strongly dependent on the chain length of the hydrocarbon tail of the surfactant molecule. Studies of the electrophoretic mobility of the particles confirm that at the c.c.c the particle mobility becomes zero and studies of the adsorption suggest that up to this point the alkyl chains lie flat on the hydrocarbon-like surface of a polystyrene particle (36).

Once the negative charges on the particle surface have been neutralised further adsorption of the surfactant occurs via the tail on to the hydrophobic patches of the surface and also by association of the hydrocarbon chains. Detailed adsorption studies have been reported on polystyrene latices (36). The additional adsorption provides a positive charge to the particles and restabilization occurs. The sharpness of this phenomenon is clearly illustrated by the experimental data given in Figure 11 and has also been examined theoretically (37). At much higher additions of surfactant the electrolyte concentration is appreciably increased and compression of the electrical double layer occurs leading to a second coagulation region. Under these conditions with the ionic groups of the surfactant exposed to the solution phase the particles are well-wetted and the coagula usually sink if the density of the particles is greater than that of the media. The conditions of the first coagulation region, however, i.e. zero surface potential with hydrocarbon chains orientated towards the solution phase leads to particles which are easily dewetted and frequently flotation is observed (34).

In the experiments on polystyrene latices using cationic surface active agents, the c.c.c. shifted systematically with hydrocarbon chain length indicating a typical Traube's Rule effect. In the coagulation of PTFE latices by cationic hydrocarbon surfactants, however, a different behaviour is observed in that only very small differences are observed in the magnitude of the c.c.c. with variation of chain length (38). This effect is illustrated by the data given in Table IV. Again, it appears to demonstrate the lack of affinity of hydrocarbon chains for fluorocarbon surfaces and suggests that in this case the chains do not lie on the PTFE surface.

Table IV

Coagulation of Polystyrene and PTFE Latices by Cationic Surfactant

Surfactant	c.c.c./mol dm^{-3} Polystyrene Latex	c.c.c./mol dm^{-3} PTFE Latex
$C_8H_{17}\overset{+}{N}Me_3.\overset{-}{Br}$	1.59×10^{-4}	4.5×10^{-5}
$C_{10}H_{21}\overset{+}{N}Me_3.\overset{-}{Br}$	1.99×10^{-5}	3.1×10^{-5}
$C_{12}H_{25}\overset{+}{N}Me_3.\overset{-}{Br}$	2.93×10^{-6}	2.9×10^{-5}

Nonionic Surfactants. It has been shown that nonionic surfactants such as dodecyl hexaoxyethylene glycol monoether, $C_{12}H_{25}(CH_2CH_2O)_6OH(C_{12}E_6)$ can adsorb onto polystyrene latex particles, below the Krafft point, to give a monolayer on the surface in which the alkyl chains are adsorbed on to the surface and the ethylene oxide groups extend into the solution phase (20). The latter are extensively hydrated and ultracentrifugation studies have indicated that the amount of water contained in the volume between the surface and that bounded by the spherical surface at a distance, δ, from the particle surface (Figure 9) could be as high as 70%.

The effect of the extensively hydrated layer is that it forms a steric barrier which prevents the particles entering into a primary minimum, since it rises very steeply as soon as the distance of separation of the native particle surfaces is of the order of 2δ. The effect is illustrated qualitatively in Figure 12; details of its theoretical treatment can be found elsewhere (20,39). The overall stabilisation which generally ensues from this type of adsorbed layer is termed steric stabilisation (40).

The c.c.c. values for polystyrene latices at various extents of surface coverage by $C_{12}E_6$ have been studied in some detail (41) and some data from these studies are shown in Figure 13. From this figure it can be seen that as the surface coverage increases so the amount of lanthanum nitrate required to produce coagulation increases until once a monolayer has been formed coagulation could not be produced even with 0.3 mol dm^{-3} lanthanum nitrate using a 600 Å diameter particle. However, the protective action was particle size dependent and $C_{12}E_6$ was less effective with a particle of diameter 0.37 μm.

It should be noted, however, in connexion with the use of nonionic surfactants that flocculation is usually observed at a temperature just below that of the Cloud point of the surfactant. The exact temperature at which this occurs, however, can depend on the type of salt used and the presence of other surfactants.

An interesting feature occurs in the use of nonionic surfactants with PTFE latices. Experiments indicate that PTFE particles adsorb ethylene oxide type surfactants (42). Since as indicated earlier hydrocarbon chains are reluctant to adsorb on polytetrafluoroethylene surfaces the circumstantial evidence is strong that the initial adsorption occurs via interaction of the ethylene oxide group with the ionic groups on the PTFE latex particles. Indeed experimentally a small drop is noted in the c.c.c. value at low concentrations of $C_{12}E_6$ which supports this idea. Subsequent adsorption of the nonionic surfactant can then nucleate around the hydrocarbon chain of the head-group "down" molecules to give eventually monolayer coverage and enhanced stability.

Heterocoagulation

In the previous section the strong affinity of cationic surfactants for negatively charged polystyrene latices was noted. This concept of a single positive ion interacting with a negative charge on a surface can be extended to the interaction between a positively charged particle and a negatively charged particle. Moreover, the particles of different charges can have the same chemical composition or be composed of different materials. Thus mixing of latices containing particles of opposing charge can lead to coagulation and this phenomena is usually termed HETEROCOAGULATION.

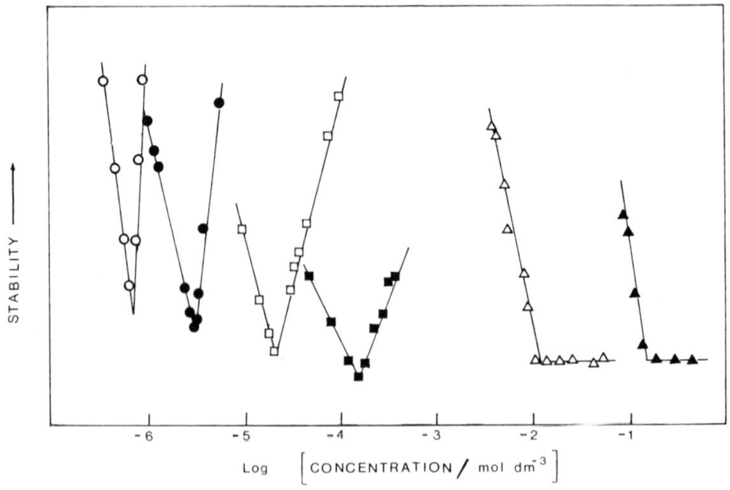

Figure 11. Influence of cationic surfactants on the stability of polystyrene latices in comparison with potassium bromide (▲). Alkyl trimethyl ammonium bromides of various chain lengths: (△) C_4; (■) C_8; (□) C_{10}; (●) C_{12}; (○) C_{16}.

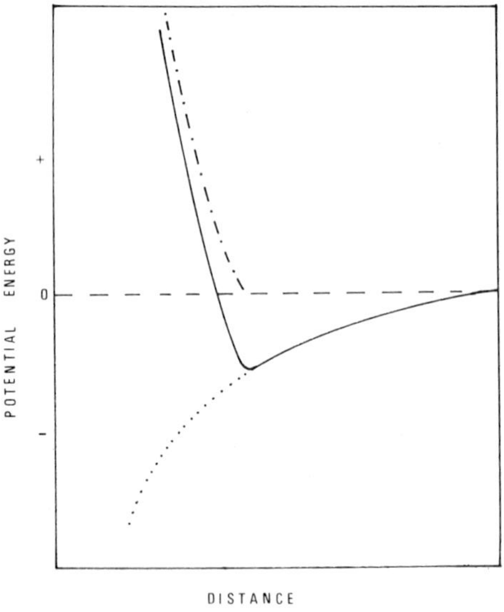

Figure 12. Diagram illustrating the form of the potential energy curve for two particles with adsorbed layers interacting in the absence of electrostatic repulsion: (– · –) steric interaction only; (· · ·) van der Waals interaction only; (——) steric plus van der Waals interaction.

An illustration of the effect is shown in Figure 14 where the change in stability of a polystyrene latex with a diameter of 52.7 nm is shown as a function of the ratio $N_-/(N_+ + N_-)$ where N_- = the number of negatively charged particles per unit volume and N_+ = the number of positively charged particles per unit volume. As can be seen, the system becomes completely unstable when this ratio reaches ca. 0.25.

When the particles are of different sizes complete coverage of the bigger particles by the smaller can occur (44). This is demonstrated by the scanning electron micrograph shown in Figure 15.

Surface Coagulation

There are a number of cases with polymer colloid systems where the coagulation process can occur at the liquid-air interface under conditions of electrolyte concentration which are far removed from those required to produce coagulation in the bulk solution. The mechanism appears to be connected with dewetting of the particle at the water-air interface either as a consequence of desorption of stabilising surfactant or the fact that the particle surface is not homogeneous. PTFE latices are particularly prone to surface coagulation and this may be partly due to their non-spherical shape. The latter is a consequence of crystallinity and because of the polymer chain folding which occurs may mean that the ionic surface groups at the chain-ends are concentrated on some surfaces, whilst the other surfaces are devoid of stabilising entities. Consequently, the latter have a high contact angle against water. One method of preventing surface coagulation is to store the latex in containers without a water-air interface. This subject has been dealt with in some depth by Heller and his collaborators (45).

The Effects of Adsorbed Macromolecules

Space prohibits a detailed discussion of this topic but a few general points can be made following the comments made in the previous section. In dealing with "hairy" particles of the type shown in Figure 2 where polyelectrolyte molecules are chemically linked to the surface, adsorption can also occur by several mechanisms including, ionic bonding - particularly via charges of opposite sign, by hydrogen bonding, by coupling with multivalent inorganic ions or by hydrophobic bonding of the hydrophobic regions of the macromolecule to the surface.

The net result is shown schematically in Figure 16. Instead of the array of surface charges leading to a well defined surface charge density and surface potential there is now a distribution of charges in space which contribute to the electrical double layer surrounding the particle. At low electrolyte concentrations the latter will extend into the space beyond the polyelectrolyte

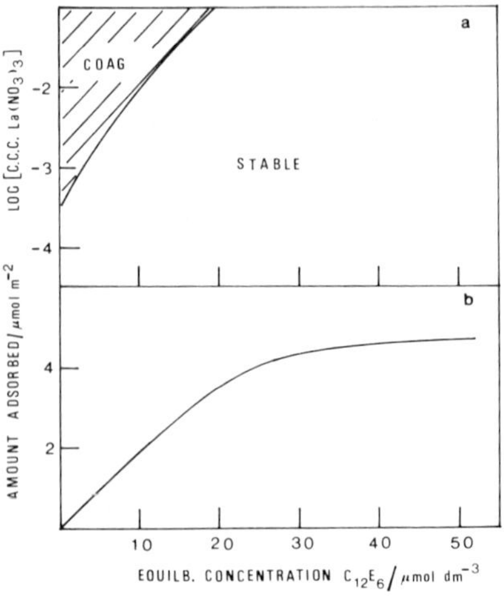

Figure 13. (a) Diagram showing the coagulation and stability regions for a polystyrene latex (particle diameter = 60 nm) in the presence of lanthanum nitrate and $C_{12}E_6$. (b) Adsorption isotherm for $C_{12}E_6$ on the surface of polystyrene latex particles.

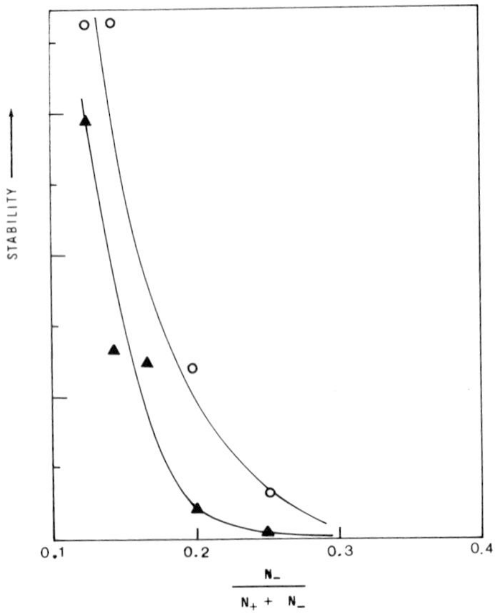

Figure 14. Changes in stability of an anionic polystyrene latex (particle diameter = 52.7 nm) admixed with a cationic latex (particle diameter = 43.4 nm). N_+ and N_- = the number concentration of the cationic and anionic latices, respectively; (–▲–) 10^{-3} mol dm^{-3} sodium chloride solution; (–○–) 2×10^{-3} mol dm^{-3} sodium chloride solution.

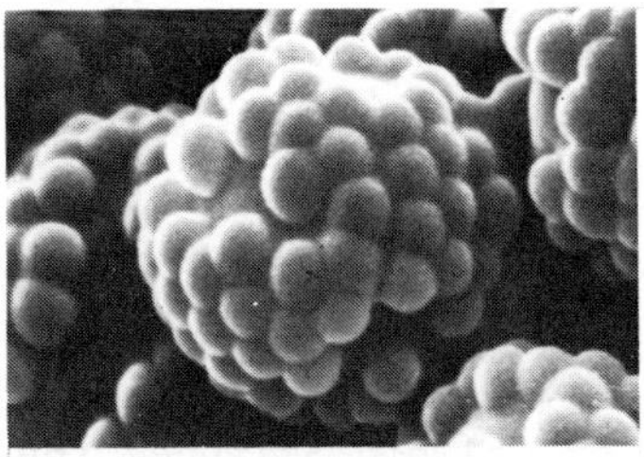

Figure 15. Scanning electron micrograph illustrating the heterocoagulation of cationic latex particles (diameter = 0.43 μm) onto a negatively charged latex particle (diameter = 2.14 μm)

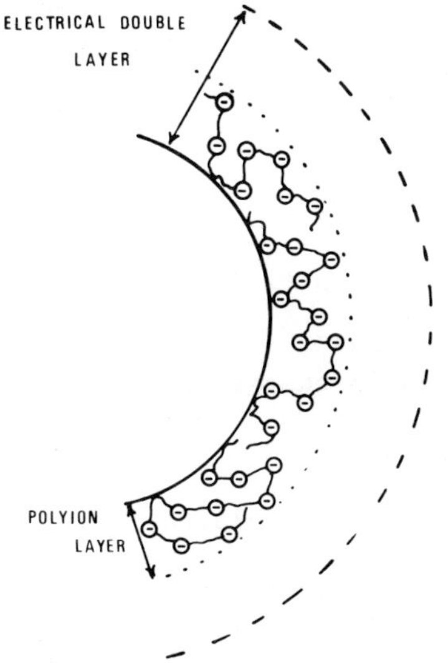

Figure 16. Schematic of a latex particle with an adsorbed layer of polyelectrolyte: (· · ·) the extension in space of the adsorbed layer and (– – –) the extension in space of the electrical double layer

layer so that the V_R term will be significant. However, with increase in electrolyte concentration and compression of the electrical double layer, the V_R term can become small or zero. However, under these conditions the particle will still be coated with an extensively hydrated layer of polymeric molecules which provide a steric stabilizing barrier. Hence, this type of system provides a two-tier mechanism of stabilization against electrolyte additions and the classical protective agents for colloidal particles such as gelatin, gum arabic etc., almost certainly act in this way. At the present meeting, the papers by Bassett (46) and by Korner (47) consider polymer colloids of this type.

Floc Structure

In the previous sections we have considered various ways of achieving the colloid stability of polymer latices and also of obtaining instability, i.e. either coagulation or flocculation. In general, in the academic literature on Colloid Science, the latter states are regarded as an end-product and the main emphasis in research has been directed towards understanding the factors which control stability. However, in the technological application of latices, it is often material in the flocculated state which is utilised in order to obtain the desired flow properties. In fact, in order to focus attention on the question of floc structure and pursue this topic further, we must ask a series of questions:-

a) why does aggregation occur?
b) how does the aggregation process proceed?
c) is the aggregated state reversible or non-reversible?
d) what are the forces acting between the particles in an aggregate?
e) what type of aggregate will be produced?
f) what is the arrangement of particles in an aggregate?
g) what is the magnitude of the bonding forces in a floc?
h) what is the mechanical strength of the floc?
i) what is the free volume in the floc?
j) how do the rheological properties of a flocculated system depend on floc structure?

In a schematic manner, some of these points can be illustrated by the diagram given in Figure 17 where I have depicted the consequences of adding an additive, X, to a stable latex. To some extent the question a, b, c and d have been discussed in the previous sections. For example, if X is a simple salt with particles of size ca.0.3μm or less, irreversible compact coagula are obtained (30) whereas, with larger particles, chains appear to dominate (31). With X = a cationic surface active agent or a hydrolysed aluminium ion, large compact clumps are formed (34). If X is a nonionic polymeric molecule, then frequently loose open structures are obtained in which the

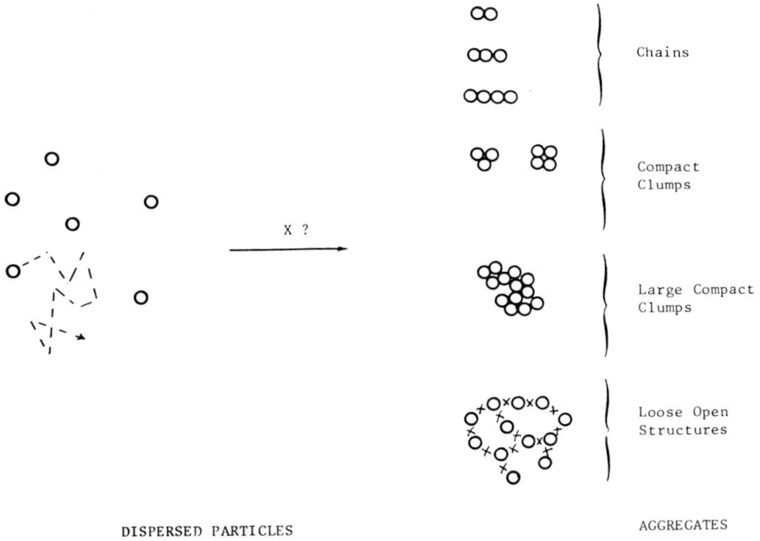

Figure 17. Effect of an additive **X** *on the stability of a latex and the type of aggregates formed according to the nature of* **X**

electrostatic repulsive forces between the particles still exist to keep them apart but the polymeric molecules link the individual particles to form a three-dimensional network. Such a system can consequently have a large volume. A prerequisite for such network formation is that the length of the polymer molecule, in its randomly coiled solution form, is sufficient to "bridge" the potential energy barrier between two individual particles. Thus the ratio of the size of the particle to the root mean square end to end distance of the polymer is an important factor. With small particles, diameter ca.50 nm, bridging occurs easily and since they are small, or similar in dimensions, to the polymers flocculation occurs easily; moreover, the number concentration of particles is high and hence a large number of bridges are formed. With large particles, diameter > 1 μm, the polymer molecule is much smaller than the particle and hence it forms an adsorbed layer which often confers a degree of steric stability (48).

Examples of this type of behaviour have been observed in a rheological study of polystyrene latices in the presence of ethyl hydroxy ethyl cellulose (48,49). The effect of particle size is illustrated in Figure 18a and 18b. With the smaller particles the viscosity rises very rapidly at quite low volume fractions and with the larger particle latex (diameter - 224 nm) a slower rise is observed. Systems of this type also show a pronounced dependence of their rheological properties on the shearing stress since as this is increased "bridges" in the network are destroyed and the latex flows as a Newtonian system. Since the breakdown of the "bridges" and their reformation on removal of the shearing stress is a kinetic process, the rheology of such systems often shows a distinctive time dependence, i.e. thixotropy. An example of this type of phenomena is given in Figure 19, where both the shearing stress against rate of shear behaviour extrapolated to zero time is given together with the steady state curve (e.g. infinite time).

Conclusion

In this presentation I have given a brief review of the factors controlling the stability and instability of polymer latices. There is little doubt that as a consequence of the ready availability of dispersions of spherical, monodisperse particles, polymer latices, our knowledge of the behaviour of colloidal dispersions has progressed rapidly over the past fifteen years. However, many phenomena remain to be investigated in quantitative detail and we must remember that the small energy changes involved in these systems, by comparison with molecular reactions, make many of the phenomena very subtle.

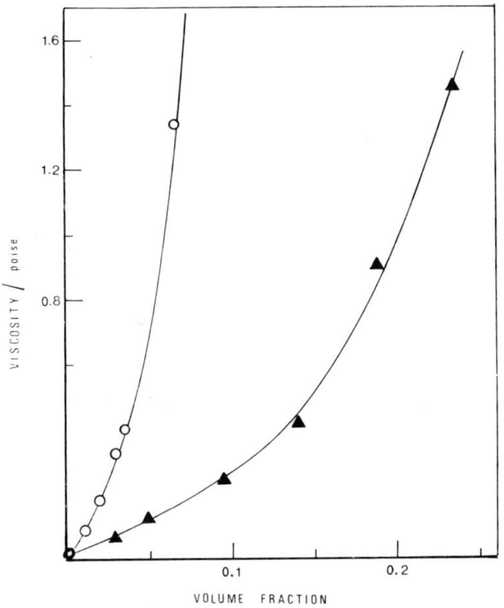

Figure 18. Viscosity vs. volume fraction for polystyrene latices in the presence of an adsorbed layer of ethyl hydroxy ethyl cellulose (full coverage): (–○–) particle diameter = 58 nm; (–▲–) particle diameter = 224 nm. Measurements made at pH 9 in 10^{-3} mol dm^{-3} sodium chloride solution at 25°C.

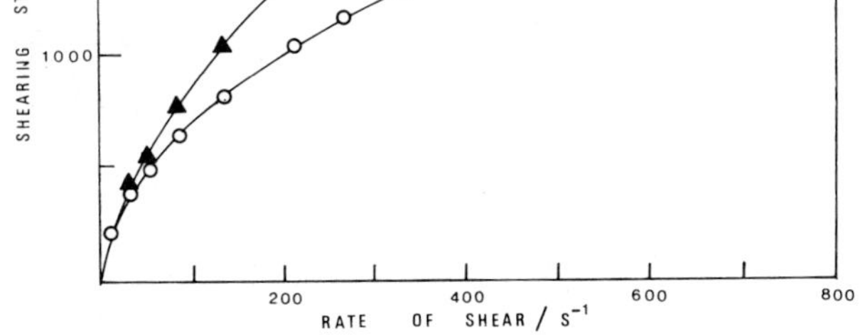

Figure 19. Shearing stress vs. rate of shear for a 1.95% solution of ethyl hydroxy ethyl cellulose in 10^{-3} mol dm^{-3} sodium chloride at pH 9 and 25°C: (–▲–) zero time data increasing rate of shear; (–○–) infinite time data decreasing rate of shear.

Literature Cited

1. Schulze, H. J.Prakt.Chem. 1882, 25, 431; 1883, 27, 320.
2. Linder, S.E.; Picton, H. J.Chem.Soc. 1892, 61, 148; 1897, 71, 568.
3. Hardy, W.B. Proc.Roy.Soc. 1900, 66, 110.
4. Hardy, W.B. Z.Phys.Chem. 1900, 33, 385.
5. Verwey, E.J.W. Chem.Rev. 1935, 16, 388.
6. Kruyt, H.R.; Overbeek, J.Th.G. Trans.Faraday Soc. 1940, 36, 110.
7. Derjaguin, B.V. Trans.Faraday Soc. 1940, 36, 203.
8. Langmuir, I. J.Chem.Phys. 1938, 6, 873.
9. London, F. Z.Phys. 1930, 63, 245.
10. Hamaker, H.C. Physica 1937, 4, 1058.
11. Derjaguin, B.V.; Landau, L. Acta Physicochim.U.R.S.S. 1941, 14, 633.
12. Verwey, E.J.W.; Overbeek, J.Th.G. "Theory of the Stability of Lyophobic Colloids"; Elsevier: Amsterdam, 1948.
13. Goodwin, J.W.; Ottewill, R.H.; Pelton, R.; Vianello, G.; Yates, D.E. Brit.Polymer J. 1978, 10, 173.
14. Reerink, H.; Overbeek, J.Th.G. Disc.Faraday Soc. 1954, 18, 74.
15. Ottewill, R.H. "Colloid Science", Ed. Everett, D.H. Chemical Society, London, 1973, 1, 173.
16. Overbeek, J.Th.G. J.Colloid and Interface Science, 1977, 58,
17. Ottewill, R.H.; Shaw, J.N. Disc.Faraday Soc. 1966, 42, 154.
18. Force, C.G.; Matijević, E. Kolloid-Z.u.Z.für Polymere 1968, 224, 51.
19. Ottewill, R.H.; Rance, D.G. Croatica Chemica Acta 1977, 50, 65.
20. Ottewill, R.H.; Walker, T. Kolloid-Z.u.Z.für Polymere, 1968, 227, 108.
21. Storer, C.S. Ph.D. Thesis, University of Bristol, 1968.
22. Pelton, R. Ph.D. Thesis, University of Bristol, 1976.
23. Neiman, R.E.; Lyashenko, O.A. Colloid J.U.S.S.R. (Eng.trans.) 1962, 24, 422.
24. Bibeau, A.A.; Matijević, E. J.Colloid and Interface Science, 1973, 43, 330.
25. Matijević, E.; Janauer, G.E.; Kerker, M. J.Colloid Science, 1964, 19, 333.
26. Matijević, E.; Force, C.G. Kolloid-Z.u.Z.für Polymere, 1968, 225, 33.
27. Matijević, E. J.Colloid and Interface Sci. 1977, 58, 374.
28. Kratohvil, S.; Matijević, E. J.Colloid and Interface Science, 1976, 57, 104.
29. Ottewill, R.H.; Rance, D.G. Croatica Chemica Acta, 1979, 52, 1.
30. Mardle, R. B.Sc. Thesis, University of Bristol, 1980

31. Cornell, R.M.; Goodwin, J.W.; Ottewill, R.H. J.Colloid and Interface Science, 1979, 71, 254.
32. Ottewill, R.H. "Surface Active Agents", Society of Chemical Industry, London, 1979.
33. Rance, D.G. Ph.D. Thesis, University of Bristol, 1976.
34. Connor, P. Ph.D. Thesis, University of Bristol, 1968.
35. Ottewill, R.H.; Chemistry and Industry, 1980, 377.
36. Connor, P.; Ottewill, R.H. J.Colloid and Interface Science, 1971, 37, 642.
37. Ottewill, R.H.; Rastogi, M.C.; Watanabe, A. Trans.Faraday Soc. 1960, 56, 854.
38. Richardson, R.A. B.Sc. Thesis, University of Bristol, 1979.
39. Napper, D.H. J.Colloid and Interface Science, 1977, 58, 390.
40. Heller, W.; Pugh, T.L. J.Chem.Phys., 1954, 22, 1778.
41. Ottewill, R.H.; Walter, T.W. J.Chem.Soc.Faraday Trans.I, 1974, 70, 917.
42. Bee, H. B.Sc. Thesis, University of Bristol, 1978.
43. Cheung, W.K. Ph.D. Thesis, University of Bristol, 1979.
44. Goodwin, J.W.; Ottewill, R.H. Disc.Faraday Soc. 1978, 65, 338.
45. Heller, W.; Peters, J. J.Colloid and Interface Science, 1970, 32, 592.
46. Bassett, D.R.; Derderian, E.J.; Johnston, J.E.; MacRury Organic Coatings and Plastics Chem. 1980, 43, 426.
47. Buscall, R.; Corner, T. Organic Coatings and Plastics Chem. 1980, 43, 203.
48. Ho, C.C. Ph.D. Thesis, University of Bristol, 1971.
49. Perks, K.B. Ph.D. Thesis, University of Bristol, 1971.

RECEIVED April 6, 1981.

3

Well-Characterized Monodisperse Polystrene Latexes as Model Colloids

J. W. VANDERHOFF

Emulsion Polymers Institute and Department of Chemistry, Lehigh University, Bethlehem, PA 18015

An ideal model colloid should comprise monodisperse spherical particles stabilized with a known number of chemically bound surface groups. Monodisperse polystyrene latexes (1) are a good starting point for the preparation of such an ideal model colloid. Their particle size can be varied systematically over a wide range, and their bound surface groups are the endgroups of the polymer molecules, so that variation of polymer molecular weight and latex particle size can give different surface charge densities. Moreover, the preparation of these latexes using persulfate initiator gives sulfate endgroups, which can be hydrolyzed to hydroxyl groups, which, in turn, can be oxidized to carboxyl groups, to give particles stabilized with the same number of three different surface groups. The purpose of this paper is to describe the preparation and characterization of such model colloids, starting with the emulsion polymerization of styrene using persulfate initiator.

The preparation of a latex by emulsion polymerization comprises two stages: (i) particle nucleation; (ii) particle growth. For the latex to be monodisperse, the particle nucleation stage must be short relative to the particle growth stage. Despite many investigations, there is disagreement as to the *locus* of particle nucleation: (i) monomer-swollen emulsifier micelles; (ii) adsorbed emulsifier layer; (iii) aqueous phase; (iv) monomer droplets. Whatever the *locus* of particle nucleation, the decomposition of persulfate is considered to produce sulfate ion-radicals in the aqueous phase. These ion-radicals are repelled by the negatively charged surface of a monomer-swollen emulsifier micelle, polymer particle or monomer droplet and therefore remain in the aqueous phase and add hydrophobic monomer units to form an oligomeric radical. When this oligomeric radical attains a critical chain length, it becomes surface-active and adsorbs on the surface of a micelle, polymer particle, or monomer droplet. If it cannot adsorb on a suitable surface, it continues to grow until it exceeds its solubility in water and therefore precipitates to nucleate a latex particle. Initiation in micelles is significant

0097-6156/81/0165-0061$05.75/0

only in the beginning of the polymerization, and the monomer droplets usually are not a significant locus of particle nucleation. Therefore, once the particle nucleation stage is completed, the monomer-swollen polymer particles are the principal locus of polymerization. The adsorption of an oligomeric sulfate radical to initiate a polymeric radical and the termination of this polymeric radical by adsorption of another oligomeric sulfate radical should give two surface sulfate groups for each polymer molecule formed assuming that no transfer reactions occur to introduce a different endgroup, that termination inside the particle occurs by combination rather than by disproportionation, and that none of the sulfate endgroups become buried inside the particle.

Polystyrene latexes have been prepared using persulfate initiator for many years, but only recently have methods been developed to determine the number and loci of the sulfate surface groups. To determine these surface groups, the latex is cleaned to remove the adsorbed emulsifier and solute electrolyte, then the surface sulfate groups in the H^+ form are titrated conductometrically with base. The latexes can be cleaned effectively by ion exchange (2-5) or serum replacement (6); dialysis is not effective in removing the adsorbed emulsifier and solute electrolyte (3,5,6).

In ion exchange, the aqueous phase ions are replaced with H^+ and OH^- ions. If the aqueous phase ions are in equilibrium with the adsorbed ions, their removal from the aqueous phase causes desorption of the adsorbed ions to maintain the equilibrium until all of the adsorbed ions have been removed. In practice, this removal is quantitative (2-5). Ion exchange is rapid and easily carried out; however, commercial ion exchange resins contain leachable polyelectrolytes which adsorb on latex particle surfaces; these polyelectrolytes can be removed only by an arduous purification process (2-5).

In serum replacement (6), the latex is confined in a cell with a semi-permeable membrane, e.g., Nuclepore filtration membrane, and water is pumped through the latex to literally replace the serum. The removal of adsorbed ions is quantitative provided the adsorption-desorption equilibrium is maintained. The Na^+ and K^+ ions are replaced by H^+ ions by pumping dilute hydrochloric acid through the latex followed by water to remove the excess acid. Serum replacement takes longer than ion exchange, but avoids the arduous resin purification step; moreover, the serum is recovered quantitatively in a form suitable for analysis.

In dialysis, the latex is confined with a semi-permeable membrane (e.g., in a dialysis bag), and the aqueous phase ions are removed by diffusion across this semi-permeable membrane. The rate of this diffusion can be increased by changing the water outside the bag more often or by using dialysis membranes of greater surface area (e.g., hollow fiber dialysis). In principle, the adsorbed ions should be removed quantitatively provided the adsorption-desorption equilibrium is maintained. In practice, however, the removal of adsorbed ions as well as the replacement of Na^+ and K^+ ions by H^+ ions is incomplete (2,3,6). The dialysis membrane must be considered a third phase in this process; moreover, the rate of dialysis decreases drastically with decreasing concentration gradient across the membrane (7).

The cleaned latex in the H^+ form is titrated conductometrically or potentiometrically with base to determine the surface charge (2,5,8). Conductometric titration is the more sensitive method when the overall conductance is very low as in a sample cleaned by ion exchange or serum replacement; potentiometric titration is the more sensitive when the conductance is relatively high. Figure 1 shows a typical conductometric titration curve (8). The solid line shows the experimental curve and the dashed line the theoretical curve. The theoretical curve lies slightly above the experimental curve; moreover, the slope of its decending leg is much greater. The difference in the descending slopes results from the distribution of the H^+ counterions in the double layer; some are held close to the particle surface and therefore do not contribute much to the overall conductance. The lower the electrolyte concentration and thus the greater the expansion of the electric double layer, the smaller the slope of the descending leg. The higher position of the theoretical curve results because the conductometric titration involves ion exchange as well as neutralization, which decreases the overall conductance.

Characterization by Ion Exchange and Conductometric Titration

Earlier work (2,3,5,9) described the characterization of monodisperse polystyrene latexes of 25-447nm particle diameter prepared using persulfate initiator, bicarbonate buffer, and four different emulsifiers. The latexes were ion exchanged using rigorously purified Dowex 50W(H^+)-Dowex 1 (OH^-) mixed resins and titrated conductometrically; this process was repeated until the surface charge was constant. These surface charges shown in Table I (3,5,9) correspond to 1.0-1.3 sulfate endgroups per polystyrene molecule (molecular weight determined by osmometry). To determine if any sulfate groups were buried inside the particle, the polymer was recovered from the ion-exchanged latex, dissolved in dioxane-water mixture, ion exchanged again, and titrated conductometrically. The total number of sulfate endgroups was in the range 1.0-1.6 per polystyrene molecule (except for one latex which had a value of 2.07), suggesting that there are other endgroups besides sulfates.

One possibility is hydroxyl endgroups, which may be formed by a side reaction of sulfate ion-radicals to form hydroxyl radicals (9) or hydrolysis of the surface sulfate groups. To determine if hydroxyl groups were present, the ion-exchanged latexes were oxidized by heating with persulfate and 10^{-5} silver ion at 90°, then ion exchanged and titrated conductometrically to determine the carboxyl groups. Table II (9) shows that some sul-

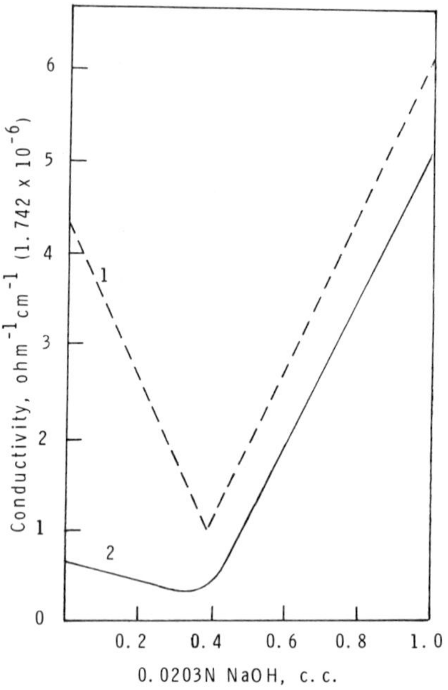

Figure 1. Conductometric titration of ion-exchanged 234-nm-diameter monodisperse polystyrene latex: (1) theoretical curve calculated assuming 100% dissociation; (2) experimental curve (8).

fate groups were hydrolyzed and oxidized to carboxyl groups during this process, but the carboxyl titration values corrected for this hydrolysis accounted satisfactorily for the requisite two endgroups per polystyrene molecule.

The extent of the side reaction of sulfate ion-radicals with water to produce hydroxyl radicals was postulated to increase with decreasing pH. Therefore, polymerizations were carried out using persulfate initiator but with the pH of the polymerization adjusted to values in the range of pH range 2-8 (9). Table III (9) shows that, at the lowest pH, the endgroups were about 90% hydroxyls and 10% sulfates; at pH 7-8, they were all sulfates.

These results showed that monodisperse polystyrene particles prepared using persulfate initiator and bicarbonate buffer contain both sulfate and hydroxyl surface groups. Some sulfate groups are on the particle surface, while others are buried inside the particle. All of the hydroxyl groups are on the particle surface. These results also showed that careful control of the pH during polymerization can produce latex particles stabilized with only surface sulfate groups.

Characterization by Serum Replacement and Conductometric Titration

For serum replacement (6), the latex is confined in a cell with a uniform-pore-size Nuclepore filtration membrane. Distilled, deionized water is pumped through the latex until the conductance of the effluent stream is about the same as that of the distilled, deionized water. This serum replacement removes the adsorbed emulsifier and solute electrolyte quantitatively and allows recovery of the serum in a form suitable for further analysis; however, it does not replace the Na^+ and K^+ counterions of the surface groups with H^+ ions. To do this, dilute hydrochloric acid (ca. 10^{-4}N) is pumped through the latex, followed by distilled, deionized water to remove the excess acid. The latex is then titrated conductometrically to determine the surface charge.

The pore size of the Nuclepore filtration membranes used is 0.50-0.75 times the particle diameter of the latex to be cleaned. The latex in the cell is agitated close to the membrane to prevent clogging by deposited particles. Even so, clogging is observed at higher pumping pressures and latex solids contents, e.g., for 10-15% solids latex at 5 psi pressure, while 5-10% solids latex at 2 psi gives no clogging. Suction filtration can also be used to increase the rate of serum replacement.

Seven polystyrene latexes prepared with persulfate initiator and bicarbonate buffer were characterized to demonstrate the efficacy of this method (6). Three were monodisperse latexes prepared using conventional emulsifiers; four were prepared using sodium styrene sulfonate or sodium vinyltoluene sulfonate as comonomeric emulsifiers. Each latex was subjected to serum replacement with

Table I
Monodisperse Polystyrene Latexes
Surface Charge and Number of Sulfate Endgroups/Polymer Molecule (3,5,9)

Latex	Particle Diameter, nm	Surface Charge, μC/cm^2	Sulfate Endgroups/Molecule Surface	Sulfate Endgroups/Molecule Total
A-1	25	0.5	12.4*	13.0*
A-2	88	3.3	0.98	1.57
A-3	234	2.0	0.87	1.59
B-1	158	4.2	1.26	1.21
B-2	248	5.7	1.02	1.23
C-1	254	5.4	0.95	0.94
D-1	109	1.6	0.95	----
D-2	187	1.8	0.90	----
D-3	285	4.9	1.03	1.63
D-4	447	8.1	0.93	2.07

A = Aerosol MA (bis-1,3-dimethylbutyl sodium sulfosuccinate)
B = Sipon WD (sodium lauryl sulfate)
C = Aerosol OT (di-2-ethylhexyl sodium sulfosuccinate)
D = potassium oleate
* μeq/gm polymer

Table II
Monodisperse Polystyrene Latex A-2
Oxidation of Sulfate and Hydroxyl Endgroups
10^{-5}N silver nitrate; 6 hours at 90°)

Potassium Persulfate, %*	Number of Endgroups, μeq/gm Sulfate	Carboxyl	Hydroxyl
None (original)	22.5	none	----
10	11.7	22.0	11.2
15	8.0	24.7	10.2
30	6.4	26.3	10.2

* based on polymer

	Endgroups/Polymer Molecule
surface sulfate	0.98
total sulfate	1.57
sulfate + hydroxyl	2.06

Table III
Monodisperse Polystyrene Latexes
Effect of pH of the Polymerization Medium (9)

pH	Particle Diameter, nm	Endgroups/Polymer Molecule Surface Sulfate	Total Sulfate	Sulfate + Hydroxyl
1.9	143	0.19	0.20	1.84
3.3	132	0.55	1.00	1.75
4.4	133	0.62	1.05	2.00
5.0	128	0.71	1.10	2.00
6.4	130	0.46	0.55	2.01
7.8	129	1.40	2.07	2.07
11.6	162	0.62	1.00	1.42

water and then titrated conductometrically. Then, this sample was ion exchanged to remove material not removed by serum replacement and titrated again. Then the sample was subjected to serum replacement with dilute hydrochloric acid to replace the Na^+ and K^+ counterions with H^+ ions and titrated. This sample was also ion exchanged and titrated again. These results were compared with those determined by ion exchange alone. Also, for comparison, latex samples were dialyzed for one month and then titrated. These samples were then ion exchanged and titrated again.

Table IV shows the results of these experiments. Serum replacement with water gave low values for the surface charge because of incomplete replacement of the Na^+ and K^+ ions by H^+ ions. However, serum replacement with water, hydrochloric acid, and water gave values equal to, or slightly less than, those obtained by ion exchange, demonstrating the efficacy of this new cleaning method. The values for the dialyzed samples were also significantly lower than those obtained by ion exchange.

Dialysis

Table IV shows that dialysis is ineffective in cleaning the latexes for characterization. Earlier work (3,5) also showed that dialysis is ineffective in removing the adsorbed emulsifier and replacing the Na^+ and K^+ counterions with H^+ ions. Others have also found that dialysis does not remove emulsifier completely. Brodnyan and Kelley (10) found that aqueous solutions of C14-tagged sodium lauryl sulfate equilibrated upon dialysis, but only 9.5% and 22% of the emulsifier was removed from latexes dialyzed under the same conditions. Matijevic et al. (11) dialyzed a butadiene-styrene copolymer latex prepared using rosin acid soap for 160 days and removed only about 50% of the emulsifier. As mentioned above, Edelhauser (7) showed that the concentration gradient across the dialysis membrane must exceed a critical value to make the dialysis proceed at a practical rate. In contrast, Ottweill and Shaw (12) found from electrophoretic mobility measurements and desorption of radioactive emulsifier that all of the emulsifier was removed by dialysis or at least that a constant surface charge was obtained.

Endgroups Produced by Persulfate Initiator

The foregoing latexes prepared using persulfate initiator and bicarbonate buffer contained surface sulfate groups and, in some cases, surface hydroxyl groups. None of these latexes contained carboxyl groups. These latexes had final pH values of 7-8. If the bicarbonate buffer was omitted, the final pH of the latex was 2-3 because of the bisulfate ion generated by the persulfate decomposition.

In contrast, others have found carboxyl groups in polystyrene latexes prepared using persulfate initiator. Ottewill and Shaw (12) cleaned latexes prepared with various emulsifiers by dialysis and found carboxyl groups by variations of electrophoretic mobility with pH, infrared spectra of the dried polymer, and potentiometric titration of dried polymer redispersed in water. Later, Ottewill et al. (13) reported carboxyl groups in polystyrene latexes prepared with hydrogen peroxide initiator and sodium laurate emulsifier as well as in emulsifier-free latexes prepared using persulfate; in addition, Krieger et al., Fitch et al., Homola et al., and Gultepe and Everett (13) reported carboxyl groups in latexes prepared using different methods. In some cases, the latexes were cleaned by dialysis and, in other cases, by ion exchange, but some of the ion exchange resins were not purified.

These results are in disagreement with the foregoing findings that only sulfate and hydroxyl groups are produced by persulfate initiator, particularly with bicarbonate buffer. There are several possible sources of the carboxyl groups found by others: (i) hydrolysis of sulfate groups followed by oxidation; (ii) dissolution of carbon dioxide in the latex; (iii) oxidation of surface sulfate groups, particularly in the presence of small concentrations of heavy metal ions; (iv) hydrolysis of ester groups for copolymers containing methacrylate and acrylate esters; (v) side reactions of initiating species. Another possibility is the contamination of the latex by oxidation products of the dialysis membranes (14).

Table IV
Surface Characterization of Polystyrene Latexes (6)

Treatment	LS-1102-A	LS-1047-E	LS-1134-B	S-10	S-13	V-4	V-6
H_2O	0.55	1.42	0.74	2.09	1.30	0.62	0
H_2O + I.E.	1.20	2.31	2.23	10.9	18.5	4.98	50.3
H_2O + HCl + H_2O	1.13	2.32	2.34	11.0	18.2	4.95	50.7
H_2O + HCl + H_2O + I.E.	1.19	2.32	2.34	10.9	18.2	4.96	50.9
I.E.	1.35	2.56	2.34	11.4	20.3	5.29	52.4
Dialyzed	1.01	1.48	1.43	4.04	5.08	1.66	14.7
Dialyzed + I.E.	1.35	2.06	2.20	9.81	18.3	5.12	51.3

Table V
Effect of Polymerization Recipe on Surface Groups (15.16)

Latex	Emulsifier	Initiator	Buffer	Surface Charge μeq/gm Sulfate	Carboxyl	pH
41	Aerosol MA	$K_2S_2O_8$	$NaHCO_3$	6.99	none	7.0
W23	Aerosol MA	$K_2S_2O_8$	$NaHCO_3$	7.71	none	8.3
520	none	$K_2S_2O_8$	none	9.34	1.45	2.0
520'	none	$K_2S_2O_8$	$NaHCO_3$	6.45	3.07	2.6
520"	none	$K_2S_2O_8$	$NaHCO_3$	9.10	none	8.0
42	Aerosol MA	$K_2S_2O_8$	$NaHCO_3$, KH_2PO_4	6.95	1.41	7.0
43	Aerosol MA	$K_2S_2O_8$	$NaHCO_3$, $Na_2B_4O_7$	6.80	4.60	8.7
43'	Aerosol MA	$K_2S_2O_8$	$NaHCO_3$, $Na_2B_4O_7$	6.79	3.48	8.0
43"	none	$K_2S_2O_8$	$NaHCO_3$, $Na_2B_4O_7$	9.87	3.43	8.7
W1	Aerosol MA	$K_2S_2O_8$	NaOH, KH_2PO_4	4.01	9.70	7.0
W2	Aerosol MA	$K_2S_2O_8$	$NaHCO_3$, NaOH, KH_2PO_4	3.12	3.57	8.7
44	Aerosol MA	$K_2S_2O_8$	$NaHCO_3$, $AgNO_3$	5.63	3.15	8.5
44'	Aerosol MA	$K_2S_2O_8$	$NaHCO_3$, $AgNo_3$	3.50	1.21	8.0
44"	none	$K_2S_2O_8$	$NaHCO_3$ $AgNO_3$	8.42	2.90	8.0
W21	Aerosol MA	$(NH_4)_2S_2O_8$	NH_4OH	7.28	none	2.7
W22	Aerosol MA	$(NH_4)_2S_2O_8$	NH_4OH	7.89	none	2.3
716D	Aerosol MA	$(NH_4)_2S_2O_8$	none	5.01	2.36	2.2

Table VI
Effect of Ageing on Ion Exchanged (H^+ Form) Latex 41

	Surface Charge, μeq/gm		
Ageing Time, days	Sulfate	Carboxyl	Total
0	7.02	3.06	10.88
1	7.79	2.92	10.71
4	6.86	4.04	10.90
8	6.97	4.73	11.70
11	6.26	5.22	11.48
15	4.39	6.88	11.27
22	1.08	7.13	8.21
32	none	9.73	9.73
---	none	10.83	10.83

Table VII
Effect of Ageing on Ion Exchanged (H^+ Form) Latex W21

	Surface Charge, μeq/gm		
Ageing Time, days	Sulfate	Carboxyl	Total
0	6.47	3.64	10.11
1	6.63	3.82	10.47
4	5.30	6.14	11.44
8	5.54	4.34	9.78
11	4.84	5.69	10.31
15	3.45	5.14	8.59
22	1.27	6.33	7.60
32	none	9.01	9.01
---	none	9.29	9.29

Table VIII
Effect of Ageing Conditions on
Ion Exchanged (H^+ Form) Latex 43' (16)

		Surface Charge, μeq/gm		
Treatment	pH	Sulfate	Carboxyl	Total
initial	8.0	---	---	---
ion exchanged	3.5	6.79	3.48	10.27
ion exchanged; 14 days at ca. 25°	---	none	10.47	10.47
pH adjusted to 3.35 with 0.1N HCl; 142 days at ca. 25	3.35	6.78	3.64	10.42

Table IX
Effect of Ageing Conditions on
Ion Exchanged (H^+ Form) Latex 520" (16)

Ageing Temp., °C	Time, days	Surface Charge, μeq/gm		
		Sulfate	Carboxyl	Total
----	none	8.57	none	8.57
ca. 25	30	5.35	2.73	8.08
ca. 25	44	none	8.22	8.22
90	4.0	none	none	0.00
90	8.8	none	none	0.00

To determine the effect of different polymerization conditions on the polymer endgroups produced, polymerizations were carried out using the standard bicarbonate buffer as well as other variations. Table V (15,16) shows that the use of the persulfate-bicarbonate combination with and without emulsifier gave latexes of final pH 7-8 with only sulfate groups. The addition of 10^{-5} silver ion gave a latex of pH 8.5, but with weak-acid groups, presumably because of oxidation of the sulfate groups. The omission of the bicarbonate buffer gave latexes of lower pH with weak-acid groups. The formation of weak-acid groups in these latexes is not merely an effect of pH during the polymerization: potassium dihydrogenphosphate, sodium borate, and sodium hydroxide, even in combination with sodium bicarbonate, gave both strong-acid and weak-acid groups, and pH values of 7-9. Also, the ammonium persulfate-ammonium hydroxide system gave latexes with pH values of 2-3, but only strong-acid groups. The omission of the ammonium hydroxide did not change the pH, but gave weak-acid groups.

Preparation of Model Colloids

Earlier work (3) has shown that cleaned monodisperse polystyrene latexes stabilized with surface sulfate (and perhaps a few hydroxyl) groups can be used as model colloids. For example, the distribution of H^+ ions in the electric double layer as determined by conductometric titration has been correlated with the particle diameter determined by ultracentrifugation (3). The conductometric titration gives two measures of the concentration of H^+ ions: the initial conductance of the latex and the amount of base required for neutralization. The number of H^+ ions determined by conductance is always smaller than the number determined by titration. This difference is attributed to the distribution of the H^+ ions in the electric double layer: those closest to the particle surface contribute least to the overall conductance. This distribution is expressed as the apparent degree of dissociation α, which is defined as the ratio H^+ ions (conductance)/H^+ ions (titration). For ion-exchanged latexes, the values of α are usually small, e.g., 0.21 for Latex A-2 described in Table I. The value of α has been used as a measure of the division of H^+ ions between the Stern layer and the diffuse double layer, e.g., 79% in the Stern layer and 21% in the diffuse double layer. This assumption allows calculation of the sedimentation diameter of these particles upon ultracentrifugation. Table X (3) compares the particle diameters determined by ultracentrifugation with that determined by electron microscopy.

Table X
Particle Diameters of Latex A-2 (3)

Sample	Particle Diameter, nm
ultracentrifugation - uncleaned latex in H_2O	85.4
ultracentrifugation - ion-exchanged latex in H_2O	75.4
ultracentrifugation - ion-exchanged latex in Aerosol MA solution	82.2
ultracentrifugation - ion-exchanged latex in electrolyte solution	79.0
sedimentation diameter from Booth's equation	76.0
electron microscopy	88.0 (σ=8.0)

The diameter of the original latex diluted in water is in good agreement with the diameter determined by electron microscopy. However, the diameter determined by ultracentrifugation of the ion-exchanged latex in electrolyte solution (79.0 nm) is most representative of the actual diameter measured by this method because the electric double layer effects are suppressed. This value is 3.2 nm smaller than the diameter determined for the ion-exchanged latex diluted in Aerosol MA solution, which gives 1.6 nm for the thickness of the adsorbed emulsifier layer. However, the 85.4 nm diameter of the original latex diluted in water is 3.2 nm larger than that of the ion-exchanged latex diluted in Aerosol MA solution, which suggests that the original latex was stabilized by a bimolecular layer of emulsifier. Moreover, the sedimentation diameter calculated from the most representative 79.0 nm diameter, assuming 79% of the H^+ ions are in the Stern layer and 21% in the diffuse double layer, is 76.0 nm, in good agreement with the 75.4 nm diameter determined by ultracentrifugation in deionized water.

Later work (8,15) showed that the value of α increased with increasing electrolyte concentration and that it could be correlated with the electrophoretic mobility. Table IX shows that the electrophoretic mobility measured using the Micromeritics Mass Transport cell increased with increasing α as observed for H^+, Na^+, and Ba^{++} counterions. These results also show that the distribution of the counterions in the electric double layer is critically dependent upon the nature of the counterion, e.g., most of the H^+ ions are close to the particle surface while most of the Na^+ ions are further away from the surface.

Table XI

Electrophoretic Mobility and Apparent Degree of Dissociation (15)

357 nm-diameter polystyrene latex (0.35% solids)

Counterion	Electrophoretic Mobility, μm cm/volt sec	α
H^+	1.41	0.063
Na^+	5.12	0.70
Ba^{++}	1.62	0.10

It was also shown earlier (9) that monodisperse polystyrene particles stabilized only with surface sulfate groups can be prepared by polymerization at pH 7-8 using potassium persulfate initiator and sodium bicarbonate buffer, followed by cleaning with ion exchange resin. Observation of monodisperse polystyrene latexes during storage showed that two of these latexes lost their strong-acid surface groups after a few years; when these latexes were ion-exchanged, they showed no measurable surface charge by conductometric titration, and the particles of one showed no movement at the stationary level in microcapillary electrophoresis. Both latexes were stable despite their zero charge; moreover, oxidation with persulfate and Ag^+ ion followed by ion exchange and conductometric titration showed the presence of carboxyl groups, suggesting that the surface sulfate groups had hydrolyzed completely to hydroxyl groups.

The first deliberate attempts to hydrolyze the surface sulfate groups to hydroxyls gave inconclusive results: continued heating at pH 3-4 hydrolyzed only part of the sulfate groups; the original latex contained mostly sulfate groups and a few hydroxyl groups and the final latex mostly carboxyl groups and a few sulfate groups (17). Higher concentrations of hydrochloric acid caused the latex to flocculate despite the addition of stabilizing emulsifier. Later, it was found (16) that allowing the latex in the H^+ form to stand at room temperature resulted in the quantitative disappearance of the sulfate groups and the appearance of a corresponding number of carboxyl groups. Tables VI and VII show the results for latexes prepared using the potassium persulfate-sodium bicarbonate and ammonium persulfate-ammonium hydroxide systems, respectively. In both cases, the strong-acid concentration decreased and the weak-acid concentration increased with ageing time at room temperature until only weak-acid groups were found, in about the same number as the original strong-acid groups were found. At intermediate times, the total number of surface groups was less than at the beginning or the end; this difference was attributed to hydroxyl groups which had not yet been oxidized. Table VIII shows for a latex prepared using the potassium persulfate-sodium bicarbonate-sodium borate system that the same degree of hydrolysis and oxidation cannot be obtained by addition of hydrochloric acid; ageing the latex at pH 3.35 (the lowest pH that the latex could withstand without flocculation) gave no change in the number of surface sulfate and carboxyl groups; however, when the ion-exchanged latex (H^+ form) was aged for 14 days at room temperature, the sulfate groups disappeared completely and were replaced with the same number of carboxyls. Table IX shows for an emulsifier-free latex prepared using the potassium persulfate-sodium bicarbonate system that ageing the ion-exchanged latex at 90° caused hydrolysis of the surface sulfate groups to hydroxyls without oxidation to carboxyls. Ageing for 44 days at room temperature gave all carboxyl groups while ageing for 4 days at 90° gave a latex with no measurable charge by conductometric titration, i.e., the slope of the conductometric titration curve was the same as that for excess sodium hydroxide. The absence of titrable groups was confirmed by back-titration with hydrochloric acid. Since the final number of weak-acid groups was the same within experimental error as the initial number of strong-acid groups, it was assumed that the zero-charge sample was stabilized with the same number of hydroxyl groups. This latex was stable and in 0.2% concentration showed the same conductance as pure water; it showed a slight negative charge upon electrophoresis in distilled water.

Other experiments (16) showed that this hydrolysis begins shortly after the latex is ion-exchanged. If the latex is titrated immediately after ion exchange, no weak-acid groups are found; however, if it is allowed to stand in the H^+ form for some time, the hydrolysis becomes significant. Moreover, it is the surface pH, not the pH of the medium, that is responsible for the hydrolysis. Because of the low value of α for H^+ counterions, the H^+ ion concentration near the surface is far higher than could be achieved by addition of hydrochloric acid; thus the hydrolysis is more effective. Generally, the hydrolysis proceeds more rapidly at temperatures higher than room temperature. Some latexes prepared using persulfate initiator cannot be hydrolyzed completely, suggesting that some of their strong-acid surface groups are sulfonates rather than sulfates. The rate of hydrolysis can be increased by increasing the surface area of the container, e.g., by adding Pyrex glass beads to the Pyrex glass vessel containing the latex; the reason for this increase in the rate of hydrolysis has not yet been determined. Generally, the hydrolysis and oxidation of the ion-exchanged latex in the H^+ form gives the same results as the earlier oxidations using persulfate ion and Ag^+ ion (9).

Thus the monodisperse polystyrene latex stabilized with strong-acid surface groups can be hydrolyzed to form a latex stabilized with the same number of nonionic hydroxyl groups, which in turn can be oxidized to form a latex stabilized with the same number of weak-acid carboxyl groups, thus offering model colloids with identical characteristics except for the type of chemically bound surface groups --- strong-acid, weak-acid, nonionic.

The electrophoretic mobility of these new model colloids has been measured as a function of pH (18). Figure 2 shows that the electrophoretic mobility of all three samples is negative and constant over the pH range 5-10, independent of the type of surface group. The presence of surface sulfate groups would have been expected to give a strong negative charge; however, it would have been expected that the charge of the carboxyl-stabilized particles would have increased with increasing pH in the range 5-10 and that of the hydroxyl-stabilized particles would have been close to zero. At pH values less than 5, the electrophoretic mobility varied as expected, decreasing to positive values for the hydroxyl- and carboxyl-stabilized particles and remaining negative for the sulfate-stabilized particles. Similarly, at pH values greater than 10, the variation of electrophoretic mobility was as expected. The results shown in Figure 2 are for two different latexes --- an emulsifier-free latex prepared using the potassium persulfate-sodium bicarbonate system and a Dow monodisperse polystyrene latex. Similar measurements have been made by several different investigators using three different instruments --- the Rank Brothers microcapillary electrophoresis, the Pen Kem Lazer Zee microcapillary electrophoresis, and the Pen Kem Model 3000, which measures in a automated manner according to the Brownian motion of the particles. The results all showed the same constant electrophoretic mobility over the pH range 5-10, independent of the type of surface group.

The charge on colloidal particles may arise from several sources: (i) dissociation of ionogenic groups, e.g., proteins or the chemically bound surface sulfate groups of polystyrene latex particles; (ii) adsorption of potential-determining ions, e.g., Ag^+ or I^- ions on silver iodide particles; (iii) dissociation of adsorbed species, e.g., sodium lauryl sulfate adsorbed on polystyrene latex particles; (iv) charged subsurface groups which induce a dipole interaction, e.g., possibly the sulfate groups buried inside polystyrene latex particles; (v) adsorption of ions from the medium, e.g., H^+ or OH^- ions from pure water and Na^+ or Cl^- from the samples run at constant ionic strength (19); (vi) contact electrification of the two phases in contact with one another, e.g., rubbing a glass rod with a piece of fur or injection of electrons into polystyrene (20).

In the present case, it would have been expected that the chemically bound surface groups would have determined the particle charge; however, the electrophoresis measurements show that this is not the case. The chemically bound surface sulfate groups are dissociated, independent of pH; the carboxyl groups are undissociated at low pH and dissociated at high pH; and the hydroxyl groups are not dissociated at all. There are no potential-determining ions in the ion-exchanged latexes, at least not in the same sense as the Ag^+ and I^- ions are potential-determining in silver iodide sols. Moreover, the emulsifier has been removed by ion exchange. These results cannot be explained by in-

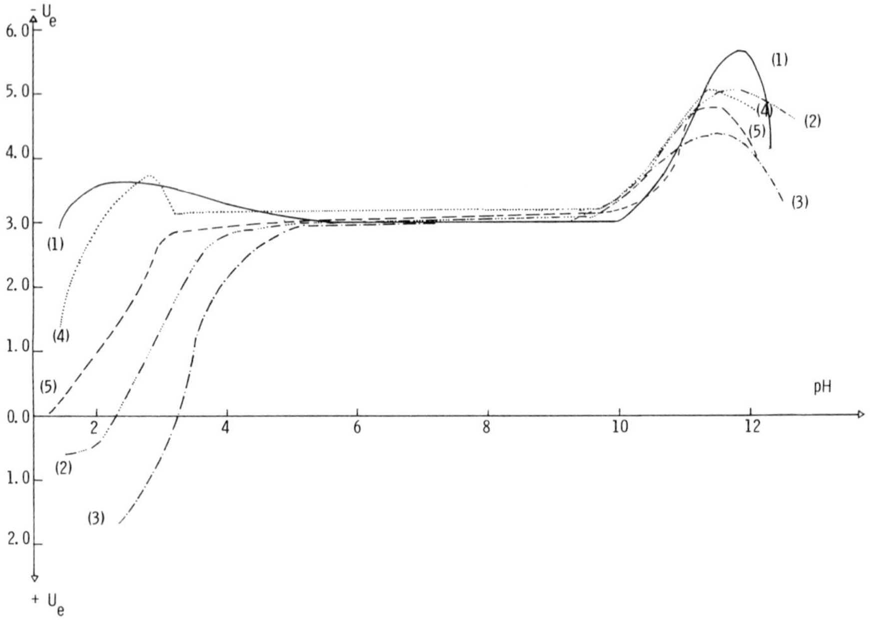

Figure 2. Variation of electrophoretic mobility in distilled water with pH for polystyrene latex particles with different surface groups: (1) 520′ sulfate; (2) 520′ carboxyl; (3) 520′ hydroxyl; (4) LS-1010-E sulfate; (5) LS-1010-E hydroxyl.

duced dipole interactions due to the buried sulfate groups, e.g., one of the latexes shown in Figure 2 initially contained 13.3 μeq surface sulfate/gm polymer and finally 13.7 μeq surface carboxyl/gm polymer, the same values within experimental error, while the buried sulfate amounted to only 3.26 μeq/gm polymer. At present, the best possibilities to explain these results are adsorption of ions from the medium or injection of electrons into polystyrene. Experiments are in progress to distinguish between these two mechanisms. Meanwhile, these results show that the type and concentration of the chemically bound surface groups do not correlate with the electrophoretic mobility; there is another effect which overrides the differences between strong-acid sulfate, weak-acid carboxyl, and nonionic hydroxyl groups. Moreover, these discrepancies between two such independent methods of measurement are observed only when such well-characterized model colloids are used for the measurements.

Application to Industrial Latexes

The foregoing methods developed for the preparation and characterization of monodisperse polystyrene latexes to be used as model colloids can also be applied to the characterization of industrial latexes. The recipes used for the preparation of these industrial latexes are complex, and most contain a small amount of a functional monomer, e.g., acrylic acid, 2-sulfoethyl methacrylate, or N-methylolacrylamide. These functional monomers are often predominantly water-soluble, so that their use may have several results; (i) the monomer may polymerize in the aqueous phase to form a water-soluble polymer that remains in the serum; (ii) it may polymerize in the aqueous phase to form a polymer that adsorbs on the particle surface, thus forming a stronger adsorptive bond than the corresponding small-molecule emulsifier; (iii) it may copolymerize at the particle surface, thus forming a chemically bound surface group; (iv) it may polymerize inside the particle, thus burying the functional group; (v) it may not polymerize at all. Characterization of industrial latexes show that, usually, as many as three of these five possibilities are found.

The loci and concentration of these functional groups often determine the latex performance in a given application. Therefore, it is important to know the distribution of functional groups between the serum, particle surface, and particle interior as a function of the type and concentration of the functional monomer and the technique of polymerization. Thus characterization methods developed to determine the loci of these functional groups are useful in research to develop new latexes and modifications of older latexes, in development to ensure that the scale-up does not result in a change in the loci of the functional groups, and in production to ensure batch-to-batch uniformity of the product.

This paper gives four examples of the characterization of industrial latexes using the foregoing methods as well as other methods that are applicable to a given system: (i) styrene-butadiene-acrylic acid and -methacrylic acid copolymers; (ii) ethyl acrylate-methyl methacrylate-acrylic acid copolymer; (iii) polyvinyl acetate prepared using polymethacrylic acid emulsifier; (iv) vinyl acetate-butyl acrylate copolymers of varying compositions prepared using batch and starved semi-continuous polymerization. Series of 54-59:41:5-0 styrene-butadiene-acrylic acid and 53-59:41:6-0 styrene-butadiene-methacrylic acid copolymers were prepared by semi-continuous polymerization at 90° using a 20nm-diameter seed latex (21). The functional monomers were dissolved in the monomer mixture, and aqueous phase contained 0.017M persulfate initiator, 0.007M sodium lauryl sulfate emulsifier, and an amount of sodium hydroxide equivalent to part of the vinyl carboxylic acid, expressed as the "degree of neutralization" ranging from 0 to 100%. After polymerization, the latexes were steam-stripped to remove the residual monomer, and the pH was adjusted to 8. The final latexes contained 45% solids and comprised monodisperse particles of 180nm diameter. The acrylic acid and methacrylic acid concentrations were varied in the ranges 0-5% and 0-6%, respectively, at 25% degree of neutralization. Also, the degree of neutralization was varied from 0% to 100% at 2.5% acrylic acid and 3.0% methacrylic acid.

The latexes were ion-exchanged with Dowex 50W(H^+) resin and the Dowex 50W(H^+)-Dowex 1 (OH^-) mixed resin in combination with the Dowex 50W(Na^+)-Dowex 1 (OH^-) resin, and the ion-exchanged samples were titrated conductometrically. The samples treated were the latex, the aqueous serum, the latex particles separated from the serum, and the latex particles swollen or dissolved in 80:20 dioxane-water mixture. The total oxygen content was determined by neutron activation and the total sulfur content by X-ray fluorescence. Material balances of acrylic or methacrylic acid found in the serum, on the particle surface, and inside the particle agreed with the amount added to within 5-10%.

For the styrene-butadiene-acrylic acid copolymers, acrylic acid was found in the aqueous serum, on the particle surface, and buried inside the particles. At 25% degree of neutralization, a constant 66% of the acrylic acid was found incorporated into the particle, independent of concentration. However, the surface concentration of acrylic acid groups was one for every 500$Å^2$ of surface area, independent of concentration. Thus increasing the concentration of acrylic acid merely increased the amount buried inside the particle and did not increase the surface concentration significantly. At 2.5% acrylic acid, the amount of acrylic acid incorporated into the particle decreased with increasing degree of neutralization and could be related to the distribution of acrylic acid between the monomer-polymer and aqueous phases. The acrylic acid found in the serum was in the form of a low-molecular-weight polymer that was removed by ion exchange.

For the styrene-butadiene-methacrylic acid copolymers, methacrylic acid was also found in the serum, on the particle surface, and buried inside the particles. At 25% degree of neutralization, less methacrylic acid was found in the serum and on the particle surface than with acrylic acid, i.e., more was buried inside the particle. At 3.0% methacrylic acid, the amount incorporated into the particle was fairly constant, independent of the degree of neutralization. The different distributions of methacrylic and acrylic acids were explained by their different distributions between the monomer-polymer and aqueous phases. Thus these characterization results show the effect of vinyl carboxylic acid type and concentration on the loci of the carboxyl groups. Similar correlations could be made with other systems.

A 62:35:3 ethyl acrylate-methyl methacrylate-acrylic acid copolymer latex was prepared by continuous addition of the monomer mixture over a 4-hour period at 80° (22). The emulsifier was a sodium lauryl ether sulfate-nonylphenol polyoxyethylene adduct (20 moles ethylene oxide) mixture, the initiator a potassium persulfate-sodium hydroxulfite mixture, and the buffer a sodium bicarbonate-potassium hydroxide mixture. The final latex of pH 6.5 contained 40% solids, and the T_g of the copolymer was 13°.

The latex was cleaned by ion exchange and serum replacement, which gave the cleaned latex plus six serum fractions. The cleaned latex and the serum samples were analyzed by conductometric titration. Also, the amount of anionic emulsifier in the serum was determined by Hyamine 1622 colorimetric titration and thin-film chromatography, and the amount of nonionic emulsifier by iodine-iodide colorimetric titration and thin-film chromatography.

The material balance of the strong-acid groups showed a theoretical total of 0.194 meq/gm polymer from the sodium lauryl ether sulfate, potassium persulfate, and sodium hydrosulfite, in comparison with a measured total of 0.205 meq/gm (0.026 on particle surface; 0.179 in serum) by serum replacement and a total of 0.215 meq/gm by ion exchange with Dowex 50W(H^+). The material balance of the acrylic acid showed that 29.9% was on the particle surface, 28.6% in the aqueous serum, and 41.5% buried inside the particle. The sodium lauryl ether sulfate found in the serum amounted to 78% of that added by Hyamine 1622 titration and 88% by thin-film chromatography. The nonylphenol polyoxyethylene adduct amounted to 113% by iodine-iodide titration and 91% by thin-film chromatography.

Thus these characterization results not only give the distribution of the acrylic acid between the aqueous serum, particle surface, and particle interior, but also account satisfactorily for the total number of strong-acid groups arising from the anionic emulsifier and initiator. In addition, both the sodium lauryl ether sulfate and the nonylphenol polyoxyethylene adduct used in the polymerization were recovered from the fractions obtained by serum replacement.

A polyvinyl acetate latex prepared by semi-continuous polymerization at 55° using a polymethacrylic acid-nonylphenol-polyethoxylate phosphate ester emulsifier and sodium persulfate-sodium formaldehyde sulfoxylate initiator (23). The latex was cleaned by ion exchange and serum replacement using both Nuclepore and Pellicon membranes, and the cleaned latex and serum fractions were analyzed by conductometric titration. In addition, the dried films were extracted with water and organic solvents, and the extracts were analyzed by infrared spectroscopy and thermogravimetric analysis.

It was found that 70-80% of the polymethacrylic acid was adsorbed on the particle surface and the remainder was in the serum. That polymethacrylic acid on the particle surface was strongly adsorbed, but there was little or no grafting to the surface (unlike polyvinyl alcohol, a significant proportion of which is grafted to the particle surface). The strong adsorption of the polymethacrylic acid was attributed to hydrophobic bonding between the α-methyl groups of the polymethacrylic acid and the acetoxy groups of the polyvinyl acetate. The polymethacrylic acid was considered to stabilize the particles by both electrostatic and steric stabilization. Infrared spectroscopy and thermogravimetric analysis showed the presence of polymethacrylic acid in the serum as well as the absence of polyvinyl acetate or polyvinyl alcohol.

Thus these characterization results demonstrate that polymethacrylic acid does not function in the same manner as polyvinyl alcohol in the emulsion polymerization of vinyl acetate and that the adsorbed polymethacrylic acid can be separated from the bulk polymer, thus distinguishing it from grafted polymer.

Vinyl acetate-butyl acrylate copolymers (0-100% butyl acrylate) were prepared by both batch and starved semi-continuous polymerization using sodium lauryl sulfate emulsifier, potassium persulfate initiator, and sodium bicarbonate buffer. This copolymer system was selected, not only because of its industrial importance, but also because of its copolymerization reactivity ratios, which predict a critical dependence of copolymer compositional distribution on the technique of polymerization. The butyl acrylate is so much more reactive than the vinyl acetate that batch polymerization of any monomer ratio would be expected to give a butyl acrylate-rich copolymer until the butyl acrylate is exhausted and polyvinyl acetate thereafter.

The latexes were cleaned by ion exchange and serum replacement, and the number and type of surface groups were determined by conductometric titration. The molecular weight distributions of the polymers were determined by gel permeation chromatography. The stability of the latexes to added electrolyte was determined by spectrophotometry. The compositional distribution was determined by dynamic mechanical spectroscopy (Rheovibron) and differential scanning calorimetry, and the sequence distribution by C13 nuclear magnetic resonance.

The results showed that all batch polymerizations gave a two-peaked copolymer compositional distribution, a butyl acrylate-rich fraction, which varied according to the monomer ratio, and polyvinyl acetate. All starved semi-continuous polymerizations gave a single-peaked copolymer compositional distribution which corresponded to the monomer ratio. The latex particle sizes and type and concentration of surface groups were correlated with the conditions of polymerization. The stability of the latex to added electrolyte showed that particles were stabilized by both electrostatic and steric stabilization with the steric stabilization groups provided by surface hydrolysis of vinyl acetate units in the polymer chain. The extent of this surface hydrolysis was greater for the starved semi-continuous sample than for the batch sample.

Thus the combination of surface characterization methods with polymer characterization methods gives new insights into the properties of this important industrial copolymer system and demonstrates the effectiveness of these methods in distinguishing between different techniques of polymerization.

Summary

The foregoing results show that the surface of monodisperse polystyrene latex particles can be cleaned by ion exchange or serum replacement to remove the adsorbed emulsifier and solute electrolyte, and the number of chemically bound surface groups arising from the persulfate initiator can be determined quantitatively by conductometric titration. Moreover, monodisperse polystyrene latexes prepared using persulfate initiator and bicarbonate buffer are good starting points for the preparation of ideal model colloids. Polymerization of the latexes at pH 7-8 gives only surface sulfate groups. Storage of these latexes in the H^+ form gives complete hydrolysis to the all-hydroxyl form, and storage at room temperature gives the all-carboxyl form. These latexes with the same number of three different surface groups ---- sulfate, carboxyl, hydroxyl ---- have been used as model colloids for measurements of electrophoretic mobility. Surprisingly, the electrophoretic mobility is negative and constant over the pH range 5-10, independent of the type of surface group. The mechanisms proposed to explain these apparently anomalous results are adsorption of ions from the medium or injection of electrons into the polystyrene particles. These characterization methods were applied to typical industrial latexes, e.g., styrene-butadiene-acrylic acid and -methacrylic acid copolymers, an ethyl acrylate-methyl methacrylate-acrylic acid copolymer, polyvinyl acetate prepared using polymethacrylic acid-nonylphenol-polyoxyethylate phosphate ester emulsifier, and vinyl acetate-butyl acrylate copolymers of 0-100% butyl acrylate.

Abstract

Monodisperse polystyrene latexes prepared with persulfate initiator are stabilized by the sulfate endgroups of the polymer molecules. These latexes can be characterized by removing the adsorbed emulsifier and solute electrolyte by ion exchange or serum replacement and determining the surface sulfate groups by conductometric titration with base. However, these latexes sometimes contain surface hydroxyl groups formed by hydrolysis of sulfate groups or a side reaction of sulfate ion-radicals to form hydroxyl radicals. These hydroxyl groups can be determined by oxidation to the carboxyl form, followed by conductometric titration. The purpose of this paper is to describe the preparation of model colloids as well as characterization of industrial latexes. Model colloids with the same number of three different types of surface groups can be prepared to give particles stabilized only with sulfate endgroups, hydrolysis to give particles stabilized only with hydroxyl groups, and oxidation to give particles stabilized only with carboxyl groups. Examples of industrial latexes characterized by these methods include styrene-butadiene-acrylic acid and -methacrylic acid copolymers, ethyl acrylate-methyl methacrylate-acrylic acid copolymers, polyvinyl acetate prepared with polymethacrylic acid emulsifier, and vinyl acetate-butyl acrylate copolymers prepared by batch and semi-continuous polymerization.

Literature Cited

1. For a brief history of the early preparation of monodisperse polystyrene latexes, see Vanderhoff, J.W., Preprints, A.C.S. Division Org. Coatings Plastics Chem., 1964, 24, (2), 223.
2. van den Hul, H.J. and Vanderhoff, J.W., J. Colloid Interface Sci., 1968, 28, 336.
3. Vanderhoff, J.W., van den Hul, H.J., Tausk, R.J.M., and Overbeek, J. Th. G., "Clean Surfaces: Their Preparation and Characterization for Interfacial Studies," G. Goldfinger, editor, Marcel Dekker, New York, 1970, p. 15.
4. McCann, G.D., Bradford, E.B., van den Hul, H.J., and Vanderhoff, J.W., "Polymer Colloids," R. M. Fitch, editor, Plenum Press, New York, 1971, p. 29.
5. van den Hul, H.J. and Vanderhoff, J.W., J. Electroanal. Chem., 1972, 37, 161.
6. Ahmed, S.M., El-Aasser, M.S., Pauli, G.H., Poehlein, G.W., and Vanderhoff, J.W., accepted for publication, J. Colloid Interface Sci., Feb. 1980, Vol. 73, No. 2.
7. Edelhauser, H., J. Polymer Sci., 1969, C27, 291.
8. Wu, W.C., El-Aasser, M.S., and Vanderhoff, J.W., "Emulsions, Latices, and Dispersions," P. Becher, and M.N. Yudenfreund, editors, Marcel Dekker, New York, 1978, p. 71.
9. van den Hul, H.J. and Vanderhoff, J.W., Brit. Polymer J., 1970, 2, 121.

10. Brodnyan, J.G. and Kelley, E.L., J. Colloid Sci., 1965, 20, 7.
11. Force, C.G., Matijevic, E., and Kratohvil, J.P., Kolloid Z. Z. Polymers, 1968, 223, 31.
12. Shaw, J.N. and Ottewill, R.H., Nature, 1965, 208, 681; ibid. Kolloid Z. Z. Polymere, 1967, 215, 161.
13. Ottewill, R. H., Krieger, I. M., Fitch, R. M., Homola, A., Everett, P. H., and Gultepe, M. E., Discussions and Preprints NATO/ASI "Polymer Colloids," Trondheim, Norway, June 30 - July 11, 1975.
14. Yates, D.E., Ottewill, R.H., and Goodwin, J.W., J. Colloid Interface Sci., 1977, 62, 356.
15. Wu, W.C., Ph.D. thesis, Lehigh University, 1977.
16. Kamel, A., El-Aasser, M.S., and Vanderhoff, J.W., unpublished research results, 1979.
17. Ahmed, S.M., M. S. Research Report, Lehigh University, 1976.
18. Kamel, A., Ma, C., Micale, F.J., El-Aasser, M.S., and Vanderhoff, J.W., unpublished research results, 1980.
19. Proposed by Micale, F.J.
20. Proposed by Fowkes, F.M.
21. Vanderhoff, J.W., van den Hul, H.J., and Hamburg[illegible]., Polymer Preprints, 1975, 16, (1), 155.
22. El-Aasser, M.S., Ahmed, S.M., Peohlein, G.W., [illegible]vira, X., Tabernero, J.T., de la Morena, P., and Vander[illegible]off, J.W., "Polymer Colloids II", R.M. Fitch, editor, Plenum Press, New York, 1980, p. 361.
23. Daniels, W.D., Enos, C.T., Iacoviello, J.G., Frost, J.H., Ahmed, S.M., and Vanderhoff, J.W., "Polymer Colloids II", R.M. Fitch, editor, Plenum Press, New York, 1980, p. 477.
24. El-Aasser, M.S., Misra, S.C., Pichot, C., and Vanderhoff, J.W., J. Polymer Sci., Polymer Letters, 1979, 17, 567.
25. El-Aasser, M.S., Makgawinata, T., Pichot, C., and Vanderhoff, J.W., accepted for publication, J. Polymer Sci.

RECEIVED April 6, 1981.

4

Characterization of Latex Particles by Light Scattering

R. L. ROWELL and J. R. FORD

Department of Chemistry, University of Massachusetts, Amherst, MA 01003

This work is directly applicable to spherical particles with a circumference in the range of 1 to 10 times the wavelength of light used for observation. The general ideas extend to a wide variety of related problems.

There are two basic approaches to the characterization of latex particles by light scattering. In the quantum approach (1, 2), which has a comparatively recent historical development under the names of Rayleigh linewidth, quasi-elastic scattering, intensity fluctuation spectroscopy, photon correlation spectroscopy and dynamic light scattering, the physical effect is a Doppler shift of the light by the moving particle. The available information is obtained through an exponential correlation function so that the method is best suited to a characteristic size and considerably less sensitive to particle size distribution. The other approach is based upon wave theory and may involve either turbidity or angular scattering as discussed below. The wave theory has long been known as Mie scattering (3, 4) and has more recently become known as Lorenz-Mie scattering in recognition of historical origins (5). It represents the exact solution to the problem of electromagnetic scattering by a dielectric sphere using Maxwell's equations and is well-known with diverse applications (5, 6).

The turbidity experiment is simple and rapid but there is less variation in the turbidity spectrum than in the angular scattering pattern so that the method has been developed mainly for particle size (7).

The angular scattering approach is the principal aim of this work. The properties of the Lorenz-Mie intensity coefficients are treated in some detail in order to illustrate their utilization for the determination of particle size distribution, refractive index and number concentration. In a related paper the internal structure of polymer latex spheres is considered (8).

0097-6156/81/0165-0085$05.00/0

The angular scattering information is embodied in the Lorenz-Mie theory intensity coefficient i_1 for V_v polarization and i_2 for H_h polarization of scattering. Although high speed computers are required for the computation of i_1 and i_2, the problem is well-known and a broad literature is available for checking (5, 6). The Lorenz-Mie theory size parameter $\alpha=2\pi r/\lambda$ may be visualized as the circumference of the particle measured in units of the wavelength of light used for observation.

The theory may be understood by representing the intensity coefficients as three-dimensional graphs giving intensity as a function of size parameter α and scattering angle θ measured from the forward direction. In the work that follows computations have been carried out for each of the three-dimensional surfaces that are to be represented below over the size parameter α, ranging from 1.0 in steps of 0.1 to 9.9. Calculations have been carried out over the angular range from 0 to 180° in steps of 1°. Each surface, then, consists of a matrix of 90 by 181 or 16,290 data points. The 16,290 data points have been plotted by computer in the form of a three-dimensional graph arranged to give one a perspective view of the behavior of the intensity coefficients as a function of both size parameter α and angle Θ.

Intensity Patterns for m = 1.200

Consider first a series of eight views of the same data matrix for the intensity coefficient i_1 for a relative refractive index m of 1.200. Figures 1 through 8 represent different perspective viewpoints of the same three-dimensional matrix viewed from the front, right-hand side, back, etc., so as to reveal details of the simple behavior of complicated mathematical functions.

In Figure 1, we are looking at the angular scattering pattern from small size in the foreground to large size in the background. Two striking features stand out. The first is the strong forward scattering represented by the smooth rising surface at the right-hand edge of the figure. The second striking feature is the remarkable regularity of the light scattering functions. The angular maxima and minima that appear in rectangular coordinate or in polar coordinate representation appear in the three-dimensional representation as a series of regularly migrating trenches or valleys. The graphical representation shows clearly that the angular minima have their origin in the backscattering region and migrate regularly to the forward direction. For example, in Figure 1, counting along the left-hand edge, one can distinguish seven minima. Each of the seven minima originates in the backward direction and moves both towards the forward direction and towards larger particle size as the size coordinate increases. Generally, the migration of the trenches or the hills and valleys is smooth except at the larger particle size where an undulation occurs along the ridges as well. The undulation is especially pronounced for larger particles and at the backward scattering angles. Before leaving Figure 1, we note one other obvious feature in the wide angle scattering pattern. For the largest size shown, one can count eight minima. Since this is one more than the number of minima that have originated in the backward direction clearly an additional minima must have crept in somehow at the higher angles and it is not completely clear from this view what has happened.

The second view of the same surface, shown in Figure 2, is from the forward direction and the α coordinate face has been exposed to reveal the features of the underneath part of the surface. Note the appearance of two distinct and sharp minima occurring in the region of large particle size at high scattering angle.

The third view of the same matrix in Figure 3 is from a point underneath the surface which reveals that the valleys show a ripple structure like that of the ridges seen from above. It is clear from this view that the sharp minima are very close to the backward direction and are at very large particle size.

In Figure 4 we have moved round to the back of the matrix and are looking down from the wide angle scattering pattern of large size toward the region of smaller size. It is clear from this view that there are eight minima along the uppermost scattering pattern in the matrix. However, this particular perspective is not the best one for distinguishing subtleties of the various undulations.

Figure 5 is from above the surface from the vantage point of highest angle and highest size parameter. This particular perspective clearly reveals the migration and fine details of the minima. We count along the θ coordinate at highest α and find the eight minima. Then we count along the α coordinate at $\theta=180^{o}$ and identify the seven minima that were seen before and discover that the eighth minimum has occurred in the region of largest size parameter and is plainly clear in this particular view. The exact correspondence between angular minima and size minima is clearly shown here.

Figure 5 also reminds us of the interference pattern that is generated by throwing two stones, a large one and a small one, simultaneously at different points upon the smooth surface of water. A cross-ripple interference pattern, generated by the wavelets emanating from the two stones, is very similar to the ripple pattern that is generated in the perspective view shown here. The major pattern, like the large-stone ripple pattern, is due to particle size while the minor pattern, like the cross-ripple interference from the smaller stone, may be attributed to refractive index as shown in subsequent figures.

In Figure 6 we have moved around to the high angle face of the matrix and are looking from the backward direction to the forward direction. The regular and pronounced periodicity of

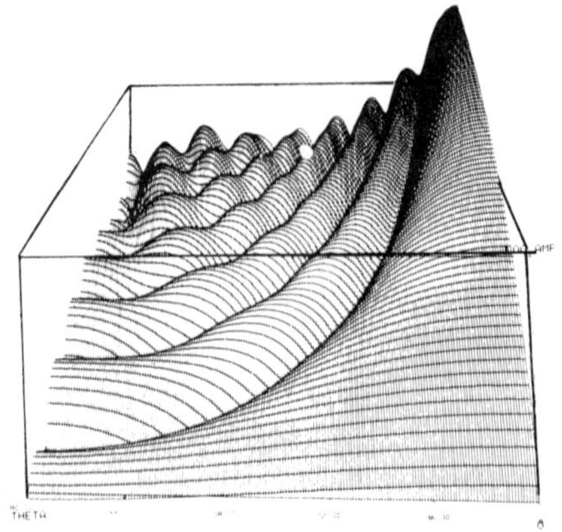

Figure 1. Log i_1 *vs.* α, θ *for* m = *1.200; view from small* α

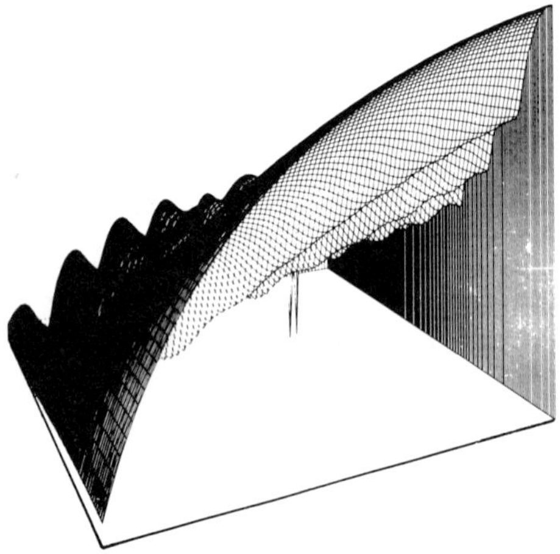

Figure 2. Log i_1 *vs.* α, θ *for* m = *1.200; view from zero angle*

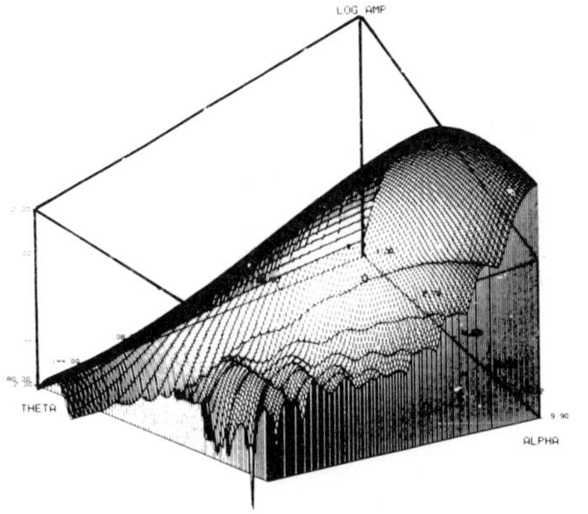

Figure 3. Log i_1 *vs.* α, θ *for* m = *1.200; view from small* α, *small* θ, *and beneath matrix*

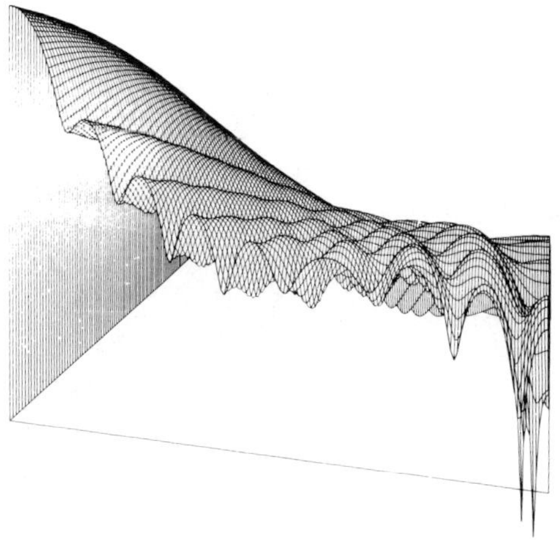

Figure 4. Log i_1 *vs.* α, θ *for* m = *1.200; view from large* α

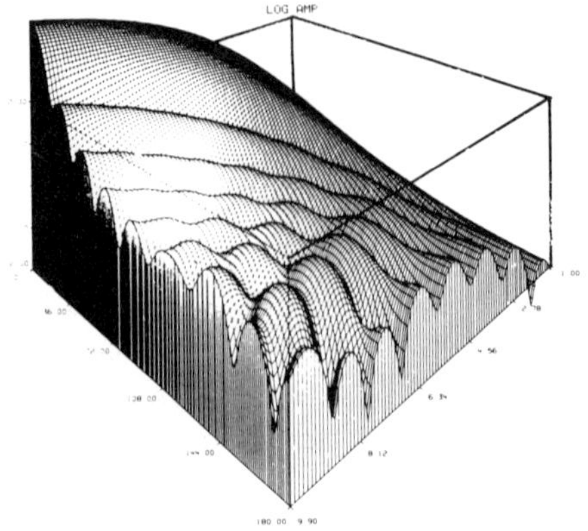

Figure 5. Log i_1 *vs.* α, θ *for* m = *1.200; view from large* α*, large* θ

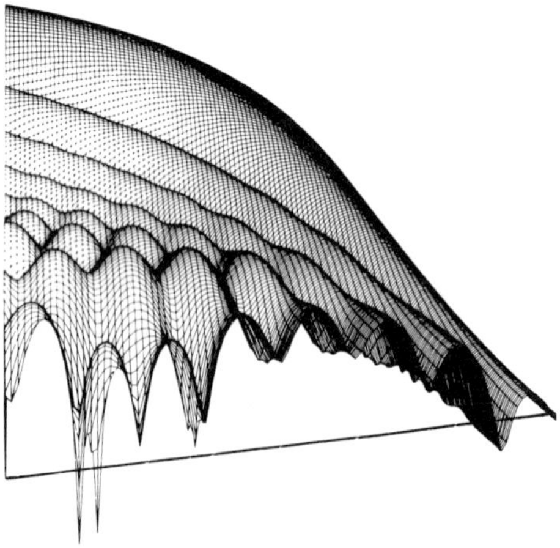

Figure 6. Log i_1 *vs.* α, θ *for* m = *1.200; view from* $\theta = 180°$

minima that occurs along the α coordinate at $\theta = 180^o$ is the most obvious feature but there appears to be a longer term undulation that displays only about 1/2 a wavelength over the entire coordinate displayed. It is as though the waveforms of the minima are superimposed upon a lower frequency modulation governing the general variation in the intensity surface. This feature is more clearly shown in Figure 7.

The next view, in Figure 7, continues our movement around the matrix and gives us an "aerial" perspective on the regular migration of the minima and the cross-ripple structure that occurs in the region of large particle size and high scattering angle. The exact correspondence of the eight minima generated along the coordinate of $\theta = 180^o$ and the eight minima occurring along the full angular coordinate at highest α is clearly shown here. It is also clear that both the trenches or valleys and the corresponding ridges develop increasingly larger undulations as particle size increases.

The final view, Figure 8 in the series, is from the perspective of small size and forward scattering. We are looking over the forward lobe towards the ripple structure at high angles and large particle size to the rear. This particular perspective will be held as a constant in order to understand the effects of refractive index and polarization on the wide angle light scattering patterns. Each of the seven following views in Figures 9 through 15 will be from this same perspective. Let us keep this view as a frame of reference and remember that it is the intensity surface for a relatively moderate relative refractive index of m = 1.200, which is typical of many aqueous polymer latex systems.

Effect of Refractive Index

In Figure 9 we observe the effect of refractive index. The i_1 intensity surface has been computed for a relative refractive index of 1.486. General features of the previous surface are retained. That is to say, there is a strong forward lobe of scattering and a remarkable regularity in the trench-like structure giving rise to the angular minima. The migration of the valleys and ridges to the forward direction seems to hold in this surface as well but the cross-ripple pattern pointed out in the earlier series is much more pronounced in this surface. It is possible to identify eleven minima along the α coordinate at $\theta = 180^o$ in contrast to the eight observed for the lower refractive index of 1.200. In addition to the increased variation in pattern or structure, a comparison of Figure 9 with Figure 8 shows that the whole scale of scattering intensity is increased. Two important effects of refractive index, then, are an increase in the general level of scattering and an increase in the structure of scattering. Notice along the α coordinate corresponding to $\theta = 0^o$ the regular increase of the forward scattering

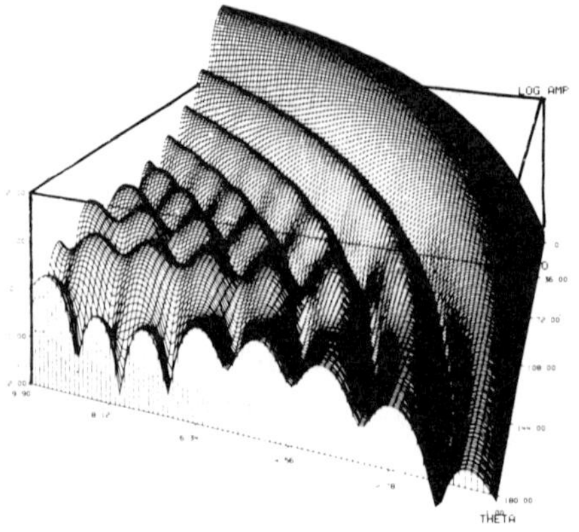

Figure 7. Log i_1 *vs.* α, θ *for* m = *1.200; "aerial" view from* $\theta = 180°$

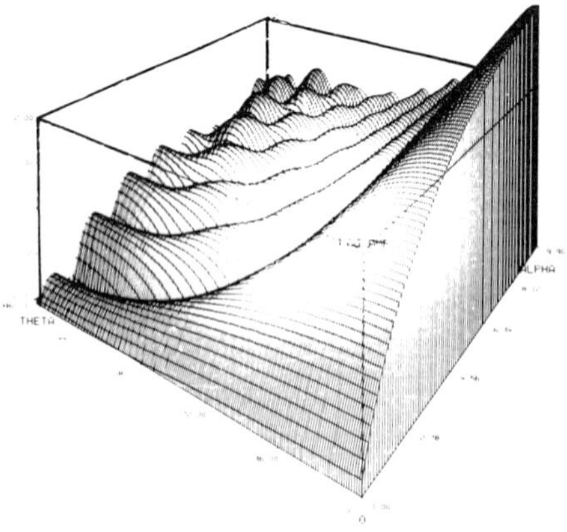

Figure 8. Log i_1 *vs.* α, θ *for* m = *1.200; view from small* α*, small* θ

shows a ripple structure. The ripple in the forward scattering seems to have some relation to the angular pattern.

In the next view, Figure 10, we use the computer to show us quantitatively the effect of refractive index difference. In this surface, we have taken the two previous surfaces and subtracted them element by element and then made a plot of the difference. The difference surface gives us a map of the sensitivity to m in the α-θ domain. Two important features are to be noted. The first of these is that the surface is relatively flat in the region of the small size and forward angle. This means that the wide angle light scattering is relatively insensitive to rather large differences in refractive index in this region. Secondly, over much of the rest of the matrix, there are large oscillations in the pattern. This means that there is a tremendous sensitivity to refractive index in the wide angle scattering pattern. It is important to stress that the z-axis spans 4 orders of magnitude and that some of the spikes are well overscale.

In Figure 11, we again look at the difference matrix but here the difference is between surfaces that have been computed for refractive indexes of 1.486 and 1.510. Again, two features are very obvious. The flat region has a much greater extent and the regions where there are large differences in the intensity functions have shrunk to very small domains which seem to occur in small, regularly spaced regions of resonance. Notice that the resonant regions follow the locus of the migrating minima, the trenches that were discussed above, and seem to be pinpointed by the cross-ripple pattern mentioned above. The sensitivity in the resonance regions is very large.

Figures 10 and 11 illustrate a very important point. Light scattering may be either extremely sensitive or extremely insensitive to refractive index depending on the region of the α-θ domain under consideration. Experimentally, the θ domain is readily selected by choice of scattering angles. The α domain is also selectable by choice of the wavelength of the illuminating radiation. Developments in laser technology are increasing the flexibility in the choices of wavelength available.

Effect of Polarization and Refractive Index

In Figure 12 we examine the effect of polarization on the wide angle scattering pattern. We are now looking at the Mie theory intensity coefficient i_2 corresponding to H_h polarization. This surface has been computed for a relative refractive index of 1.200. Again, we note that the minima that occur in the wide angle scattering pattern have their origin in the backward direction and move toward the forward direction as particle size increases. From this view we can count the eight minima that occur along the backward direction so that the number of minima

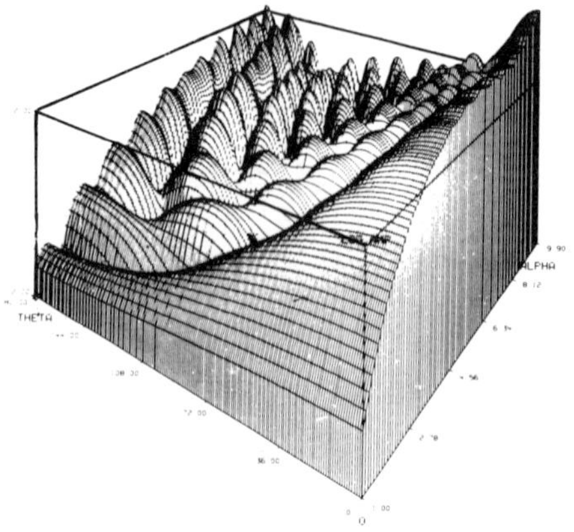

Figure 9. *Log* i_1 *vs.* α, θ *for* m = *1.486; view from small* α, *small* θ

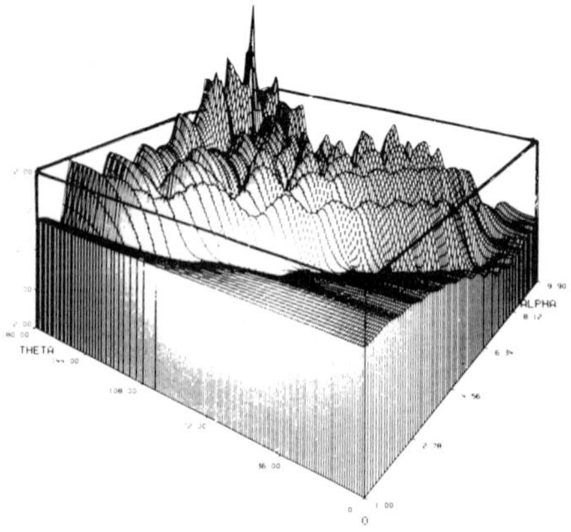

Figure 10. *(Log* i_1, m = *1.200) – (Log* i_1, m = *1.510) vs.* α, θ*; view from small* α, *small* θ

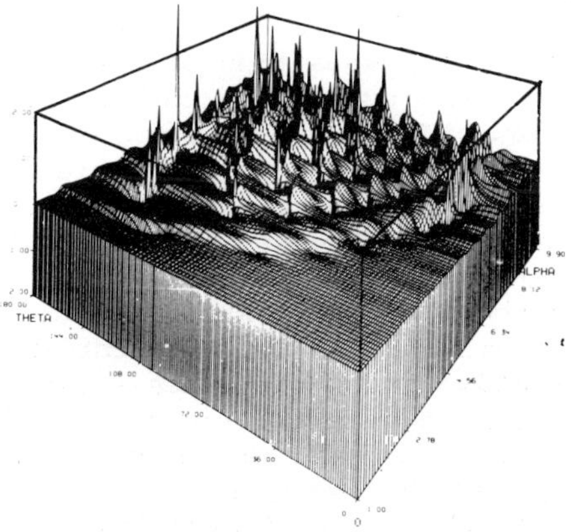

Figure 11. *(Log* i_1, m = *1.510)— (Log* i_1, m = *1.486) vs.* α, θ*; view from small* α*, small* θ

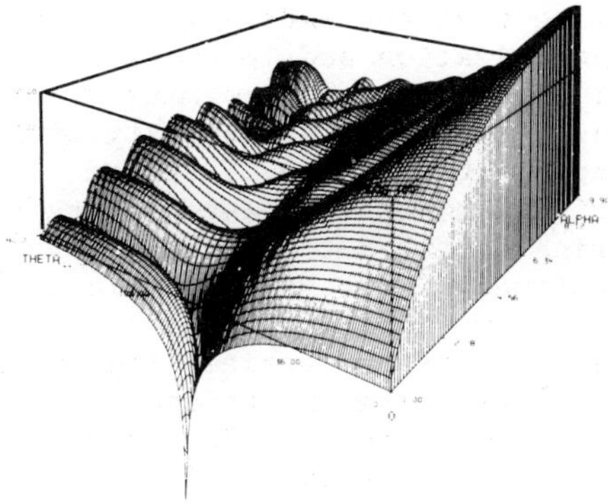

Figure 12. Log i_2 *vs.* α, θ *for* m = *1.200; view from small* α*, small* θ

in backscattering is independent of polarization. The most pronounced feature of this surface is the deep minimum occurring at 90° which is the Rayleigh minimum for very small particles. We notice that as particle size increases the Rayleigh minimum fades smoothly into the overall structure of the surface. Comparison of Figure 12 with Figure 8 suggests that the main effect of polarization is the superposition of the Rayleigh minimum on a regular pattern determined by size and refractive index.

In the next view, Figure 13, we examine both the effect of polarization and refractive index. The surface is again for the component i_2 but is for a refractive index of 1.486. It is clear from this view the same general features that we had observed before are apparent here as well. We find that in the backward direction there are eleven minima, the same number as for the i_1 polarization. This is consistent with our understanding that refractive index increases structure, independent of polarization. We also note that over most of the surface there is much scattering so that the increase in the general level of scattering and the superposition of the Rayleigh minimum has moderated the large dissymmetry between forward and backward scattering.

In Figure 14, we see in a quantitative way the effect of a difference in refractive index. In this view, we have used the computer to subtract the two preceding matrices point by point and make a plot of the difference surface. Again, as before, we note a region that is relatively flat at small angles and small sizes showing that the wide angle scattering is insensitive to both polarization and refractive index in this region. Over a good deal of the rest of the matrix, there is pronounced structure showing the extreme sensitivity of light scattering to refractive index. Comparison of Figure 14 with Figure 10 shows that polarization has increased the region of sensitivity of light scattering in the α-θ domain. In particular, the position and magnitude of the Rayleigh minimum is especially sensitive to refractive index.

In the final view of the series, Figure 15, we look at the effect of a very small change in refractive index on the wide angle scattering pattern for i_2. This surface has been computed for a refractive index difference 1.486 minus 1.510. Most of the surface is fairly flat showing an insensitivity to such a small refractive index difference. But there are regions, particularly at the larger scattering angles and for larger size particles where there are pronounced resonances again following the patterns of the migrating minima and pinpointed by the cross-ripple structure. These very small regions of resonance are very interesting because they show several orders of magnitude of sensitivity for very small difference in refractive index. This suggests that if the proper instrumentation were available to tune our experiments into the region of theoretical space

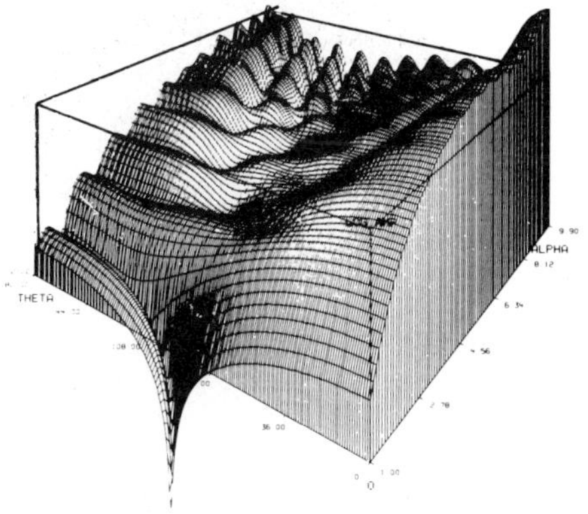

Figure 13. Log i_2 *vs.* α, θ *for* m = *1.486; view from small* α, *small* θ

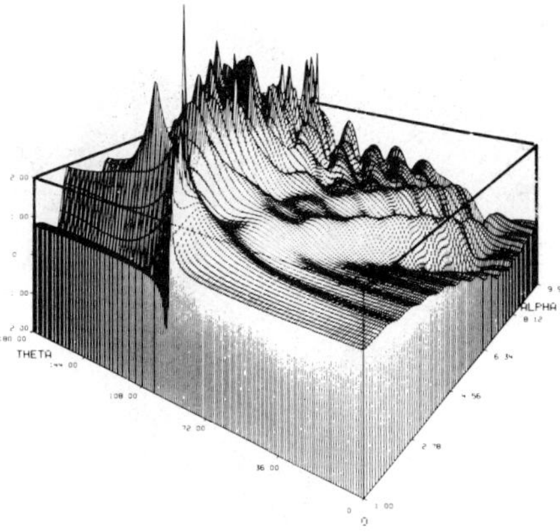

Figure 14. *(Log* i_2, m = *1.200) – (Log* i_2, m = *1.510); view from small* α, *small* θ

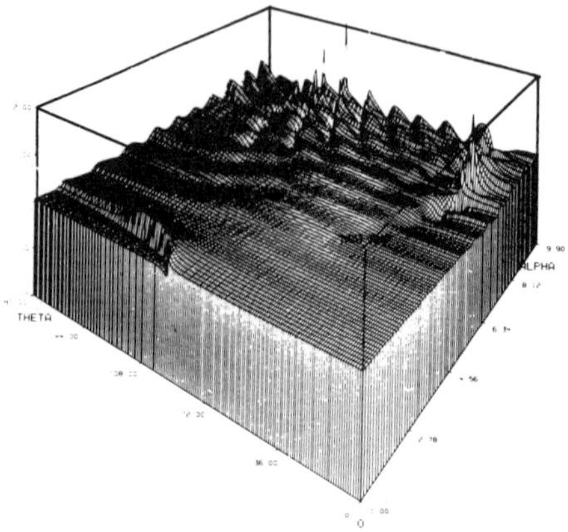

Figure 15. *(Log* i_2, m = *1.510) – (Log* i_2, m = *1.486) vs.* α, θ*; view from small* α*, small* θ

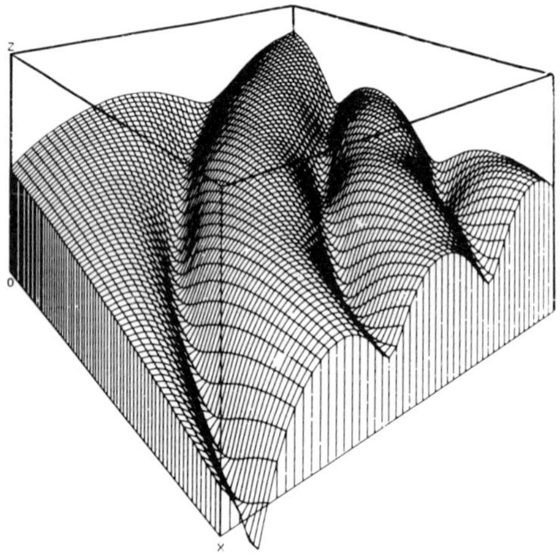

Figure 16. *Log* i_1 *vs.* $\theta = 45°(1°)115°$, $\alpha = 2.0(0.1)6.9$ *for* $m = 1.200$*; view from small* θ*, small* α

surrounding a resonance maxima or minima it would be possible to gain an enormous sensitivity to refractive index and hence particle composition. The resonances are less pronounced for i_2 than for i_1 as seen by comparing Figures 15 and 11.

The Inversion Problem

The inversion problem consists of extracting useful information from light scattering data. The details differ but the general approach has been to find a match between an experimental record and a library of machine computations. The matching procedures have without exception been one dimensional, i.e. the θ-template generated from an angular scan at constant size or the α-template generated from a growing system observed at constant θ.

Figure 16 will be used to illustrate some of the difficulties encountered in the inversion problem. The surface shows log i_1 over $\theta = 45^o(1^o)115^o$ and $\alpha = 2.0(0.1)6.9$ for $m = 1.200$. The angular coordinate is to the left with the forward direction identified by increasing intensity. The direction of increasing α-coordinate to the right may be identified by the pattern of migration of the extrema shown in Figures 1-8. One of the difficulties is that the θ-range and the α-range may have limits imposed by the characteristics of the system and the limitations of available apparatus so that only a limited pattern in the α-θ domain is relevant.

With a limited pattern such as shown in Figure 16, the problem of multivaluedness may be serious. To make an extreme illustration, if angular data were taken only over $75^o(1^o)80^o$, there would be several α values at which a fit could be obtained within the limits of ordinary experimental error. On the other hand, if angular data were taken over $45^o(1^o)115^o$, it is clear that a unique fit could be obtained with the pattern in Figure 16 leading to a characteristic α and m. In the latter case, however, there is no guarantee that some other combination of size parameters, refractive index and polarization would not lead to the same "unique" fit. Generally then, some *a priori* knowledge of the system and the patterns of light scattering theory are required.

Figure 16 may also be used to illustrate the problem of polydispersity. For simplicity, consider a small polydispersity in particle size at constant refractive index. The theoretical surface shown in Figure 16 must be modified with a running integration over the size distribution which tends to fill in the minima and round off the maxima so that some of the structure is averaged out. Graphical illustrations of the effect have been reported (9) and the effect has been exploited to obtain both particle size and distribution width (10). It should be realized that both very narrow or very broad distributions tend to average out the light scattering pattern. However, the inversion problem involves the folding or multiplication of a light scattering pattern with a polydispersity function. If the periodicity of the size distribution matches the periodicity of the light scattering pattern an enhancement will result. The relationship is the same mathematical problem as the interference between two signals. We have used polydispersity in size for illustration but the same ideas apply to polydispersity or variation in refractive index or other characteristic properties.

Since the inversion problem involves, in essence, finding a concordance between an experimental set of data and a theoretical set of data, it follows that the probability of a unique match increases with the number of data points in the set. The sensitivity surfaces show that the arrangement of data points is important as well. If the experimental data involves, for example, two angular scans collected simultaneously at two wavelengths, then the matching process is equivalent to fitting a pair of templates at a fixed separation to the theoretical surface. This amounts to the simplest possible case of a two-dimensional match or surface-to-surface fit. The advantages are obvious from the preceding discussion and it is anticipated that experimental exploitation will soon follow.

Sensitivity Maps

The intensity surfaces and difference surfaces that have been presented here may be viewed in general as sensitivity maps. A number of important uses of sensitivity maps are possible, among which are:

(1) The dependence of the light scattering patterns on a wide variety of variables may be explored, i.e. size, angle, intensity or combination of intensities such as the depolarization ratio i_2/i_1 or the polarization $(i_2-i_1)/(i_2+i_1)$.

(2) The effects of particle shape and internal structure may be explored since many theoretical solutions are available (5, 6).

(3) Examination of the loci of the minima in the preceding figures, such as Figures 1-8, show that in some regions of the α-θ domain the valleys are smooth whereas in other regions an undulation occurs. Where the valleys are smooth, the computer time involved in the inversion procedure may be drastically reduced by following a steepest descent to a valley and then seeking a concordance in the valley as in the "short fit" developed in our laboratory (11).

(4) The sensitivity to size and refractive index may be assessed and used to suit a particular application. For exam-

ple, in the study of a system of particles where only particle size is important and a variable refractive index is inconsequential, the low-angle region in the α-θ domain is suggested by the sensitivity surfaces.

(5) The remarkable sensitivity to refractive index illustrated by the regions of optical resonance in Figure 11 remains to be experimentally exploited.

Applications

The applications of wide angle light scattering are beyond the scope of this work and too numerous to be summarized here but we would like to call attention to some comprehensive reviews (12, 13), basic treatises (5, 6), conference proceedings (14, 15) and finally some representative articles, selected from our own experience, that introduce the basic size distribution information obtainable from wide angle light scattering.

The simplest demonstration of wide angle light scattering may be found in the measurements of the light scattered by single aerosol droplets of dioctylphthalate (16, 17).

Determination of the time dependence of particle size distribution and number concentration in a growing system has been evaluated on the well-known LaMer sulfur sols (18, 19, 9), an extensively studied system (5).

Recently, a four-laboratory cooperative study has led to a comparison of the determination of particle size distribution by electron microscopy, quasi-elastic light scattering and wide angle light scattering (20).

The preceding works have been selected because of their coherence with our general treatment. It is hoped that a small group of papers will serve as a readily grasped frame of reference and an introduction to the vast literature on light scattering.

Acknowledgements

We are grateful to Mr. R. N. Rowell for assistance in the preparation of magnetic tape records of the Lorenz-Mie theory computations which were carried out at the University of Massachusetts Computer Center. The tape records were used under the direction of M. L. Prueitt with his program PICTURE at the University of California Los Alamos Scientific Laboratory to generate the computer-drawn perspective graphs. The work was supported in part by grants from the National Science Foundation.

Abstract

The methodical relationship of angular light scattering to turbidity and dynamic light scattering is given. The wide angle scattering method is based upon the exact electromagnetic wave theory of angular light scattering which is reviewed and represented in graphical form in order to illustrate the regularity and simplicity of the three most pronounced general features: (1) the increase in forward scattering with increased particle size, (2) the increase in the number and magnitude of angular maxima and minima with increased particle size and (3) the dependence of intensity level and angular structure on refractive index and polarization. The angular method is used to illustrate the concept that structure or variation in the data is the key to the determination of particle size distribution and refractive index by light scattering. The inter-relationship between polydispersity in size and refractive index or other characteristic property is discussed in relationship to the inversion of light scattering data. The graphical representation gives sensitivity maps which are shown to be useful in exploring theoretical patterns, defining the limits of a problem and revealing new phenomena such as the origin of the angular extrema and the locus of the regions of optical resonance. Selected applications are given as an introduction to wide angle light scattering and as a comparison with other approaches.

Literature Cited

1. Chu, B. "Laser Light Scattering"; Academic Press: New York 1974.
2. Berne, B. J.; Pecora, R. "Dynamic Light Scattering"; Wiley: New York, 1976.
3. Mie, G. Ann. Physik 1908, 25, 377.
4. Debye, P. Ann. Physik 1909, 30, 57.
5. Kerker, M. "The Scattering of Light and Other Electromagnetic Radiation"; Academic Press: New York, 1969.
6. van de Hulst, H. C. "Light Scattering by Small Particles"; Wiley: New York, 1957.
7. Wallach, M. L.; Heller, W.; Stevenson, A. F. J. Chem. Phys. 1961, 34, 1796.
8. Ford, J. R.; Rowell, R. L.; Bassett, D. R. this volume, p.
9. Levit, A. B.; Rowell, R. L. J. Colloid Interface Sci. 1975, 50, 162.
10. Wallace, T. P.; Scott, W. B. J. Polymer Sci. Part A-2 1972, 10, 527.
11. Rowell, R. L.; Ford, J. R.; Parsons, J. W. in "Polymer Colloids II", Fitch, R. M., Ed.; Plenum Press: New York, 1980; p. 27.
12. Kratohvil, J. P. Anal. Chem. 1964, 36, 458R.
13. Kratohvil, J. P. Anal. Chem. 1966, 38, 517R.
14. Kerker, M. "Electromagnetic Scattering"; Pergamon Press: London, 1963.

15. Rowell, R. L.; Stein, R. S. "Electromagnetic Scattering"; Gordon and Breach: New York, 1967.
16. Gucker, F. T.; Egan, J. J. J. Colloid Sci. 1961, 16, 68.
17. Gucker, F. T.; Rowell, R. L. Disc. Faraday Soc. 1960, 30, 185.
18. Rowell, R. L.; Wallace, T. P.; Kratohvil, J. P. J. Colloid Interface Sci. 1968, 26, 494.
19. Rowell, R. L.; Kratohvil, J. P.; Kerker, M. J. Colloid Interface Sci. 1968, 27, 501.
20. Rowell, R. L.; Farinato, R. S.; Parsons, J. W.; Ford, J. R.; Langley, K. H.; Stone, J. R.; Marshall, T. R.; Parmenter, C. S.; Seaver, M.; Bradford, E. B. J. Colloid Interface Sci. 1979, 69, 590.

RECEIVED April 6, 1981.

5

The Molecular Weight Distributions of Emulsion Polymers

DONALD H. NAPPER, GOTTFRIED LICHTI, and ROBERT G. GILBERT

Departments of Physical and Theoretical Chemistry, The University of Sydney, N.S.W. 2006 Australia

The object of this paper is to review a general procedure (1) that allows the molecular weight distribution (MWD) of the polymer produced in an emulsion polymerization to be predicted with considerable generality. The theory to be presented permits the following microscopic kinetic processes to be comprehended: (i) the entry of free radicals into the latex particles from the continuous phase; (ii) the exit (or desorption) of free radicals from the particles; (iv) chain transfer, whether to monomer or added chain transfer agent; (v) bimolecular termination, whether by combination and/or disproportionation. The procedure allows all of these kinetic events to be acting simultaneously. It could be readily extended to encompass the effects of retarders. Chain transfer to polymer, which may be important in some industrial emulsion polymerizations and which results in long chain branching, is specifically excluded from the present treatment, although an extension of the theory should permit this kinetic process to be incorporated.

The prediction of the MWD of emulsion polymers proved to be a relatively intractable problem even after the advent of the Harkins-Smith-Ewart theory. Perhaps the most successful early attack on the problem was that of Katz, Shinnar and Saidel (2). They considered only two microscopic events: entry and bimolecular termination by combination. Their theory resulted in a set of partial integrodifferential equations, whose numerical solution provided the lower moments of the molecular weight distribution function. Other attempts to predict the MWD of emulsion polymers include those of Parts and Watterson (3), Sundberg and Eliassen (4), Min and Ray (5) and Gardon (6).

Overall Strategy

The overall strategy to be described in detail below is straightforward (1). We consider first a set of chains, all of which began growing at the same instant. We shall refer to these chains as being 'distinguished': they are distinguished from all

0097-6156/81/0165-0105$05.00/0

other chains generated in the latex by having commenced growth at the same instant. Particles containing distinguished free radicals will themselves be referred to as being 'distinguished'.

Next, we follow the growth histories of each distinguished chain by following the time evolution of the distinguished particles. This permits us to specify the number concentration of the different types of distinguished particles at any time.

Finally, the time of growth of each chain is found by determining how many distinguished latex particles stopped growing at any particular instant. This is obtained readily from the product of the number concentration of the different types of distinguished particles and the (known) rate coefficient for the appropriate kinetic event. Of course, the growth time of each chain determines the molecular weight of the polymer produced on termination.

The foregoing procedure is that applicable to chain stoppage by a first order process (e.g., transfer, exit) but an extension permits bimolecular termination, whether by combination or disproportionation, to be encompassed. For bimolecular termination, two free radicals are involved in the production of the dead polymer. It is therefore necessary to know the growth histories of both free radicals involved in the bimolecular event. This gives rise to the concept of doubly distinguished particles, i.e., particles that contain two free radicals, the first of which began growing at one arbitrarily chosen instant and the second of which commenced growing at another arbitrarily chosen instant. In this way the growth histories of both distinguished free radicals can be followed and proper allowance made for the termination mechanism. Again, the growth times determine the molecular weight of the polymer produced by the emulsion polymerization process.

It is necessary in a compartmentalized reaction to keep track of not only the distinguished free radicals in a latex particle but also the other free radicals in the particle (i.e., the nondistinguished radicals). This suggests the notation N_i' and N_i'', where N = relative number of latex particles, i denotes the total number of free radicals in the latex particle and the number of prime superscripts specifies the number of distinguished chains still growing in the particle. The nondistinguished free radicals determine the range of microscopic kinetic processes that the distinguished chain can undergo. For example, the distinguished chain in an N_1'-type particle cannot undergo bimolecular termination whereas that in an N_2'-type particle can.

For the purposes of this article, we shall make two simplifying assumptions in that the discussion will be confined to the polymer produced in the steady state domain (i.e., Interval II) of a '0-1-2' system. By a '0-1-2' system, we imply one in which particles can contain only 0, 1 or 2 free radicals, the populations of particles containing more than two free radicals being considered too small to influence the overall behaviour. It is stressed that neither of the foregoing simplifying assumptions is necessary in the general theory (1), although the inclusion of particles containing more than two free radicals usually precludes analytical formulae being obtained. Note that Interval I, where nucleation occurs, is not comprehended by this analysis.

The Smith-Ewart Equations

Like most theoretical discussions of emulsion polymerization, that of the MWD begins with the Smith-Ewart equations (7):

$$\frac{dN_i}{dt} = \rho(N_{i-1} - N_i) + k([i+1]N_{i+1} - iN_i) + c([i+2][i+1]N_{i+2} - i[i-1]N_i) \quad (1)$$

where N_i = relative number of latex particles containing i (=0,1, 2,3,...) free radicals, ρ = radical entry rate coefficient (i.e., the average number of free radicals that enter a particle in unit time), k = exit rate coefficient and 2c = bimolecular termination rate coefficient. Note that the normalization adopted is $\sum_i N_i = 1$; moreover, only if $i \geq 2$ will bimolecular events contribute terms to equations (1). Any term involving a negative subscript is ignored.

The Smith-Ewart equations can be solved using a single numerical eigenvalue determination under all conditions. Analytical solutions can also be obtained if $\bar{n}$ is not too large ($\bar{n} \leq 0.7$)(8,9,10). These solutions encompass both the steady state and the approach to the steady state. Thus the particle number concentrations N_o, N_1, N_2,...are known once ρ, k and c have been determined experimentally. As will be seen, these populations are the starting point for the MWD analysis.

The Distinguished Particle Equations

The Number of Distinguished Particles Produced. We will assume that the steady state has been established and that at some time, denoted by t = 0, a stopwatch (numbered 1) is started. The time on stopwatch 1, denoted by t, thus specifies the period that the experiment has been under way. After an arbitrary period, a second stopwatch (number 2) is started to signal the commencement of growth of the distinguished set of free radicals. Stopwatch 2 accordingly specifies how long the distinguished chains have been growing; this second time we denote by t'.

In an 0-1-2 system, the N_1'-type particles are produced at any instant by the entry of free radicals into particles containing no free radicals or by chain transfer in particles containing one free radical. The number concentration generated at any instant is readily found from the product of the respective number concentration of the different types of particles and the appropriate rate coefficient. Thus

$$N_1' = \rho N_0 + k_{tr} N_1 \qquad (2)$$

where k_{tr} = (pseudo-first-order) transfer rate coefficient ($=k_{tr,m}C_m$, where $k_{tr,m}$ = (second order) rate constant for transfer to monomer and C_m = monomer concentration in the latex particles). For styrene, N_0 and N_1 are often comparable in magnitude whereas k_{tr} (of order $0.1s^{-1}$) may be larger than ρ (of order, say, $\leqslant 10^{-2}\ s^{-1}$)(8). The second term is then dominant.

Particles that contain one distinguished and one nondistinguished growing chain are produced at t'=0 by radical entry into particles containing one free radical or by chain transfer from either of the free radicals in particles containing two free radicals:

$$N_2' = \rho N_1 + 2k_{tr} N_2 \qquad (3)$$

For styrene, it often happens that $N_1 \gg N_2$ and the first term may then be dominant.

The Time Evolution of the Distinguished Particles. We see that N_i' must be functions both of actual time (t) and distinguished time (t'). For the steady state considered here, the t dependence can be dropped. The number concentration of distinguished particles changes in distinguished time as a consequence of two different types of events: those associated with the distinguished character (labelled with the subscript D below) of the particles and those not so associated (labelled with the subscript ND). The former lead to the loss of the distinguished character of the particles whereas the distinguished character is conserved when the latter events occur.

For N_1'-type particles, only chain transfer and exit lead to growth of the distinguished chains being stopped:

$$\left.\frac{dN_1'}{dt'}\right|_D = -(k_{tr} + k)N_1' \qquad (4)$$

The mechanism whereby exit occurs suggests that for styrene $k_{tr} \gg k$ commonly. N_2'-type particles are subject to both of the first order processes considered for N_1'-type particles and also to bimolecular termination events:

$$\left.\frac{dN_2'}{dt'}\right|_D = -(k_{tr} + k + 2c + \rho/2)\ N_2' \qquad (5)$$

Note that no coefficient of 2 precedes the first two rate coefficients since only one distinguished chain is involved. The $\rho/2$ term is an artifact of the 0-1-2 system: it indicates that half the entries into the N_2'-type particles terminate instantaneously the distinguished chains.

The analogous equations for nondistinguished chains are

$$\left.\frac{dN_1'}{dt'}\right|_{ND} = \rho\, N_1' + (k+\rho/2)\, N_2' \tag{6}$$

$$\left.\frac{dN_2'}{dt'}\right|_{ND} = \rho N_1' - (k+\rho/2)\, N_2' \tag{7}$$

These equations specify that entry into an N_1'-type particle produces an N_2'-type particle whereas an exit event (and only the nondistinguished chain is involved here) results in the reverse transformation.

The overall time evolution of the singly distinguished particles is obtained by summing the changes that dissipate the distinguished character of the particles with those that conserve it:

$$\frac{dN_1'}{dt'} = -(\rho+k_{tr}+k)\, N_1' + (k+\rho/2)\, N_2' \tag{8}$$

$$\frac{dN_2'}{dt'} = \rho N_1' - (\rho+2k+k_{tr}+2c)\, N_2' \tag{9}$$

These equations are similar to the Smith-Ewart equations presented above and can be solved in an analogous manner. The boundary conditions, i.e., the values of N_1' and N_2' at $t'=0$, required to solve these equations have been presented above (equations (2) and (3)).

The solution of equations (8) and (9) yields the number concentration of singly distinguished particles at any time t'. These can be used to calculate the rate of stoppage of chains if termination is a first order event.

First Order Chain Stoppage. The rate of stoppage of distinguished chains is obtained from the product of the populations of singly distinguished particles with the respective rate coefficient. Thus for stoppage in the absence of combination and disproportionation, we have from equations (4) and (5)

$$s_1^{tt} = (k_{tr}+k)\, N_1' \tag{10}$$

$$s_2^{tt} = (k_{tr}+k) + \rho/2)\, N_2' \tag{11}$$

where s_i specifies the distribution of singly distinguished particles in state i whose distinguishing chain ceases growth at time t' after its creation. The total distribution is obtained by summing s_1 and s_2.

Integration for Comparison of Theory with Experiment. The sum of equations (10) and (11) yields the MWD for one arbitrarily chosen starting time. In any experimental determination of the MWD, measurements are made not on just one set of distinguished chains but on all the distinguished chains (i.e., the totality of chains) generated in the particles from the start of the experiment (t=0) to the time (t=t*) when the system is sampled for polymer (i.e., t* is the final time displayed on stopwatch 1). This requires integration over all possible starting times:

$$S^{tt}(t^*,t') = \int_0^{t^*} (s_1^{tt} + s_2^{tt})\, dt \qquad (12)$$

The molecular weight M is readily obtained from the growth time t' since M=at', where a is the increase in the molecular weight of a free radical growing in a latex particle per second ($a = k_p C_m M_0$, where k_p = propagation rate constant and M_0 = molecular weight of monomer). S^{tt} is the target of the analysis. S^{tt}/t^* gives the average number of chains of each molecular weight generated per second in each latex particle. Note that the original normalization adopted for N_i ensures that the calculated rate of production of chains refers to a single latex particle.

The Doubly Distinguished Particle Equations

The foregoing procedure must be extended to incorporate bimolecular events. Two free radicals are involved in the termination step and the growth histories of both chains must be known to calculate the MWD.

The Number Concentration of Doubly Distinguished Particles. In the 0-1-2 system, doubly distinguished particles (N_2'') can only be formed from singly distinguished particles by entry of a free radical into a N_1'-type particle or chain transfer (involving the nondistinguished chain) in an N_2''-type particle. Therefore

$$N_2'' = \rho N_1' + k_{tr} N_2' \qquad (13)$$

Note that the values of N_1' and N_2' are known from the solution of equations (8) and (9). We will assume that at the instant when the doubly distinguished particles are formed, a third stopwatch (numbered 3) is started and that stopwatch 2 is stopped. The time which is displayed on stopwatch 3, t'', is the time of co-existence of the two distinguished free radicals or the growth time of the shorter of the two radicals.

Note that equation (13) shows that there is a hierarchy of differential equations: solution of the Smith-Ewart equations provides the boundary conditions for the singly distinguished particle equations; these in turn provide the boundary conditions for the doubly distinguished particle equations.

The Time Evolution of the Doubly Distinguished Particles. In the 0-1-2 system, all events associated with the doubly distinguished particles lead to the loss of the particles. Entry, bimolecular combination, transfer (from either distinguished chain) and exit (again from either distinguished chain) all may occur:

$$\frac{dN_2''}{dt''} = -(\rho+2c+2k_{tr}+2k)\ N_2'' \qquad (14)$$

This clearly specifies an exponential decay in N_2'', the initial value of N_2'' being given by equation (13). The value of N_2'' at any time t'' is thus known.

Termination. Termination of the doubly distinguished chains is given by

$$s_2^{bt} = 2cN_2'' \qquad (15)$$

Integrations for Comparison with Experiment. The equations (13)-(15) hold irrespective of whether termination occurs by combination or disproportionation. Further development, however, demands a different approach for the two different types of bimolecular termination.

Combination Alone. The polymer sampled from an emulsion polymerization terminated solely by combination will contain distinguished sets of chains spanning all possible starting times and all possible co-existence times. It is thus necessary to integrate over all possible values of these variables. Accordingly, the target of the analysis, S^{bc}, is given by

$$S^{bc}(t^*,M) = \int_0^{t^*} dt \int_0^{M/2a} dt''2c_cN_2'' \qquad (16)$$

where the superscript bc denotes bimolecular combination, t* = time of the experiment and $2c_c$=combination rate coefficient. Note that the maximum possible co-existence time of two chains that combine is M/2a. The molecular weight M is related in this case to the times displayed on stopwatches 2 and 3 by M=a(t'+2t''), the factor of two indicating that the two chains grew concurrently for a time t''; this relationship permits the variable t' to be eliminated from equation (16).

Disproportionation Alone. Each disproportionation event results in the production of two chains: one shorter and one longer. To obtain the distribution function of the longer chains (S_ℓ^{bd}), it is necessary to integrate over all possible starting times and all possible growth times of the shorter chains:

$$S_\ell^{bd}(t^*,M) = \int_0^{t^*} dt \int_0^{M/2a} dt'' \; 2c_d \; N_2'' \qquad (17)$$

where $2c_d$ = disproportionation rate coefficient. Note that the molecular weight in this instance is given by $M = a(t'+t'')$ and that the upper limit to the growth time of the second distinguished chain is just M/a. Similarly, it is necessary for the shorter chains to integrate over all possible starting times and all possible growth times of the longer chains:

$$S_s^{bd}(t^*,M) = \int_0^{t^*} dt \int_0^{t^*} dt' \; 2c_d \; N_2'' \qquad (18)$$

where $M = at''$. The upper limit to the growth time of the longer chain is clearly the time of the experiment t*.

The total distribution function for the chains produced by disproportionation is obtained by summing the number distributions of the longer and shorter chains:

$$S^{bd} = S_\ell^{bd} + S_s^{bd} \qquad (19)$$

The Total Chain Production

The foregoing analysis was expressly elaborated to treat a diverse range of microscopic kinetic processes. Each termination process, however, was regarded as acting alone. It is possible to handle any combination of termination mechanisms by simple addition of the number distribution functions for the first order and second order events:

$$S = S^{tt} + S^{bc} + S^{bd} \qquad (20)$$

Polydispersity Index

One advantage of the procedure delineated above is that it permits the complete molecular weight distribution function to be calculated, sometimes analytically, whatever the termination mechanism. Of course, the lower moments of the distribution function can also be readily calculated:

$$\langle M_w \rangle = \int_0^\infty SM^2 dM \Big/ \int_0^\infty SMdM$$

$$\langle M_n \rangle = \int_0^\infty SMdM \Big/ \int_0^\infty SdM$$

allowing the commonly used polydispersity index $\langle M_w \rangle / \langle M_n \rangle$ to be found.

Results

First Order Stoppage Alone. If stoppage is determined solely by a first order process, such as transfer, the foregoing analysis predicts a nearly exponential distribution function. The polydispersity index must then be very close to 2.00. The same result is obtained for bulk and solution polymerizations dominated by chain transfer. Compartmentalization thus has no major effect on the polydispersity of the polymer produced, as was recognized by Gerrens (11), if the stoppage process is dominated by chain transfer. This contrasts with the significant effects of compartmentalization if bimolecular events dominate termination.

Bimolecular Termination Alone. The effect of compartmentalization in an emulsion polymerization is to broaden significantly the MWD of the polymer produced if termination is dominated by bimolecular events. This was clearly established by Katz, Shinnar and Saidel (2) for termination by combination but also holds for disproportionation. We note in passing that one theory (6) of emulsion polymerization claims that compartmentalization decreases the polydispersity of the polymer produced at any instant; there is, however, no sound theoretical basis for this claim.

The origin of the broadening of the MWD by compartmentalization is readily located. The width of the MWD reflects the different environments in which the polymer chains are created, grow and cease growth. If the average number of free radicals per particle ($\bar{n}$) is large, the environments of the chains in the compartmentalized system will be similar to those in the bulk and the polydispersity index will be close to the bulk value. As $\bar{n}$ decreases, however, the environments differ more widely from those in the bulk system. Of particular importance are the growing chains in particles containing only one free radical. These chains cannot undergo bimolecular termination and so grow unhindered until a second free radical enters the latex particle. This environment is manifestly different from that in other growing latex particles and from that in the bulk system. This argument also explains why compartmentalization has no effect on the MWD if termination is by chain transfer because the chains containing one free radical can still undergo the transfer process, just as they do in the bulk system.

The limiting behaviour in the steady state for bimolecular termination of emulsion polymerizations is summarized as follows:

Combination

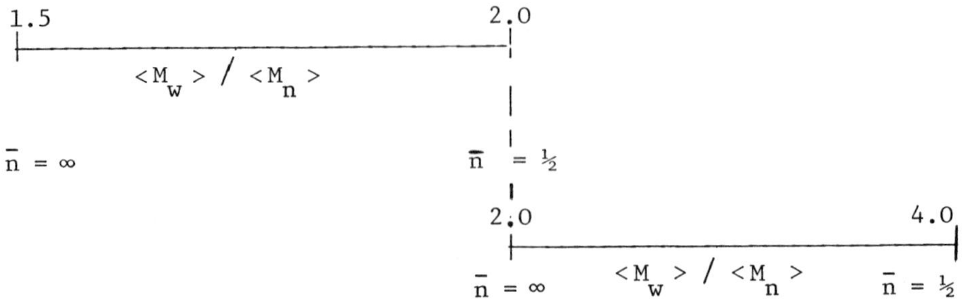

Disproportionation

Combination Alone. Figure 1 displays the dependence of the polydispersity index on $\bar{n}$ if termination is solely by combination. Note that these results have been calculated using the full analysis (1), not just the 0-1-2 system. Other theories given in the literature imply different values for $\langle M_w \rangle / \langle M_n \rangle$ if $\bar{n} = \frac{1}{2}$: these range from 1.0(6) to 2.5 (5). The limiting value of 2.00 was also obtained by Katz, Shinnar and Saidel (2) by an entirely different method from that elaborated above and is unquestionably correct. It can be established by the following reasoning. If $\bar{n} = \frac{1}{2}$, each particle contains at most one free radical. Growing chains in the latex particles can thus either grow or be terminated instantaneously by entrant free radicals. These mutually exclusive kinetic events immediately prescribe the Flory 'most probable' distribution function for the growing chains (12); this is an exponential distribution function with a polydispersity index of 2.00 (13). The termination process occurs instantaneously via entrant free radicals of (near) zero molecular weight. These radicals do not perturb significantly the distribution of chain lengths in converting growing chains to dead polymer. Indeed, termination in this instance is equivalent to chain transfer, which gives an identical value for the polydispersity index.

Disproportionation Alone. Figure 2 displays the polydispersity index as a function of $\bar{n}$ for termination by disproportionation. Again, these have been calculated by a full analysis (1). As with termination by combination, the polydispersity shows a rapid decrease in the $\bar{n}$ range of 0.5 - 1.0.

The reason for the upper limit of $\langle M_w \rangle / \langle M_n \rangle = 4.00$ follows immediately from the discussion given in the preceding section for termination by combination. Again, the growing chains have a Flory 'most probable' distribution with $\langle M_w \rangle / \langle M_n \rangle = 2.00$.

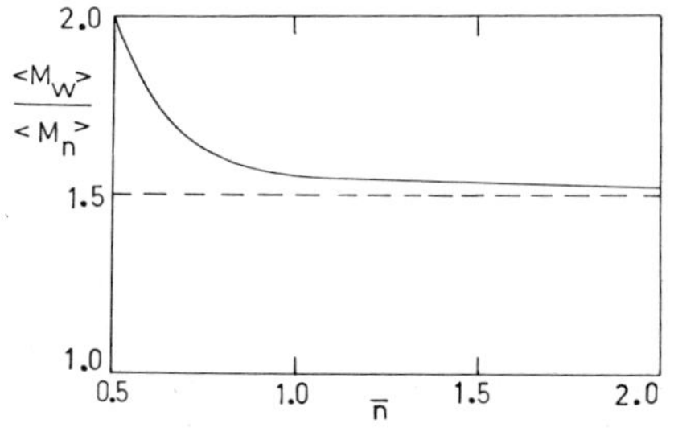

Figure 1. Polydispersity index of the polymer produced in Interval II of an emulsion polymerization terminated solely by combination as a function of the average number of free radicals per particle

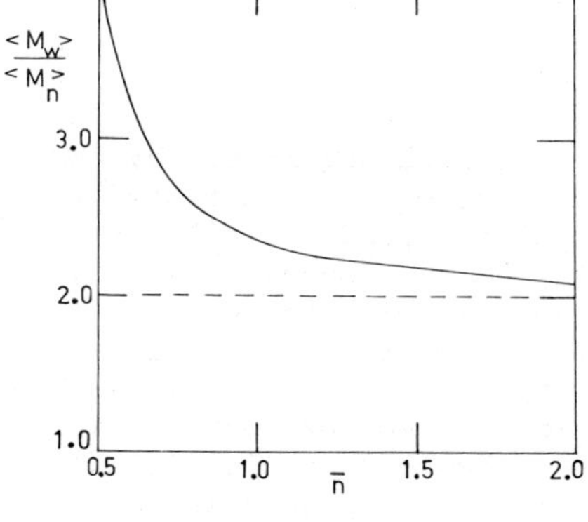

Figure 2. Polydispersity index of the polymer produced in Interval II of an emulsion polymerization terminated solely by disproportionation as a function of the average number of free radicals per particle.

Termination in this case produces a number, equal to the number of growing chains, of low molecular weight dead chains. This does not affect $<M_w>$ but halves $<M_n>$. Hence $<M_w>/<M_n>$ is doubled to 4.00.

A Combination of Termination Mechanisms

Displayed in Figure 3 are the results for the polydispersity index for an emulsion polymerization system in which chain stoppage occurs by a combination of chain transfer (k_{tr} = 1 reciprocal time unit) and disproportionation (c_d = 100 reciprocal time units). These results were obtained by varying ρ and thus $\bar{n}$. The results suggest how it might be possible to tailor a distribution to some desired polydispersity index.

Comparison with Experiment

The experimental results on the polymer produced in emulsion polymerizations published thus far are both confusing and contradictory. Several factors may be responsible for this: first, many surfactants behave as chain transfer agents, which has often not been recognized; second, measurements have often been made on samples that contain polymer from Intervals I, II and III, which leads to a significant increase in the polydispersity index because $<M_n>$ is sensitive to the presence of lower molecular weight species; third, direct measurements of the MWD have only recently become possible with the advent of gel permeation chromatography.

For many monomers (e.g., styrene, methyl methacrylate, vinyl acetate), it seems likely that the molecular weight of the polymer produced by emulsion polymerization is in many instances dominated not by bimolecular termination as in the bulk but rather by chain transfer to monomer. This is most convincingly demonstrated by the average molecular weight of the polymer produced (4). $<M_w>$ would be of order tens of millions if chain stoppage were dominated by bimolecular termination but only several millions or less if chain transfer to monomer is operative. Experimental results usually confirm the latter order of magnitude, which is itself an order of magnitude larger than that for polymer produced in bulk or in solution. The kinetic results of Hawkett *et al.* (8) allow the process of chain transfer to monomer to be placed in perspective. The experimental results show that in these systems a free radical entry event occurs only once every few minutes, even at the highest initiator concentration. An exit event takes place every fifteen minutes. Chain transfer to monomer, however, occurs every ten seconds. This is clearly the dominant chain stoppage event and should result in polymer with a polydispersity index of 2.00. This result demonstrates that compartmentalization can cause a fundamental change in the chain stopping mechanism.

Piirma and co-workers (14, 15) have performed careful studies on the polystyrene produced by emulsion polymerization using gel

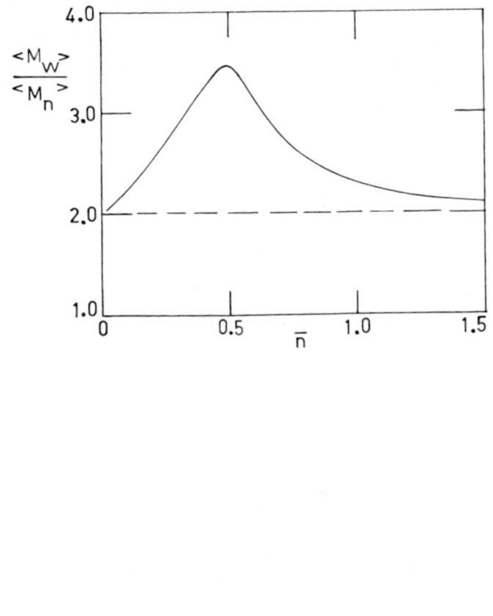

Figure 3. Polydispersity index of the polymer produced in Interval II of an emulsion polymerization as a function of the average number of free radicals per particle. Chain stoppage occurs by chain transfer (k_{tr} = *1 reciprocal time unit) and disproportionation* (c_d/k = *100).*

permeation chromatography. Their results broadly support the contention that chain transfer to monomer controls the MWD. Note that if $\bar{n} = \frac{1}{2}$, the polydispersity index has a value of 2.00 irrespective of whether termination is dominated by transfer to monomer or by combination, which is the dominant bimolecular termination event for styrene in bulk or solution polymerizations. This ambiguity can usually be removed by consideration of the (weight) average molecular weight of the polymer produced, as noted above.

Note that if it is possible to cause free radical entry events to occur much more frequently than chain transfer to monomer events (e.g., for styrene, at a rate greater than 1 per second), then the polymer will be controlled primarily by bimolecular termination. Note, too, that if chain stoppage is dominated by chain transfer, the molecular weight of the polymer produced would be independent of particle volume. Morton et al. (16) have obtained data for styrene that support this conclusion.

Conclusions

It is possible to predict the MWD of polymer produced by an emulsion polymerization with significant generality. The results show that not only is the instantaneous average molecular weight significantly larger but also that the MWD of the polymer produced in compartmentalized reactions is relatively broader than the corresponding polymer produced in a bulk or solution polymerization provided bimolecular termination is dominant. If, however, chain transfer to monomer is the dominant chain stopping mechanism, compartmentalization has no effect on the MWD of the polymer produced. The latter appears to be the case for some common monomers (e.g., styrene) in many emulsion polymerizations. Thus compartmentalization can cause a fundamental change in the dominant chain stoppage mechanism.

Legend of Symbols

a = molecular weight increase per unit time of a growing chain.
c = pseudo-first-order bimolecular termination rate coefficient.
c_c = rate coefficient for termination by combination.
c_d = rate coefficient for termination by disproportionation.
C_m = monomer concentration in latex particles.
k = rate coefficient for exit.
k_p = propagation rate coefficient.
k_{tr} = pseudo-first order rate coefficient for chain transfer.
M_o = molecular weight of monomer.
$\bar{n}$ = average number of free radicals per particle.
N_i = relative number of latex particles containing i free radicals.
N_i' = relative number of latex particles containing one distinguished free radical and (i-1) nondistinguished radicals.
N_i'' = relative number of latex particles containing two distinguished free radicals and (i-2) nondistinguished radicals .

s_i^{tt} = distribution of singly distinguished particles in state i whose distinguishing chain ceased growth at time t' after its creation due to first order stoppage events.
s_2^{bt} = distribution of doubly distinguished particles whose distinguishing chains ceased growth at time t'' after the creation of the second chain due to mutual termination.
S^{tt} = distribution of nongrowing chains produced by stoppage events.
S^{bc} = distribution of nongrowing chains produced by combination.
S^{bd} = distribution of nongrowing chains produced by disproportionation.
S_ℓ^{bd} = contribution to S^{bd} by longer chains.
S_s^{bd} = contribution to S^{bd} by shorter chains.
t* = time the experiment has run.
t' = growth time of distinguished chains in singly distinguished particles.
t'' = simultaneous growth times of distinguished chains in doubly distinguished particles.
ρ = entry rate coefficient.

Acknowledgements

We thank the Australian Research Grants Committee for their support of these studies.

Literature Cited

1. Lichti, G.; Gilbert, R.G.; Napper, D.H. J. Polym. Sci. Polm. Chem. 1980, 18, 1297.
2. Katz, S.; Shinnar, R.; Saidel, G.M. Adv. Chem. Ser. 1969, 91, 145.
3. Watterson, J.G.; Parts, A.G. Makromol. Chem. 1971, 146, 1.
4. Sundberg, D.C.; Eliassen, J.D. in "Polymer Colloids", Fitch, R.M., Ed.: Plenum: New York, 1971; p.153.
5. Min, K.W.; Ray, W.H. J. Macromol. Sci. Rev. Macromol. Chem. 1974, C11, 177.
6. Gardon, J.L. J. Polym. Sci. A-I, 1968, 6, 665.
7. Smith, W.V.; Ewart, R.H. J. Chem. Phys. 1943, 16, 592.
8. Hawkett, B.S.; Napper, D.H.: Gilbert, R.G. J. Chem. Soc. Faraday Trans. I 1980, 76, 1323.
9. Hawkett, B.S.; Napper, D.H.; Gilbert, R.G. J. Chem. Soc. Faraday Trans. I 1977, 73, 690.
10. Brooks, B.W. J. Chem. Soc. Faraday Trans. I 1980, 76, 1599.
11. Gerrens, H. Fortschr. Hochpolymer-Forsch. 1959, 1, 234.
12. Flory, P.J. "Principles of Polymer Chemistry"; Cornell University: Ithaca, 1953; chap. 8.
13. Peebles, L.H. "Molecular Weight Distributions in Polymers"; Interscience: New York, 1971; p.10.
14. Piirma, I.; Kamath, V.R.; Morton, M. J.Polym. Sci. Chem. Ed. 1975, 13, 2087.

15. James, H.L.; Piirma, I. in "Emulsion Polymerization", Piirma, I. and Gardon, J.L., eds., ACS Symp. Series 24, American Chemical Society: Washington, D.C.; 1976; p. 197.
16. Morton, M.S.; Kaizerman, S.; Altier, M.W. J. Colloid Sci. 1954, 9, 300.

RECEIVED April 6, 1981.

6

On the Optimal Reactor Type and Operation for Continuous Emulsion Polymerization

MAMORU NOMURA
Department of Industrial Chemistry, Fukui University, Fukui, Japan
MAKOTO HARADA
Institute of Atomic Energy, Kyoto University, Uji, Japan

Continuous emulsion polymerization processes are presently employed for large scale production of synthetic rubber latexes. Owing to the recent growth of the market for polymers in latex form, this process is becoming more and more important also in the production of a number of other synthetic latexes, and hence, the necessity of the knowledge of continuous emulsion polymerization kinetics has recently increased. Nevertheless, the study of continuous emulsion polymerization kinetics has, to date, received comparatively scant attention in contrast to batch kinetics, and very little published work is available at present, especially as to the reactor optimization of continuous emulsion polymerization processes. For the theoretical optimization of continuous emulsion polymerization reactors, it is desirable to understand the kinetics of emulsion polymerization as deeply and quantitatively as possible.

The present review paper, therefore, refers firstly to the particle formation mechanism in emulsion polymerization, the complete understanding of which is indispensable for establishing a correct kinetic model, and then, deals with the present subject, that is, what type of reactor and operating conditions are the most suitable for a continuous emulsion polymerization process from the standpoint of increasing the volume efficiency and the stability of the reactors.

Although the early literature described the application of a tubular reactor for the production of SBR latexes(1), the standard continuous emulsion polymerization processes for SBR polymerization still consist of continuous stirred tank reactors(CSTR's) and all of the recipe ingredients are normally fed into the first reactor and a latex is removed from the last one, as shown in Figure 1. However, it is doubtful whether this conventional reactor combination and operation method is the most efficient in continuous emulsion polymerization. As is well known, the kinetic behavior of continuous emulsion polymerization differs very much according to the kind of monomers. In this paper, therefore, the discussion about the present subject will be advanced using the

0097-6156/81/0165-0121$05.75/0

styrene emulsion polymerization system because the basic mechanism and the mathematical reaction model of this system are fairly well established.

In continuous emulsion polymerization of styrene in a series of CSTR's, it was clarified that almost all the particles formed in the first reactor(2,3). Since the rate of polymerization is, under normal reaction conditions, proportional to the number of polymer particles present, the number of succeeding reactors after the first can be decreased if the number of polymer particles produced in the first stage reactor is increased. This can be realized by increasing emulsifier and initiator concentrations in the feed stream and by lowering the temperature of the first reactor where particle formation is taking place(3). The former choice is not desirable because production cost and impurities which may be involved in the polymers will increase. The latter practice could be employed in parallel with the technique given in this paper. Our final goal in the present paper is to devise an optimal type of the first stage reactor and its operation method which will maximize the number of polymer particles produced in continuous emulsion polymerization. For this purpose, we need a mathematical reaction model which explains particle formation and other kinetic behavior of continuous emulsion polymerization of styrene.

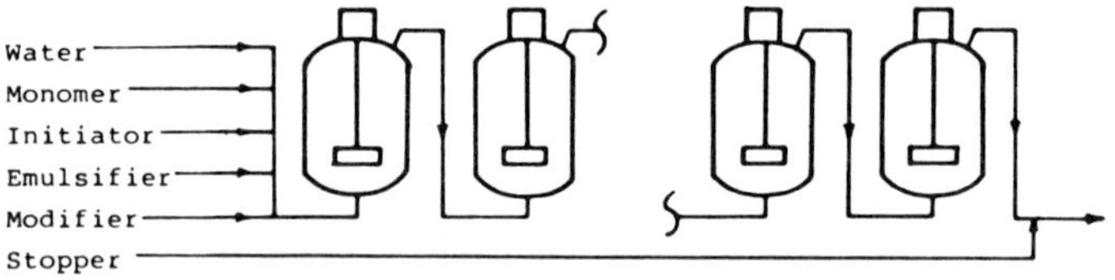

Figure 1. Flow diagram of typical continuous emulsion polymerization reactor system

Mathematical Reaction Model

Basic Equations. Mathematical reaction models for continuous emulsion polymerization of styrene proposed to date are roughly classified into three groups according to the particle formation mechanism which they adopted. Dickinson(4), and Min and Ray(5) took into consideration the homogeneous particle formation which Fitch and Tsai(6) made quantitative. Gershberg and Longfield(3), Degraff and Poehlein(7), and Omi and coworkers(8) form one group, where they employed the Smith-Ewart second idealized situation for particle formation(9), that is, the free radicals in the water phase are captured in proportion to the surface area of micelles and polymer particles. On the other hand, Nomura and coworkers(10) presented a different expression which involves a concept of radical capture efficiency of micelles, although adopting the micelle hypothesis. The kinetic behaviors of styrene emulsion polymerization can be predicted sufficiently well, as a whole, by micellar particle formation and without using a complicated reaction model with a number of kinetic parameters. Therefore, a simple reaction model developed by Nomura and coworkers(11) (referred to as the Nomura and Harada model) will be used for the discussion of the present subject because their reaction model also includes the Smith-Ewart second idealized situation in particle formation(9), and the Gershberg model(3) is a special case of the Nomura and Harada model, as shown later.

The Nomura and Harada model proceeds as follows, using the elementary reactions and their rate expressions shown in Table I (11).

Table I Elementary Reactions and Their rates

Reaction	Reaction type	Reaction rate	
Initiation of radicals	$I \rightarrow 2R^*$	$r_i = 2k_d fI$	(A)
Initiation of particle from micelle	$R^* + m_s \rightarrow N^*$	$k_1 m_s R^*$	(B)
Initiation	$R^* + N \rightarrow N^*$	$k_2 N R^*$	(C)
Termination	$R^* + N^* \rightarrow N$	$k_2 N^* R^*$	(D)
Propagation in particle	$P^*_j + M \rightarrow P^*_{j+1}$[a]	$k_p M_p N^*$	(E)

[a] P^*_j is a polymer radical containing j monomer units in an active particle N^*_j.

(I) Particle Formation:

$$\frac{dN_T}{dt} = R_i - \frac{N_T}{\theta} \tag{1}$$

where R_i is the rate of particle formation, N_T is the number of polymer particles per unit volume of water and θ is the mean residence time in the first reactor.

Since the number of radicals flowing out of the first reactor in the exit stream can be neglected under normal reaction conditions, we have:

$$R_i = r_i \left[\frac{k_1 m_s R^*}{k_1 m_s R^* + k_2 N_T R^*}\right]$$

$$= r_i \left[\frac{1}{1 + (k_2/k_1) M_m (N_T/S_m)}\right] = \frac{r_i}{1 + \varepsilon(N_T/S_m)} \tag{2}$$

here, $\varepsilon = (k_2/k_1) M_m$

where r_i is the rate of radical production in the water phase, m_s is the micelle concentration, M_m is the aggregation number per micelle, S_m is the number of emulsifier molecules forming micelles in unit volume of water and R^* is the radical concentration in the water phase and given by;

$$\frac{dR^*}{dt} = r_i - k_1 m_s R^* - k_2 N_T R^* - \frac{R^*}{\theta}$$

Considering that the term R^*/θ can be neglected and steady state hypothesis is applicable to R^*, Eq.(3) leads to:

$$R^* = \frac{r_i}{k_1 m_s [1 + (\varepsilon N_T/S_m)]} \tag{3}$$

According to the Gershberg model where it was assumed that the radicals in the water phase enter both micelles and particles in proportion to their total surface areas, R_i is expressed as:

$$R_i = r_i\left[\frac{A_m}{A_m + A_p}\right] = r_i\left(\frac{a_s S_m}{a_s S_F}\right) = r_i\left(\frac{S_m}{S_F}\right) \quad (4)$$

where A_m is the surface area of micelles, A_p is the surface area of polymer particles, a_s is the surface area occupied by an emulsifier molecule and S_F is the emulsifier concentration in the feed stream.
Considering that $A_m = \pi d_m^2 m_s$ and $A_p = \pi d_p^2 N_T$ where d_m and d_p are the average diameters of a micelle and a polymer particle, respectively, Eq.(4) can be also expressed by the following equation:

$$R_i = r_i\left[\frac{\pi d_m^2 m_s}{\pi d_m^2 m_s + \pi d_p^2 N_T}\right] = \frac{r_i}{1 + (d_p/d_m)^2 M_m (N_T/S_m)} \quad (5)$$

Equation(2) can include this expression by supposing that $k_1 = \pi d_m^2$ and $k_2 = \pi d_p^2$. We have, therefore, the following expressions for ε according to the mechanism of radical entry into micelles and particles.

<u>Case A:</u> If k_1 and k_2 do not depend on the diameters of micelle and particle, we have:

$$\varepsilon = (k_2/k_1)M_m = \text{constant} = K_o \quad (6)$$

<u>Case B:</u> If Fick's diffusion theory can be applied to the entry of radicals into micelles and particles, we have:

$$k_1 = 2\pi D_w d_m \quad \text{and} \quad k_2 = 2\pi D_w d_p \quad (7)$$

$$\varepsilon = (d_p/d_m)M_m = K_1 d_p \quad (8)$$

where D_w is the diffusion coefficient of radical in the water phase.
<u>Case C:</u> This situation corresponds to the Smith-Ewart second idealized situation for particle formation and to the Gershberg model , as shown by Eq.(5). In this case, we have:

$$k_1 = \pi d_m^2 \quad \text{and} \quad k_2 = \pi d_p^2 \quad (9)$$

$$\varepsilon = (d_p/d_m)^2 M_m = K_2 d_p^2 \quad (10)$$

<u>Case D:</u> Ugelstad states that radical entry into micelles and particles is, under some conditions, proportional to their volumes. If these conditions are fulfilled, we have:

$$k_1 \propto d_m^3 \quad \text{and} \quad k_2 \propto d_p^3 \quad (11)$$

$$\varepsilon = (d_p/d_m)^3 M_m = K_3 d_p^3 \quad (12)$$

(II) The number of active particles containing a radical is given by:

$$dN^*/dt = k_1 m_s R^* + k_2 N_o R^* - k_2 N^* R^* - N^*/\theta$$

$$= r_i [1 - \frac{2N^*/N_T}{1+(S_m/\varepsilon N_T)}] - N^*/\theta \tag{13}$$

(III) Monomer conversion is expressed by:

$$dX_M/dt = (\frac{k_p [M_p] M_g}{M_F N_A}) N^* - X_M/\theta \tag{14}$$

(IV) Emulsifier balance:
Since the depletion of emulsifier micelles occurs only because they break up and their molecules are adsorbed onto the surface of growing particles. The balance on the micelles in the first stage reactor is given by the following equation if the reaction is started so that the emulsifier concentration in the feed stream does not change with time:

$$S_m = S_F - A_p/a_s \tag{15}$$

where,

$$A_p = (36\pi)^{1/3} \bar{v}_p^{2/3} N_T \tag{16}$$

When monomer droplets exist,

$$\bar{v}_p = \frac{(M_F X_M)(1+\gamma)}{N_T \rho_p} \qquad (X_M \leqq 0.43) \tag{17}$$

When monomer droplets disappear,

$$\bar{v}_p = (\frac{M_F}{N_T \rho_p}) \qquad (X_M > 0.43) \tag{18}$$

where $\bar{v}_p$ is the average volume of a monomer-swollen particle, γ is the monomer-polymer weight ratio in the particles, M_F is the monomer concentration in the feed stream and ρ_p is the density of monomer-swollen particles.

Gershberg and Longfield evaluated A_p in a different manner, taking into account the particle size distribution. The fraction of polymer particles which stayed for time $\tau \sim \tau + d\tau$ in a parfectly mixed stirred tank reactor can be given by:

$$\frac{dN_i}{N_T} = \frac{1}{\theta} e^{-\tau/\theta} \tag{19}$$

and then,

$$A_p = \pi d_p^2 N_T = \pi \int_0^{N_T} d_{pi}^2 dN_i = (36\pi)^{1/3} \int_0^{N_T} \bar{v}_{pi}^{2/3} dN_i \tag{20}$$

If the volume of a micelle is neglected, the volume of a particle which stayed for time τ in the reactor can be approximately given by:

$$\bar{v}_{pi} = \mu\tau \tag{21}$$

where μ is the volumetric growth rate of a particle and defined by

$$\mu = \left(\frac{k_p [M_p] M_g}{N_A \rho_p}\right) \bar{n} (1 + \gamma) \qquad (22)$$

where $\bar{n}$ is the average number of radicals per particle and defined by $N^* = \bar{n} N_T$.

Introducing Eqs.(19) and (21) into Eq.(20) and integrating Eq.(20) lead to:

$$A_p = \Gamma(5/3)(36\pi)^{1/3}(\mu\theta)^{2/3} N_T \qquad (23)$$

It is found from Eqs.(14),(17),(22) and $N^* = \bar{n} N_T$ that at steady state $\mu\theta = \bar{v}_p$. Therefore, the value of A_p given by Eq.(23) is greater than that given by Eq.(16) by a factor of $\Gamma(5/3) = 0.903$. Thus, Eqs.(16) and (23) seem to give almost the same accuracy in calculating A_p, considering that competitive growth rate of polymer particles with different size is not necessarily well understood(12).

Comparison between Experimental Results and Model Predictions. As will be shown later, the important parameter ε which represents the mechanism of radical entry into the micelles and particles in the water phase does not affect the steady-state values of monomer conversion and the number of polymer particles when the first reactor is operated at comparatively shorter or longer mean residence times, while the transient kinetic behavior at the start of polymerization or the steady-state values of monomer conversion and particle number at intermediate value of mean residence time depend on the form of ε. However, the form of ε influences significantly the polydispersity index $\bar{M}_w/\bar{M}_n$ of the polymers produced at steady state. It is, therefore, preferable to determine the form of ε from the examination of the experimental values of $\bar{M}_w/\bar{M}_n$. The effect of radical capture mechanism on the value of $\bar{M}_w/\bar{M}_n$ can be predicted theoretically as shown in Table II, provided that the polymers produced by chain transfer reaction to monomer molecules can be neglected compared to those formed by mutual termination. Degraff and Poehlein(7) reported that experimental values of $\bar{M}_w/\bar{M}_n$ were between 2 and 3, rather close to 2, as shown in Figure 2. Comparing their experimental values with the theoretical values in Table II, it seems that the radicals in the water phase are not captured in proportion to the surface area of a micelle and a particle but are captured rather in proportion to the first power of the diameters of a micelle and a particle or less than the first power. This indicates that the form of ε would be Case A or Case B. In this discussion, therefore, Case A will be used as the form of ε for simplicity.

Let us determine the value of ε by comparing the transient kinetic behavior of monomer conversion in continuous emulsion polymerization of styrene with the model prediction by the Nomura and Harada model. It is reported in the literature that sustained

Table II Effect of radical capture mechanism on the form of ε and theoretical values of $\bar{M}_w/\bar{M}_n$.

$k_1 \propto d_m^n$, $k_2 \propto d_p^n$

n	0	1	2	3
ε	Case A	Case B	Case C	Case D
$\bar{M}_w/\bar{M}_n$	2.0	2.4	4.8	>4.8

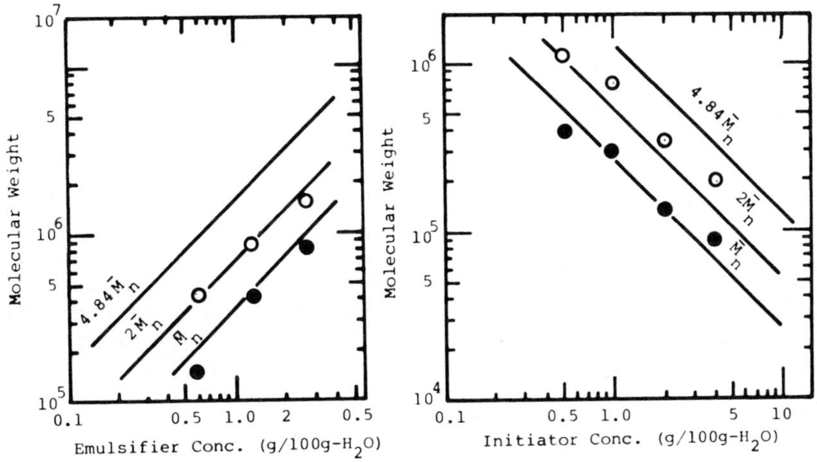

*Figure 2. Experimental data of average molecular weight of polystyrene formed in a CSTR: (●) $\bar{M}_n$; (○) $\bar{M}_w$; (——) theory (7) ((*left*) θ = 58.5 min; I_o = 0.8 g/kg H_2O; reaction temp.* T *= 70°C; (*right*) θ = 7.2 min;* S_o *= 27.9 g/kg H_2O; reaction temp. T = 70°C)*

oscillations in monomer conversion and the number of polymer particles when a CSTR is operated at a high temperature(3)or extremely high conversion range(13). This sort of oscillatory response did not appear at a moderate temperature such as 50°C, as shown in Figure 3, even when a CSTR is used as the first stage reactor where particle formation mainly takes place. Figure 4 shows a typical example of the course of continuous emulsion polymerization of styrene, which was started with the following procedure. The reactor is filled with the desired amount of all recipe ingrediants except initiator, and the continuous run is started by simultaneously pouring the given amount of an aqueous initiator solution and pumping monomer and an aqueous initiator and emulsifier solution into the first reactor. The solid lines show the model predictions calculated using the Nomura and Harada model with $\varepsilon = 1.28 \times 10^5$ the value of which was obtained in batch experiments(10). The dotted lines are those calculated using Eq.(4) in place of Eq.(2) as the expression for particle formation, the other equations and calculation conditions being the same as those in the above calculations. This situation is identical to the Gershberg and Longfield model(3)(or the Nomura and Harada model with ε of Case C), except that Eq.(16) was used instead of Eq.(23). It is clear from Figure 4 that the Nomura and Harada model with the constant value of $\varepsilon = 1.28 \times 10^5$ predicts well the kinetic behavior of continuous emulsion polymerization of styrene, including the behavior in transient period, and that the Gershberg and Longfield model which supposes that radicals in the water phase enter the micelles and particles in proportion to their surface areas does not explain transient behavior. The reason why both models coincide with each other at steady state will be shown in the succeeding section.

If Case B is rather likely for ε, as the diffusion theory and the examination of experimental value of $\bar{M}_w/\bar{M}_n$ predict, one must introduce the concept of radical capture efficiency of a micelle relative to a polymer particle, α in the form of αk_2 where k_2 is the radical capture coefficient for a polymer particle. The approximate value of α is estimated to be 0.01 for emulsion polymerization of styrene because the value of ε is 1.28×10^5, the value of d_p/d_m is at the highest 10 and the value of M_m is about 100.

Pre-Reactor Principle

Let us consider the steady state characteristics of continuous emulsion polymerization of styrene in the first stage reactor. The steady state value of the number of polymer particles formed in the first stage reactor can be calculated using the following equations. From Eqs.(1) and (2), we have:

$$N_T = \frac{r_i \theta}{1 + \varepsilon N_T / S_m} \qquad (24)$$

The value of S_m is given by the following equation obtained from Eqs.(15) and (16) and $\bar{v}_p = \mu\theta$:

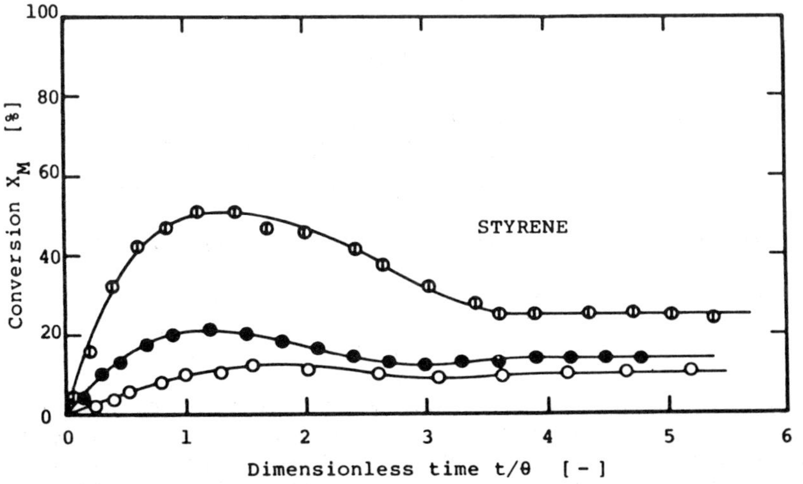

Figure 3. Polystyrene conversion transient at start-up in a CSTR (S_F = 12.5 g/L H_2O; I_F = 1.25 g/L H_2O; M_F = 0.5 g/cc H_2O; T = 50°C; mean residence time θ: (⦶) 155 min; (●) 67.5 min; (○) 38.6 min)

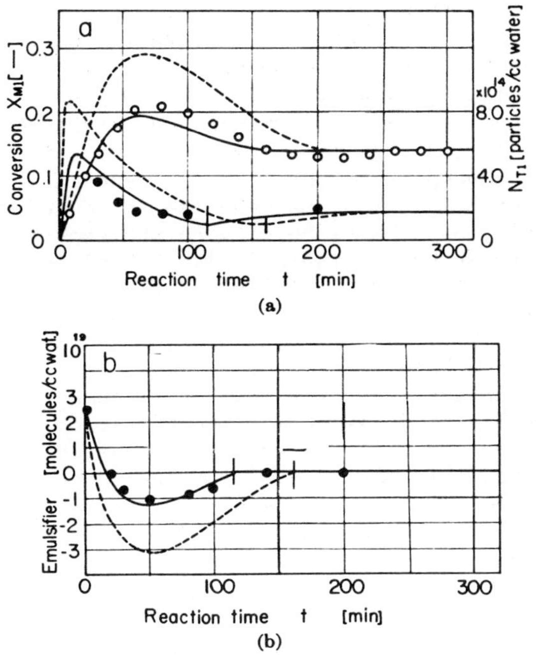

Figure 4. Typical course of continuous emulsion polymerization of styrene: (a) (○) conversion; (●) particle number (calculated: (– – –) Smith and Ewart; (——) Nomura et al. (b) emulsifier balance S_F = 12.5 g/L H_2O; I_F = 1.25 g/L H_2O; M_F = 0.50 g/cc H_2O; 50°C; experimental: θ = 67 min)

$$S_m = S_F - A_p/a_s = S_F - (36\pi)^{1/3}(\mu\theta)^{2/3} N_T/a_s = S_F - \beta N_T \tag{25}$$

where, $\beta = (36\pi)^{1/3}(\mu\theta)^{2/3}a_s^{-1}$

Combination of Eqs.(24) and (25) leads to:

$$N_T = \frac{(S_F - r_i\theta\beta) - \sqrt{(S_F - r_i\theta\beta)^2 - 4(\beta-\varepsilon)r_i\theta S_F}}{2(\beta-\varepsilon)} \tag{26}$$

At limiting values of θ, Eq.(26) gives the following values of N_T which do not depend on ε,

when $\theta \to 0$, $$N_T = r_i\theta \tag{27}$$

when $\theta \to \infty$, $$N_T = S_F/\beta = 0.21(\mu\theta)^{-2/3}(a_sS_F)^{1.0} \tag{28}$$

This is the reason why the steady state value of the number of polymer particles coincide with each other, as shown in Figure 4, regardless of the form of ε when the first stage reactor is operated at comparatively longer residence time. On the other hand, if Eq.(23) is used instead of Eq.(16) for calculating A_p value, we have:

$$N_T = S_F/\Gamma(5/3)\beta = 0.23(\mu\theta)^{-2/3}(a_sS_F)^{1.0} \tag{29}$$

This equation was derived by Gershberg and Longfield(3) at $\theta \to \infty$.

Figure 5 represents a typical example of the variation of the number of polymer particles with mean residence time θ. The solid line shows the theoretical value predicted by the Nomura and Harada model with $\varepsilon = 1.28 \times 10^5$. The dotted line is that predicted by the Gershberg model (or the Nomura and Harada model with Case C for ε), where Eq.(23) was used instead of Eq.(16) for A_p. The value of N_T produced at longer mean residence time differs, therefore, by a factor of $\Gamma(5/3)$ between the solid and dotted lines in Figure 5. From the comparison between the experimental and theoretical results shown in Figure 5, it is confirmed that the steady state particle number can be maximized by operating the first stage reactor at a certain low value of mean residence time θ_{max}, which is considerably lower than that in the succeeding reactors. This is so-called "pre-reactor principle". It is, therefore, desirable to operate the first reactor at such mean residence time as producing something like a maximum number of polymer particles in order to increase the rate of polymerization in the succeeding reactors. This will result in a decrease in the number of necessary reactors and hence, in the capital cost.

θ_{max} where maximum number of polymer particles can be formed is predicted theoretically by differentiation of Eq.(26) by θ. Thus, we have:

$$\theta_{max} = 0.25\mu^{-4/7}(r_i/\varepsilon)^{-3/7}(a_s^2S_F)^{3/7} \tag{30}$$

On the other hand, θ_{max} has the following relation with t_{cl}, the disappearance time of micelles in batch operation conducted with the same recipe as that in continuous operation(11).

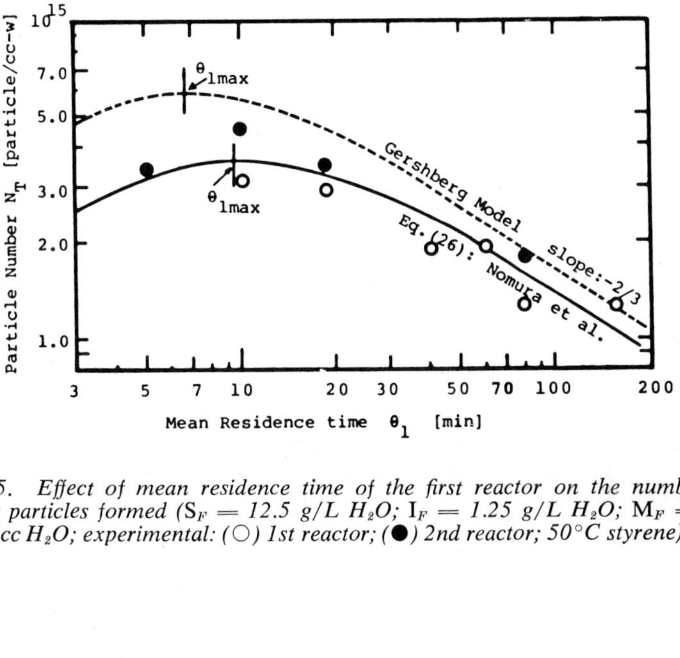

Figure 5. Effect of mean residence time of the first reactor on the number of polymer particles formed (S_F = 12.5 g/L H_2O; I_F = 1.25 g/L H_2O; M_F = 0.5 g/cc H_2O; experimental: (○) 1st reactor; (●) 2nd reactor; 50°C styrene)

$$\theta_{max} = 0.83\, t_{cl} \tag{31}$$

It is apparent from Eq.(30) that the higher the temperature of the first stage reactor and the value of r_i, the smaller the value of θ_{max}. This may be the reason why Degraff and Poehlein could not find θ_{max} in their experiments. Equation(31) suggests that one can estimate the value of θ_{max} simply by determing t_{cl} by measuring surface tension with the use of, for example, a du-Nöuy tensiometer.

The maximum number of polymer particles produced in the first stage reactor being operated at θ_{max} can be obtained by introducing Eq.(30) into Eq.(26). Thus,

$$N_{Tmax} = 0.32\, a_s^{3/7} (r_i/\varepsilon)^{2/7}\, \mu^{-2/7} S_F^{5/7} \tag{32}$$

On the other hand, the number of polymer particles formed in batch operation with the same recipe as in continuous operation is given by:

$$N_{TB} = 0.56\, a_s^{3/7} (r_i/\varepsilon)^{2/7}\, \mu^{-2/7} S_F^{5/7} \tag{33}$$

Equations(32) and (33) give:

$$N_{Tmax} = 0.57\, N_{TB} \tag{34}$$

This means that as long as a CSTR is used as the first stage reactor and all the recipe ingrediants are fed into the first stage reactor, one cannot have more than 57% of the number of particles produced in a batch reactor with the same recipe as in continuous operation. The validity of these expression is clear from the comparison between the experimental and theoretical values shown in Figure 5. From Figure 5, it is found that the optimum mean residence time of the first stage reactor is about 10 minutes under these reaction conditions. Equation(30) predicts 10.0 minutes, while experimental value is 10.4 minutes where the number of polymer particles is about 60% of that produced in a batch reactor. The observed disappearance time of micelles t_{cl} in a batch reactor with the same recipe as in this continuous operation was 12.8 minutes and 10.4 minutes is about 80% of 12.8 minutes, as Eq.(31) predicts. Thus, it proves that Eqs.(31), (32) and (34) are valid.

θ_{max}, N_{Tmax} and N_{TB} which are predicted by the Gershberg and Longfield model without taking into account the radical capture efficiency of micelles(or the Nomura and Harada model with Case C for ε) are given, for comparison, as follows:

$$\theta_{max} = 0.53 \left(\frac{a_s S_F}{r_i}\right)^{3/5} \mu^{-2/5} \tag{35}$$

$$N_{Tmax} = 0.21 (r_i/\mu)^{2/5} (a_s S_F)^{3/5} \tag{36}$$

$$N_{TB} = 0.37 (r_i/\mu)^{2/5} (a_s S_F)^{3/5} \tag{37}$$

On the other hand, Eqs.(36) and (37) give the same relationship as Eq.(34). Equation(35) underestimates θ_{max}, while Eq.(36) overestimates the value of N_{Tmax}, as shown in Figure 5.

Another method to increase the number of polymer particles produced in the first stage reactor with initiator and emulsifier concentrations fixed is to employ a plug flow type reactor such as a tubular reactor for the first stage. The minimum residence time of a plug flow reactor θ_p necessary to produce the same number of polymer particles as in a batch reactor is t_{cl}. Thus, from Eq.(31) We have:

$$\theta_p = t_{cl} = 1.2\,\theta_{max} \tag{38}$$

This means that one can increase the number of polymer particles about 75% higher than that formed when a CSTR of θ_{max} is used, by employing a plug flow type reactor which is only 20% bigger in volume than a CSTR of θ_{max}.

When the first stage reactor is operated at such a low mean residence time as θ_{max} or θ_p, monomer conversion in the first stage reactor is considerably lower, say several %, and hence, almost all the monomer fed into the first stage reactor only passes through it without playing any important role. This suggests that one can further decrease the reactor volume of the first stage reactor by supplying only a small portion of the total monomer feed into the first stage and the rest of the monomer into the second stage. But how far can we decrease the monomer feed rate into the first stage reactor, and what happens, then, to particle formation behavior?

Proposition of Operation with Divided Monomer Feed

Nomura and Harada already reported an experimental and theoretical study on the effect of lowering the amount of monomer initially charged on the number of polymer particles formed in a batch reactor(14). Under usual conditions in batch operation, micelles disappear and the formation of particles terminates before the disappearance of monomer droplets in the water phase. However, if the initial monomer concentration is extremely low, micelles would exist even after the disappearance of monomer droplets and hence, particle formation will continue until all emulsifier molecules are adsorbed on the surfaces of polymer particles. This condition is quantitatively expressed by the following emulsifier balance equation.

$$S_m = S_o - (36\pi)^{1/3}(\bar{v}_p)^{2/3}\,N_T/a_s \tag{39}$$

After the time when monomer droplets have disappeared, $\bar{v}_p$ is approximately given by $(M_o/N_T\rho_p)$. Then, Eq.(39) is rewritten as:

$$S_m = S_o - (36\pi)^{1/3}(M_o/\rho_p)^{2/3}\,N_T^{1/3}a_s^{-1} \tag{39'}$$

When reaction time becomes infinite, micelles will disappear ultimately, and then, Equation(39') becomes:

$$N_T = (a_s^3 \rho_p^2 / 36\pi) S_o^3 M_o^{-2} I_o^0 \qquad (40)$$

where S_o, M_o and I_o are the emulsifier, monomer and initiator concentrations, respectively.
Figures 6, 7 and 8 show experimental verification of Eq.(40) in batch emulsion polymerization of styrene(14). The number of polymer particles was measured by electron micrscopy, not at finite but at 1 hour after the start of polymerization. Figure 6 represents the effect of lowering the initial monomer concentration, M_o on the number of polymer particles formed at fixed initial initiator and emulsifier concentrations. The number of polymer particles formed is constant even if M_o is lowered to the critical value M_c. This is because normal condition that micelles disappear before the disappearance of monomer droplets is satisfied in the range of monomer concentration above M_c. The value of M_c can be calculated by the following equation obtained by equating X_{Mc1}, the monomer conversion where micelles disappear, to X_{Mc2}, the monomer conversion where monomer droplets disappear.

$$M_c = 0.126\, a_s^{9/7} (\rho_p/(1+\gamma))(\mu\varepsilon/r_i)^{1/7} S_o^{8/7} X_{Mc2}^{-1} \qquad (41)$$

In case of styrene emulsion polymerization, X_{Mc2} is 0.43. If M_o is further decreased below M_c, the number of polymer particles formed increases with decreasing M_o, as shown in Figure 6. The solid lines in the figures represent theoretical values predicted by Eq.(40) using the following numerical constants:

$k_p = 212$ l/mol.sec, $[M_p] = 5.48$ mol/l, $\rho_p = 1.0$ g/cm^3
$a_s = 35 \times 10^{-16}$ cm^2/molecule, $X_{Mc2} = 0.43$, $r_i/\varepsilon = 2.9 \times 10^7$
molecule/cm^3.sec at $I_o = 1.25$ g/l-water

The reason why the experimental values of particle number are somewhat lower than the theoretical values seems to be that the time where the number of polymer particles was measured is not at infinite but at only 1 hour after the start of polymerization. Figure 9 shows that the number of polymer particles increases with reaction time. The solid lines represent the theoretical values predicted by the Nomura and Harada model. However, since $N_T = 0$ when $M_o = 0$, there would be an optimum value of M_o where the number of polymer particles formed becomes maximum. Unfortunately, it is difficult at present to predict the optimum value of M_o theoretically because any reaction model cannot yet explain perfectly the kinetic behavior at high monomer-conversion range. Therefore, one cannot help determining, at present, the optimum value of M_o experimentally. Figures 7 and 8 also show that Eq.(40) roughly satisfies the experimental results.

Applying the above mentioned knowledge to continuous emulsion polymerization of styrene, we can propose a very effective operation method(14). A schematic diagram of porposed process is shown

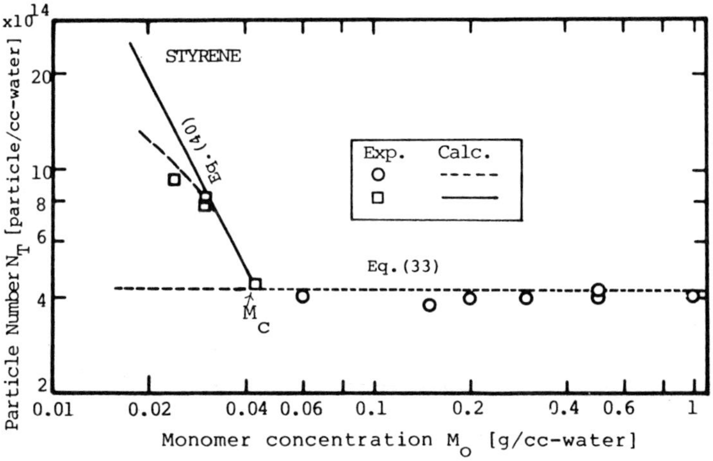

Figure 6. Effect of lowering initial monomer concentration on particle formation in batch operation ($S_o = 6.25$ *g/L* H_2O*;* $I_o = 1.25$ *g/L* H_2O*;* $T = 50°C$*)*

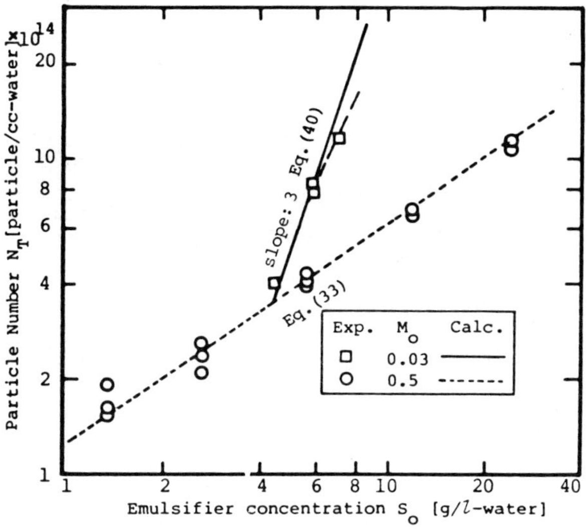

Figure 7. Effect of initial emulsified concentration when initial monomer concentration is very low ($I_o = 12.5$ *g/L* H_2O*;* M_o *g/cc* H_2O*)*

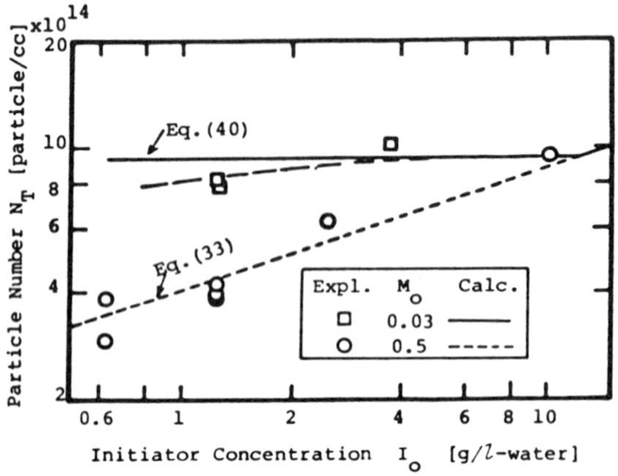

Figure 8. Effect of initiator concentration on particle formation when monomer concentration is very low ($S_o = 6.25$ *g/L* H_2O, *50°C)*

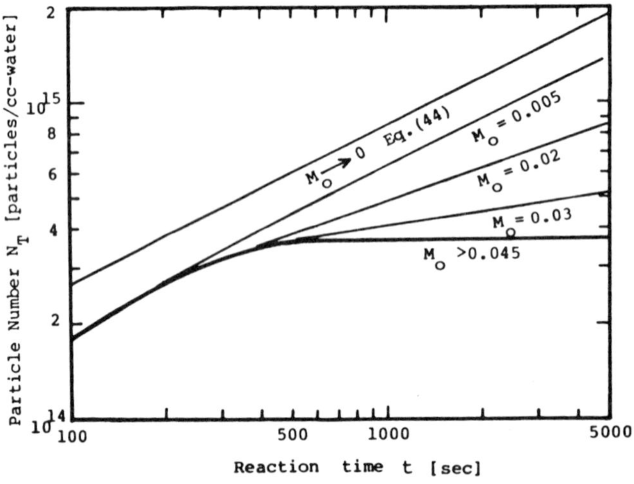

Figure 9. Variation of particle number with reaction time in batch operation (calculation conditions: $S_o = 6.25$ *g/L* H_2O; $I_o = 1.25$ *g/L* H_2O; $S_{CMC} = 0.50$ *g/L* H_2O; $\bar{n} = 0.5$*)*

in Figure 10. In this operation, a very small portion of the total monomer feed and all other recipe ingrediants are basically fed into the first stage reactor and the rest of the monomer into the second stage. This operation method is named "an operation with divided monomer feed".

(1) When a stirred tank reactor is used as the first stage. Figure 11 shows a typical example of the theoretical relationship between the steady state particle number and mean residence time in the first stage reactor. The solid line indicates the theoretical values calculated by Eq.(26) using the value of $\varepsilon = 1.28 \times 10^5$, where monomer droplets exist in the water phase. On the other hand, the broken lines represent the theoretical values predicted by Eqs.(24) and (39'), where monomer droplets disappear. When M_{F1}, the rate of monomer feed into the first stage reactor decreases to zero, the number of polymer particles formed approaches the value calculated by the following equation:

$$N_T = (r_i S_F/\varepsilon)^{1/2} \theta^{1/2} \tag{42}$$

Equation(42) is the solution of simultaneous equations(24) and (39') at $M_{F1} \rightarrow 0$. Let us consider the variation of N_T with θ in Figure 11 when M_{F1} is, for example, 0.03 g/cc-water. The stady state particle number produced in the first stage reactor firstly increases along the solid line with increasing the mean residence time of the first stage reactor θ, begins to decrease, passing through N_{Tmax} at θ_{max}, and then, shifts to the broken line at θ_c where monomer droplets just disappear. With an increase in the value of θ above θ_c, particle number again increases gradually along the broken line. As is clear from Figure 11, a CSTR is less efficient in the ability of particle formation in the range of shorter mean residence time even if the operation method with divided monomer feed is employed, comparing a plug flow type reactor with the residence time of θ_p which produces the same number of polymer particles as in a batch reactor. However, the efficiency of a CSTR in particle formation is greatly improved when the operation method with divided monomer feed is employed in the range of longer mean residence time, as shown in Figure 11.

(2) When a plug flow type reactor is used as the first stage. The theoretical relationship between N_T and θ is shown in Figure 12 when a plug flow type reactor is used for the first stage. Figure 12 can be obtained by replacing t by θ in Figure 9, because the kinetics of continuous emulsion polymerization of styrene in a plug flow type reactor is the same as in a batch reactor. The solid line corresponding to the condition $M_{F1} \rightarrow 0$ is obtained by Eq.(44) which can be derived as follows.

In a plug flow type reactor, the rate of particle formation is expressed by(10):

$$\frac{dN_T}{d\theta} = R_i = \frac{r_i}{1 + \varepsilon N_T/S_m} \simeq \left(\frac{r_i S_F}{\varepsilon N_T}\right) \tag{43}$$

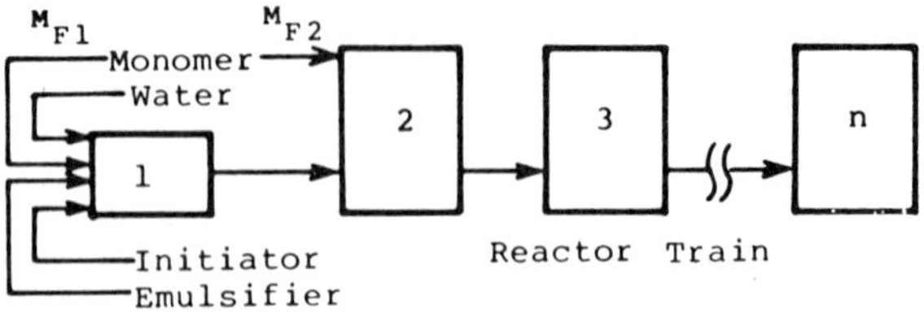

Figure 10. Schematic of proposed process and its operation method

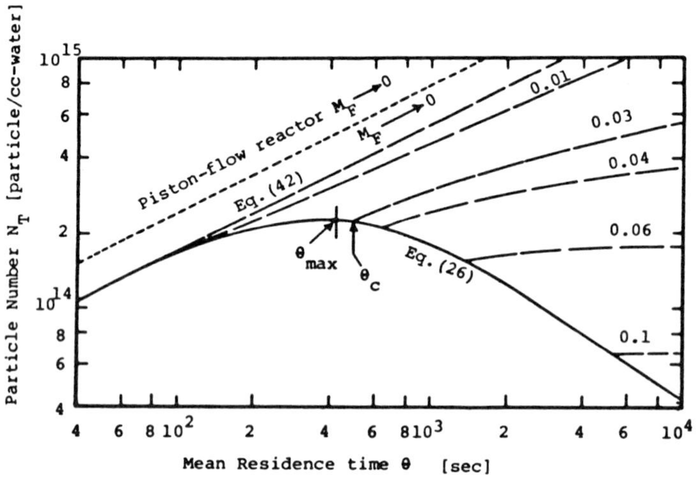

Figure 11. Effect of mean residence time and monomer concentration on steady-state particle number when a CSTR is used for the first stage (S_F = 6.25 g/L H_2O; I_F = 1.25 g/L H_2O; 50°C)

Since monomer droplets do not exist in the reactor in this case, Equation(39') can be used for S_m. Thus, combining Eqs.(39') and (43) and integrating at the condition $M_{F1} \rightarrow 0$, we have:

$$N_T = \sqrt{2}\,(r_i S_F/\varepsilon)^{1/2}\theta^{1/2} \qquad (44)$$

Theoretical values calculated by Eq.(44) are also shown in Figure 11 by a dotted line, for comparison. It is found from the comparison between Figures 11 and 12 that a plug flow type reactor with a divided monomer feed is much better in the efficiency of particle formation than a CSTR with a divided monomer feed and a plug flow type reactor with all recipe ingrediants fed together into the first stage reactor. The validity of the theoretical predictions represented in Figure 12 was experimentally verified using a plug flow type reactor with residence time of $\theta = 20$ minutes, which was made of a coiled glass tube, and is shown in Figure 13 where the solid line is the theoretical values calculated by Eqs.(39) and (43). It is seen from Figure 13 that the steady state particle number produced at $M_{F1} = 0.01$ g/cc-water is about 1.5 times that at $M_{F1} = 0.2$ g/cc-water. It is clear that the volume of the reactor operated at $M_{F1} = 0.2$ g/cc-water is bigger than that operated at $M_{F1} = 0.01$ g/cc-water because the residence times of both reactors are fixed at $\theta = 20$ minutes. From this theoretical and experimental results, therefore, a plug flow type reactor with a divided monomer feed is recommended for the first stage reactor(pre-reactor), because the volume of the reactor can be decreased by decreasing monomer concentration in a feed stream. Nevertheless, the steady state particle number attained in this reactor can be increased.

Further, continuous emulsion polymerization in a plug flow type reactor has another advantage. In continuous emulsion polymerization in a CSTR, sustained oscillations of monomer conversion , particle number and average molecular weight of polymers formed are sometimes observed(15). This phenomenon is regarded as disadvantage from the standpoint of the stability of continuous reactors and the quality of products. However, it was experimentally found that sustained oscillation can be avoided when particle formation is conducted in a plug flow type reactor, as shown in Figure 14. Figure 14b displays a course of continuous emulsion polymerization of vinyl acetate obtained when a plug flow type reactor with the residence time of 8 minutes is used as a seeder in front of a CSTR with the mean residence time of 12 minutes, and all the recipe ingredients are fed into the plug flow type seeder. Experimental data points represent the monomer conversion in the CSTR and reach a steady state value very smoothly. On the other hand, Figure 14a shows the course of continuous emulsion polymerization of vinyl acetate conducted in a CSTR with the mean residence time of 20 minutes, other reaction conditions being identical to those in Figure 14b(15-16).

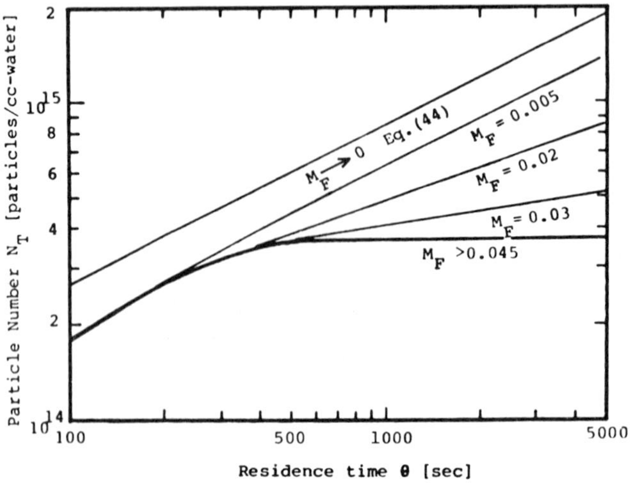

Figure 12. Effect of mean residence time and monomer concentration on steady-state particle number when a plug flow-type prereactor is used for the first stage (calculation conditions: $S_F = 6.25$ g/L H_2O; $I_F = 1.25$ g/L H_2O; $S_{CMC} = 0.50$ g/L H_2O; $\bar{n} = 0.5$)

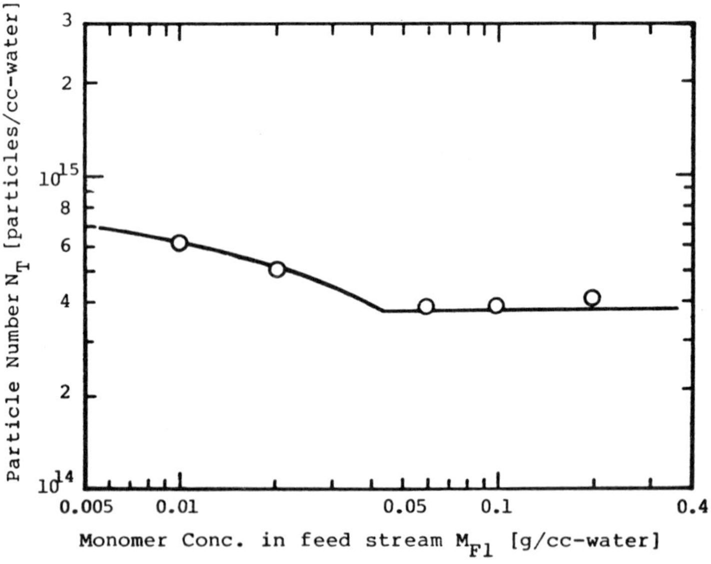

Figure 13. Effect of monomer concentration fed into a piston flow prereactor on steady-state particle number produced (reaction conditions: $S_F = 6.25$ g/L H_2O; $I_F = 1.25$ g/L H_2O; residence time $\theta_1 = 20$ min)

In some cases, the following operation method is more efficient than that stated above (14). Figure 15 shows a schematic diagram of the proposed process and operation method which is the same as that represented in Figure 10, except the method of water feed. In this operation method, water feed is also divided. A small amount of water which just dissolves initiator and emulsifier is fed into the first stage reactor along with all the initiator and emulsifier, and the rest of water into the second stage reactor.

The number of polymer particles formed in the first stage reactor can be generally expressed by the following form:

$$N_T = K S_F^a I_F^b \theta^c \qquad (45)$$

When 1/f of the total water feed is supplied into the first stage reactor and its mean residence time is kept constant by adjusting its reactor volume, the number of polymer particles formed in the first stage reactor is:

$$N_{T1} = K(fS_F)^a (fI_F)^b \theta^c = K f^{a+b} S_F^a I_F^b \theta^c \qquad (46)$$

However, the concentration of polymer particles will be diluted by f times with the rest of water fed into the second stage. Therefore, the number of polymer particles in the second stage reactor N_{T2} is given by:

$$N_{T2} = f^{a+b-1} K S_F^a I_F^b \theta^c \qquad (47)$$

Since the value of f is greater than unity, it is concluded that:

(1) when $a+b>1$, one can increase the number of polymer particles formed and furthermore, decrease the volume of the reactor by employing a divided water feed along with a monomer divided feed,

(2) when $a+b=1$, the number of polymer particles does not change. However, the volume of the first stage reactor can be decreased in proportion to a decrease in the volume of the water fed into the first stage, and

(3) when $a+b<1$, one can decrease the volume of the first stage reactor. However, the number of polymer particles also decreases with a decrease in the volume of the first stage reactor. The operation method with a divided water feed is not necesarily useful.

In case of continuous emulsion polymerization of styrene, it seems that $a+b=1$ holds in a wide range of operation conditions, considering Eqs. (27), (28), (42) and (44). The operation method with a divided water feed would, therefore, be useful, although the validity of above discussion is not yet proved experimentally.

To Conclude

Optimal reactor type and its operation method for the first stage in continuous emulsion polymerization was discussed in this paper. It was clarified theoretically and experimentally uaing a

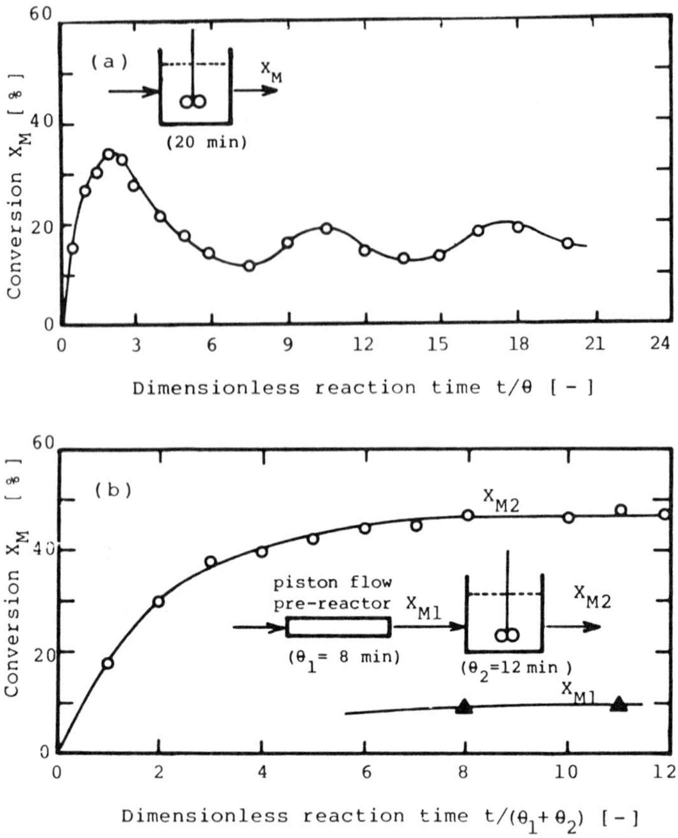

Figure 14. Effect of prereactor type on vinyl acetate conversion transient at start-up (15, 16) (S$_F$ = *2.0 g/L H$_2$O;* I$_F$ = *1.25 g/L H$_2$O;* M$_F$ = *0.20 g/cc H$_2$O; θ = 20 min, 50°C)*

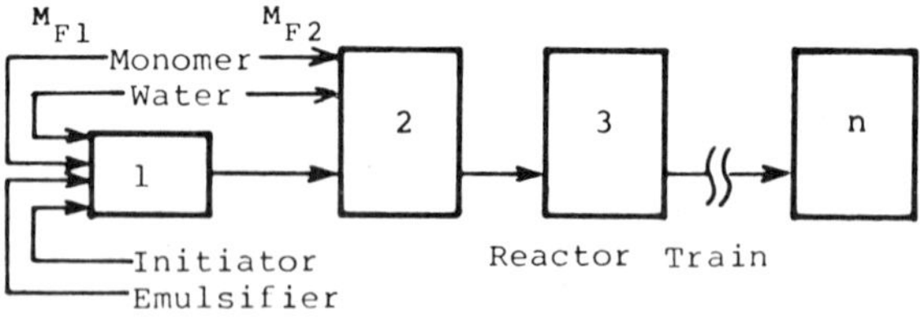

Figure 15. Schematic of proposed process and its operation method

styrene emulsion polymerization system that the most suitable reactor was a plug flow type reactor with a divided monomer feed and in some cases, with a divided water feed, additionally. The basic principles developed in this paper would be useful in designing the first stage reactor(pre-reactor) and its operation method for continuous emulsion polymerization of other monomers.

Literature Cited

1. Feldon, M.; McCann, R. F.; Laundrie, R. W. India Rubber World 1953, 128, 51.
2. Poehlein, G. W.; Dpughherty, D. J. Rubber Chem. Technol. 1977, 50, 601.
3. Gershberg, D. B.; Longfield, J. E. Simp. Polym. Kinetics and Catalyst System Preprints 10, 45th A. I. Ch. E. Meeting, New York 1961.
4. Dickinson, R. F. Ph.D. dissertation, Dept. of Chem. Eng., Waterloo Ontario, Canada, 1976.
5. Min, K. W.; Ray, W. H. J. Macromol. Chem. 1974, C11(2) 177.
6. Fitch, R. M.; Tsai, C. H. "Polymer Colloids" Fitch, R. M. Ed., Plenum Press: New York, 1971; Ch.6.
7. Degraff, A. W.; Poehlein, G. W. J. Polym. Sci. 1971 A-2, 9, 1955.
8. Omi. S.; Ueda, T.; Kubota, H. J. Chem. Eng. Jpn. 1969, 2, 123.
9. Smith, W. V.; Ewart, R. H. J. Chem. Phys. 1948 16, 592.
10. Harada, M.; Nomura. M.; Kojima, H.; Eguchi, W.; Nagata, S. J. Appl. Polym. Sci. 1972, 16, 811.
11. Nomura, M.; Harada, M.; Kojima, H.; Eguchi, W.; Nagata, S. J. Appl. Polym. Sci. 1971, 15, 675.
12. Vanderhoff, J. W.; Vitkuske, J. F.; Bradford, E. B.; Alfrey, T, J. Polym. Sci. 1956, 20, 225.
13. Ley, G.; Gerrens, H. Makromol. Chem. 1974, 175, 563.
14. Harada, M.; Nomura, M.; Eguchi, W.; Nagata, S. Kobunshi Kagaku 1972, 29, 844.
15. Nomura, M.; Sasaki, S.; Fujita, K.; Harada, M.; Eguchi, W. ACS Preprints of Organic Coatings and Plastics Chemistry 1980, 43, 834.
16. Nomura, M. Unpublished data.

RECEIVED April 6, 1981.

7

Emulsifier-Free Emulsion Copolymerization of Styrene with Acrylamide and Its Derivatives

HARUMA KAWAGUCHI, YOSHISHIGE SUGI, and YASUJI OHTSUKA

Department of Applied Chemistry, Keio University, Yokohama, Japan

Emulsifier-free latices are useful not only for industrial purposes but also for studies on colloidal properties (1, 2) and medical applications (3, 4). Various methods have been tried to prepare characteristic emulsifier-free latices (5-8). Among them, copolymerization of hydrophobic monomers with hydrophilic comonomers has been the most applicable one (7, 8). There have been many studies on the effects of ionic comonomers on the kinetics of aqueous copolymerization and the properties of the resulting latices, but nonionic hydrophilic comonomers have rarely been used for these purposes.

This paper deals with the copolymerization of styrene with acrylamide and its derivatives in emulsifier-free aqueous media. It is expected that the effects of acrylamides on the nucleation and stabilization of particles differ from those of ionic comonomers. The reaction mechanism, the characteristics of the latices prepared, and the effect of the properties of acrylamides on them are discussed.

Experimental

Materials. Ion-exchanged and distilled water was used in all the polymerizations. Four kinds of acrylamides were used as comonomers (M2). Acrylamide (AA, Wako Chemicals Co.) and methacrylamide (MA, Tokyo Kasei Co.) were recrystallized from benzene. N-(Hydroxymethyl)acrylamide (HMA, Tokyo Kasei Co.) was recrystallized from ethyl acetate. N,N-Dimethylacrylamide (DMA, Tokyo Kasei Co.) and styrene (St, Kashima Kagaku Yakuhin Co.) were distilled at $54^0C/3.5$ mmHg and $40^0C/14.5$ mmHg, respectively. In some copolymerizations cross-linking reagents were added to reduce the formation of water-soluble polymer. N,N'-Methylenebisacrylamide (MBA, Nakarai Chemicals Co.) and N-allylacrylamide (AAA, Polysciences, Inc.) were used as received. Divinylbenzene (DVB, Tokyo Kasei Co.) was treated with 10 % sodium hydroxide and dried. Two kinds of initiators were used: Potassium persulfate (KPS, Taisei Kagaku Co.) was recrystallized from water and azobis(isopropyl-

0097-6156/81/0165-0145$05.00/0

amidine hydrochloride) (AIPA, Wako Chemicals Co.) was used without further purification.

Procedures. A standard recipe for the latex preparation is shown below: (St + M2) 20 g, (water + buffer) 160 g, and initiator 5 mmole/l. The weight fraction of M2 in monomer charge (f) was varied from 0.01 to 0.50. Polymerizations were carried out at 55^0C or 70^0C and pH 2.5 or 9.0 under nitrogen. Samples were withdrawn from the reaction mixture at various time intervals and the polymer was precipitated in an excess of acetone. The conversion and polymer composition were determined by gravimetric means and by elemental analysis, respectively. The M2 fraction in instantaneously-formed copolymer (Fi) was calculated from eq. 1:

$$Fi = Fc + c(dFc/dc) \qquad (1)$$

where Fc = the M2 fraction in accumulated copolymer at conversion c.

The diameter of latex particles was measured from their transmission electron micrographs which were obtained by use of a Hitachi electron microscope HU-12AF. The uniformity ratio of particle size (U) was calculated from eq. 2:

$$U = Dw / Dn \qquad (2)$$

where $Dw = \Sigma NiDi^4 / \Sigma NiDi^3$, $Dn = \Sigma NiDi / \Sigma Ni$, and Ni = the number of particles with diameter Di.

Some polymer precipitates were extracted with water and benzene by use of a Soxhlet extractor for 72 hrs and the extracts were elementally analyzed. A few latices were centrifuged with a Kokusan H-502 centrifuge at 25000g for 1 hr to determine the amount and composition of the polymer which dissolved in the latex serum.

The partition coefficient, the ratio of the M2 concentration in St to that in water, was determined from the absorbance (at 200 mμ for DMA, 201 mμ for AA, 203 mμ for HMA, and 205 mμ for MA) of each phase equilibrated at 70^0C. The solubility of St in the aqueous solution of AA was determined from the absorbance of the solution at 290 mμ.

Results and Discussion

The dependence of the conversion and polymer-composition on the reaction time is shown in Figure 1 for the copolymerization of St with AA at f = 0.4, pH 9.0 and 70^0C. The polymerization course was found to consist of three stages: At first AA polymerizes preferentially but the preference decreases rapidly with increasing conversion (0 - 1.0 hr); St polymerizes exclusively (1.0 - 1.5 hrs); and again AA polymerizes preferentially (1.5 hrs - end). A similar copolymerization mode was observed in an aqueous copolym-

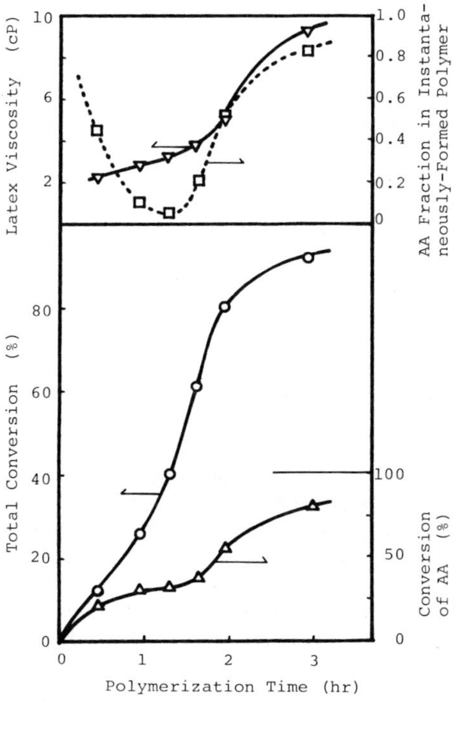

Figure 1. Copolymerization of St with AA under the conditions of the weight fraction of acrylamide in monomer charge (f) *= 0.9, pH = 9.0, and 70°C*

erization of St with 4-vinylpyridine (8). In the following sections the details of each reaction stage are discussed not only for the copolymerization of St with AA but also for those with HMA, DMA, and MA.

The First Stage. The preferential polymerization of AA at the initial stage of copolymerization means that the main reaction locus is the aqueous phase just as Juang and Krieger pointed it out for the aqueous copolymerization of St with sodium styrenesulfonate (SSS) (9). In the St-SSS system, SSS polymerized preferentially up to a few percent conversion under the condition of SSS/St (w/w) = 0.014. Copolymerization of hydrophobic monomer with a large amount of hydrophilic comonomer was considered to yield a greater amount of information with respect to the reaction mode. By use of a relatively large amount of AA or its derivatives the characteristic reaction mode of the copolymerization of St with acrylamides could be clarified.

Some polymer-composition vs. conversion curves were obtained for the copolymerizations with different f's (Figure 2), and all of them seem to intersect the ordinate at 1.0. From the initial slope of the curves and the monomer ratio in the aqueous phase the monomer reactivity ratio was calculated, but the calculation resulted in a negative r_2. Therefore, it was concluded that the copolymerization could not be regarded as a homogeneous one even just after the beginning of the reaction. The first stage was considered to be a transitional stage to establish the particle formation.

According to the homogeneous nucleation mechanism by Fitch (10), in the emulsifier-free aqueous medium, the growing oligomers coil to primary particles and such primary particles agglomerate to form secondary particles which then begin to imbibe St monomer and consequently polymerization of St in the secondary particles proceeds. Almost the same process is expected to take place in St-AA copolymerization system. Judging from the rapid decrease in the AA fraction in instantaneously-formed polymer (Figure 1) in spite of the extraordinarily large amount of AA compared with that of St in the aqueous phase and no appriciable predominance of St in copolymerizability (11), localization of St in the vicinity of the active sites might occur earlier than expected, that is, even the growing oligomers might give favorable sites for St to be localized in concentrated state to accerelate the polymerization of St and the particle nucleation. According to the data on the extraction of AA-St copolymer, the copolymer molecules containing more than 15 % St are insoluble in water. An alternative mechanism of particle formation, the micellization-type one (12), can not be neglected at present.

The gradual increase in the number of particles must play an important role to decrease the AA fraction in polymer rapidly because as the number of particles increases the coalescence of a growing radical with a particle becomes more frequent and decreases the occasion for the radical to react with AA in the aqueous phase.

The total conversion where the latex particles came to appear first on the electron micrographs were more than 20 % for the copolymerization at f = 0.5, 12 % at f = 0.4, ca. 3 % at f = 0.3, and less than 1.2 % at f = 0.2. A common fact in four copolymerizations is that the AA fraction in instantaneously-formed polymer drops below 0.60 - 0.65 at the presented conversions. The fraction continues to decrease further to less than 0.05.

The Second Stage. The second stage is defined as the stage where St polymerizes exclusively. To clarify the main reation locus at this stage, some quantitative analysis was done for the copolymerization at f = 0.4. Using the AA conversion at which AA polymerization first levels off (35 %), the solubility of St in the aqueous solution of AA (3.92 mmole/l), and the partition coefficient of AA between St and water (0.093), the molar concentration ratio of AA to St in the aqueous phase was calculated to be about 120 whereas that in the particles was calculated to be 0.008. From the monomer reactivity ratio reported by Minsk and his coworkers (13), the AA fraction in the polymer instantaneously formed in the particles is calculated to be 0.03, which is very close to the experimental value. Therefore, it is concluded that during the second stage polymerization in the aqueous phase is negligible, or the main reaction locus is the particles. This must result from fast diffusion of radicals into the preexisting particles. The number of particles in this stage was found to be almost constant as the slope of lines in Figure 3 was almost 1/3 although the copolymerizations at higher f's resulted in some decrease in the number of particles with increaseing conversion.

Exclusive polymerization of St continues until the St conversion attains about 50 %. This critical St conversion was constant regardless of f and it corresponds to the equilibrium concentration of poly-St in the particle. Therefore, the second stage is considered to come to end when all of the St droplets disappear.

The Third Stage. The beginning of the third stage is characterized by gradual acceleration of AA polymerization, accompanied with an increase in the latex viscosity as shown in Figure 1. These phenomena must result from alteration of the main reaction locus from the particles to the aqueous phase.

Disappearance of St droplets leads to a decrease in the concentration of St in the aqueous phase as well as in the particles. The decrease in the concentration of St in the aqueous phase causes growing radicals to keep their hydrophilicity for a long time. The radicals can react with many AA molecules before they diffuse into the particles. Some radicals would not lose their solubility in water even after they propagated to be polymer molecules and such kind of molecules would cause an increase in the latex viscosity. Judging from the dependence of the latex viscosity on the reaction time, most of the polymer dissolved in water is considered to form during the third stage. To reduce the

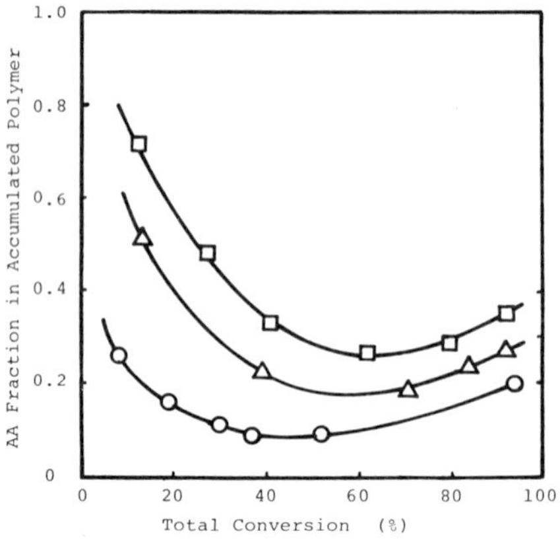

Figure 2. Dependence of polymer composition on conversion in copolymerization of St with AA at pH 9.0 and 70°C (f: (○) 0.2; (△) 0.3; (□) 0.4)

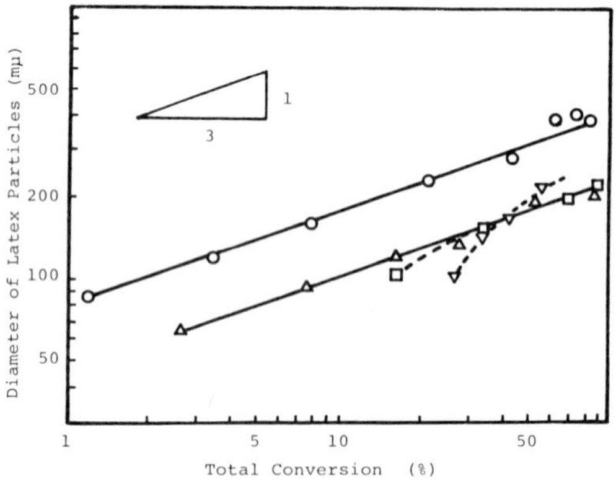

Figure 3. Dependence of particle size on conversion in copolymerization of St with AA at pH 9.0 and 70°C (f: (○) 0.2; (△) 0.3; (□) 0.4; (▽) 0.5)

formation of the polymer dissolved in water, a few cross-linking reagents (MBA, AAA and DVB) were used but no successful results were obtained: addition of MBA or AAA (5 wt % to AA) increased the amounts of coagulum and use of DVB (also 5 wt %) did not change the shape of time-conversion curve and distribution of AA units significantly.

Copolymerization of St with Acrylamide Derivatives. The dependence of the M2 fraction in instantaneously-formed polymer on the total conversion is compared among four kinds of acrylamides-St copolymerizations at f = 0.4 (Figure 4) (14). HMA and DMA resemble AA concerning the copolymerization mode, that is, their reaction course is divided into three stages. On the other hand MA copolymerizes in a different mode.

The partition coefficients of acrylamides between St and water at 70^0C are shown below: AA (0.093); HMA (0.089); DMA (0.293); and MA (0.159). It is expected that use of an acrylamide derivative which has a larger partition coefficient causes shorter duration of the first stage in its copolymerization with St. DMA is less hydrophilic than AA and HMA and the first stage of St-DMA copolymerization ends quickly compared with others. Consequently the DMA conversion where the DMA polymerization levels off is lower than the corresponding values for other acrylamides as shown in Figure 5 (14). DMA units seem to be less effective for stabilization of the latex particles and the number of particles in St-DMA copolymer latex is much less than those in St-AA or St-HMA copolymer latices. As mentioned in the preceding section the second stage resembles the steady stage in usual emulsion polymerization from the viewpoint of kinetics. Therefore, the duration of the second stage, that is, the duration of levelling-off of acrylamides conversion is expected to be inversely proportional to the number of particles. The results shown in Figure 5 satisfy the expectation qualitatively. The levelling-off of DMA conversion at the second stage might be surprizing because an appreciable amount of DMA should exist in the particles. This phenomenon can be explained by considering low copolymerizability of DMA with St. The reactivity ratios of St and acrylamides obtained by Saini and his coworkers (11) support this consideration.

The partition coefficient of HMA is very close to that of AA. But it was found that HMA units in copolymer is less effective to make the copolymer water-soluble than AA units, that is, St-HMA copolymer loses its water-solubility by containing a smaller amount of St than St-AA copolymer does (Figure 6). The average molar fraction of AA in the water-soluble AA-St copolymer was 0.86 while the corresponding value for HMA was 0.92. Such lower ability of HMA to make the copolymer water-soluble may be attributed to negligible hydrogen-bonding of -NH- groups of HMA due to steric hindrance and some loss of hydrogen-bondable -OH groups due to undesirable side-reactions. This property of HMA would cause the first stage of St-HMA copolymerization to end at a slightly lower

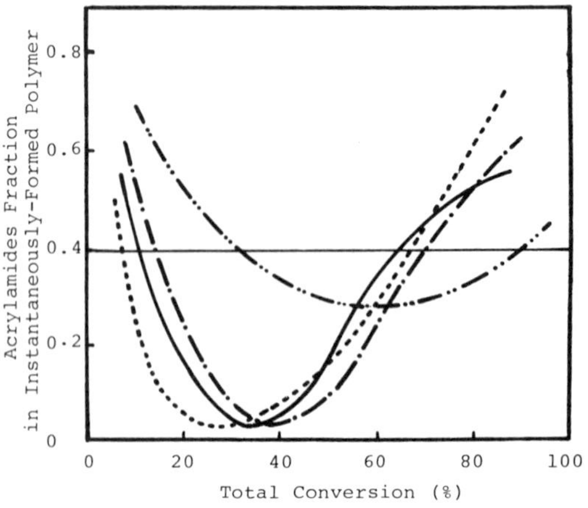

Journal of Applied Polymer Science

Figure 4. Dependence of polymer composition on conversion in copolymerization of St with AA's at f = *0.4* (14) *((— · —) AA; (——) HMA; (- - -) DMA; (— · · —) MA)*

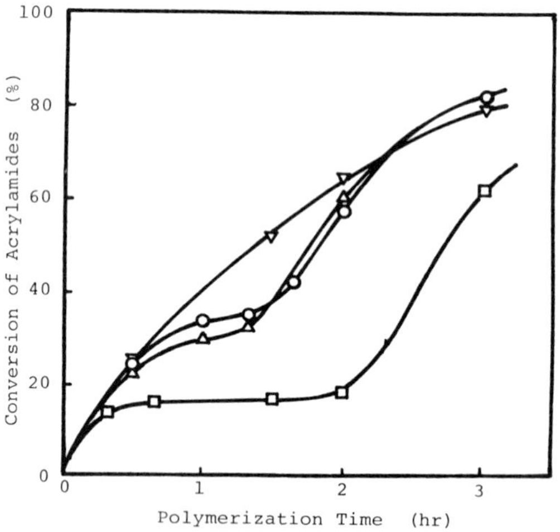

Journal of Applied Polymer Science

Figure 5. Conversion of AA's as a function of time in copolymerization with St at f = *0.4* (14) *((○) AA; (△) HMA; (□) DMA; (▽) MA)* conversion, and the amount of the polymer dissolved in water to decrease. Under the same reaction conditions the number of particles formed in St-HMA copolymerization was about the same with that in St-AA latex but the distribution of particle-size of the former was narrower than the latter. Some discussions for the preparation of monodisperse St-HMA latices are presented in the following section.

No clear evidence was obtained about the alteration of the main reaction locus for the copolymerization of St with MA. This would be attributed to the difference in the partition coefficient between MA and AA (see above) and the difference in the number of latex particles formed in their copolymerization systems with St. (The number of particles in St-AA copolymer latex was about a third of that in St-AA latex. The small number for St-MA system is considered to result from the lower particle-stabilizing ability of MA due to its lower hydrophilicity.) These factors would alter the balance of the polymerization in two reaction loci, that is, the aqueous phase and the particles, and consequently serve to change the reaction mode.

Preparation of Monodisperse and Clean Latices. In the polymerization described above, conditions with relatively high f's were adopted to make clear the behavior of M2 in the reaction course. The resulting latices contain a fairly large amount of polymer dissolved in the serum. It was also clarified that addition of an excess amount of M2 monomers makes the particle-size distribution widespread. Some polymerizations of St with a small amount of HMA were carried out to obtain monodisperse latices free from the polymer dissolved in the serum.

The dependence of the particle-size and uniformity ratio on the charged amount of HMA is shown in Figure 7. Judging from the results in Figure 7 the uniformity ratio would take the minimum at the f close to 0.1. Monodisperse but much larger particles were obtained at the same monomer feed ratio but changing the mode of monomer charge. The aqueous prepolymerization of HMA followed by addition of St resulted in the formation of the latices with large particles and without significant broadening of the particle-size distribution. In the cases shown in Figure 7 St was charged after 1 hr polymerization of HMA whose conversion at this time was about 60 %.

Conclusions

Copolymerization of styrene with acrylamides (acrylamide, N-(hydroxymethyl)acrylamide, N,N-dimethylacrylamide, and methacrylamide) were carried out in emulsifier-free aqueous media. When either of the former three acrylamides were used, the copolymerization course was divided into three srages on the basis of the main reaction locus. At first acrylamides polymerized preferentially in the aqueous phase. After the particle formation styrene

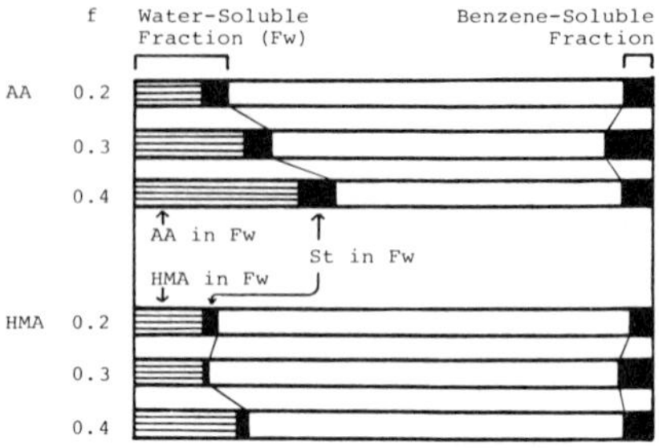

Figure 6. Extraction of St–AA and St–HMA copolymers (fractions are given in weight basis)

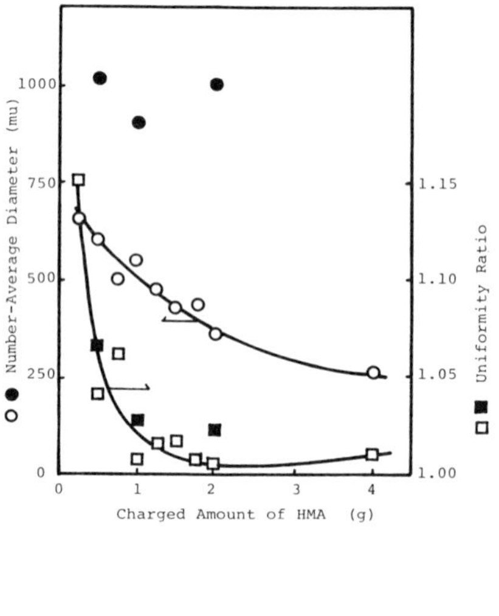

Figure 7. Dependence of particle size and uniformity on charged amount of HMA at pH 9.0 and 70°C (solid marks: St was charged after 1 h prepolymerization of HMA)

polymerized exclusively in the particles even if a significant amount of acrylamides remaining in the aqueous phase. The decrease in the concentration of styrene after disappearance of styrene droplets caused the transfer of the main reaction locus from the particles to the aqueous phase and consequently led to the formation of an appreciable amount of the polymer dissolved in the aqueous phase. In the copolymerization of styrene with methacrylamide no distinct alteration of the main reaction locus was detected. The hydrophilicity and reactivity of acrylamides affected the polymerization mode and also the latex properties, e. g., the particle size, the distribution of acrylamides units in the latex, etc. Polymerizations of styrene in the presence of a small amount of N-(hydroxymethyl)acrylamide gave monodisperse and clean latices with the diameter of 350 mμ to 1100 mμ, which depended on the mode of monomer charge.

Acknowledgement

The authors gratefully acknowledge the assistance of Mr. S. Nakata for elemental analysis and Mr. K. Fujita, Electron Microscope Laboratory, for TEM work.

Literature Cited

1. Vincent, B.; Young, C. A. Faraday Discuss. Chem. Soc., 1978, 65, 297
2. Healy, T. W.; Homola, A.; James, R. O.; Hunter, R. J. Faraday Discuss. Chem. Soc., 1978, 65, 156
3. Rembaum, A.; Yen, P. S.; Molday, R. S. J. Macromol. Sci-Chem., 1979, A13, 603
4. Norde, W.; Lyklema, J. J. Colloid & Interface Sci., 1979, 71, 350
5. Matsumoto, T.; Ochi, A. Kobunshi Kagaku, 1965, 22, 481
6. Goodall, A. R.; Hearn, J.; Wilkinson, M. C. Br. Polym. J., 1978, 10, 141
7. Liu, L-J.; Krieger, I. M. "Emulsions, Latices, and Dispersions" ed. by Becker, P.; Yudenfreund, M. N.; Decker: New York, 1978, p. 41
8. Ohtsuka, Y.; Kawaguchi, H.; Hayashi, S. Polymer, to be published.
9. Juang, M. S.; Krieger, I. M. J. Polym. Sci., Polym. Chem. Ed., 1976, 14, 2089
10. Fitch, R. M.; Prenosil, M. P.; Sprick, K. J. "New Concepts in Emulsion Polymerization" ed. by Hwa, J. C. H.; Vanderhoff, J. W.; Interscience, New York, 1969, p.95
11. Saini, G.; Leoni, A.; Franco, S. Makromol. Chem., 1971, 144, 235; ibid., 146, 165; ibid., 147, 213
12. Goodall, A. R.; Wilkinson, M. C.; Hearn, J. J. Polym. Sci., Polym. Chem. Ed., 1977, 15, 2193

13. Minsk, L. K.; Koflarchik, C.; Darlak, R. S. J. Polym. Sci., Polym. Chem. Ed., 1973, 11, 353
14. Kawaguchi, H.; Sugi, Y.; Ohtsuka, Y. J. Appl. Polym. Sci., to be published.

RECEIVED April 6, 1981.

8

Polyelectrolyte-Stabilized Latices

RICHARD BUSCALL and TERENCE CORNER

I. C. I. Corporate Laboratory, P.O. Box No. 11, The Heath, Runcorn, Cheshire, England

The stability of electrostatically charged sols has been studied extensively and is now reasonably well understood. More recently the stabilising action of adsorbed or chemically anchored non-ionic polymers has received much attention. There has been however little systematic work on polyelectrolyte stabilisers apart from a number of investigations of the flocculation of particles bearing adsorbed biopolymers, usually proteins, by simple salts (2). These have shown that polyelectrolyte covered particles can be more stable with respect to the addition of salt than simple charged systems, and the extra stability has been ascribed to the polymeric nature of the surface layer. The precise mechanism by which polyelectrolytes stabilise dispersions in the presence of high concentrations of salt has however remained unclear.

The current theories of steric stability (3-6) predict that provided the particles are well-covered and the polymer is well-anchored particles bearing non-ionic polymers should flocculate at or near the Θ-point of the stabilising chains. The available experimental date (3, 7, 8) confirm this result in as much as critical flocculation temperatures and pressures have been found to correlate tolerably well with the relevant Θ-points for a wide range of systems. Where the correlation has been less than satisfactory the discrepancy has often been understandable in terms of multiple anchoring, selective adsorption of lyophobic blocks, or other specific effects (9, 10).

The theories of polymer solutions upon which steric-stability theories are based are usually formulated in terms of a portmanteau interaction parameter (for example Flory's χ-parameter and the excluded volume integral) which does not preclude electrostatic interactions, particularly under conditions where these are short range. It is thus appropriate to consider whether polyelectrolyte-stabilisation can be understood in the same broad terms as stabilisation by non-ionic polymers. It was this together with the fact that polyelectrolyte solutions containing simple salts show phase-separation behaviour reminiscent of that of non-ionic

0097-6156/81/0165-0157$05.00/0

polymers in mixed solvents (11) that prompted the present investigation into the preparation of polyelectrolyte stabilised latices.

Polyacrylic acid was chosen as the stabilising moiety because there is a substantial body of literature on its solution properties and because Napper (12) has demonstrated a correlation between critical flocculation temperature and θ-temperature for particles stabilised by copolymers of unionised PAA. Also, since PAA is a weak acid the degree of ionisation can be varied by titration with base.

Experimental

I Materials, Double distilled water and absolute ethanol were used in all polymerisations. Styrene monomer was washed with 10% w/w aqueous sodium hydroxide solution and then distilled under a nitrogen atmosphere with reduced pressure to remove inhibitor and impurities. 4.4'-Azobis (4-cyanovaleric acid) (ADIB), recrystallised from absolute ethanol to remove any peroxide impurities, and benzoyl peroxide (BzP) of reagent grade were utilised.

The polyacrylic acids were supplied by Allied Colloids Ltd, Bradford, England.

II Latex Preparation, The polyelectrolyte stabilised latices were prepared by the polymerisation of styrene monomer in a solution of polyacrylic acid in an alcohol-water mixture adjusted so that sufficient alcohol was present to solubilise the styrene. The agigation speed was fixed at 350 revolutions per min for all experiments although it was later found not to be critical. A typical example of the method is as follows.

A solution of 5g PAA of molecular weight 1.0×10^6 in 500 cm^3 ethanol and 280 cm^3 water was charged into a reaction flask, set in a water bath at 78°C. Nitrogen was bubbled through the solution for 2 hr after which 0.2g BzP was added and the resultant solution allowed to degas for 8-12 hr. After this time a solution of 1g BzP in 50 cm^3 styrene and 20 cm^3 ethanol was charged into the reaction flask. Polymerisation was allowed to proceed for 8 hr at 78°C under a stream of nitrogen to give a polystyrene (PST) latex with a mean particle diameter of 0.6 μm and a coefficient of variation of <5%. The conversion of styrene to latex was shown to be 100% by analysing the supernatant, obtained by centrifuging the latex, by UV spectroscopy.

III Latex Clean-Up, Typically 20-50g of latex was subjected to a series of centrifugation/redispersion cycles in 900 cm^3 of 1:1 volume ratio alcohol:water mixtures (3 times) followed by centrifugation/redispersion in water (6 times). This procedure was shown to be more than adequate for the removal of unreacted styrene, polyelectrolyte and initiator residues etc, by the analysis of the supernatants obtained after each successive centrifugation.

IV Particle Size, The modal diameters of latices were determined by measuring the diameters of particles on electron micrograph negatives obtained by conventional procedures. In order to determine the mean, and the standard deviation from the mean, between 200 - 300 particles were measured. Latices prepared from polyacrylic acid in the manner described above were found to have coefficients of variation of <10% in all cases and sometimes <5%. Errors in particle size were estimated to be ± 10%.

V Surface Characterisation, A quantitative analysis of the amounts of PAA associated with the latices was obtained by conductometric titration and a qualitative analysis of its behaviour by potentiometric titration. Titrations were performed under a nitrogen atmosphere using a Radiometer Autoburette system and a M64 pH meter together with a Wayne Kerr B642 Bridge.

VI Flocculation Studies, Critical flocculation temperatures were measured using suspensions of 0.6% v/v latex containing the appropriate amount of sodium chloride. Differing degrees of neutralisation were produced by the addition of NaOH to stock solutions and checked by conductometric titration with base in the absence and presence of added HCl. Flocculation temperatures (CFT) were determined by visual observation at a cooling/heating rate of $\sim 0.5^{o}C\ s^{-1}$. The flocculation was generally reversible provided that the latices had not been left in the flocculated state for more than a few minutes or so.

Results and Discussion

I Latex Preparation, In the present work it was found that the largest changes in particle size were obtained by variation of either the styrene concentration, the ratio of alcohol to water or by the choice of alcohol. In Tables I and II it can be seen that increases in either the styrene concentration or the ratio of ethanol to water caused the formation of latex particles of increased particle size. The coefficients of variation for the latices were always less than 10%.

Table I Dependence of particle diameter on styrene concentration

Styrene Concentration (Mole dm^{-3})	0.008	0.080	0.240	0.470	0.640	4.200
Particle Diameter μm	No Latex	0.5	0.7	0.8	1.5	Swollen Gel

(Note. The results presented in Table I were obtained using solutions of 5g PAA of molecular weight 1 x 10^6 in 500 cm^3 ethanol and 280 cm^3 water containing 1g BzP. All of the polymerisations were performed at 78°C. Conversions to PST in the 8 hours allowed for the experiments was ∿100% in all cases except that using the highest styrene concentration. In this case a sticky precipitate was produced instead of a latex and the experiment was terminated after 1 hour).

Table II Dependence of particle diameter on ethanol : water ratio

Ethanol : Water Ratio	5:1	5:2.5	5:2.7	5:2.8
Particle Diameter (μm)	1.5	0.8	0.6	0.5

(Note. The results given in Table II were obtained from a series of experiments in which 5 cm^3 styrene was polymerised in solutions of ethanol (50 cm^3) and water containing 0.5 g PAA of molecular weight 27,000 and 0.1g BzP at 78°C for 6 hours after which conversions to PST ∿100% in all cases).

The effect of replacing ethanol by methanol was investigated by carrying out pairs of experiments at 60°C under identical conditions apart from the alcohol used. In all cases it was observed that the use of methanol instead of ethanol resulted in latex dispersions whose ultimate particle size were 50 - 60% of the size obtained with ethanol as the co-solvent. However, little change in particle size was obtained when the PAA molecular weight was varied between 5,000 and 1 x 10^6 or the BzP concentrations from 0.6% to 6.0%. When the BzP was replaced by ADIB or the PAA by polymethacrylic acid, an increase in the particle size of 10 - 20% was obtained. Finally, when the concentration of PAA was varied it was observed that the particle size increased with increasing PAA concentration up to ∿4.0% after which a reduction in size was observed (see Figure 1).

The essentially aqueous dispersion polymerisations described here were investigated because the authors thought that they represented the aqueous analogues of NAD polymerisations in which methyl methacrylate is polymerised in solutions of degraded rubber in aliphatic hydrocarbons (1). Although the detailed mechanism by which particles are produced and stabilised during these aqueous dispersion polymerisations cannot be deduced from the somewhat limited data which has been obtained, it seems reasonable to postulate what the mechanism is likely to be. With the exception of the results obtained when the PAA concentration was varied, the influence of the various experimental parameters on the ultimate particle sizes of dispersions is analogous to that found in the aforementioned NAD systems. Also it is not possible to distinguish between these aqueous and non-aqueous dispersion polymerisations by visual observation. The onset of opalescence as particle nucleation begins, followed by a gradual change to a milky latex, is the same in both instances. Therefore the authors put forward the following mechanism in order to explain the observed results. In the early stages of the polymerisation soluble styrene oligomers form together with small amounts of PAA-g-PST graft copolymer. Then, as the PST precipitates, aggregates of monomer swollen PST form onto which PAA-g-PST graft copolymer adsorbs. Colloidally stable PST particles produced in this way grow until the supply of styrene is exhausted. Increasing the styrene concentration, or increasing the ratio of alcohol to water, or using ethanol instead of methanol increases the solvency of the reaction medium towards PST. Therefore the tendency of any PAA-g-PST graft copolymer to associate with growing PST during particle formation or to anchor to particles which have already been produced, will be reduced. The result is the formation of fewer particles which grow to a larger size. The graft copolymer is probably formed by the addition of styrene to free radical sites on the PAA produced by hydrogen abstraction reactions. Free radicals produced from the decomposition of BzP are very reactive and are known to abstract hydrogen atoms readily whereas free radicals produced from ADIB do so to a much lesser extent.Also, PMA is less prone to undergo hydrogen abstraction reactions than PAA. Thus, the replacement of BzP by ADIB and PAA by PMA results in less graft copolymer and so fewer particles are produced which grow to a larger size if a similar amount of styrene is utilised. This mechanism also explains the manner in which the rate of conversion and particle size change during the polymerisation. In Figure 2 it can be seen that the percentage conversion and the partical size increase slowly at first and then increase rapidly, the former to a level much higher than that observed in an equivalent solution polymerisation.

In the early stages of the dispersion polymerisations reported here the rate of polymerisation must be very similar to that of the equivalent solution polymerisation, hence the similarity in the percentage conversion versus time curves for the initial 'slow' period. However, when the growing PST radicals become insoluble in the reaction medium, coagulation to form particles occurs in the manner described above. Thereafter polymerisation occurs in a heterogeneous system. The rate of polymerisation in the monomer swollen particles will be greater than in solution due to the higher concentration of both styrene and PST radicals in the particles. The latter is due to the lower termination of polymeric radicals in what is essentially a gel phase. Thus, after coagulation to form particles, it is expected that the rate of polymerisation will increase rapidly since a changeover from a solution to a gel phase polymerisation takes place. Also, since the styrene is being polymerised at a much faster rate in the particles than in solution, the existing particles will grow at the expense of the formation of new particles, hence the observed increase in particle size following the coagulation stage.

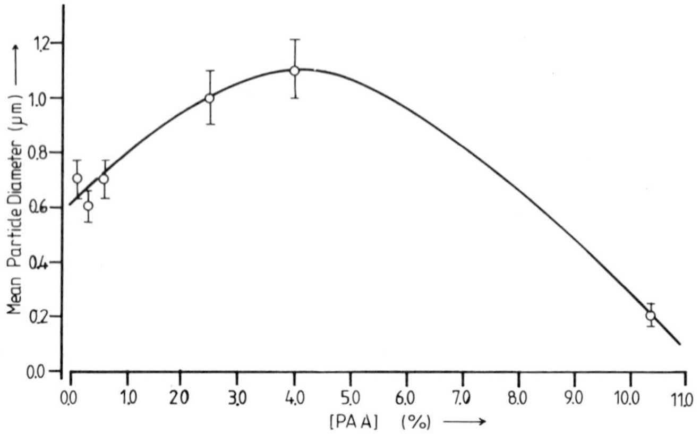

Figure 1. Effect of PAA concentration on particle size. Error bars constructed assuming particle size to be accurate to ±10%.

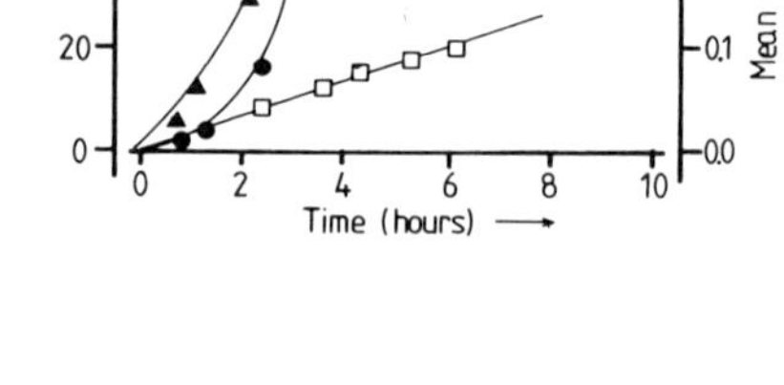

Figure 2. Percentage conversion (●) and particle size (▲) as a function of time for a dispersion polymerization and percentage conversion as a function of time for an equivalent solution polymerization (□).

II Surface Characterisation, The surface charge densities of a number of rigorously purified latices possessing narrow particle size distributions (ie with coefficients of variation of <5%) were determined by a combination of conductometric titration, to yield the total titratable charge, and electron microscopy, to yield the surface area. Values for the surface charge densities were in the range 80 - 250 μC cm^{-2}. These values are significantly higher than those obtained by emulsion polymerisation and they are equivalent to about 0.5 - 2.0 mg of PAA per m^2 of PST surface, values which are comparable to experimental adsorption densities of non-ionic water soluble polymers on PST latices (13). From the experimentally determined surface charge densities, the area of surface per molecule of PAA was determined. The values obtained for several latices are given in Table III together with values for the root mean square (rms) radii of gyration $(s^2)^{\frac{1}{2}}$ and the rms end to end distance $(r^2)^{\frac{1}{2}}$ for PAA's in solution in 1, 4 dioxane, a θ solvent for PAA (14). In their work on NAD polymerisation Osmond and Walbridge (1) showed that the average spacings of poly (lauryl methacrylate), estimated by taking the square root of the area per polymer molecule, were quite similar to their rms dimensions in solution. Inspection of Table III shows that a similar correspondence is obtained here.

Typical potentiometric titration data for the latices and also for the PAA solutions are shown in Figure 3 in the form of a plot of effective pK (15) against degree of ionisation (α'). The curve obtained for the latex is displaced upwards from that for the PAA solution and also exhibits a point of inflection in the region of α' ∿0.5. These differences must be associated with differences in the means by which the PAA expands upon ionisation. It has been suggested that the latices are stabilised by PAA-g-PST graft copolymer formed in-situ. If the proposed mechanism is correct it is likely that some of the PST chains are oligomeric rather than polymeric in nature, and, if sufficiently short, it is reasonable to suppose that these will tend to remain in solution rather than become buried in the particle. However, after preparation, the latices are transferred from alcohol/water mixtures to a poorer solvent for the styrene segments, namely water. At this stage any oligomeric styrene will tend to either associate by hydrophobic bonding, or adsorb onto the latex surface (see Figure 4). In either case the effect would be to cause the stabilising chains to adopt a more compact conformation than would be the case for chemically unmodified PAA. It is suggested that the titration behaviour of the latices can be explained in these terms. Thus as a result of the locally higher segment density in the surface layer, the stabiliser behaves as a weak acid than PAA itself. However, as ionisation proceeds the electrostatic energy builds up to a point where it is sufficient either to break up the aggregates of styrene oligomers or to cause desorption of the short chains. After this point ionisation

Table III Analysis of some conductometric titration data for the latices

Particle Radius (μm)		0.24	0.15	0.15	0.20	0.12	0.20
PAA Mol. Wt. ($x10^{-4}$)		1.0	3.0	3.0	23.0	100.0	100.0
Surface Charge Density ($\mu C\ cm^{-2}$)		100	116	115	128	244	159
Adsorption Density ($mg\ m^{-2}$)		0.75	0.87	0.86	0.96	1.80	1.20
Average Spacing per PAA molecule (nm)		2.8	7.2	7.6	20.0	30.0	37.4
rms Dimensions for PAA in Solution (nm) (Ref 11)	$(\overline{r^2})^{\frac{1}{2}}$	3.9	10.9	10.9	31.9	66.5	66.5
	$(\overline{s^2})^{\frac{1}{2}}$	1.6	4.5	4.5	13.0	27.2	27.2

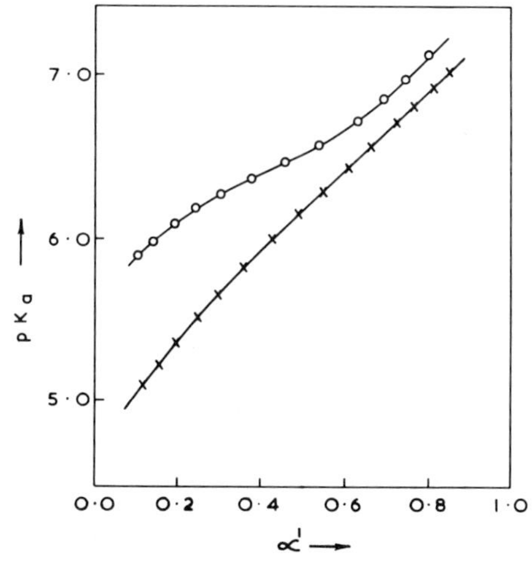

Figure 3. Potentiometric titrations—dependence of apparent pK (pKa) on α' in 10^{-2}M NaCl at 25°C for a PAA solution (×) and for a latex (○).

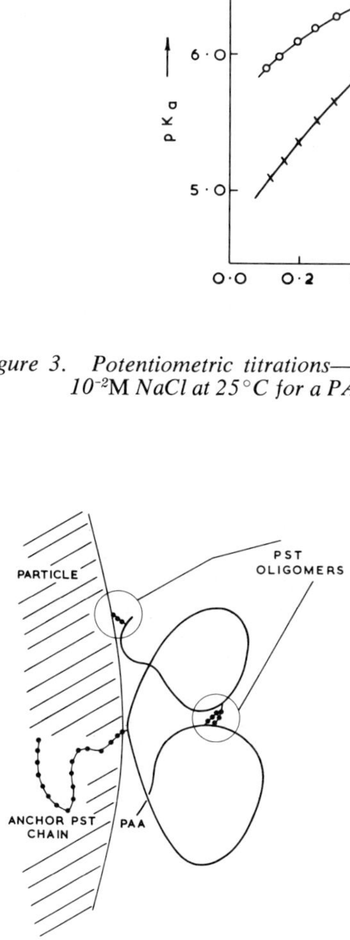

Figure 4. Schematic of the surface layer of a latex particle

proceeds in a manner similar to PAA (cf the similarity in slopes at values of α' >0.5). Some support for this picture was obtained by performing titrations in the presence of 30% v/v ethanol. The addition of ethanol would be expected to reduce the tendency for any hydrophobic segments to associate or adsorb and was indeed found to remove the inflection from the titration curves (see Figure 5).

III Flocculation Studies, The stability of the latices in the presence of added electrolyte has been studied by determining the critical flocculation temperatures (CFT) at various degrees of neutralisation in the presence of various concentrations of added sodium chloride. In Figure 6 the CFT for three sodium chloride concentrations are plotted against the degree of neutralisation (α). It can be seen that the latices flocculate both on heating and cooling. Now it is well known that particles stabilised by non-ionic polymers flocculate at or near the θ point of the stabilising polymer (3). It has also been shown that a similar correlation holds for PAA stabilised particles in aqueous media (16). In principle every polymer-solvent system has two θ points, although it is not always possible to observe both experimentally. In the case of PAA only one θ point has been reported, this corresponding to an upper critical solution temperature (UCST). Therefore it was expected that the latices should flocculate on cooling. That flocculation was also observed on warming implies that the stabilising polymer has an experimentally accessible lower critical solution temperature (LCST). Phase separation at a LCST is thought to be dominated by the free volume dissimilarity between polymer and solvent (17). As a result of the polymerisation mechanism by which the latices are produced it is likely that oligomeric styrene units are attached to the stabilising polymer. It is suggested that the presence of these hydrophobic groups and their ordering effect upon water in the region of the stabiliser chains causes a lowering of the LCST. It can be inferred from potentiometric titration data that the amount of oligomeric styrene associated with the stabilising chains could well be as high as 20 mol % . Solution studies of a number of polyelectrolyte copolymers have shown that molecules containing similar levels of hydrophobic groups are often only soluble in water when highly ionised. It is therefore likely that the grafted stabiliser chains are only marginally soluble in the aqueous medium.

Conclusions

Polystyrene latices can be prepared by the dispersion polymerisation of styrene in alcohol/water mixtures containing polyelectrolytes. The experimental data obtained lend support to the hypothesis that the polymerisation mechanism which operates is analogous to that which occurs in the non aqueous dispersion polymerisation of methyl methacrylate in solutions of degraded

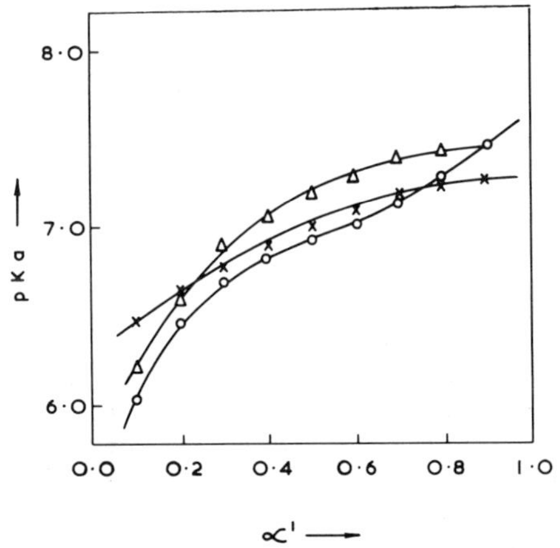

Figure 5. Potentiometric titrations—dependence of apparent pK on α′ in 10^{-2}M NaCl at 25°C for a PAA solution in EtOH/H_2O (△) and for a latex in EtOH/H_2O (×) and in H_2O (○).

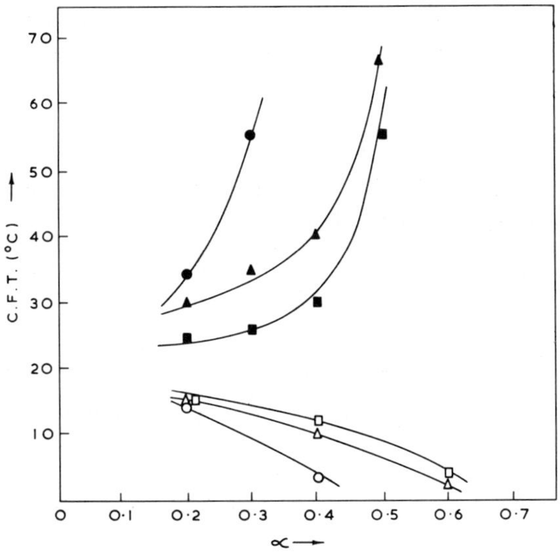

Figure 6. Critical flocculation temperatures (CFT's) for a latex as a function of the degree of neutralization (α) in 0.83M, 1.10M, and 1.38M NaCl on heating (●, ▲, ■) and on cooling (○, △, □), respectively.

rubber in aliphatic hydrocarbons. Further support for this hypothesis comes from the results obtained by analysis of the poly acrylic acid associated with the latices by conductometric and potentiometric titration, and from an evaluation of the stability of the latices in the presence of added 1:1 electrolyte. These results are indicative of the latices being stabilised by a layer of poly(acrylic acid-g-styrene) graft copolymer.

Abstract

Polyacrylic acid stabilised latices have been prepared by aqueous dispersion polymerisation. The method used is analogous to the non-aqueous dispersion (NAD) polymerisation methods originally used to prepare polymethyl methacrylate particles in aliphatic hydrocarbons (1). In effect the components of a NAD polymerisation have been replaced as follows: aliphatic hydrocarbon by aqueous alcohol, and degraded rubber, the stabiliser, by polyacrylic acid (PAA). The effect of various parameters on the particle size and surface charge density of the latices is described together with details of their colloidal stability in the presence of added electrolyte.

Acknolwedgements

The authors wish to thank A Price and I McGowan for carrying out much of the experimental work which is reported here and the Management of ICI for permission to publish this paper.

Literature Cited

1. Barrett, K.F.J, Ed. "Dispersion Polymerisation in Organic Media"; Wiley; London; 1975; p.80.

2. Van der Scheer, A; Tanke, M. A; Smolders, C. A. Faraday Disc. Chem. Soc., 1978, 65, 264.

3. Napper, D. H. J. Colloid Interface Sci. 1977, 58, 390.

4. Hesselink, F. Th. J. Polym. Sci. Poly. Symposia, 1977, 61, 439.

5. Dolan, A. K; Edwards, S. F. Proc. Roy. Soc. London, 1975, A-343, 427.

6. Gerber, P. R; Moore, M. A. Macromolecules, 1977, 10, 476.

7. Croucher, M. D; Hair, M. L. Macromolecules, 1978, 11, 874.

8. Everett, D. H; Stageman, J. F. Faraday Disc. Chem. Soc. 1978, 65, 230.

9. Dobie, J. W; Evans, R; Gibson, D. V; Smitham, J.B; Napper, D. H. J. Colloid Interface Sci., 1973, 45, 557.

10. Lambe, R; Tadros, Th. F; Vincent, B. J. Colloid Interface Sci., 1978, 66, 77.

11. Flory, P. J; Osterheld, J. E. J. Phys. Chem., 1954, 58, 653.

12. Evans, R; Davison, J. B; Napper, D. H. J. Polym. Sci., Poly. Letters, 1972, 10, 449.

13. Garvey, M. J; Tadros, Th F; Vincent, B. J. Colloid Interface Sci., 1976, 55, 440.

14. Brandup, J; Immergut, E. H., Ed. "Polymer Handbook'; Wiley; New York, 1966; p 53.

15. Conio, G; Patrone, E; Russo, S; Trefiletti, V. Makromol. Chem., 1976, 177, 49.

16. Buscall, R. J. Chem. Soc., Faraday Trans. In Press.

17. Patterson, D. Macromolecules, 1969, 2, (6), 672.

RECEIVED April 6, 1981.

9

Mechanical and Chemical Stability of Polymer Latices

D. C. BLACKLEY

National College of Rubber Technology, The Polytechnic of North London, Holloway, London, N7 8DB, England

The purpose of this paper is to summarise results which have recently been obtained for the effects of various soaps and surfactants upon the mechanical and chemical stability of natural rubber latex, and to indicate the inferences which have been drawn in the course of endeavouring to interpret these observations. Some of these results have already been published and discussed in detail elsewhere; it is intended that the others will be published and discussed in detail elsewhere in due course.

Experimental procedure

All the results described in this paper were obtained using a high-ammonia centrifuged natural rubber latex concentrate. As yet, no investigations have been made using low-ammonia centrifuged natural rubber latex concentrates. Although generally similar effects will be expected in the case of low-ammonia concentrates, it is possible that the presence of secondary preservatives, such as sodium pentachlorophenate or zinc oxide, will cause some perturbation of the observed effects. It is also the case that no investigations have as yet been made using natural rubber latices from which some of the non-rubber constituents have been removed, nor have any investigations as yet been made using synthetic rubber latices. It seems likely that further progress in understanding the interesting effects which have been observed so far will require that measurements now be made upon rubber latices which have a more closely-defined composition than the high-ammonia centrifuged natural rubber latex concentrate which is widely used in industry. In particular, there is now a need to elucidate the effects of proteins and protein-degradation products in influencing the mechanical stability of natural rubber latex, as well as for further elucidation of the role of indigenous and added soaps and surfactants.

Mechanical stabilities were determined at 35°C and 55% total solids content by means of a Klaxon high-speed-stirring mechanical stability test apparatus. The tests were carried out according to the procedure specified in either BS 1672:Part 2:1954 or

0097-6156/81/0165-0171$05.00/0

BS 1672:1972. Each mechanical stability was determined in duplicate, and the average of the two values reported. The reproducibilities of the determinations were satisfactory in all cases.

The ability of a soap or surfactant to enhance the chemical stability of natural rubber latex was assessed by ascertaining its effect upon the mechanical stability of natural rubber latices whose stabilities had been reduced by various chemical modifications. Natural rubber latices of reduced stability were produced in three different ways as follows:

(i) by the addition of sufficient potassium chloride to reduce the mechanical stability time of the latex to approximately half its initial value;

(ii) by the addition of sufficient acetic acid to reduce the mechanical stability time of the latex to approximately half its initial value;

(iii) by de-ammoniation to ca. pH 8.5 by aeration.

In the case of (i), the potassium chloride was added to the latex as a 10% aqueous solution. The amount of potassium chloride which was required in order to effect the desired reduction of stability was equivalent to approximately 1 part by weight per 100 parts of latex solids. In the case of (ii), the acetic acid was added to the latex as a 5% aqueous solution. The required amount was approximately 0.2 parts by weight per 100 parts of latex solids. In the case of (iii), de-ammoniation was effected by blowing air across the surface of the latex whilst the latex was gently warmed and stirred.

As far as possible, fatty-acid soaps and sulphate/sulphonate surfactants of high purity were used. The full details of the grades and preparative procedures are given elsewhere. The ethylene oxide condensates were commercial materials produced by reacting a mixture of cetyl and oleyl alcohols with various amounts of ethylene oxide. The average mole ratio of ethylene oxide to fatty alcohol in these condensates ranged from 2 to 60. In endeavouring to interpret the effects of these condensates upon the stability of natural rubber latex, it must be borne in mind that commercial ethylene oxide condensates are usually heterogeneous with respect to the length of the polyethylene oxide chain. Furthermore, the extent of the heterogeneity increases as the overall mole ratio of ethylene oxide to hydrophobe base increases.

Wherever possible, the soaps and surfactants were added to the natural rubber latex as dilute aqueous solutions. The cases where this was not possible were (a) ethylene oxide-fatty alcohol condensates of low ethylene oxide:fatty alcohol mole ratio, and (b) sparingly-soluble fatty-acid soaps such as lithium laurate and calcium soaps. The former were added as pastes with water, the latter as dry powders. In all cases, the latex samples were allowed to mature for about three days at room temperature before their mechanical stabilities were determined. This allowed some opportunity for the attainment of adsorption equilibrium.

Effects of added fatty-acid soaps upon mechanical and chemical stability of natural rubber latex (1,2,3)

Saturated straight-chain fatty-acid soaps (1). Figure 1 shows the effects of increasing levels of various potassium saturated straight-chain fatty-acid soaps upon the mechanical stability of natural rubber latex. For convenience of making comparisons between the various soaps, the levels of added soap are expressed as moles per 100 g. of latex solids.

These results show two features of especial interest:

(i) Very large enhancements of mechanical stability can be brought about by the addition of amounts of fatty-acid soap which are small relative to the amounts of soap indigenously present in the latex. An indication of the concentration of indigenous soaps present in the latex can be gained from the difference between the VFA and KOH numbers of the latex, for there are good reasons(4) for supposing that the concentration of indigenous soaps is equivalent to approximately one third of this difference. Using this index, the conclusion is reached that the addition of, say, potassium laurate in an amount equivalent to ca. only 5-10% of the indigenous soaps can enhance the mechanical stability by a factor of 3.

(ii) The ability of an added fatty-acid soap to enhance the mechanical stability of natural rubber latex depends very much upon the chain length of its alkyl group. For any given molal level of addition of soap, the optimum enhancement is observed when the alkyl chain of the soap contains approximately 11 carbon atoms. In Figure 2, the data of Figure 1 are re-presented in such a way as to demonstrate this point more clearly than is evident from Figure 1. In this diagram, the mechanical stability at various molal levels of addition of soap has been plotted as a function of the number of carbon atoms in the alkyl chain of the soap. The results plotted in this way give a pictorial indication of the molecular efficiencies of the various soaps in enhancing the mechanical stability of natural rubber latex.

The results summarised in Table I show the effect of equal parts by weight of each of the potassium fatty-acid soaps upon the mechanical stability of each of the three chemically-destabilised latices. For convenience in making comparisons, estimates of the corresponding results for unmodified natural rubber latex are also included. It is clear from these results that the ability of added potassium fatty-acid soaps to enhance the stability of chemically-destabilised natural rubber latex roughly parallels their abilities to enhance the mechanical stability of unmodified natural rubber latex.

In endeavouring to explain these observations - in particular, the observation that the molal efficiency of a straight-chain fatty-acid soap in enhancing the mechanical stability of natural rubber is a maximum when the alkyl chain of the soap contains approximately 11 carbon atoms - we have come to the conclusion

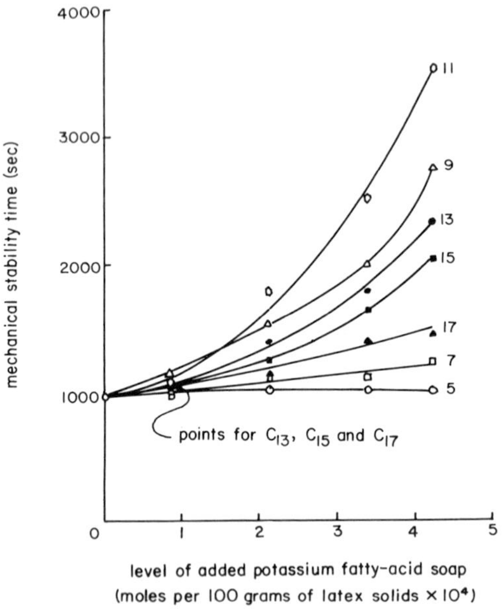

Plastics and Rubber: Materials and Applications

Figure 1. Effect of added straight-chain potassium fatty-acid soaps upon mechanical stability of natural rubber latex (1). Numbers appended to curves are number of carbon atoms in alkyl chain of soap.

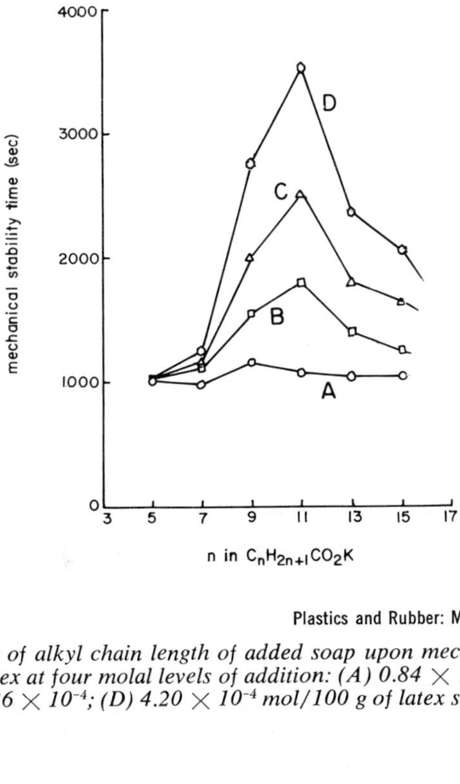

Plastics and Rubber: Materials and Applications

Figure 2. Effect of alkyl chain length of added soap upon mechanical stability of natural rubber latex at four molal levels of addition: (A) 0.84 × 10^{-4}; (B) 2 × 10^{-4}; (C) 3.36 × 10^{-4}; (D) 4.20 × 10^{-4} mol/100 g of latex solids (1) that added fatty-acid soaps enhance the mechanical and chemical stability of natural rubber latex principally by making the indigenous soaps more effective as stabilisers, rather than by increasing the total amount of adsorbed soap present in the latex and thereby increasing the surface potential at the interface between rubber and water. According to our present view, the existence of the optimum soap alkyl chain length arises from the balance of the following two opposing tendencies as the alkyl chain length of the added soap is increased:

(i) an increasing tendency for the added soap anion to be adsorbed at the rubber-water interface; and

(ii) a decreasing ability of the added soap anions to make the indigenous soap anions more effective as stabilisers.

Insofar as some of the added soaps may be very efficient in "activating" the indigenous soaps, in the sense that a small number of molecules of added soap is able to activate a large number of molecules of indigenous soap, then we have a ready explanation for the ability of some added soaps to effect large enhancements of mechanical stability even although they are present in an amount which is small compared to the concentration of indigenous soaps in the latex.

There then arises the question of the mechanism by which small amounts of added soap might make the indigenous soaps more effective as stabilisers. Our present suggestion is that the indigenous soap anions are mainly present as coherent clusters or "islands" adsorbed at the rubber-water interface, and that mechanical destabilisation occurs by way of mechanically-induced interactions between those regions of the surfaces of contiguous rubber particles which are relatively bare of adsorbed soap anions. The bare regions through which the initial interaction occurs may be essentially those which must exist if (a) the adsorbed soap anions are clustered and (b) (as is known to be the case) the latex particles are not saturated with soap. Alternatively, the extent of the bare regions in those parts of the particle surfaces where interaction is occurring may be augmented by electrostatic repulsions between the clusters of soap anions on different particles as they approach one another. These repulsions are envisaged as causing the clusters to move away from the regions of impending contact. (In this latter connection, it is interesting to note that Napper(5) has postulated that colloid particles which are stabilised by an adsorbed steric stabiliser can become destabilised by the stabiliser molecules moving laterally around the surface of the particle whilst remaining attached to the particle. We also note that carboxylate ions which are chemically combined at the polymer-water interface are known to be considerably more effective in conferring mechanical stability upon a latex than are carboxylate ions which are held at the interface by adsorption. Presumably this is because the latter are able to move laterally in the particle surface, whereas the former are not.) We propose that a given number of adsorbed soap anions is more effective in conferring mechanical stability if able to move independently of each other than if aggregated into coherent clusters. This is either because the normal equilibrium distribution of bare patches is different, or because the electrostatic repulsions between particles carrying the adsorbed anions are greater, or because the surface osmotic pressure arising from the presence of the adsorbed anions is greater. A plausible activation mechanism is then one in which the added soap anions adsorb at the rubber-water interface, mix with the indigenous adsorbed soap anions, and thereby encourage the coherent clusters to disperse. According to this view, the soaps of short alkyl chain length are relatively inefficient as enhancers of mechanical and chemical stability because they are not readily adsorbed at the rubber-water interface. On the other hand, the soaps of long alkyl chain length are inefficient because, although they are strongly adsorbed, they are similar in nature to the indigenous soaps, and therefore have little effect upon the coherence of the soap-anion clusters. A soap such as potassium laurate is very efficient as an enhancer of mechanical and chemical stability because on the one hand the length of its alkyl chain is sufficient to ensure that the soap anion is reasonably strongly adsorbed at the rubber-water interface, and on the other hand the alkyl chain is sufficiently short seriously to disrupt the coherence of the clusters of indigenous adsorbed soap anions when the added anions mix with them.

Straight-chain C_{18} carboxylate soaps of various hydrophobe structures (2). Figure 1 and Table I show that potassium stearate is relatively inefficient as an enhancer of the mechanical and chemical stability of natural rubber latex. This observation was rather surprising in view of the widely-held belief that water-soluble stearate soaps are efficient stabilisers for anionic latices at alkaline pH. It was therefore of interest to investigate to what extent this inefficiency is a general feature of potassium carboxylate soaps which contain 18 carbon atoms in all. Experiments with a range of potassium straight-chain C_{18}-carboxylate soaps have shown that the ability of such soaps to enhance the mechanical stability of natural rubber latex depends markedly upon the chemical structure of the C_{17} hydrophobe chain; thus, for instance, potassium oleate is much more effective in enhancing stability than is potassium stearate. The results for effects upon mechanical stability are summarised in Figure 3. They show that, broadly speaking, the more the structure of the added soap deviates from that of potassium stearate, the greater is the ability to enhance mechanical stability. The effects of the C_{18} soaps upon chemical stability (as assessed by the methods used in this work) broadly parallel the effects of the soaps upon mechanical stability. These observations are consistent with the view that stability is enhanced by reduction of the coherence of clusters of indigenous adsorbed soap anions, if it is also assumed that the indigenous soap anions have a chemical structure which is similar to that of the stearate anion.

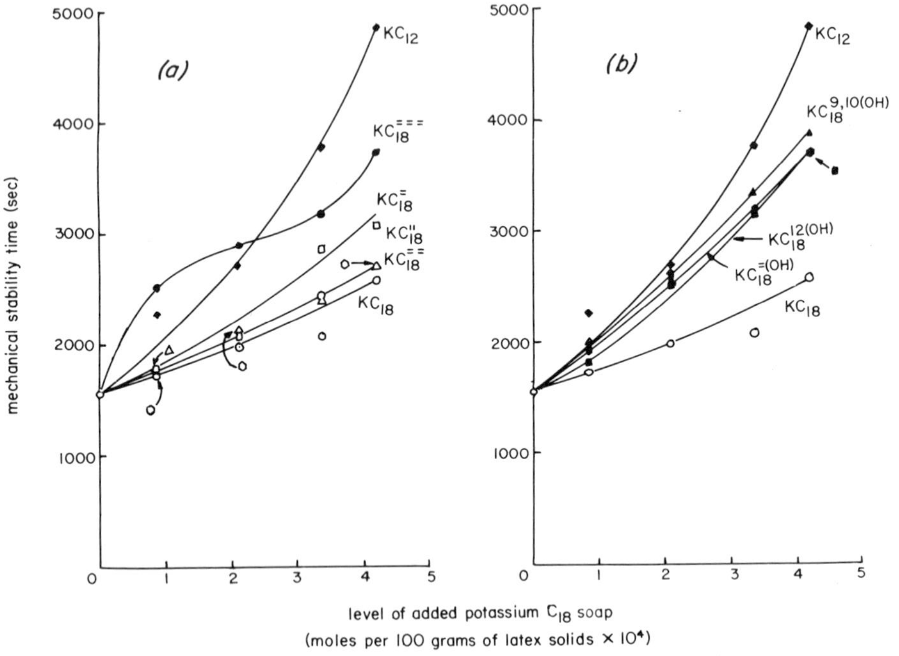

Plastics and Rubber: Materials and Applications

Figure 3. Effect of various straight-chain potassium C_{18} carboxylate soaps upon mechanical stability of natural rubber latex (2): (KC_{18}) potassium stearate; ($KC_{18}{}^{=}$) potassium oleate; ($KC_{18}{}''$) potassium elaidate; ($KC_{18}{}^{==}$) potassium linoleate; ($KC^{===}$) potassium linolenate; ($KC_{18}{}^{12(OH)}$) potassium 12-hydroxystearate; ($KC_{18}{}^{9,10(OH)}$) potassium 9,10-dihydroxystearate; ($KC_{18}{}^{=(OH)}$) potassium ricinoleate. Results also are included for potassium laurate (KC_{12}).

Counterion effects (3). Experiments which have been carried out using lithium, sodium, potassium, ammonium and morpholinium laurates have shown that the effects of these five laurates upon mechanical and chemical stability are broadly similar, although such differences as are observed are statistically significant. The results for effects upon mechanical stability are summarised in Table II. That lithium laurate behaves similarly to, say, potassium laurate is perhaps surprising, in that it is known that a lithium salt is more effective in reducing the mechanical stability of natural rubber latex than is the corresponding potassium salt (6). The inference has been drawn that the counterion of the carboxylate soap has a negligible effect upon the ability of the soap to enhance mechanical stability, relative to the effect of the anion, at least for those cations for which specific adsorption effects are absent.

Table II shows that morpholinium laurate is markedly less effective in enhancing mechanical stability than are the other laurates which have been investigated. This is attributed to specific counterion adsorption, with a consequent reduction of the effective surface potential at the rubber-water interface.

The abilities of the five laurates to protect natural rubber latex against chemical destabilisation appear to be broadly parallel to their effects upon mechanical stability.

We have also recently discovered that added calcium laurate is able markedly to enhance the mechanical stability of natural rubber latex (7). This observation is surprising, partly because of the low solubility of calcium laurate in water, and partly because calcium ions are known to be powerful destabilisers of natural rubber latex (6). It indicates that the stabilising effect of the laurate anion is much greater than the destabilising effect of the calcium cation.

It is important to point out that our investigation of counterion effects in carboxylate soaps has so far been concerned almost exclusively with laurate soaps. Laurate soaps were chosen partly because they are generally convenient to handle in that many of them are readily soluble in water to give solutions of low viscosity, and partly because, as has been shown above, laurate soaps are very effective in enhancing the mechanical and chemical stability of natural rubber latex. It must therefore be borne in mind that the conclusions which have been drawn from this investigation concerning effects attributable to counterion variation in laurate soaps may not be generally valid for carboxylate soaps as a family.

Effects of added sulphate and sulphonate surfactants upon the mechanical and chemical stability of natural rubber latex (8)

The effects of a range of sodium *n*-alkyl sulphates and sodium *n*-alkyl sulphonates upon the mechanical stability of natural rubber latex are summarised in Figures 4 and 5 respectively. As in the case of added potassium fatty-acid soaps, small additions of

Table I: Effect of 0.1 part by weight per 100 parts latex solids of various potassium fatty-acid soaps upon mechanical stability of unmodified and chemically-destabilised natural rubber latices (1)

number of carbon atoms in alkyl chain of soap	mechanical stability time (seconds)			
	unmodified latex	latex containing 1 part by weight potassium chloride per 100 parts of latex solids	latex containing 0.2 part by weight acetic acid per 100 parts of latex solids	latex de-ammoniated to pH 8.5 by aeration
5	-	500	754	530
7	1310*	910	1200	980
9	3190*	1360	2195	1580
11	3540	1210	2620	2367
13	2060	1140	2330	1685
15	1650	860	1768	1420
17	1320	740	1300	1240
MST of latex in absence of fatty-acid soap	965	455	490	350

* obtained by extrapolation

Table II: Effects of added laurate soaps of various counterions upon mechanical stability of natural rubber latex (3)

level of added soap (moles per 100 g of latex solids)	mechanical stability time (seconds) in presence of laurate soap of counterion indicated				
	lithium	sodium	potassium	ammonium	morpholinium
0.84×10^{-4}	2280	2460	2540	2080	1650
2.10×10^{-4}	2670	3120	3080	2900	1950
3.36×10^{-4}	3390	3640	3880	3280	2580
4.20×10^{-4}	3990	4080	4220	3640	3090

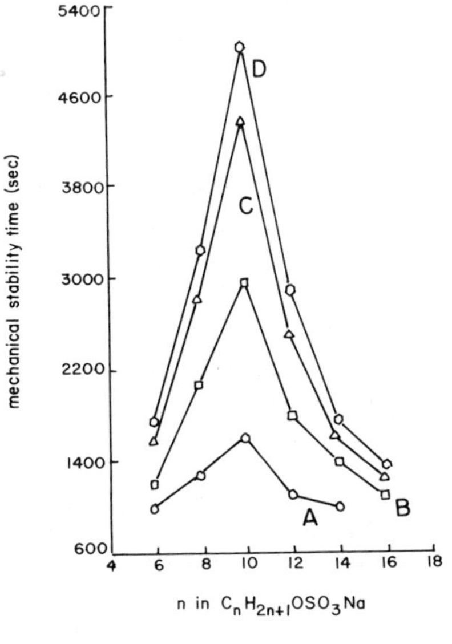

Figure 4. Effect of alkyl chain length of added sodium n-*alkyl sulfate upon mechanical stability of natural rubber latex at four levels of addition: (A) 20; (B) 60; (C) 100; (D) 120 mg/100 g of latex solids* (8)

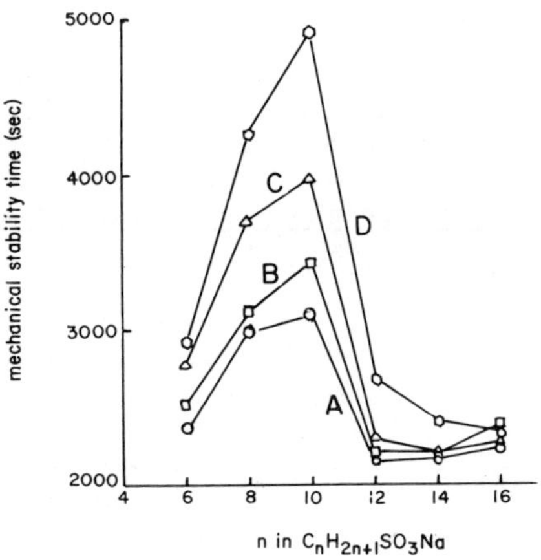

Figure 5. Effect of alkyl chain length of added sodium n-*alkyl sulfonate upon mechanical stability of natural rubber latex at four levels of addition: (A) 20; (B) 40; (C) 60; (D) 80 mg/100 g of latex solids* (8)

both n-alkyl sulphates and n-alkyl sulphonates can cause marked enhancements of the mechanical stability of natural rubber latex, and the magnitude of the enhancement at any given level of addition depends upon the length of the alkyl chain in the surfactant. As in the case of the potassium fatty-acid soaps, an optimum alkyl chain length is observed as the alkyl chain length of the surfactant is increased. The optimum alkyl chain length in the case of the sulphates and sulphonates is approximately 10 carbon atoms, and this is very similar to the optimum alkyl chain length of approximately 11 carbon atoms observed for the carboxylates. The explanation which we offer for the existence of optimum alkyl chain lengths for the sulphates and sulphonates is similar to that proposed for the carboxylates. The ability of these surfactants to enhance the mechanical stability of natural rubber latex is attributed primarily to their being adsorbed at the rubber-water interface, and there mixing with the anions of the indigenous soaps, thereby altering the physical nature of the monolayer of adsorbed anions.

The abilities of the sulphates and sulphonates to protect natural rubber latex against chemical destabilisation are again broadly parallel to their abilities to enhance mechanical stability.

Some data are also available for the effect of the counterion of a dodecyl sulphate upon its ability to enhance the mechanical stability of natural rubber latex. As in the case of the laurates the lithium, sodium, potassium and ammonium salts are similar in behaviour, but the morpholinium salt is slightly less effective. Again, the latter effect is attributed to specific adsorption of the morpholinium cation. Calcium and magnesium dodecyl sulphates are also effective in enhancing mechanical stability, their abilities being similar to that of morpholinium dodecyl sulphate.

Effects of added n-alkyl triethyl ammonium bromides upon the mechanical stability of natural rubber latex (9)

Some interesting results have recently become available for the effects of a range of n-alkyl triethyl ammonium bromides upon the mechanical stability of natural rubber latex. The number of carbon atoms in the alkyl group varied from 6 to 18. Figure 6 summarises the results. It is usually believed that the addition of cationic surfactants to an anionic latex such as natural rubber latex invariably leads to a reduction in colloid stability, the effect being attributed to adsorption of the cations with consequent partial neutralisation of the particle charge and reduction of the counterion cloud surrounding the particles. Whilst the results summarised in Figure 6 show that this expectation is often realised, they also show that small additions of an n-alkyl triethyl ammonium bromide whose alkyl group contains between approximately 8 and 12 carbon atoms can bring about a small increase in mechanical stability. In such cases, the increase in stability cannot possibly be attributed to enhance-

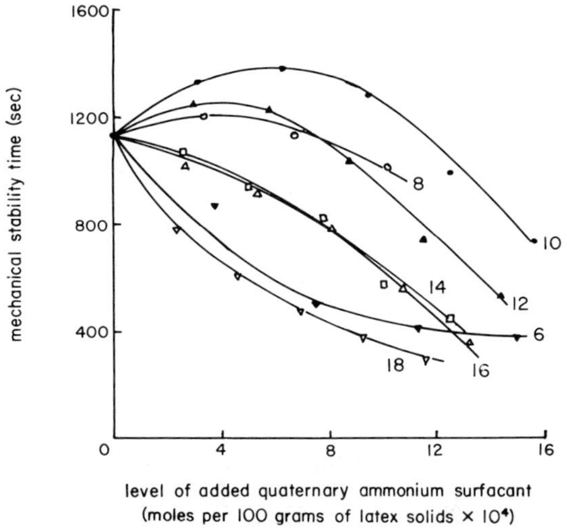

Figure 6. Effect of added n-*alkyl triethyl ammonium bromides upon mechanical stability of natural rubber latex (9). Numbers appended to curves are numbers of carbon atoms in* n-*alkyl chain of surfactant.*

ment of the charge carried by the particles, since presumably the particle charge can only be reduced by the addition of a cationic surfactant. It appears that in these cases the destabilising effect of the reduced particle charge is more than offset by some other effect which accompanies the adsorption of the surface-active cation at the particle surface and which causes the stability to increase. One possible effect which meets these requirements is the mixing effect, with reduction of lateral coherence within the adsorbed layer of soap anions, which has been postulated above. To this extent, the observations on the effects of added cationic surfactants are consistent with the suggestions which have been made above, and so add weight to their credibility.

Effects of ethylene oxide-fatty alcohol condensates upon the mechanical and chemical stability of natural rubber latex (10)

The results summarised in Figure 7 show that small additions of ethylene oxide-fatty alcohol condensates to natural rubber latex generally cause the mechanical stability of the latex to fall. This phenomenon is attributed to the displacement of adsorbed proteinaceous molecules by the condensate molecules. Although the latter are more surface active than the former, they are presumably less effective in conferring mechanical stability upon the rubber particles, perhaps because, unlike the proteinaceous molecules, they are not ionised.

Figure 7 also shows that, with increasing additions of ethylene oxide-fatty alcohol condensates, the mechanical stability passes through a minimum and then increases. However, only if the overall mole ratio of ethylene oxide to fatty alcohol in the condensate exceeds about 30 does the mechanical stability increase above that of the initial latex, at least at levels of addition which would normally be used in practice. An interesting fact emerges when, as in Figure 8, mechanical stability is plotted as a function of the level of addition of condensate expressed as moles of ethylene oxide units added per unit mass of latex solids. This is that, for condensates for which the overall mole ratio of ethylene oxide to fatty alcohol is within the approximate range 6-30, the ability to enhance the stability of natural rubber latex after the minimum in stability has been passed depends primarily upon the total number of ethylene oxide units which have been added to the latex as the condensate. The length of the blocks in which the units are added is of secondary importance. We interpret this observation as implying that, for these condensates, the effect upon mechanical stability is determined primarily by the binding of water to the ethylene oxide units which are anchored to the rubber-water interface by the fatty-alcohol moiety of the condensate. In the case of condensates for which the overall mole ratio of ethylene oxide to fatty alcohol exceeds ca. 30, the effect upon mechanical stability is much greater than would be expected on the basis of the total amount of ethylene oxide which has been added to the latex, as evidenced by the

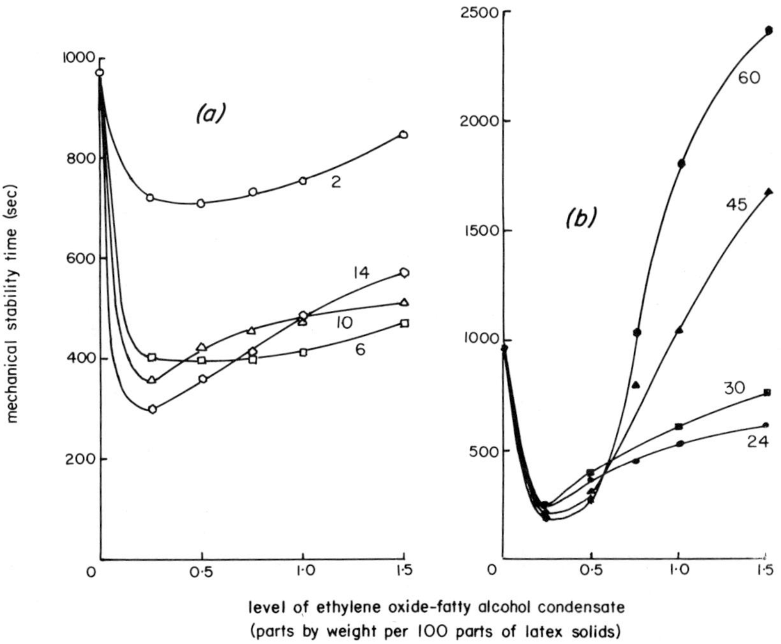

Plastics and Rubber: Materials and Applications

Figure 7. Effect of added ethylene oxide–fatty alcohol condensates upon mechanical stability of natural rubber latex (10). Levels of condensate are expressed in parts by weight. Numbers appended to curves indicate overall mole ratio ethylene oxide:fatty alcohol in condensate.

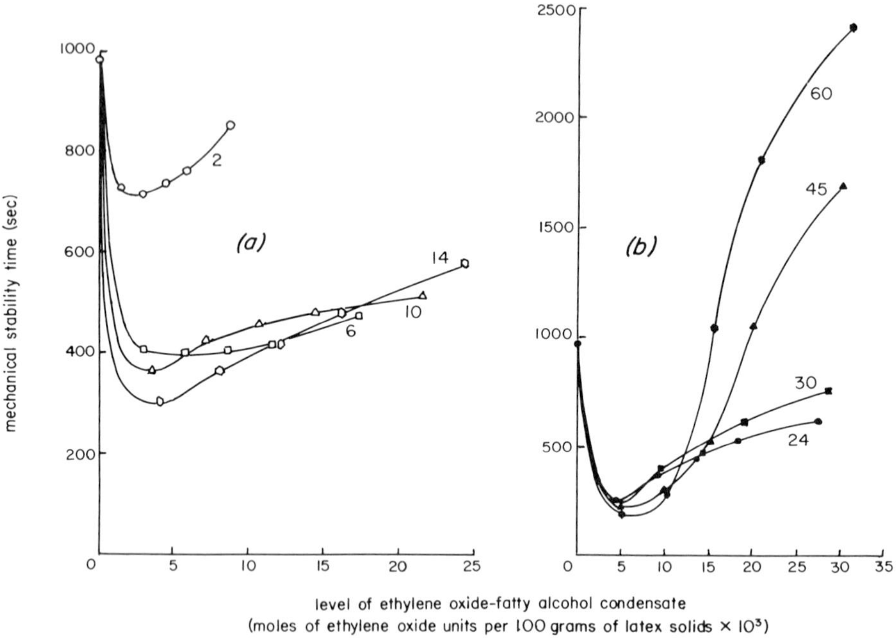

Plastics and Rubber: Materials and Applications

Figure 8. Effect of added ethylene oxide–fatty alcohol condensates upon mechanical stability of natural rubber latex (10). Levels of condensate are expressed in moles of ethylene oxide units. Numbers appended to curves indicate overall mole ratio ethylene oxide:fatty alcohol in condensate.

Table III: Effect of 1 part by weight per 100 parts latex solids of various ethylene oxide-fatty alcohol condensates upon mechanical stability of unmodified and chemically-destabilised natural rubber latices (10)

overall mole ratio of ethylene oxide to fatty alcohol	mechanical stability time (seconds)			
	unmodified latex	latex containing 1 part by weight potassium chloride per 100 parts of latex solids	latex containing 0.2 part by weight acetic acid per 100 parts of latex solids	latex deammoniated to pH 8.5 by aeration
2	757	710	623	520
6	415	247	325	500
10	479	364	461	590
14	482	575	558	630
24	531	850	728	708
30	612	1041	802	785
45	1050	1993	1130	1256
60	1810	2578	2003	2015
MST of latex in absence of ethylene oxide-fatty alcohol condensate	970	440	499	374

results for condensates containing between 6 and 30 moles of ethylene oxide per mole of fatty alcohol. We suggest that in such cases steric stabilisation makes an important contribution to the mechanical stability in addition to stabilisation by hydration.

Data for the abilities of these condensates to protect natural rubber latex against chemical destabilisation (as assessed by the methods used in this work) are given in Table III. The effects upon chemical stability broadly parallel effects upon mechanical stability. An interesting and significant observation is that the chemically-destabilised latices to which the ethylene oxide condensates had been added were in most cases more mechanically stable than the unmodified latex to which the same level of the same condensate had been added. Thus it appears that the ability of an ethylene oxide condensate to enhance the mechanical stability of natural rubber latex is itself enhanced by the presence of factors (such as increased ionic strength) which normally tend to reduce the stability of the latex. In the case where that factor is increased ionic strength, a possible explanation is that the adsorption tendency of the condensate is increased by a salting-out effect. But whatever the explanation, we note that our data provide objective support for the view, widely held in the latex-using industry, that ethylene oxide condensates of suitable composition are very efficient enhancers of the chemical stability of natural rubber latex.

Literature cited

1. Blackley, D.C.; Nor Aisah, bt. A.A.; Twaits, R., Plastics and Rubber : Materials and Applications, 1979, 4, 77

2. Blackley, D.C.; Azas, M., Plastics and Rubber : Materials and Applications, 1980, 5, 57

3. Blackley, D.C.; Haynes, A.C., Plastics and Rubber : Processing and Applications, in press

4. Calvert, K.O., Plastics and Rubber : Materials and Applications, 1977, 2, 59

5. Napper, D.H., J. Colloid Interface Sci., 1977, 58, 390

6. Blackley, D.C.; Loha, S.; Twaits, R., unpublished results

7. Blackley, D.C.; Pieris, H.S.M., unpublished results

8. Blackley, D.C.; Emengo, F.N., unpublished results

9. Blackley, D.C.; Zahari bin Hashim, unpublished results

10. Blackley, D.C.; Salleh, b. N.B.; Twaits, R., Plastics and Rubber : Materials and Applications, 1977, 2, 117

RECEIVED April 6, 1981.

10

Stabilization in Nonaqueous Radical Dispersion Polymerization with AB Block Copolymers of Polystyrene and Poly(dimethyl siloxane)

J. V. DAWKINS, G. TAYLOR[1], S. P. BAKER, and R. W. R. COLLETT[2]

Department of Chemistry, Loughborough University of Technology, Loughborough, Leicestershire LE11 3TU, England

J. S. HIGGINS

Department of Chemical Engineering and Chemical Technology, Imperial College, London SW7 2BY, England

Nonaqueous polymer dispersions are prevented from flocculation by surrounding each particle by a surface layer of adsorbed polymeric stabilizer. Such dispersions are conveniently prepared by polymerizing in the presence of a preformed block or graft copolymer a monomer dissolved in a diluent which is a precipitant for the polymer. The stabilizing copolymer contains A blocks which are insoluble in the dispersion medium and act as anchors for the B blocks which are swollen by the diluent and which extend away from the particle surface. Most of the work on the preparation of nonaqueous dispersions has been concerned with the radical initiation of acrylic monomers in the presence of copolymer stabilizers having the A block the same as the acrylic polymer in the particles (1). Our work has involved the preparation of nonaqueous polystyrene dispersions in aliphatic hydrocarbons stabilized with well-defined AB block copolymers of polystyrene (PS) and poly(dimethyl siloxane) (PDMS) (2, 3), giving particles stabilized by surface layers of silicone. The preparation of dispersions of other polymers in the presence of the same PS-PDMS block copolymers in aliphatic hydrocarbons is of interest because effective anchoring of the copolymer may be influenced by the degree of compatibility between the A block and the polymer molecules in the particles. The present paper describes the properties of poly(methyl methacrylate) (PMMA) particles in aliphatic hydrocarbons, together with initial results on the radical dispersion polymerization of vinyl acetate (VA).

Experimental

Dispersion Polymerization. AB block copolymer stabilizers were prepared using anionic polymerization techniques by the addition of hexamethylcyclotrisiloxane to "living" polystyryl-

[1]Current address: Unilever Research, Port Sunlight, Wirral, Merseyside, U.K.
[2]Current address: Berger, Newcastle upon Tyne, U.K.

0097-6156/81/0165-0189$05.00/0

lithium. Polymerizations were performed under conditions of rigorous purity using an inert gas blanket technique or a high vacuum procedure as described elsewhere (4, 5). Block copolymers prepared over a range of molecular weights and compositions were characterized by gel permeation chromatography (GPC), osmometry and silicon analysis. The samples had a narrow molecular weight distribution ($\bar{M}_w/\bar{M}_n$ typically <1.25), which when used as stabilizers would be expected to give a constant layer thickness around the particles. The anchor/soluble balance (ASB) for the anchoring A blocks and stabilizing B blocks was calculated from the number average molecular weights of the PS blocks and PDMS blocks, $\bar{M}_n$(PS)/$\bar{M}_n$(PDMS). Block copolymers containing protonated polystyrene blocks PS(H) and deuterated polystyrene blocks PS(D) were prepared.

Polymer dispersions in aliphatic hydrocarbons were prepared by radical polymerization, e.g. with azobisisobutyronitrile as initiator in the temperature range 323-343 K, with the monomer concentration in the range 10-20 weight per cent and with the PS-PDMS stabilizer having a concentration of about 5 weight per cent. The PS-PDMS samples had $\bar{M}_n$ (PS) in the range 8.8-61.0 x 10^3 and $\bar{M}_n$ (PDMS) in the range 2.4-48.0 x 10^3 with the ASB generally between 0.5 and 4.0. Seeding techniques were frequently used in the dispersion polymerization. Dispersions were washed by repeated centrifuge/diluent exchange cycles to remove unadsorbed stabilizer and unconverted monomer, which also served to exchange a diluent for a different one.

Dispersion Properties. Particle diameters were estimated from transmission electron micrographs. The surface coverage of a particle defined as the area A occupied or stabilized by a given PDMS chain was estimated from silicon analysis on dry particles and the particle diameter D. Dispersion stability was assessed visually with the dispersion contained in a cylindrical glass cell with a light beam arranged so that light scattered by the dispersion at about 45° from the transmitted beam could conveniently be observed by the human eye. The relative viscosity η_r of a dispersion was determined with an Ostwald-Fenske capillary viscometer at 298 K. The viscometer was treated with a solution of chlorotrimethyl silane to prevent adhesion of particles to the walls. The volume fraction of the polymer particle cores Φ_o was calculated from the total polymer content of the dispersion, the PDMS content and the density of the core. Small angle neutron scattering experiments on dispersions containing PS(D) blocks were carried out using the D11 and D17 spectrometers at ILL Grenoble. From the radial distribution of scattered intensities, data of the normalized scattering intensity I(K) were computed as a function of the scattering wave vector K.

Results and Discussion

PMMA Particles. It has been shown that the value of the surface coverage A for a given PS-PDMS block copolymer was constant for PMMA particles having D in the range 96-480 nm (4). This suggests that "total" coverage of the surfaces of the PMMA particles may be assumed for these dispersions. Data for A as a function of $\bar{M}_n$ (PDMS) are shown in Figure 1, suggesting that the area stabilized by a given PDMS chain is similar on both PMMA and PS particle surfaces (6). Furthermore, the value of A appears to be independent of the molecular weight of the PS anchor block, suggesting that the PS anchor block does not extend significantly into the dispersion medium. We may conclude that the PDMS chains may be regarded as being terminally anchored at the particle surface, so that the data in Figure 1 confirm that the area which one chain is capable of stabilizing increases with increasing $\bar{M}_n$ (PDMS).

Viscosity data for the dispersions were plotted according to Equation 1

$$\Phi_o/\ln \eta_r = (1/k_1 f) - (k\,\Phi_o/k_1) \tag{1}$$

in which k_1 is the Einstein coefficient for solid spheres, f is a factor representing an increase in the Einstein coefficient because of the adsorbed surface layer of PDMS and k is a crowding factor. From the reciprocal of the intercept of this plot Goodwin (7) proposed that the thickness of the stabilizing surface layer δ may be calculated with Equation 2

$$k_1 f = k_1\,[1 + (2\delta/D)]^3 \tag{2}$$

in which k_1 was confirmed to be close to 2.5 for spherical particles free from aggregation. Our values of δ (6) are shown as a function of $\bar{M}_n$ (PDMS) in Figure 2. We may conclude that the thickness of the adsorbed surface layer is similar on both PMMA and PS particles.

In Figure 3 the scattering function I(K) is plotted against the wave vector K for dispersions containing PS(D) blocks in n-heptane (8). The intensity scattered from the PS particles relative to the background of undeuterated particles is an order of magnitude less than that from the PMMA particles with corresponding concentrations of PS(D) blocks. This suggests that there are larger scattering objects in the PMMA dispersions and hence aggregates of PS(D) blocks. Estimates of the radius of gyration and molecular weight of PS(D) blocks in PS particles from the scattering intensity suggest that the PS blocks with $\bar{M}_n$ (PS) = 23400 are molecularly dispersed. We may conclude that the PS(D) blocks cluster into domains in PMMA particles.

Flocculation studies were performed by lowering the temperature of PMMA dispersions in a binary mixture of n-heptane and

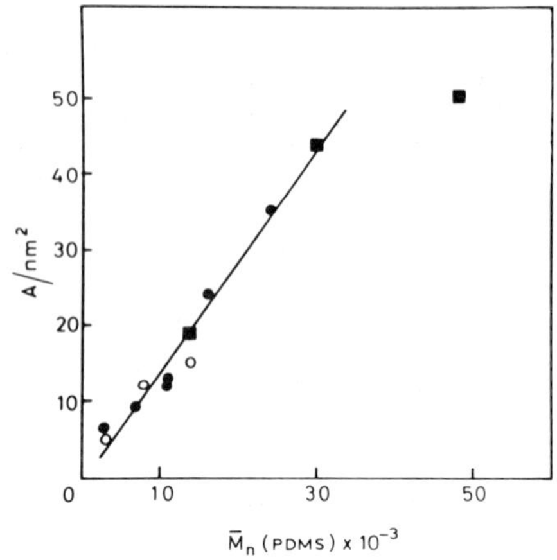

J. C. S. Faraday I

Figure 1. Dependence of surface coverage on $\overline{M}_n$ (PDMS): (●) PMMA particles with $\overline{M}_n$ (PS) < 20000; (■) PMMA particles with $\overline{M}_n$ (PS) > 30000; (○) PS particles (6)

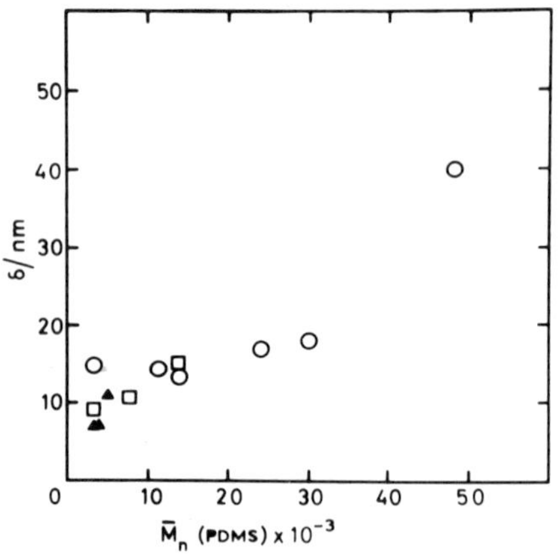

J. C. S. Faraday I

Figure 2. Dependence of surface layer thickness on $\overline{M}_n$ (PDMS): (○) PMMA particles; (□, ▲) particles (6) ethanol (51/49, v/v). The flocculation temperature was studied as a function of the molecular weight and composition of PS-PDMS, the particle size of the dispersion and the surface coverage of the particles (9). Values of the flocculation temperature in Table I are in the range 338.2–340.5 K which may be compared with values of the theta temperature of 339 ± 1 K and 341.2 ± 2.0 K determined by two separate procedures for PDMS homopolymer dissolved in the same binary mixture. We may conclude that the mechanism of steric stabilization (10) operates for the PMMA dispersions.

Table I. Stabilizing Copolymers, Particle Sizes and Surface Coverage of PMMA Dispersions in Flocculation Experiments *(9)*

Block Copolymer		D	A	Flocculation Temperature
$\bar{M}_n$ (PS)	$\bar{M}_n$ (PDMS)	(nm)	(nm^2)	(K)
12700	3200	110	6.4	339.6
8800	11200	250	12.6	340.1
44000	13700	670	19.4	339.0
16400	16100	330	24.6	340.4
12700	23800	74	35.4	340.5
44000	29800	69	44.5	340.4
33400	48000	95	51.3	338.2

Colloid and Polymer Science

The stability of PMMA dispersions was examined further by redispersion of the particles in cyclohexane at 333 K. Above 307 K cyclohexane is a good solvent for PS and PDMS, and if the PS-PDMS block copolymer was not firmly anchored, desorption of stabilizer by dissolution should occur at 333 K, followed by flocculation of the PMMA dispersion. However, little change in dispersion stability was observed over a period of 60 h. Consequently, we may conclude that the PS blocks are firmly anchored within the hard PMMA matrix. Experiments suggesting that grafting of PS-PDMS to PMMA is unlikely, have been reported previously (4). The stability of a PMMA dispersion, with a block copolymer having $\bar{M}_n$ (PS) = 12700 and $\bar{M}_n$ (PDMS) = 23800, was also examined after redispersion in n-dodecane by heating above the glass transition temperature of PMMA. A tendency towards flocculation occurred above 393 K. Solubility studies (3) of PS in alkanes suggest that PS with $\bar{M}_n$ = 12700 will dissolve in n-dodecane at 393 K. Therefore, it is proposed that since the results in Figure 3 indicate that the PS blocks are not compatible with PMMA, the PS blocks above 393 K diffuse through the soft PMMA matrix to the particle surface where the blocks are dissolved by n-dodecane so that rapid flocculation occurs. This suggests that for polymer particles having a glass transition temperature above the dispersion polymerization

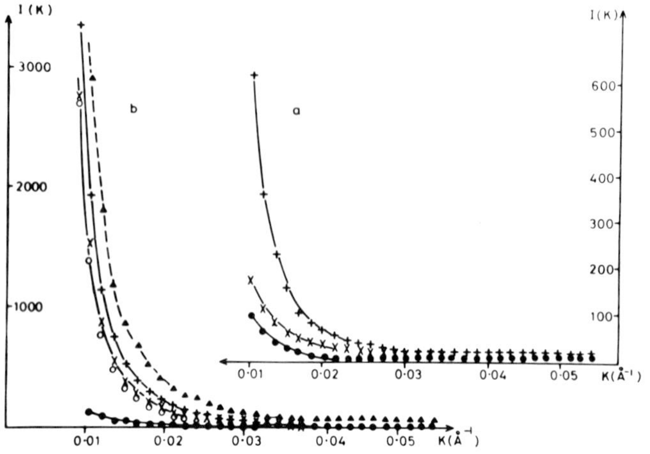

Polymer

Figure 3. Scattering intensity from PS(D) blocks in PS particles (a) and PMMA particles (b): (+, ×, ▲, ○) different PS(D)/PS(H) ratios between 1/2 and 1/12 in the particles; (●) undeuterated particles (8)

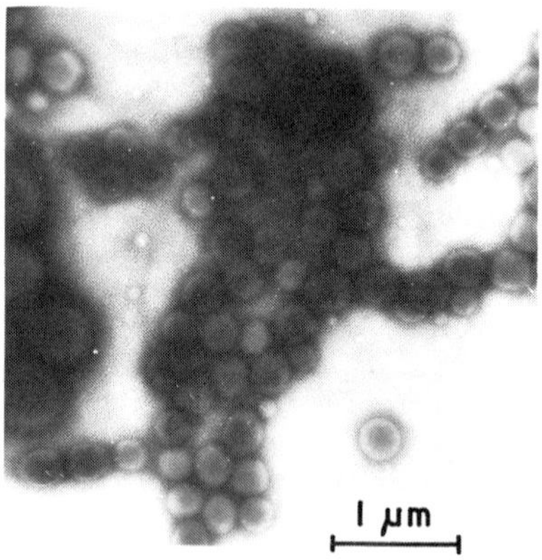

Figure 4. Electron micrograph of poly(vinyl acetate) particles

temperature the PS-PDMS block copolymer becomes entrapped within a hard polymer matrix despite the polymer incompatibility effect and stabilizes polymer particles after removal of excess block copolymer by redispersion. The preparation of stable polyacrylonitrile dispersions confirms this view. Nonaqueous dispersions of PMMA and polyacrylonitrile stabilized by BAB block copolymers of PS and PDMS have also been reported by Everett and Stageman (11, 12).

PVA Particles. Initial studies (13) on polymers with glass transition temperatures below the dispersion polymerization temperature suggest that stable dispersions of poly(vinyl acetate) (PVA), and also of poly(ethyl acrylate), may be prepared in the presence of excess PS-PDMS. An example is shown in Figure 4 in which PVA particles are superimposed on a background consisting of micelles of PS-PDMS. PVA particles prepared with PS-PDMS having $\bar{M}_n$ (PS) ∿ 10000 showed a tendency to flocculate at ambient temperature during redispersion cycles to remove excess block copolymer, particularly if the dispersion polymerization had not proceeded to 100% conversion of monomer. It is well documented that on mixing solutions of polystyrene and poly(vinyl acetate) homopolymers phase separation tends to occur (14, 15), and solubility studies (3) of PS in n-heptane suggest that PS blocks with $\bar{M}_n$ (PS) ∿ 10000 will be close to dissolution when dispersion polymerizations are performed at 343 K. Consequently, we may postulate that for soft polymer particles the block copolymer is rejected from the particle because of an incompatibility effect and is adsorbed at the particle surface. If the block copolymer desorbs from the particle surface, then particle agglomeration will occur unless rapid adsorption of other copolymer molecules occurs from a reservoir of excess block copolymer.

With a careful redispersion technique stable dispersions free of excess block copolymer are produced for PVA particles with the anchoring PS block having $\bar{M}_n$ (PS) = 33400. This suggests that more effective anchoring occurs when the solubility of the block copolymer in the dispersion medium is reduced. PVA dispersions stabilized with block copolymers having $\bar{M}_n$ (PS) = 33400 were redispersed in n-dodecane and retained their stability on heating to 393 K. In view of the incompatibility effect of PS in the soft PVA matrix and the tendency of the PS blocks towards dissolution at elevated temperatures, effective anchoring of the PDMS chains may require covalent grafting of the copolymer to the particles by reaction between PVA radicals and PS-PDMS during dispersion polymerization. Literature values for chain transfer constants for the reaction between PVA radicals and PS suggest that the grafting reaction is more likely than the chain transfer reaction between PVA radicals and PVA homopolymer which produces branched PVA at high conversions of monomer (16). In order to confirm the existence of covalent grafting of PS-PDMS to the particles, we have isolated block copolymer from a washed and dried sample of a PVA dispersed phase. Acetonitrile was used as solvent in a Soxhlet extraction for 60 h in order to remove PVA homopolymer. The residue was washed, dried and characterized by IR spectroscopy and GPC. The molecular weight distribution of the residue was broader than that of the original block copolymer, and the residue displayed an intense IR absorption at about 1730 cm^{-1} (17).

Conclusions

Our results demonstrate that PS-PDMS block copolymers stabilize both hard and soft polymer particles during nonaqueous dispersion polymerization. The dispersions have been characterized by electron microscopy, silicon analysis and viscometry. It was observed that the surface coverage and surface layer thickness of the stabilizing PDMS blocks were similar for PS and PMMA particles and that the PDMS chains are in a somewhat extended chain conformation. The anchoring of the PS blocks depends on the degree of compatibility between PS and the core polymer and on the incidence of radical grafting reactions during dispersion polymerization. From the results of the flocculation experiments, it is suggested that hard particles are stabilized because the PS blocks although incompatible with the PMMA chains are trapped within the PMMA matrix. The phase separation of the PS blocks into domains in the particles may be identified by small angle neutron scattering.

Acknowledgements

This work was supported by the Science Research Council and by an S.R.C. CASE award in collaboration with Dow Corning.

Literature Cited

1. Barrett, K.E.J., Ed. "Dispersion Polymerization in Organic Media"; Wiley: New York, 1975.
2. Dawkins, J.V.; Taylor, G. In "Polymer Colloids II"; Fitch, R.M., Ed.; Plenum: New York, 1980; p. 447.
3. Dawkins, J.V.; Taylor, G. Eur.Polym.J. 1979, 15, 453.
4. Dawkins, J.V.; Taylor, G. Polymer 1979, 20, 599.
5. Dawkins, J.V.; Taylor, G. Makromol. Chem. 1979, 180, 1737.
6. Dawkins, J.V.; Taylor, G. J.C.S. Faraday I 1980, 76, 1263.
7. Goodwin, J.W. In "Colloid Science"; Everett, D.H., Ed.; Specialist Periodical Report, The Chemical Society: London, 1975; Vol. 1, Chapter 7.
8. Higgins, J.S.; Dawkins, J.V.; Taylor, G. Polymer 1980, 21, 627.
9. Dawkins, J.V.; Taylor, G. Colloid Polym.Sci. 1980, 258, 79.
10. Napper, D.H. J.Colloid Interface Sci. 1977, 58, 390.
11. Everett, D.H.; Stageman, J.F. Colloid Polym.Sci. 1977, 255, 293.

12. Everett, D.H.; Stageman, J.F. Faraday Disc. Chem. Soc. 1978, 65, 230.
13. Dawkins, J.V.; Collett, R.W.R. unpublished work, 1979.
14. Dobry, A.; Boyer-Kawenoki, F. J.Polym.Sci. 1947, 2, 90.
15. Kern, R.J.; Slocombe, R.J. J.Polym.Sci. 1955, 15, 183.
16. Brandrup, J.; Immergut, E.H. Eds. "Polymer Handbook"; Wiley-Interscience: New York, 1976.
17. Dawkins, J.V.; Gatehouse, P. unpublished work, 1980.

RECEIVED April 6, 1981.

11

The Formation of Coagulum in Emulsion Polymerization

J. W. VANDERHOFF

Emulsion Polymers Institute and Department of Chemistry, Lehigh University, Bethlehem, PA 18015

In emulsion polymerization, a water-immiscible monomer is emulsified in water using an oil-in-water emulsifier and polymerized using a water-soluble or oil-soluble initiator. The product is a latex, i.e., a colloidal sol comprised of submicroscopic polymer spheres suspended in water. The emulsion polymerization may be carried out by: (i) batch polymerization in which all ingredients are added to the reactor and the mixture is heated under agitation to the polymerization temperature: (ii) semi-continuous polymerization in which the monomer as well as other ingredients are added continuously or in increments over the course of the polymerization to remove the heat of reaction which would otherwise exceed the cooling capacity of the reactor; (iii) continuous polymerization in which all ingredients are added continuously to a tubular reactor or a continuous stirred-tank reactor, either singly or in series, and partially or completely polymerized latex is removed continuously.

All three types of emulsion polymerization can be carried out using seeded emulsion polymerization, i.e., by adding monomer, initiator, and emulsifier to a previously-prepared small-particle-size latex, the particles of which grow in size without initiation of a new crop of particles. The purpose of seeded emulsion polymerization is to avoid the uncertainties of the particle initiation stage, obtain better batch-to-batch reproducibility, and give a stable latex of the desired particle size.

The usual description of these different polymerization processes suggests that all produce stable latexes and various hypotheses have been advanced to explain the stability of these latexes to such factors as added electrolyte, mechanical shear and freezing and thawing. In the literature, there is little mention of the fact that many of these polymerizations produce varying amounts of coagulum, i.e., polymer recovered in a form other than that of a stable latex. This coagulum is produced in all sizes of polymerization reactors, ranging from the smallest laboratory

0097-6156/81/0165-0199$05.00/0

reactor to the largest production reactor. It may be a mere nuisance in small laboratory polymerization reactors, but in large-scale polymerizations it may prevent the scale-up of a commercially-acceptable latex or exact a heavy economic penalty in longer cycle times and reduced yields.

The formation of coagulum is observed in all types of emulsion polymers: (i) synthetic rubber latexes such as butadiene-styrene, acrylonitrile-butadiene, and butadiene-styrene-vinyl pyridine copolymers as well as polybutadiene, polychloroprene, and polyisoprene; (ii) coatings latexes such as styrene-butadiene, acrylate ester, vinyl acetate, vinyl chloride, and ethylene copolymers; (iii) plastisol resins such as polyvinyl chloride; (iv) specialty latexes such as polyethylene, polytetrafluoroethylene, and other fluorinated polymers; (v) inverse latexes of polyacrylamide and other water-soluble polymers prepared by inverse emulsion polymerization. There are no major latex classes produced by emulsion polymerization that are completely free of coagulum formation during or after polymerization.

The coagulum formed during polymerization may take many forms and is commonly referred to by many names, often colloquial, e.g., reactor fouling, filterable solids, button, sediment, silt, grit, seeds, sand, waste, scrap, or worse. In this discussion, the term "coagulum" will be used to denote any polymer recovered in a form other than stable latex.

Formation of Coagulum

The coagulum formed in a latex can be divided into three main types: (i) coagulum formed during polymerization and recovered from the latex afterwards by filtration or sedimentation; (ii) coagulum deposited on the reactor surfaces during polymerization, on the walls, roof, bottom, agitator, baffle, thermowell, or cooling coil; (iii) coagulum formed in the latex after polymerization, during storage or transportation.

The coagulum formed during polymerization may range from a single lump of polymer with little fluid latex to tiny sand-like grains in an otherwise stable latex. It is unusual for a latex to coagulate completely during polymerization, to solidify in the reactor and cause the agitator to seize; nevertheless, it occurs often enough to warrant development of means to remove the solidified polymer. Usually, the latex coagulates only partially, to form lumps of coagulum which can be removed by filtration or sedimentation or which remain in the reactor after the latex is drained; the latex is usually relatively stable after removal of the coagulum. The lumps of coagulum filtered from the latex may be solid or porous. The solid lumps may be soft and sticky or hard, friable, and granular according to the polymer and its degree of plasticization with monomer. The porous lumps are usually relatively hard and friable. Occasionally, the coagulum is too fine in particle size to be removed easily by filtration or sedimentation; in this case, it remains dispersed in the latex, but manifests itself as surface roughness or haziness when a film is cast from the latex.

The coagulum deposited on the reactor surfaces may be lumps and accretions on the reactor walls or roof and on the agitator shaft and blades, baffles, thermowell, or cooling coil; it may be deposited on the reactor as a surface layer or "skin". The lumps and accretions on the various reactor surfaces may be sticky or hard and granular according to the polymer composition and its degree of plasticization with monomer. Similarly, the surface layer on the reactor surface may display the properties of the polymer if it is relatively thin and unplasticized; a thicker film plasticized with monomer may be soft and sticky. Often, coagulum may also be found on the bottom of the reactor where it is deposited when the latex is drained from the reactor. Generally, the coagulum left in the reactor resembles the lumps of coagulum filtered from the latex; however, in some cases, it is different, suggesting that it was formed by a different mechanism.

The coagulum formed in the latex after polymerization during storage or transfer may result from flocculation of the latex by mechanical shear, e.g., by pumping the latex from the reactor to the storage tank and from the storage tank to the tank car or truck, transportation to the customer's plant, and pumping from the tank car or truck into the customer's storage tank. Coagulum may also be formed by temperature variations within the latex during storage or transportation. Of course, freezing often causes flocculation of latexes; however, the stability of latexes varies with temperature above the freezing point, and a latex that is stable at room temperature may slowly flocculate when the temperature is raised or lowered sufficiently. This slow flocculation may not be noticeable over a short time, but may result in the formation of significant coagulum over a longer time. Indeed accelerated storage stability tests, e.g., heating the latex at 50°, are often used in industry to predict storage stability at ambient conditions. Also, some latexes may contain microscopic polymer particles that slowly settle or cream during storage. The critical particle size for settling or creaming can be estimated using Overbeek's criterion (1), i.e., a particle that settles or creams at a rate of only 1 mm in 24 hours according to Stokes law will never settle or cream in practice because of the Brownian motion of the particles and the chance thermal convection within the sample. For example, for polystyrene particles (density = 1.050 gm/cm^3) the critical size for settling in water is 0.65 μm, which has been confirmed by many observations of the storage stability of monodisperse polystyrene latexes. The settling or creaming of these microscopic particles may form a dense layer which will not pass through a filter and thus appear as coagulum.

The formation of coagulum during or after polymerization presents serious problems in industrial latex production for several reasons: (i) it decreases the yield of the polymer by the amount of coagulum formed; (ii) it increases the "down time" of the reactor, thus lengthening the cycle time unpredictably; (iii) it necessitates cleaning the reactor between successive polymerization batches; (iv) it increases the batch-to-batch variation in latex quality and properties; (v) the coagulum must be disposed of usually by incineration or burial in sanitary landfill; (vi) the swelling of coagulum with monomer may pose a health hazard, particularly for monomers that have been shown, or are suspected, to be toxic, e.g., vinyl chloride and acrylonitrile. These disadvantages emphasize the importance of determining the mechanism of coagulum formation and developing methods to reduce or eliminate it.

Mechanism of Formation of Coagulum

Two mechanisms proposed for the formation of coagulum are: (i) a failure of the colloidal stability of the latex during or after the polymerization, to cause flocculation of the particles and eventually form microscopic and macroscopic coagulum; (ii) polymerization of the monomer by a mechanism other than that of emulsion polymerization, to give polymer of different form than latex particles.

Latex Stability Failure

The failure of latex stability, and the resultant flocculation of the latex particles, may cause the formation of coagulum that is recovered from the latex after polymerization as well as a buildup on the reactor surfaces. Moreover, the inherent instability of the latex may also cause flocculation during storage or transportation.

The experimental factors that affect the stability of the latex during and after polymerization are the recipe used for the polymerization, the type and intensity of agitation during and after the polymerization, the temperature of polymerization and storage, and the age and storage conditions of the latex. The recipe used in the polymerization included the mode of polymerization, the monomer-water ratio, the solubility of the monomer in water, the emulsifier type and concentration, initiator type and concentration, total electrolyte concentration, and impurities present in the system.

The stability of latexes during and after polymerization may be assessed at least qualitatively by the theoretical relationships describing the stability of lyophobic colloids. The Verwey-Overbeek theory (2) combines the electrostatic forces of repulsion between colloidal particles with the London-van der Waals forces of attraction. The electrostatic forces of repulsion arise from the surface charge, e.g., from adsorbed emulsifier ions, surface sulfate endgroups introduced by persulfate initiator, or ionic groups introduced by using functional monomers. These electrostatic repulsion forces are greatest closest to the particle surface and decrease exponentially with increasing distance from the particle surface. The rate of exponential decrease is dependent upon the electrolyte concentration: the higher the concentration and the higher the valence of the counterions, the more rapid the decrease. The London-van der Waals forces of attraction arise from the difference in dielectric constant between the particles and the medium. These attractive forces are greatest closest to the particle surface and decrease exponentially with increasing distance from the particle surface. Unlike the electrostatic repulsion forces, the London-van der Waals forces are only slightly affected by electrolyte concentration and other parameters of the system. The net overall interaction between colloidal particles is described by the sum of the potential energies of repulsion and attraction. The usual result for stable colloidal sols is a deep potential energy well of attraction at very small interparticle distances, with the potential energy of interaction rising sharply to a repulsion peak at slightly greater distances, followed by a slow decrease to zero with increasing interparticle distance. For relatively large particles, the slow decrease from the repulsion peak may give a shallow secondary minimum of attraction at relatively large interparticle distances. Thus the overall interaction will result in flocculation if the energy of collision is great enough to surmount the repulsion peak and the interparticle distance decreases to that corresponding to the deep primary attraction potential energy well. Once flocculated in the primary potential energy well, the particles are difficult to separate and will remain in the flocculated state. An increase in electrolyte concentration or valence of the counterion lowers the repulsion peak and thus favors flocculation. Similarly, particles flocculated in the secondary attraction minimum may be redispersed without difficulty.

In addition to electrostatic repulsion and London-van der Waals attraction, steric stabilization (3) must be taken into account. Steric stabilization results from the adsorption of an uncharged hydrated species on the particle surface, which hinders the approach of other particles and thus acts as a mechanical barrier to flocculation. Steric stabilizers include such uncharged water-soluble polymers as methylcellulose or nonionic emulsifiers such as nonylphenol-ethylene oxide adducts. The potential energy of interaction between sterically-stabilized colloidal particles is repulsion at very small interparticle distances, followed by a rapid decrease to a shallow attraction well at somewhat larger distances, an increase to a relatively low repulsion peak at slightly greater distances, and a slow decrease to zero with increasing distance. Thus sterically-stabilized particles tend to flocculate in the shallow potential energy well and are relatively easy to redisperse.

It should be noted, however, that the same uncharged polymer molecules that give rise to steric stabilization also may cause flocculation of colloidal particles by "bridging" (4). Steric stabilization results from the adsorption of an uncharged hydrated polymer molecule on a single particle, perhaps with loops of the polymer chain extending into the aqueous phase. Flocculation may result from the adsorption of the same molecule on more than one particle to form a "bridge" between them; a sufficient number of such bridges would cause complete flocculation of the latex. The result of addition of such polymers ---- steric stabilization or flocculation ---- is determined by the specific system, the relative concentration of polymer and particles, and the method of mixing the polymer with the latex. Such flocculation is also observed with polymer moleclues containing ionic groups. It should be emphasized that the use of functional monomers such as acrylic acid or 2-sulfoethyl methacrylate may form such polymers in the aqueous serum of the latex.

The flocculation of colloidal particles can be divided into two main types: (i) diffusion-controlled flocculation; (ii) agitation-induced flocculation. The diffusion-controlled flocculation has been analyzed theoretically by von Smoluchowski (5). Each particle is considered as a center to which other particles diffuse by Browian motion. Thus the rate of flocculation is proportional to the square of the number of particles. The original treatment assumed that all interparticle collisions were effective in causing flocculation. Later modifications (6) assumed that the potential energy barrier between particles resists flocculation and that only those collisions with sufficient energy to overcome this barrier will cause flocculation. The agitation-induced flocculation has also been analyzed theoretically (7). In this case, the rate of flocculation depends upon the collision radii of the particles and the rate of shear. A combination of the two theories (8) shows that diffusion-controlled flocculation is predominant at particle sizes of ca. 0.1μm, but agitation-induced flocculation becomes equivalent at sizes of ca. 1μm and becomes predominant at larger particle sizes. This finding applied to latexes suggests that the initial stages of flocculation are diffusion-controlled, but as the aggregates grow in size, they eventually reach the size where agitation-induced flocculation becomes competitive, and grow much more rapidly when the agitation-induced flocculation becomes predominant. Thus an inherent instability of a latex during polymerization may result in an autoaccelerating flocculation of the particles as the polymerization proceeds.

The formation of coagulum can often be correlated with: (i) conversion; (ii) agitation rate; (iii) mode of addition of ingredients. Generally, the amount of coagulum formed increases with increasing conversion, but this observation may be the result of an initial stability failure which becomes more obvious as the polymerization proceeds. Since latexes can be flocculated by mechanical shear, the agitation rate used must be sufficient to give good mixing and heat transfer, yet not so great as to cause flocculation. Generally, for batch polymerizations, the agitation rate should be vigorous to form the monomer emulsion, moderate from the initiation of polymerization to the disappearance of the monomer droplets, and mild (only sufficient to maintain good heat transfer and mixing) thereafter. The mode of addition of ingredients also affects the formation of coagulum, e.g., continuous addition of monomer often reduces the amount of coagulum compared to that formed in batch polymerization.

Different Mechanism of Polymerization

Polymerization of monomer by a different mechanism also causes the formation of coagulum. Coagulum found in the latex may result from polymerization in large monomer droplets or a separate layer of monomer. In most polymerizations, monomer emulsion droplets are 1-10μm in diameter. The chance entry of a radical into these droplets causes their polymerization to form microscopic (as opposed to submicroscopic) particles; however, although these microscopic particles are less stable than the submicroscopic particles, the presence of only a few such particles is usually not sufficient to result in the formation of coagulum. In large reactors, however, there may exist large monomer drops as well as a separate layer of monomer, particularly if the agitation is not sufficient to mix all parts of the system equally well. In this case, entry of radicals into these large drops and the separate layer will cause a bulk polymerization that will certainly result in the formation of coagulum.

The coagulum deposited on the reactor surfaces may be the result of polymerization in large monomer drops or a separate monomer layer, or it may be the result of polymerization of the monomer in the vapor space above the latex or a surface polymerization on the walls and roof of the reactor. Polymerization in the vapor space of the reactor will form solid polymer in the form of particles which may stick to the reactor surfaces or fall into the latex; in the later case, these particles serve as nuclei for the formation of coagulum. Polymerization of monomer on the reactor surfaces will form solid particles that become swollen with monomer and grow by flocculation of the latex particles. The surface polymerization can be related to the smoothness of the reactor surface; the smoother the surface, the lesser the tendency for surface polymerization and formation of coagulum. In this respect, glass-lined reactors are superior to stainless steel reactors, although highly polished stainless steel surfaces are often satisfactory. Spots where the glass lining has been fractured and is missing, or scratches in the stainless steel walls, serve as nuclei for the formation of coagulum. The important parameter is the wetting of the reactor surface by the monomer-polymer phase; such wetting is facilitated by surface roughness or discontinuities and inhibited by a perfectly smooth surface.

Elimination of Coagulum

There are no general methods or approaches to eliminate the formation of coagulum in emulsion polymerizations. Each polymerization system must be considered separately, and although remedies worked out in one system may be useful in another system, these differences are such that there are only a few generalizations that can be applied. Generally, the coagulum can be reduced or eliminated by modification of the polymerization recipe and technique or by modification of the reactor. There are several approaches to reduce or eliminate coagulum by modification of the polymerization recipe and technique: (i) the substitution of a seeded system for a unseeded system often eliminates the formation of a coagulum and decreases the batch-to-batch variations in product properties; (ii) addition of stabilizing emulsifier at the appropriate conversion may stabilize particles that would otherwise flocculate; (iii) rigorous temperature control often decreases the amount of coagulum formed whereas allowing the temperature to rise slightly during the rapid exothermic polymerization gives rise to formation of coagulum; (iv) varying the mode of monomer addition may eliminate the formation of coagulum, e.g., adding the monomer continuously instead of at the beginning of the reaction; (v) variation of the agitation rate, usually a decrease at each stage of the reaction to the level just sufficient to maintain good transfer and mixing; (vi) a better understanding of the polymerization system may suggest alterations of the recipe that will be effective.

There are several approaches to reduce or eliminate the formation of coagulum by modification of the reactor design: (i) substitution of a semi-continuous process for a batch process and a continuous process for a semi-continuous process; (ii) substitution of a different reactor configuration, e.g., a tumbler type may give less coagulum than a stirred tank reactor and a tube reactor used for continuous polymerization may give less coagulum than a batch reactor; (iii) modification of the agitator and baffle system to ensure uniform agitation throughout the reactor and complete but mild mixing of the ingredients; (iv) better temperature control; (v) different mode of addition of ingredients, e.g., addition of monomer below the surface of the latex rather than by dropping it through the vapor space to the upper surface of the latex.

Despite these generalizations, the reduction or elimination of coagulum is usually best accomplished by a "systems approach", i.e., a consideration of latex properties to be achieved in the emulsion polymerization, the economics of the polymerization process, and the deliberate design of the reactor system for that particular polymerization system. Each polymerization system must be considered as a separate system and treated as such. The most effective approach to reduce or eliminate the formation of coagulum is to determine the mechanism by which it is formed and the approximate stage in the conversion at which it is formed. Once this is known, the remedies usually suggest themselves. Without this knowledge, the efforts are limited to cut-and-try methods without basis in mechanism or logic.

Abstract

Coagulum is formed in many emulsion polymerizations, from the smallest laboratory size to the largest production reactors. It is observed in many forms, from a single lump of polymer with little or no fluid latex to tiny sand-like grains suspended in an otherwise stable latex. Usually, it is found as lumps in the latex or deposited on the reactor surfaces. The type and amount of coagulum formed depends upon the polymer system and the polymerization recipe and technique. Two mechanisms are proposed for the formation of coagulum: (i) a failure of the stability of the latex, giving rise to flocculation and growth of the aggregates to macroscopic size; (ii) a different mechanism of polymerization, e.g., polymerization in large monomer drops or a separate monomer layer in the vapor space above the latex and on the reactor surfaces. The factors affecting latex stability during polymerization (electrostatic repulsion, London-van der Waals attraction, steric stabilization, flocculation by "bridging") as well as the possibility of polymerization by mechanisms other than emulsion polymerizations are discussed in terms of the parameters of polymerization. Generalizations on the formation of coagulum as a function of the parameters of the polymerization are reviewed, and recommendations are made to reduce or eliminate coagulum in specific cases. Generally, however, the reduction or elimination of coagulum depends upon the examination of the particular system and the determination of the mechanism and the time of coagulum formation, which determines the appropriate remedy.

Literature Cited

1. Overbeek, J. Th. G., "Colloid Science", Vol. 1, H. R. Kruyt, editor, Elsevier, New York, 1952, p. 80.
2. Verwey, E. J. W., and Overbeek, J. Th. G., "Theory of Stability of Lypohobic Colloids", Elsevier, New York, 1948.
3. Heller, W., and Pugh, T. L., J. Chem. Phys. 1954, 22, 1778; Ottewill, R. H., "Nonionic Surfactants", M. Schick, editor, Marcel Dekker, New York, 1967.
4. Linke, W. F., and Booth, R. B., Trans. Am. Inst. (Metall.) Engrs. 1959, 217, 364; Michaels, A. S., and Morelos, O., Ind. Eng. Chem. 1955, 47, 1801; Healy, T. W., and La Mer, V. K., J. Phys. Chem., 1962, 66, 1835; see also Audsley, A., Mineral Processing Information Note No. 5 "Flocculation of Suspensions of Solids with Organic Polymers --- A Literature Survey", Warren Spring Laboratory, Herts, England, 1965.
5. von Smoluchowski, W., Z. Physik, 1916, 17, 557, 585; ibid. Z. Physik, 1917, Chem. 92, 129.

6. Fuchs, N., Z. Physik, 1934, 89, 736.
7. von Smoluchowski, M., Z. Physik, 1917, Chem. 92, 155; Tuorila, P., Kolloidchem. Beihefte, 1927, 24, 1; Mueller, H., Kolloidchem. Beihefte, 1928, 27, 223
8. Overbeek, J. Th. G., "Colloid Science", Vol. 1, H. R. Kruyt, editor, Elsevier, New York, 1952, P. 290.

RECEIVED April 6, 1981.

12

Effect of Spray Drying Parameters on Polyvinyl Chloride Resin Particle Size Distribution and Plastisol Rheology

D. J. HOFFMANN and P. M. SAFFRON

Diamond Shamrock Corporation, T. R. Evans Research Center,
Painesville, OH 44077

PVC plastisols are dispersions of very small PVC particles in a plasticizer. These very small particles are made by spray drying PVC latex. The latex particles range in size from 0.1 to 2.0 micrometers and the distribution is extended to around 80 micrometers due to the agglomerates formed in the spray drying process. The spray dried agglomerated particles can make up a large percentage of the total resin particle size distribution. The agglomerates can have various degrees of fusion depending on the drying parameters such as temperature, latex solids concentration and atomized droplet size.

The viscosity of a plastisol is primarily determined by the particle size (PS) and particle size distribution (PSD) of the resin[1,2]. Studies have shown that agglomerates play an important role in the plastisol viscosity and viscosity aging[3,9] Although it is recognized that the drying operation is important in the determination of the resin PS and PSD, its effects are not clearly understood. We therefore undertook a study to determine the effect of drying parameters on resin PSD and subsequent plastisol viscosity and viscosity aging. A single latex was used to eleviate the confounding effects of latex parameters on spray drying. The area of latex parameter/spray drying effects will be the topic of a future paper.

Experimental

Latex Preparation & Analysis. Deionized water was used for the aqueous phase in the polymerization and commercial grade vinyl chloride monomer (99.9% pure) was used without further purification. The emulsifier was a salt of a sulfated fatty alcohol, and the emulsifier solution for metering was prepared to have 0.15 mol/L in water. The initiator system was a water soluble redox system.

The polymerization was conducted in a 50-gal. pilot plant reactor using the charge procedure described by Jones and Schaefer.[4] The polymerization was seeded with a PVC seed latex produced in a separate polymerization. During the

0097-6156/81/0165-0209$05.00/0

polymerization, initiator and emulsifier were metered into the reactor. At the end of the polymerization (noted by a pressure drop of 45 psig), the remaining monomer was stripped off. The stripped latex contained 40.6% solids and less than 2000 ppm residual vinyl chloride monomer.

An ICI-Joyce Loebl Disc Centrifuge MK III, a photosedimentometer, was used to measure the latex particle size distribution. The latex had a unimodal particle size distribution with a diameter of 1.05 micrometers (surface area average). The methods of separating latex particles by a centrifugal field and detecting the size distribution by a photocell may be found in the literature.(5,6,7)

Latex Drying. The stripped latex was dried in a pilot plant NICHOLS/NIRO spray drier using centrifugal disc atomization. Different disc speeds were obtained by changing the atomizer drive pulleys. The disc speeds could be varied between 10,000-24,000 rev/min, (tip speeds 62-150 m/sec) and were measured with a tachometer. The dryer has a 1.22 m inside diameter and is heated by a gas-fired heater. For a given dryer inlet temperature, the dryer outlet temperature is controlled by varying the feed rate to the atomizer, e.g. increasing the feed rate lowers the outlet temperature. The dried resin was collected in a cyclone.

Particle Size Analysis. Particle size distribution measurements of the dry resin were made using a Micromeritics Sedigraph 5000D particle size analyzer. This instrument employs a gravity settling technique and uses Stokes law to determine the particle size. The PSD was run over the size range of 0.36 to 100 micrometers. All measurements were made in dibutyl sebacate as the suspending medium.

Viscosity Measurements. Plastisols were prepared by mixing 60 parts plasticizer, dioctyl phthalate (DOP), with 100 parts resin using a laboratory scale paddle type mixer. The viscosity measurements were conducted on a Rheometrics Conometer equipped with a high sensitivity torque cell (10^{-6} to 10^{-9} Kg-m). Aging behavior was followed by running viscosity analyses after 2 hours, 24 hours, and 7 days. The reproducibility of the viscosity analysis was determined to be very good. Viscosity data error is less than 5% for reanalysis of a given mix. Remixing and analysis of a given sample gives an error of less than 10%.

ϕ_{max} Determinations. The maximum volume fraction of resin that forms a flowable system was determined by the following procedure:

1. A mix was prepared using resin and 37% by volume DOP.
2. Additional plasticizer was added and hand stirred until a smooth paste resulted. (In general, the pastes were highly dilatent. The addition of plasticizer was halted when a shiny texture was noticed.)

3. $\phi max = (WT_r/\rho_r)/(WT_r \rho_r + WT_p \rho_p)$

WT_r = WT. of Resin

ρ_r = Density of Resin

WT_p = Wt. of Plasticizer

ρ_p = Density of Plasticizer

The endpoint determination, while subjective was found to give values of ϕmax to within ±.02. This procedure was taken from the thesis work of G. Williams at Lehigh University.(11)

Microscopy. Scanning electron microscopy was run on resin samples using a AMR 1200 Scanning Electron Microscope. The samples were mounted on the stub using double stick tape and then sputter coated with gold.

Optical microscopy of resin and plastisols were conducted using transmitted light on a Zeiss Optical Microscope.

Results

Particle Size Analysis. The PSD for each unground resin in a plastisol formulation (using dibutyl sebacate as the plasticizer) was determined with the following outcome:

1. Particle size of the primary emulsion particles was unaffected by the spray drying operation.
2. Agglomeration increases with increasing outlet temperature, (Figure I).
3. Agglomeration increases with increasing atomizer speed, (Figure II).
4. Agglomeration increases with increasing inlet temperature, (Figure III).

The effects of dryer temperatures and atomizer speeds on the percent of agglomeration present in the plastisols are also shown in Table I.

Table I

EFFECT OF DRYER CONDITIONS ON RESIN AGGLOMERATION (%)*

DRYER	ATOMIZER TIP SPEED (M/SEC)			
INLET/OUTLET (°C)	79	98	118	138
150/60	8	17	25	35
150/75	35	45	60	70
180/60	18	20	30	35
180/75	35	50	60	70

* Agglomerates are defined as particles 4 µm.

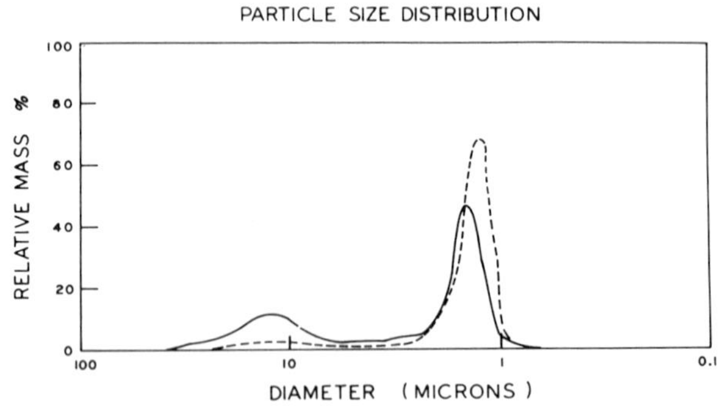

Figure 1. Variation in particle size with dryer outlet temperature (dryer conditions—inlet/outlet: (——) 150°/75°C; (– – –) 150°/60°C; atomizer tip speed: 79 m/s)

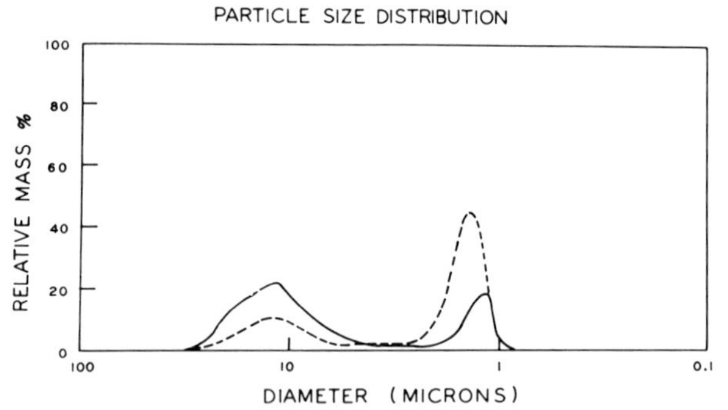

Figure 2. Variation in particle size with dryer atomizer speed (dryer conditions—(– – –) 79 m/s; (——) 138 m/s; inlet/outlet: 150°/75°C)

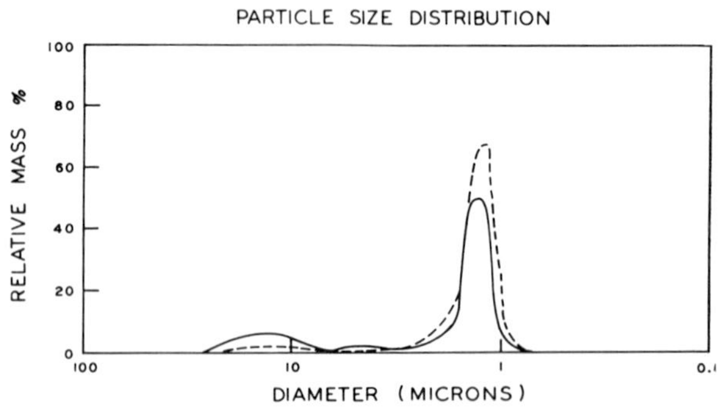

Figure 3. Variation in particle size with dryer inlet temperature (dryer conditions—inlet/outlet: (——) 180°/60°C; (– – –) 150°/60°C; atomizer tip speed: 79 m/s)

Effect of Outlet Temperature. Low (60°C) and high (75°C) dryer outlet temperatures were studied. The results are summarized in Table I and are depicted in Figure I. They indicate that resin agglomeration present in the plastisols increases with increasing outlet temperature (decreased latex feed rate to the atomizer). Increasing the outlet temperature causes two changes, both of which should increase the degree of agglomerate fusion. First, decreasing feed rates to centrifugal disc atomizers produce smaller spray droplets,(8) which dry faster and should be more highly fused because of their longer heat history. Second, the higher dryer temperature should more thoroughly fuse the agglomerates. The highly fused agglomerates do not readily break apart during plastisol mixing and result in more agglomerates in the plastisols (Figure IV, Case I). By the same reasoning, low dryer outlet temperature obtained by increasing the feed rate to the atomizer should produce larger, loosely fused agglomerates in the dried resin. These would be expected to fall apart more easily during plastisol mixing and should produce few agglomerates in the plastisol (Figure IV, Case II).

Effect of Atomizer Speed. Four different atomizer tip speeds, 79, 98, 118, and 138 M/sec., were studied. Increasing the atomizer speed increased the agglomeration present in the plastisols (Table I and Figure II). Increasing atomizer speed is known to produce smaller spray droplets.(8) This should result in more highly fused resin agglomerates, (Figure IV, Case I).

Effects of Dryer Inlet Temperature. Low (150°C) and high (180°C) dryer inlet temperatures were studies. Both Figure III and the results in Table I indicate that the amount of agglomerates present in the plastisols increases with increasing inlet temperature. The droplets should be more highly fused because of the high inlet temperature, and therefore, will produce more agglomerates in the plastisols (Figure IV, Case III).

Viscosity Analysis. Rheological analyses of the unground spray dried resins in dioctyl phthalate (DOP) plastisols gave the viscosity flow curves in Figures V-VIII. Characteristic data are presented in Tables II and III.

From the viscosity data, two trends are evident:

1. Viscosity increases with increasing outlet temperature.
2. Viscosity decreases with increasing inlet temperature.

There does not seem to be a clear cut dependence of viscosity on atomizer speed.

Analysis of the viscosity aging data, Table III, reveals one significant trend. In general, as dryer temperatures are increased, plastisol viscosity aging decreases, Figure IX. The

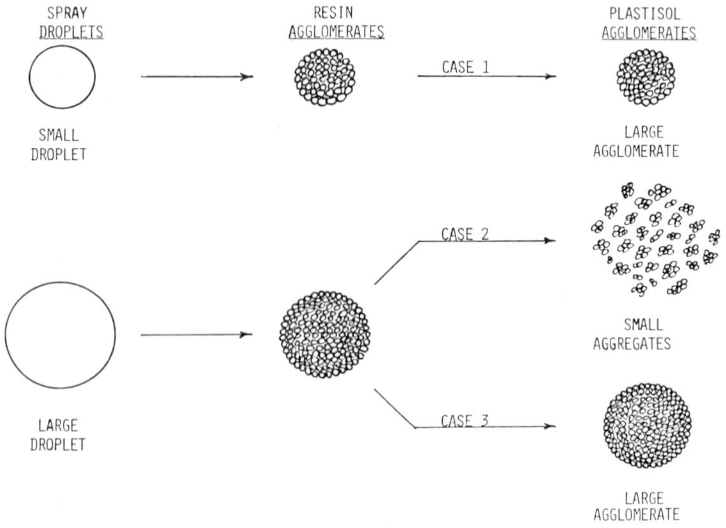

Figure 4. Relationship between spray droplet size and the thoroughness of fusion on agglomerates

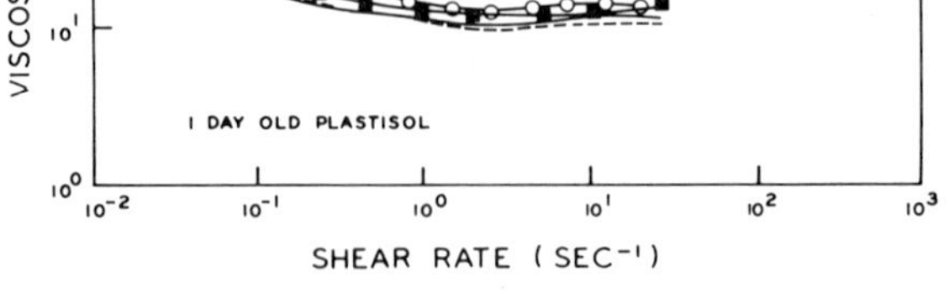

Figure 5. Effect of atomizer speed at low dryer inlet and outlet temperatures on plastisol viscosity (dryer conditions—inlet/outlet: 150°/60°C; atomizer tip speed: (-○-) 79 m/s; (——) 98 m/s; (- - -) 118 m/s; (-■-) 138 m/s)

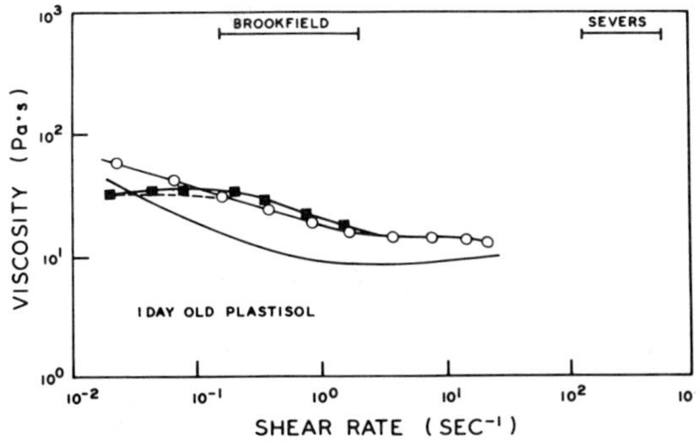

Figure 6. Effect of atomizer speed at low inlet and high outlet dryer temperatures on plastisol viscosity (dryer conditions—inlet/outlet: 150°/75°C; atomizer tip speed: (–○–) 79 m/s; (——) 98 m/s; (– – –) 118 m/s; (–■–) 138 m/s)

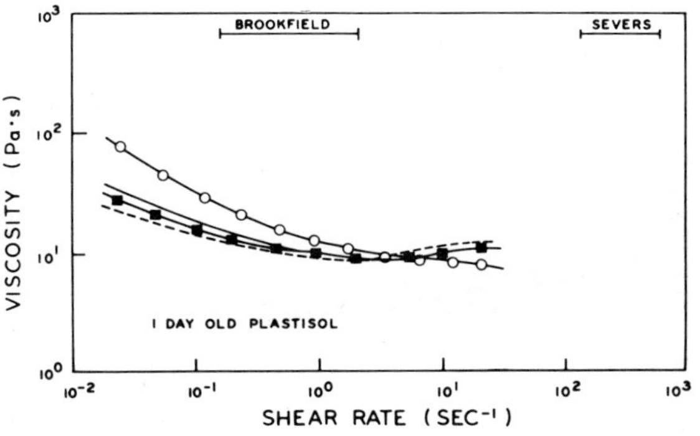

Figure 7. Effect of atomizer speed at high inlet and low outlet temperatures on plastisol viscosity (dryer conditions—inlet/outlet: 180°/60°C; atomizer tip speed: (–○–) 79 m/s; (——) 98 m/s; (– – –) 118 m/s; (–■–) 138 m/s)

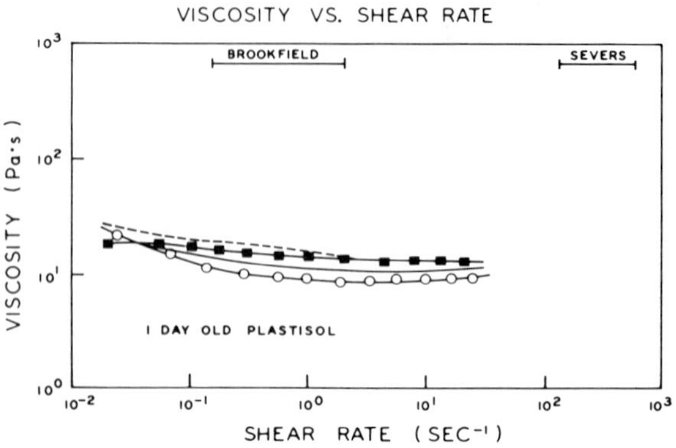

Figure 8. Effect of atomizer speed at high inlet and outlet temperatures on plastisol viscosity (dryer conditions—inlet/outlet: 180°/75°C; atomizer tip speed: (–○–) 79 m/s; (——) 98 m/s; (– – –) 118 m/s; (–■–) 138 m/s)

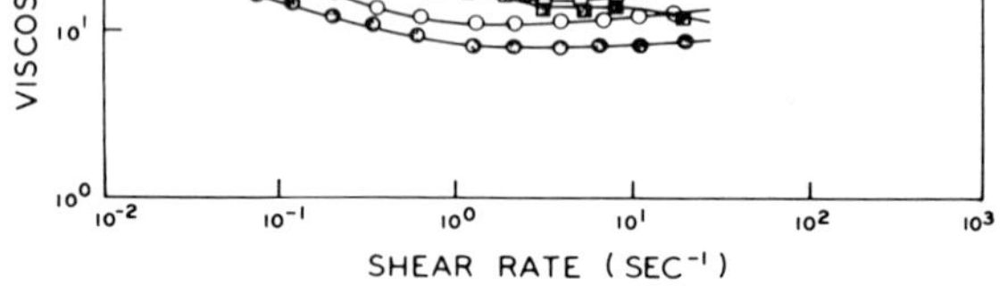

Figure 9. Influence of dryer outlet temperature on viscosity aging (dryer conditions—inlet/outlet: (●) 2 h, 180°/60°C; (○) 7 day, 180°/60°C; (■) 2 h, 180°/75°C; (□) 7 day, 180°/75°C; atomizer tip speed: 138 m/s)

dryer outlet temperature is seen to have a much greater effect on aging than inlet temperatures.

Table II

EFFECTS OF DRYER CONDITIONS ON VISCOSITY (Pa-S)*

DRYER INLET/OUTLET (°C)	ATOMIZER TIP SPEED (M/SEC) 79	98	118	138
150/60	14	11	11	12
150/75	18	22	19	21
180/60	12	9	8	10
180/75	9	11	17	15

* Viscosity data was taken at 1 sec^{-1} for day old plastisols.

Table III

EFFECTS OF DRYER CONDITIONS ON VISCOSITY AGING*

DRYER INLET/OUTLET (°C)	ATOMIZER TIP SPEED (M/SEC) 79	98	118	138
150/60	2.27	1.5	1.4	1.7
150/75	2.7	1.3	1.3	1.0
180/60	1.5	1.4	1.1	1.5
180/75	0.8	1.3	0.0	0.90

* Viscosity aging = $\frac{\eta\ (7\ day)}{\eta\ (2\ hr)}$ @ 0.1 $sec.^{-1}$

The higher dryer temperatures are felt to decrease aging due to the latex droplets being subjected to greater temperatures. This results in more densely packed, more fused agglomerates. (3) The densely packed, fused agglomerates will be less susceptible to deagglomeration in the plastisol. The higher dryer conditions also increases the amount of agglomerates which will decrease the overall exposed surface area. These two mechanisms have been shown to improve viscosity aging by minimizing the surface area available for interaction between resin and plasticizer.(9)

There also is a slight trend of decreasing aging with increasing atomizer tip speed. This again is felt to be due to the latex droplets receiving a longer heat history. Smaller latex droplets from higher tip speeds will dry faster, therefore increasing the heat history of the droplet. This results in more highly fused agglomerates. The optimum conditions for minimizing viscosity aging are at maximum atomizer speed and dryer temperatures. This, however, does not result in the lowest overall viscosity.

Microscopic Analysis. SEM Microscopy of the unground resin revealed several things. First, it confirmed the essentially unimodal nature of the resin primary particles, Figure X. Second, it showed the interior of the agglomerates at all the drying conditions to be very loosely packed and highly porous, Figure XI. Visually there was no obvious difference in degree of fusion of the agglomerates as dryer conditions were altered.

Optical microscopy analysis of the dried resins confirmed the relative extent of agglomeration of the resins in a plastisol, Figure XII. In addition, it again revealed the highly porous nature of the agglomerates. Adsorption of plasticizer into the internal pores takes place very rapidly, 1 sec, and even with high viscosity fluid (900 Pa-S, total adsorption is accomplished in 40 seconds, Figure XIII.

ϕmax Analysis. ϕmax is an experimentally determined parameter that represents the maximum volume fraction of resin that will still form a flowable system. This is the point where the system has gone through a phase inversion from a liquid filled porous solid to a solid filled liquid. The pores are due to the inter- and intra- particle voids. ϕmax gives an indication of particle/particle interactions where the interactions increase with decreasing ϕmax. This increase in interactions should cause a subsequent increase in viscosity of the plastisol.

The result of the ϕmax determinations are presented in Table IV.

Table IV

EFFECT OF DRYER CONDITIONS ON ØMAX

DRYER	ATOMIZER TIP SPEED (M/SEC)			
INLET/OUTLET (°C)	79	98	118	138
150/60	0.63	0.61	0.61	0.62
150/75	0.59	0.56	0.57	0.52
180/60	0.63	0.60	0.58	0.61
180/75	0.59	0.56	0.56	0.60

This data coupled with the viscosity and microscopic analysis illustrates a factor important to plastisol viscosity. For example, samples from dryer conditions 150/75 °C and 180/75 °C @ 138 M/sec tip speed have essentially identical particle size distributions, Figure XIV, but have different viscosities, Figure XV and Table II. Ømax for the two samples indicate a difference

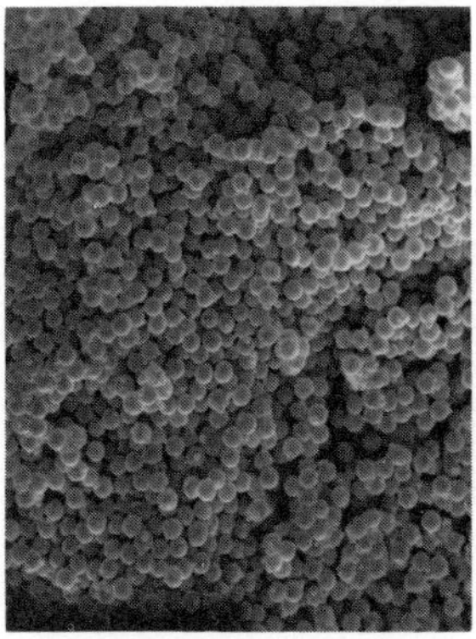

Figure 10. Microphotograph of resin showing essentially unimodal nature of latex particles

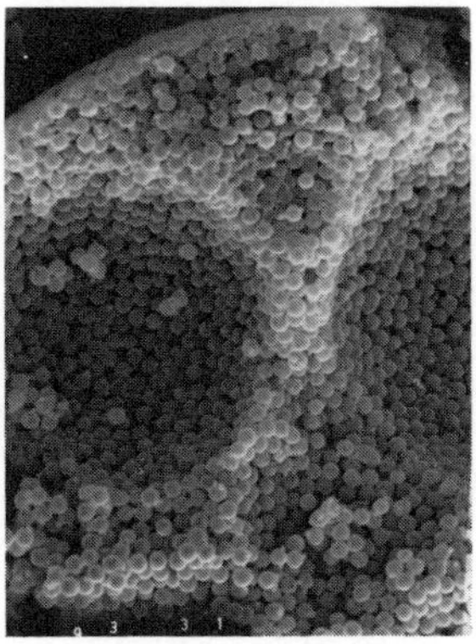

Figure 11. Microphotograph of a porous resin agglomerate

A

B

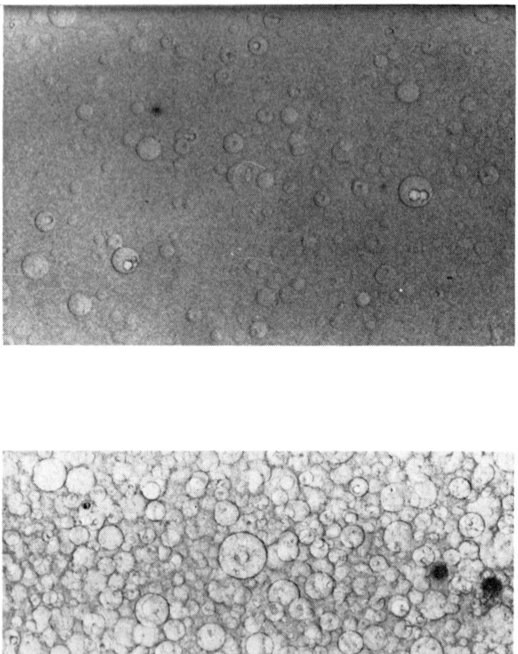

Figure 12. Microphotograph of agglomerates showing relative amount of agglomerates in plastisols (dryer conditions: (A) 150°/60°C, 79 m/s, (B) 180°/75°C, 138 m/s)

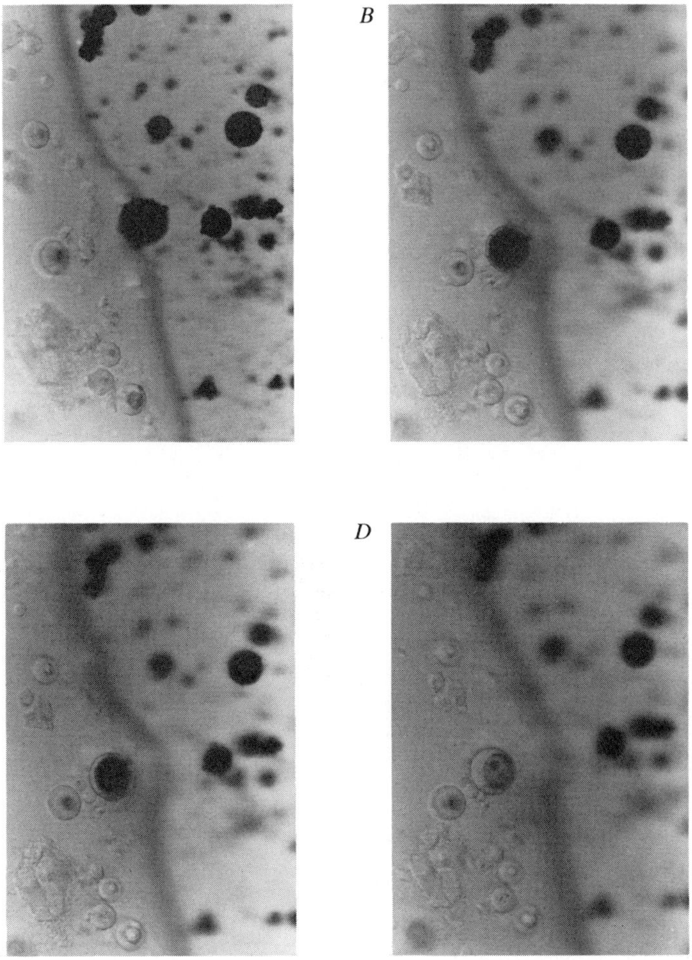

Figure 13. Absorption of high viscosity (900 Pa-S) silicone oil by porous agglomerates

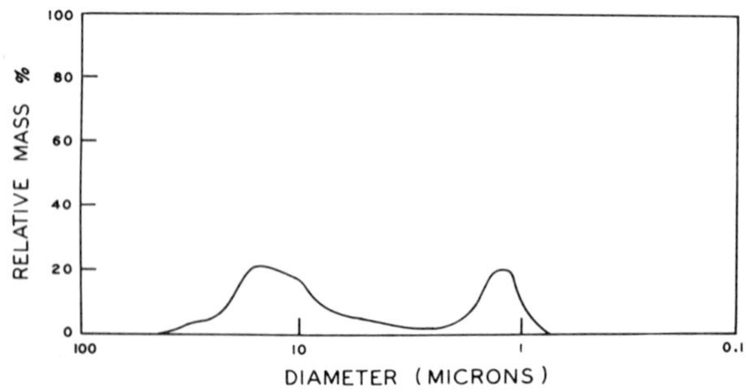

Figure 14. Particle size distribution of resins produced at two different dryer conditions (dryer conditions—inlet/outlet: 150°/75°C and 180°/75°C; atomizer tip speed: 138 m/s)

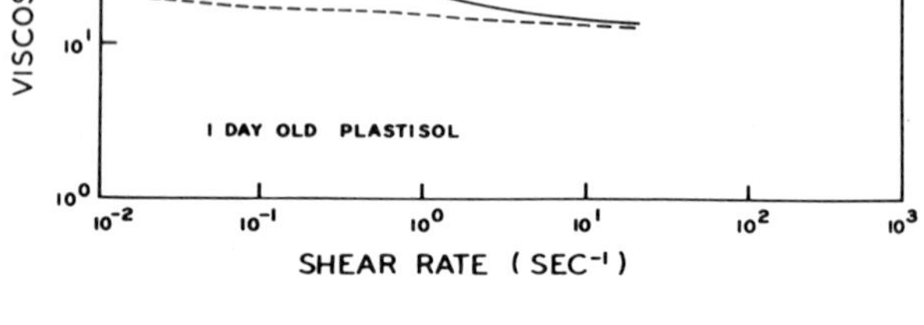

Figure 15. Illustration of different plastisol viscosities produced by resins having the same particle size distribution (see Figure 14) (dryer conditions—inlet/outlet: (——) 150°/75°C; (– – –) 180°/75°C; atomizer tip speed: 138 m/s)

where sample 180/75 °C has a lower Ømax, Table IV. A lower Ømax implies more plasticizer is required to fill voids. This effectively reduces the amount of plasticizer available for particle lubrication. This difference can account for the different viscosity behavior. The difference in Ømax is felt to be due to differences in the agglomerate internal porosity. Mukhina[3] reported a similar effect on agglomerate packing and fusion due to increased dryer outlet temperatures.

Ømax is seen to follow one major trend with respect to dryer conditions; Ømax decreases with increasing dryer outlet temperature. This trend substantiates Mukhina's work that increasing outlet temperatures results in more densely packed agglomerates.

Discussion

The spray drying process for emulsion PVC has been shown to be very important to plastisol viscosity and viscosity aging stability, both here and in the literature.[3] Mukhina et al reported that as the dryer outlet temperature is increased, the agglomerates formed become more densely packed, and more highly fused resulting in plastisols with better viscosity aging stability. The observations in this study are in agreement with Mukhina's results, Figure IX.

Viscosity aging behavior has been shown to be dependent on exposed resin surface area and agglomerate fusion.[3,9] Higher dryer temperatures will decrease the former and increase the latter, resulting in better viscosity aging stability.

Initial viscosities seem to be dependent on several resin variables within the plastisol. Resin particle size distribution is important to the viscosity in that decreased interparticle void volume by improved packing will decrease plastisol viscosities.[10,11,12,13] The intraparticle void volume within agglomerates will also affect the viscosity. Plasticizer which is absorbed by the agglomerates can not lubricate the particles, so it has been essentially removed from the continous phase. This effectively increases the solids volume loading resulting in higher viscosities. The combination of both inter- and intraparticle voids were determined by Ømax analysis. (While Ømax generally related to the plastisol viscosity, certain anomalies were noted. It is unclear what to attribute these to at this time. Differences between Ømax and plastisol mixing procedures may be one cause. Mixing can cause particle attrition resulting in less agglomerates.)

It is clear that in order to produce a low initial viscosity plastisol with good viscosity stability, the drying conditions need to be optimized for two factors:

1. Optimization of the particle size distribution in order to minimize void volume in the plastisol to get good initial viscosities.

2. Production of well fused agglomerates for good viscosity stability through the appropriate choice of temperature and atomization.

The drying conditions required for the optimization of plastisol viscosities will be dependent upon the latex PSD and percent solids. These factors need to be considered when performing drying studies and predictive work.

Literature Cited

1. O. Bjerke, "Advances in Polymer Science and Technology" (Soc. Chem. Ind., Monograph No. 26), Gordon and Breach, 1967, pp. 370-380.

2. E. A. Collins, D. J. Hoffmann and P. L. Soni, J. Colloid Interface Sci., 71, (1) 21 (1979).

3. I. A. Mukhina, B. P. Shtarkman, and I. N. Vishnevskaya, Mekh. Protsessov Plankoobrazov, Polim. Rastvorov Dispersii, Akad. Nauk. Ussr, Sb. Statei 1966, 124-127.

4. Jones, W. E. Schaefer, S. W., U.S. Patent 3,317,495 (1967).

5. Mueller, J., Stauffer, W., Moser, K., Chima, 27 (2), 82-86 (1973).

6. Provder, T., Hosworth, R. M., Am. Chem. Soc. Div. Coat. Plast. Chem. Pap., 36(2), 150-155 (1976).

7. Seidewand, R. J., Erickson, J. R., Polym. Eng. Sci., 18 (15) 1182-1185, (1978).

8. J. J. Friedman, F. A. Glucert, and W. R. Marshall, Jr. Chem. Eng. Prog., 48 (4) 81 (1952).

9. D. J. Hoffmann, E. G. Garcia, Rheology of PVC Dispersions: Effect of Time and Temperature, presented at the Third International Symposium on PVC, Cleveland, Aug. 1980.

10. C. W. Johnston and C. H. Brower, SPE J. 26 (9), 31 (1970).

11. G. Williams, J. Bergen, and G. Poehlein, J. of Rheology., 23 (5), 591, (1979).

12. J. S. Chong, E. B. Christiansen, and A. D. Baer, J. Appl. Polym. Sci. 15, 2007 (1971).

13. C. E. Chaffey, and I. Wagstaff, J. Colloid Interface Sci., 59, (1) 63 (1977).

RECEIVED April 6, 1981.

13

Surfactant Interactions in Polyvinyl Acetate and Poly (vinyl acetate–butyl acrylate) Latexes

B. R. VIJAYENDRAN, T. BONE, and C. GAJRIA

Celanese Plastics and Specialties Company, Jeffersontown, KY 40299

Recent investigations have shown that the behavior and interactions of surfactants in a polyvinyl acetate latex are quite different and complex compared to that in a polystyrene latex (1, 2). Surfactant adsorption at the fairly polar vinyl acetate latex surface is generally weak (3,4) and at times shows a complex adsorption isotherm (2). Earlier work (5,6) has also shown that anionic surfactants adsorb on polyvinyl acetate, then slowly penetrate into the particle leading to the formation of a polyelectroyte type solubilized polymer-surfactant complex. Such a solubilization process is generally accompanied by an increase in viscosity. The first objective of this work is to better understand the effects of type and structure of surfactants on the solubilization phenomena in vinyl acetate and vinyl acetate-butyl acrylate copolymer latexes.

It was reported earlier (1) that surfactant adsorption at a polymer/water interface can be related to the polarity of the polymer surface. The model used in that study was tested satisfactorily by using the available literature data on polymer polarity and sodium lauryl sulfate adsorption on latex surfaces. The second objective is to verify experimentally the predicted relationship between polymer polarity and surfactant adsorption by studying the adsorption of a non-ionic surfactant that shows a saturation type isotherm behavior on vinyl acrylic latexes of varying polarity.

In order to achieve the above objectives, three vinyl acrylic latexes of varying butyl acrylate content have been prepared and 'cleaned' for use in the study. Several anionic and nonionic surfactants commonly used in emulsion polymerization have been used to investigate the effects of surfactant structure and polymer composition on the solubilization process. Polarity of latex surface estimated from contact angle measurements have been used to study the effect of polymer polarity on surfactant adsorption.

0097-6156/81/0165-0225$05.00/0

Experimental Section

Materials: Commercial grade vinyl acetate (VA) and butyl acrylate (BA) from Celanese Chemical Company were used. Reagent grade potassium persulfate was used as the initiator. Sodium lauryl sulfate (NaLS) from BDH Chemicals after purification (7) was used as the emulsifier in the polymerization. Surfactants used in the adsorption studies were the following: Igepal CO-630 (10 mole ethylene oxide adduct of nonyl phenol) and Alipal EP-110 and EP-120 from GAF Corporation; Aerosol A-102 from American Cyanamid and BDH Chemical solium lauryl sulfate. According to the manufacturer's literature, Alipal EP-110 and EP-120 are ammonium salts of a sulfated nonyl phenoxy poly (ethyleneoxy) ethanol of molecular weights 708 and 1640 respectively. Aerosol A-102 is an anionic/non-ionic type surfactant of the composition disodium ethoxylated alcohol (C_{10} - C_{12}) half ester of sulfo-succinic acid of molecular weight of about 1800.

Emulsion Polymerization: A typical recipe is give in Table I. Emulsion polymerization was carried out at 60°C under a nitrogen atmosphere using a batch process. Theoretical solids content in all the formulations was 25%, and generally the conversions were better than 98%. A polyvinyl acetate homopolymer and two poly (vinyl acetate-butyl acrylate) copolymers having VA/BA composition of 85/15 and 70/30 were prepared according to the above procedure.

Ion Exchange and Dialysis of Latexes: The above three latexes were 'cleaned' by the ion exchange method of Vanderhoff (8) and dialysis. In general, the latexes were stable after the cleaning process. In some cases, the ion-exchange resin was found to discolor after the treatment indicating some interaction, perhaps hydrolysis of the latex, between the latex and ion-exchange resin.

Particle-Size Determination: Particle-size of the 'cleaned' latexes were determined using transmission electron microscopy after freeze-drying the samples and counting the particles with a Quantimet image analyzer. The number average particle diameters (d_n) of the homopolymer, the 85/15 VA/BA and 70/30 VA/BA latexes were found to be 0.057 μm, 0.062 μm and 0.073 μm, respectively. The polydispersity $\frac{d_w}{d_n}$ of the three samples were 1.09, 1.05 and 1.06, respectively.

Adsorption Studies: Appropriate amounts of cleaned latex and surfactants were mixed, equilibrated for 24 hours and the excess surfactant in serum analyzed after separation by centrifugation. Igepal CO-630 and Alipal surfactants were analyzed by colorimetric titration with cetyltrimethylammonium chloride in the presence of methylene blue and a chloroform layer similar to the method of Epton (9). Some non-ionic surfactants and solubilized serum components were found to have a positive interference in the colorimetric analysis of ionic surfactants.

Contact Angle Measurements and Polarity of Latex Films: Advancing contact angle measurements were made at 25°C using a Rame' Hart contact angle goniometer. Latexes 'cleaned' by the above procedure were drawn on a glass plate and dried at 40°C. An average of eight measurements was taken as the contact angle. Water and methylene iodide were used as test liquids. Polarity of the dried latex films were estimated according to the method of Kaelble (10).

Results and Discussion

Shape of Adsorption Isotherms and Latex Thickening: Adsorption isotherms of NaLS and Aerosol A-102 surfactants on a 'clean' 85/15 VA/BA copolymer latex are shown in Figure 1. It is seen that the shapes of the two isotherms are quite different. A-102 seems to exhibit a Langmuir type saturation adsorption behavior typical of surfactant adsorption at a polystyrene/water interface. NaLS seems to have a more complex adsorption isotherm, perhaps involving a multiple step process as discussed by Giles et al (11). The initial region of the isotherm upto B appears to be normal and beyond that the isotherm become linear. Giles et al have explained such a linear isotherm (classified as Type C) by penetration of substrate by solute, leading to new adsorption sites. Compared to NaLS, the size of Aerosol A-102 molecule is much larger and bulkier and thus perhaps hindered in being able to penetrate into the latex particle.

Figure 2 shows adsorption isotherms of two sulfated ethoxylate type anionics - Alipal EP-110 and Alipal EP-120 - on the 85/15 VA/BA latex surface. Again it is seen that the lower molecular weight EP-110 shows a C type isotherm similar to NaLS while the higher molecular weight EP-120 exhibits a normal saturation type isotherm.

It is well known (3,5,6) that sodium lauryl sulfate interacts with some polymers such as polyvinyl acetate causing solubilization of the insoluble polymer leading to an increase in viscosity. In Figure 3, viscosity of the homopolymer and 70/30 VA/BA at various NaLS/polymer ratio is shown. It is seen that the viscosity of the 2% latex dispersion increases with increase in NaLS/polymer ratio. Similar viscosity data for the 85/15 VA/BA was intermediate between the homopolymer and 70/30 VA/BA latexes. Surfactants that showed a normal saturation type adsorption behavior did not show any significant viscosity increase of the latex.

Edelhauser (3) has explained his results on the interaction of anionic surfactants (sodium lauryl sulfate and sodium dodecylbenzene sulfonate) with a PVAC latex as a two-step mechanism involving surface adsorption followed by surfactant penetration into

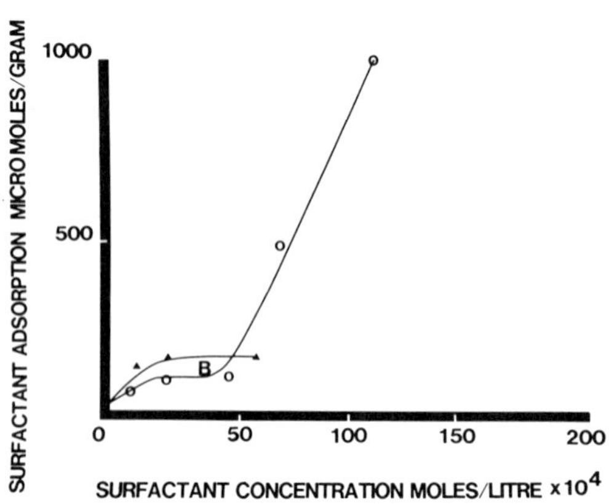

Journal of Applied Polymer Science

Figure 1. Adsorption isotherms of NaLS and Aerosol A-102 surfactants at 85/15 VA/BA latex/water interface (17) *((○) sodium lauryl sulfate; (▲) Aerosol A-102)*

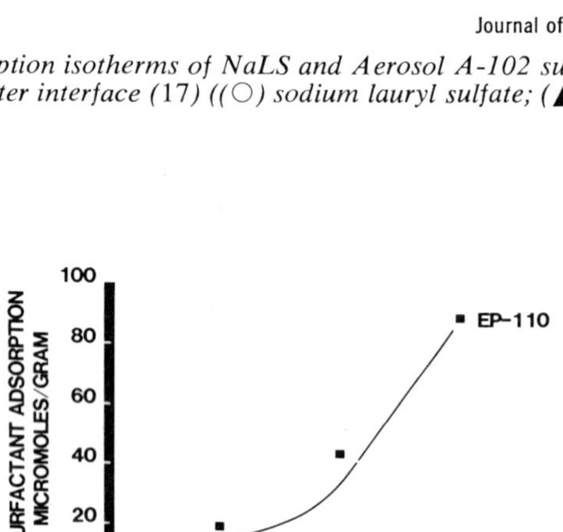

Journal of Applied Polymer Science

Figure 2. Adsorption isotherms of Alipal EP-110 and Alipal EP-120 surfactants at 85/15 VA/BA latex/water interface (17) *((■) Alipal EP-110; (▲) Alipal EP-120)*

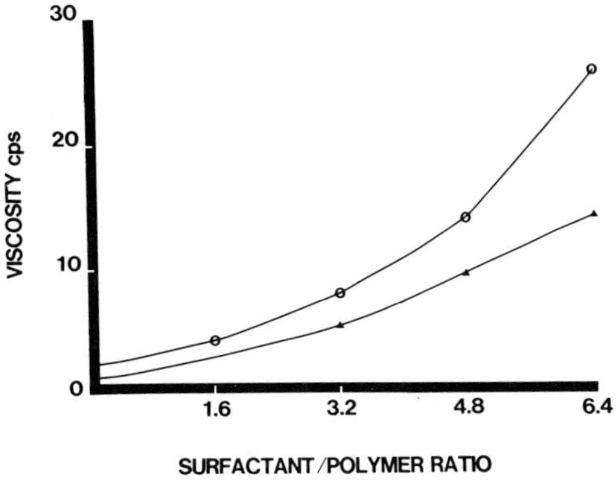

Journal of Applied Polymer Science

Figure 3. Thickening of vinyl acrylic latexes in the presence of NaLS (17) ((○) PVAC homopolymer; (▲) 70/30 VA/BA copolymer)

TABLE I

EMULSION POLYMERIZATION RECIPE (17)

MONOMERS	500 GRAMS
SURFACTANT NaLS	17.4 GRAMS
$K_2S_2O_8$	2.48 GRAMS
DISTILLED WATER	1575 GRAMS
SOLIDS	25%

Journal of Applied Polymer Science

TABLE II

ANIONIC SURFACTANTS STUDIED IN VINYL ACRYLIC LATEXES (17)

PENETRATING TYPE

(A) SODIUM LAURYL SULFATE
$C_{12}H_{25}SO_4Na$ M.WT = 288.4

(B) SODIUM DODECYL BENZENE SULFONATE $C_{12}H_{25}$  SO_3Na
M. WT = 348.5

(C) ALIPAL EP-110 M.WT = 708
C_9H_{19}  $-(OCH_2CH_2)_{8\text{-}9}O.SO_3NH_4$

NON PENETRATING TYPE

(A) AEROSOL A-102 M. WT = 1800
DISODIUM ETHOXYLATED ALCOHOL (C_{10} - C_{12}) HALF ESTER OF SULFOSUCCINIC ACID

$CH_2 \cdot COO(H_2C \cdot CH_2O)_{30}C_{11}H_{23}$
|
CH COO Na
|
SO_3Na

(B) ALIPAL EP-120 M. WT = 1640
C_9H_{19}  $-(OCH_2 \cdot CH_2)_{29\text{-}30}O.SO_3NH_4$

Journal of Applied Polymer Science

the interior of PVAC particles taking water with it and causing swelling and gradual dissolution of the disintegrated polymer chains. Based on this model, it is convenient to classify the adsorption behavior of anionic surfactants at vinyl acrylic latex surface into penetrating and non-penetrating types as shown in Table II. The smaller surfactant molecules seem to be capable of penetrating into the latex, causing disintegration and solubilization of the otherwise insoluble polymer. The larger and bulkier surfactants studied seem to adsorb only at the latex surface unable to penetrate into the particle. It appears that there exists a critical size and perhaps shape of anionic surfactant molecule, for the surfactant to be able to penetrate into the particle. It is also likely that the nature of hydrophobe adsorbing at the latex/water interface and the charge density of the polymer-surfactant complex may influence surfactant penetration and latex solubilization. More extensive work with other anionic surfactants of varying chain length and shape would be needed to better define molecular parameters that influence penetration of surfactants into a vinyl acrylic latex.

Latex thickening in the presence of penetrating type anionic surfactants such as NaLS appears to depend on polymer composition as seen in Table III. The extent of latex thickening in the presence of excess NaLS decreases with the VA content of a vinyl acetate-butyl acrylate copolymer. Polystyrene and poly acrylate copolymer latexes do not show any thickening.

Presence of non-ionic surfactants such as Igepal CO-630 seems to prevent the thickening of PVAC latex by NaLS, as shown in Table IV. This can be interpreted as to show that the presence of non-ionic surfactant at the PVAC latex surface prevents the penetration of NaLS into the particle.

Formation of solubilized surfactant-latex complexes can influence the properties and performance of vinyl acrylic latexes prepared with NaLS and other penetrating type anionic surfactants. Such complexes seem to affect glass transition temperature and film coalescence process (12).

Our results on the interactions of anionic surfactants in vinyl acrylic latexes can be summarized as follows:

> Lower molecular weight (300-800) anionic surfactants such as NaLS and Alipal EP-110 adsorb at a vinyl acrylic latex in a complex C-type adsorption isotherm, leading to penetration and solubilization of latex.
>
> Relatively higher molecular weight (1600) anionic surfactants such as Alipal EP-120 and Aerosol A-102 adsorb in a normal manner, suggesting that there exists a critical size, specific charge density and perhaps shape for the surfactant to be able to penetrate into the latex.

TABLE III

EFFECT OF LATEX COMPOSITION ON THICKENING BY SODIUM LAURYL SULFATE *(17)*

LATEX COMPOSITION	LATEX THICKENING
POLY VINYLACETATE	19.3
POLY VINYL ACETATE-BUTYL ACRYLATE (70/30)	10.0
POLYSTYRENE	1.1
POLY (METHYL METHACRYLATE-ETHYL ACRYLATE)	1.1

THICKENING $\frac{\eta}{\eta_0}$ = $\frac{\text{BROOKFIELD VISCOSITY OF 2\% LATEX IN THE PRESENCE OF NaLS}}{\text{BROOKFIELD VISCOSITY OF 2\% CONTROL LATEX}}$

Journal of Applied Polymer Science

TABLE IV

EFFECT OF NON-IONIC SURFACTANT ON THICKENING OF POLYVINYL ACETATE-SODIUM LAURYL SULFATE (NaLS) *(17)*

LATEX	LATEX THICKENING
POLYVINYL ACETATE + NaLS	19.3
POLYVINYL ACETATE + IGEPAL CO-630	1.0
POST ADDITION OF IGEPAL CO-630 TO POLYVINYL ACETATE + NaLS	17.4
POST ADDITION OF NaLS TO POLYVINYL ACETATE + IGEPAL CO-630	1.3
POLY VINYL ACETATE + MIXED IGEPAL CO-630 AND NaLS	1.4

*SEE TABLE III FOR DEFINITION OF LATEX THICKENING

Journal of Applied Polymer Science

Latex thickening, a result of surfactant penetration and solubilization of the polymer, increases with the vinyl acetate content of vinyl acetate-butyl acrylate co-polymer latexes.

Presence of non-ionic surfactants at the latex/water interface seem to interfere with the penetration of low molecular weight penetrating type anionic surfactants.

Polarity of Vinyl Acrylic Latex and Surfactant Adsorption: Contact angle measurements, dispersion and polar contribution to latex film surface tension and polarity of polymer calculated according to the method of Kaelble (10) of the three latex films are whown in Table V. It is seen that the polarity of the latex film decreases with increase in butyl acrylate content of the vinyl acrylic co-polymer. The polarity of the 70/30 (VA/BA) latex is very similar to that of the polybutyl acrylate homopolymer estimated to be about 0.21 (1).

Polarity of 0.54 for the PVAC homopolymer is considerably higher than the 0.33 value determined for the homopolymer from interfacial tension measurements on polymer melts (13). The higher polarity value found in the PVAC film strongly suggests that there was a significant amount of hydrolysis of PVAC, either during polymerization and/or cleaning of the latex resulting in a more hydrophilic PVAC surface.

Table VI shows the adsorption data of Igepal CO-630 surfactant at the three latex/water interfaces. The isotherms were of the normal type and no thickening of the latex in the presence of surfactant was observed. A typical adsorption isotherm of Igepal CO-630 on a 85/15 VA/BA latex surface is shown in Figure 4. Area per molecule was calculated according to equation - 1 (14).

$$A = \frac{9.961 \times 10^{-3}}{\Gamma \cdot \rho \cdot d} \qquad -(1)$$

where

A = Area per molecule in nm^2.

d = diameter of latex in nm.

ρ = density of polymer latex. Taken as 1.17 and 1.14 g/cc for the homopolymer and co-polymer respectively.

Γ = Moles surfactant/gram polymer latex.

Interaction parameter (I) defined as A_o/A, where A_o= limiting area per molecule of surfactant, is given in the last column of the table.

TABLE V

POLARITY OF LATEX SURFACE BY CONTACT ANGLE MEASUREMENTS (17)

LATEX	CONTACT ANGLE, θ H_2O	CONTACT ANGLE, θ CH_2I_2	DISPERSION CONTRIBUTION DYNES/CM	POLAR CONTRIBUTION DYNES/CM	POLARITY X^P =
VA/BA 70/30	72°	41°	30.5	9.3	0.23
VA/BA 85/15	67°	38°	30.0	12.2	0.29
VA/BA 100/0	44°	29°	25.4	29.8	0.54

VA = VINYL ACETATE

BA = BUTYL ACRYLATE

θ = AVERAGE OF 8 MEASUREMENTS

Journal of Applied Polymer Science

TABLE VI

ADSORPTION OF IGEPAL CO-630 ON VINYL ACRYLIC LATEX OF DIFFERENT POLARITY (17)

LATEX	X^P POLARITY	D, PARTICLE SIZE NM	ADSORPTION MOLES/CM^2 OF POLYMER	A, AREA PER MOLECULE NM^2	I, INTERACTION PARAMETER
VA	0.54	0.057	8.89×10^{-11}	1.87	0.29
VA/BA 85/15	0.29	0.062	14.7×10^{-11}	1.13	0.49
VA/BA 70/30	0.23	0.073	17.3×10^{-11}	0.96	0.57

I = INTERACTION PARAMETER = $\frac{A_o}{A}$ A_o = 0.55 NM^2 FOR IGEPAL CO-630

Journal of Applied Polymer Science

It is seen that the adsorption of Igepal CO-630 at the three latex/water interfaces decreases with increase in polarity of the vinyl acrylic latex surface. Explanation for such a decrease in surfactant adsorption at a polymer/water interface with increase in polymer polarity has been discussed in detail elsewhere (1). Briefly, increased polarity of the polymer lowers the interfacial free energy of the polymer latex/water interface and this, in turn, reduces the free energy of adsorption for a simple saturation type adsorption process of a surfactant at a latex surface in aqueous media. Such a lowering in free energy of surfactant adsorption at a polymer latex/water interface with increase in polymer polarity leads to the observed results, namely, decrease in the adsorption of Igepal CO-630 with polarity increase of the VA/BA latex particle.

The interaction parameter, as expected, decreases with increase in polarity of the latex surface (12). It shows that at saturation adsorption, the extent of interaction of Igepal CO-630 with the PVAC homopolymer and the two VA/BA co-polymer latexes is 29%, 49%, and 57% respectively of the theoretical limit corresponding to a close packed monolayer adsorption.

Figure 5 shows a plot of log A vs polarity (x^p) of latex surface. It is readily seen that the plot is quite linear and fits equation 2.in good agreement with the results on the adsorption of sodium lauryl sulfate at various latex/water interfaces (1).

$$\log A = \text{const} + k(X^p) \qquad (2)$$

Attempts to correlate the adsorption data of other surfactants such as Alipal EP-110 and NaLS on the three latex surfaces in a similar manner failed because of the more complex and specific interactions observed in these systems. Equation 2 can adequately describe the adsorption data of surfactants at polymer/water interfaces, provided that the free energy of the interface is related to the free energy of adsorption and there are no specific interactions between surfactant and interface (15).

The polarity and adsorption data discussed above reveal some interesting aspects of the surface chemistry of vinyl acrylic latex surfaces. It is quite likely that the polarity of the latex films, expecially of the two co-polymers, determined by contact angle measurements may not correspond exactly with their respective latex surfaces in the dispersed state due to reorientation of polymer chains during film formation. But the surfactant adsorption data shows clearly that the three latex surfaces in their dispersed state do exhibit varying polarity paralleling the trend found from contact angle measurements. The result also shows that the surface of the co-polymer latex surface is a mixture of vinyl acetate and acrylate units. This result is somewhat unexpected in a vinyl acrylic latex, prepared by a batch

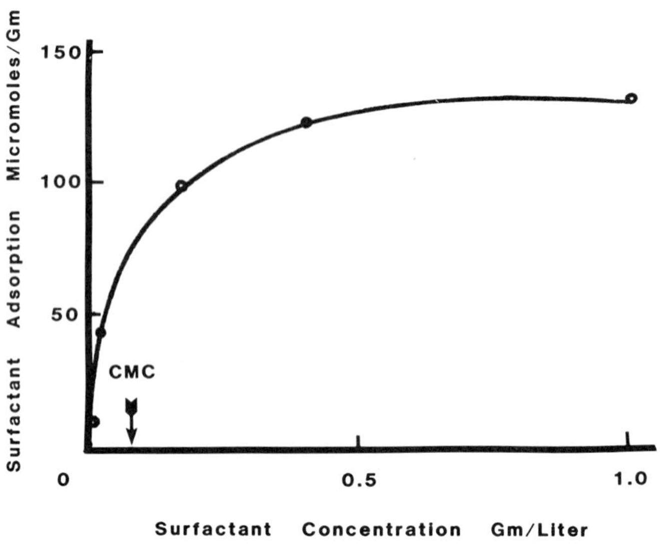

Figure 4. Adsorption isotherm of Igepal CO-630 surfactant at 85/15 VA/BA latex/water interface

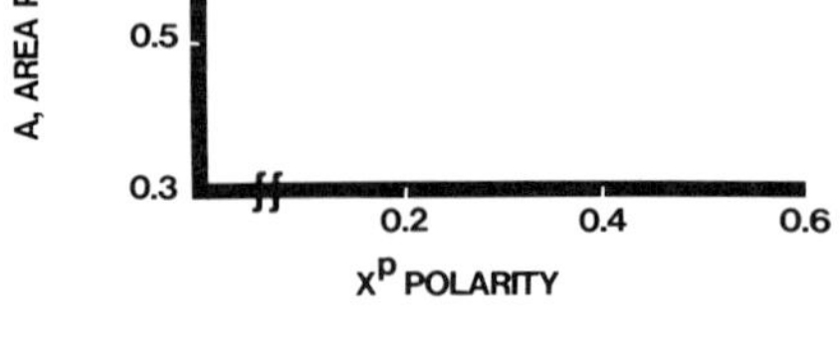

Journal of Applied Polymer Science

Figure 5. Effect of polymer polarity on the adsorption of Igepal CO-630 surfactant at vinyl acrylic latex/water interface (17) process since one would expect the more reactive butyl acrylate to predominate the core structure leaving the less reactive vinyl acetate near the latex surface. Based on the electron micrographs of a batch and semi-continuous polymerized 63/37 VA/BA latex particles, Misra etal (16) concluded that the batch sample has a heterogenous structure comprising of a relatively large butyl acrylate-rich core surrounded by a vinyl acetate-rich shell. In comparison, the semi-continuous polymerized particle was found to have a homogenous structure comprising of a much smaller butyl acrylate-rich core surrounded by a shell of vinyl acetate-butyl acrylate copolymer. It would be interesting to study the two VA/BA latex samples of Misra et al by the above methods to ascertain the differences in the surface composition of the two latexes.

In agreement with our earlier studies (1,15), the adsorption results of Igepal CO-630 on the three vinyl acrylic latexes show that the area per molecule of surfactant can be related to the polarity of polymer surface. Further, the results show that one can employ the techniques discussed above to characterize the polarity of co-polymer latex surfaces.

Acknowledgements

It is a pleasure to acknowledge Miss Diane Fife for the preparation of latexes. Thanks are also due to Celanese Plastics and Specialties Company for permission to present this work.

Abstract

Types of surfactant interactions encountered in polyvinyl acetate and poly (vinyl acetate-acrylate) latexes are: (1) simple saturation adsorption at the latex/water interface, (2) adsorption and penetration into the latex particle leading to the formation of solubilized polyelectrolyte type polymer-surfactant complex, and (3) grafting to the polymer chains, especially with non-ionic surfactants. In this paper, the effects of molecular weight and structure of anionic surfactants on their interactions with model vinyl acrylic latexes are considered. It is shown that anionic surfactants having fairly low molecular weight and perhaps simple structure readily penetrate into the latex core causing new adsoption sites to open up. Such a process seems to result in a linear C-type adsorption isotherm and solubilization of the latex. On the other hand, higher molecular weight and bulkier anionic surfactants seem to show a saturation type adsorption isotherm without any solubilization of the latex.

It is also shown that the adsorption of non-ionic surfactants at a vinyl acrylic latex/water interface that exhibit a saturation type isotherm can be related to the polarity of the polymer surface, in agreement with earlier sufactant adsorption studies.

References:

(1.) B.R. Vijayendran, J. Appl. Polym. Sci., 23, 733 (1979).
(2.) S.M. Ahmed, M.S. El-Aasser, G.W. Poehlein and J.W. Vanderhoff, Emulsion Polymers Institute Research Reports, (1979).
(3.) H.A. Edelhauser, J. Polym-Sci.: Part C., 27, 291 (1969).
(4.) B.R. Vijayendran, Division of Colloid Abstracts, Spring ACS Meeting in Honolulu, (1979).
(5.) T. Isemura and A. Imanishi, J. Polym-Sci., 33, 337 (1950).
(6.) H. Arai and S. Harin, J. Coll. and Int. Sci., 30, 372 (1969)
(7.) B.R. Vijayendran, J. Coll., and Int. Sci., 60, 418 (1977).
(8.) H.J. Van den Hull and J.W. Vanderhoff, Br. Polym. J., 2 121 (1970).
(9.) J.R. Epton, Trans. Faraday Soc., 44, 226 (1948).
(10.) D.H. Kaelble, Physical Chemistry of Adhesion, Wiley-Interscience, New York, 1971, Chap. 4.
(11.) C.H. Giles, D. Smith and A. Huitson, J. Coll and Int. Sci., 47, 755 (1974).
(12.) B.R. Vijayendran, T. Bone and C. Gajria, Vinyl Acetate Symposium Proceedings, Lehigh University, April, 1980.
(13.) S. Wu, J. Macromol Sci-Rev. Macromol Chem., C-10, 1 (1974).
(14.) S.H. Maron, M.E. Elder, and I.N. Ulevitch, J. Colloid Sci., 3, 87 (1954).
(15.) B.R. Vijayendran, J. Coll. and Int. Sci., 68, 383 (1979).
(16.) S.C. Misra, C. Pichot, M.S. El-Aasser, and J.W. Vanderhoff, J. Polym. Sci. Polym. Letters., 17, 567 (1979).
(17.) B. R. Vijayendran, T. Bone, and C. Gajria, J. of Appl. Polym. Science, 26, 1351-1359 (1981).

RECEIVED April 6, 1981.

14

Rapid Density Gradient Centrifugation of Polymer Latices

H. LANGE

Bayer AG, FE-DPP-Strukturforschung, 5090 Leverkusen, West Germany

For the development, production, and application of polymer latices the determination of the size distribution and the analysis of the chemical composition and heterogeneity of the latex particles are important. The size distribution can be determined rapidly by ultracentrifugation, electron microscopy or light scattering (1-4), but for the analysis of the composition distribution there is only the time-consuming density gradient centrifugation.

In this method, usually a density profile is formed in the ultracentrifugation cell by centrifuging solutions for instance of sucrose or Metrizamide in water or methanol until the equilibrium state between sedimentation and diffusion of the dissolved molecules has been established. Figure 1 shows schematically two of such equilibrium density profiles (dotted lines). In these density profiles the latex particles, added before starting the experiment, migrate to that position in the cell where their density coincides with the density of the surrounding medium. The position of the particles can be recorded by schlieren optics or, if there is a particle density distribution, more precisely by scanning extinction measurements normally used for the characterization of proteins. Thus the density and extinction profile in the ultracentrifugation cell yield a criterion for the density distribution and hence, because of the correlation between chemical composition and particle density, a criterion for the composition distribution or heterogeneity of the latex particles.

This standard method of density gradient centrifugation, however, has a serious disadvantage. Because of the sedimentation and diffusion equilibrium formation of the density profile the time needed for one experiment amounts to more than 15 hours.

0097-6156/81/0165-0239$05.00/0

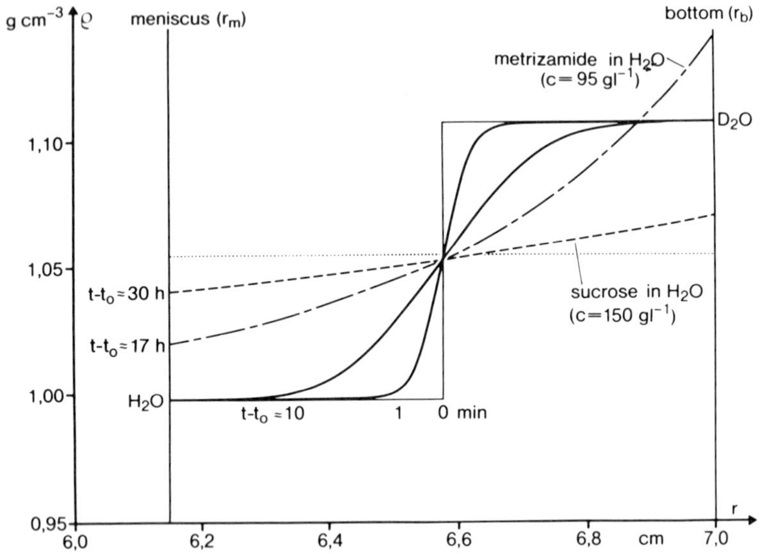

Figure 1. Density profile in the cell of the ultracentrifuge in case of standard and rapid density gradient centrifugation (rotor speed N = *40.000* min^{-1}*)*

The Rapid Method

To reduce this time a rapid method of density gradient centrifugation is developed by avoiding the equilibrium formation and generating the density profile only by diffusion. This is done by adding a layer of a dispersion medium of low density to a chemically similar medium of high density in the ultracentrifugation cell, for instance H_2O to D_2O (full lines in Figure 1).

The procedure is as follows. At the beginning of the rapid experiment a volume of H_2O with a density value of about 1.0 g cm^{-3} is added to an approximately equal volume of D_2O with a density value of about 1.1 g cm^{-3} containing the particles to be measured. Immediately after that, there is a sharp density step of 0.1 g cm^{-3} in the middle of the cell. Then this step is equalized by diffusion and after a few minutes a slowly changing continuous density profile has been established. During this time the particles migrate, just as in standard density gradient centrifugation, to that position in the cell where their density coincides with the density of the surrounding medium. In such an experiment, however, the migration time of the particles to the right position in the cell must be negligible in relation to the temporal change of the density profile. But this condition is fulfilled for nearly all latex particles if the rotor speed is high enough.

For the evaluation of the experiment two quantities are necessary: the density profile and the local position of the particles in the cell at the same time.

The density profile (upper part of Figure 2) can be determined by measuring the refractive index gradient along the cell (lower part of Figure 2) and the correlation between the refractive index and the density for the gradient system (Figure 3). This yields the density profile with an accuracy of about ± 1 per thousand.

The local position of the particles in the cell can be recorded in all cases by scanning extinction measurements. Only in the special case of chemically uniform particles can the schlieren optical method also be used.

Examples for Application

The upper part of Figure 4 shows the results of the rapid density gradient centrifugation of a latex mixture of poly(styrene) and two chemically uniform co-

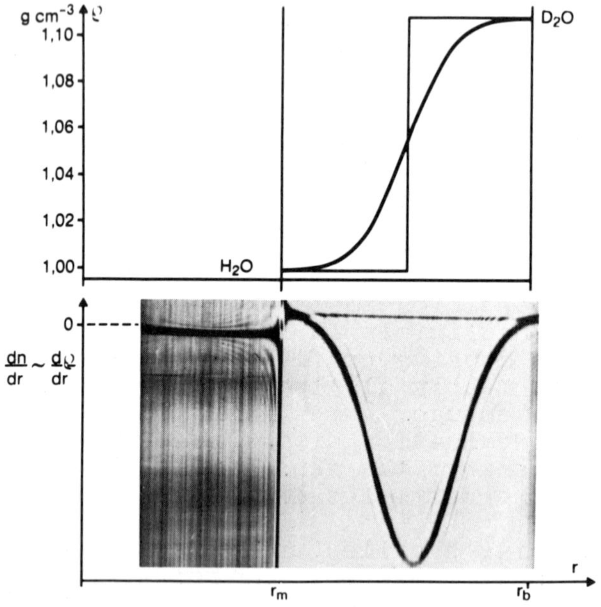

Figure 2. Refractive index gradient and density profile in the cell of the ultracentrifuge in rapid density gradient centrifugation (N = *40.000* min^{-1}, *centrifugation time* $t - t_o = 4$ *min, schlieren angle* $\alpha = 65°$)

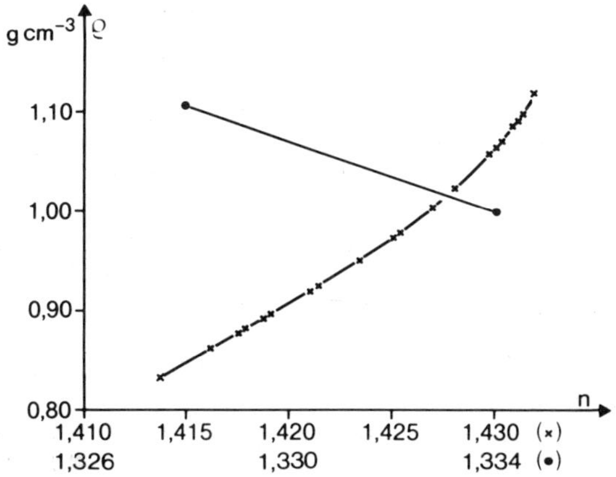

Figure 3. Relation between density and refractive index of mixtures of 3-butene-2-ol and ethylene glycol (× resp. H_2O and D_2O (●))

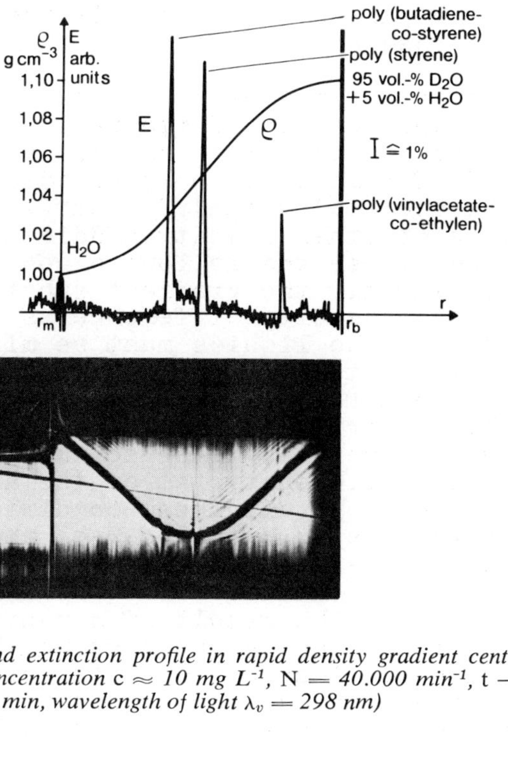

Figure 4. Density and extinction profile in rapid density gradient centrifugation of a latex mixture (concentration c ≈ 10 mg L^{-1}, N = 40.000 min^{-1}, t − t_o = 15 min, wavelength of light λ_v = 298 nm)

polymers of butadiene and styrene and of vinyl acetate and ethylene. The gradient system consists of H_2O and D_2O. ρ means the density and E the extinction profile in the cell. The three types of particles are well separated: to the left the copolymer particles of butadiene and styrene, in the middle the particles of poly(styrene) with the known density of 1.054 g cm^{-3}, and to the right the copolymer particles of vinyl acetate and ethylene. The lower part of Figure 4 shows the corresponding schlieren pattern with the three faint perpendicular schlieren lines at the positions of the particle accumulations. In this experiment the three kinds of particles are separated in less than 15 min.

In Figure 5 the results obtained by rapid density gradient centrifugation of a polymer latex of unknown composition are shown. The gradient system consists of pure H_2O and D_2O. The density distribution of the particles has two broad peaks at the density values 1.01 and 1.04 g cm^{-3}. That means the latex contains two chemically heterogeneous kinds of particles. This experiment requires only 4 min. In such a case of a broad particle density distribution, the local positions of the particles in the ultracentrifugation cell can be recorded only by scanning extinction measurements.

The gradient system of H_2O and D_2O, however, excludes the application of the rapid method to particles with densities lower than 1.0 and higher than 1.1 g cm^{-3}, for instance poly(butadiene) with 0.9 and poly(vinyl acetate) with 1.2 g cm^{-3}. To investigate such particles other density gradient systems are necessary. For these the following conditions must be fulfilled. Firstly the gradient substances must be liquids with a density difference as large as possible. Secondly, the liquids must be miscible without considerable heat of mixing to avoid thermal disturbances. Furthermore the liquids must be nonsolvents for the latex particles and the difference in their refractive indices has to be so small that the standard optical schlieren system of the analytical ultracentrifuge can be used. These conditions are fulfilled by the system 3-butene-2-ol and ethylene glycol. It yields a density range from about 0.83 to 1.1 g cm^{-3}.

In Figure 6 the result of a rapid density gradient centrifugation is shown using this system. It deals with a graft copolymer latex of butadiene and styrene acrylonitrile. The gradient media are different mixtures of 3-butene-2-ol and ethylene glycol.

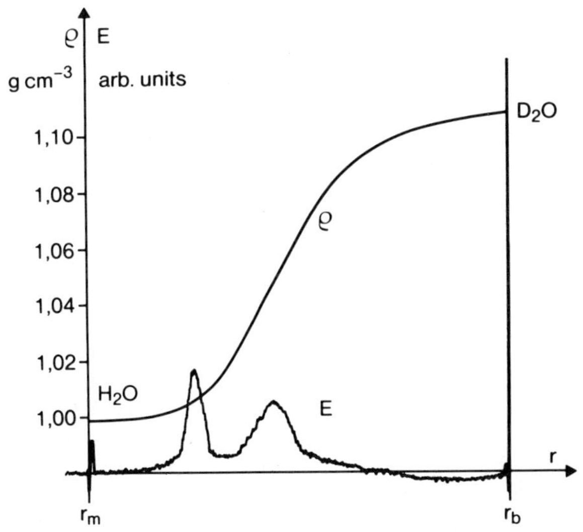

Figure 5. Density and extinction profile in rapid density gradient centrifugation of a polymer latex of unknown composition (c ≈ *5 mg* L^{-1}, N = *40.000* min^{-1}, t − t_o = *4 min*, λ_v = *546 nm)*

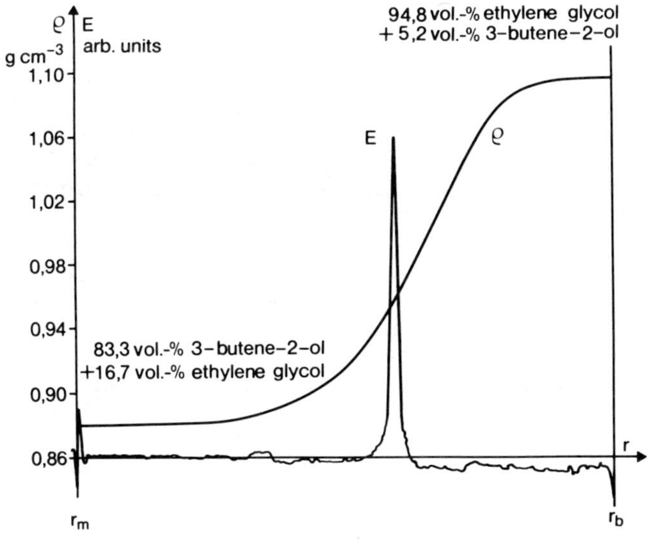

Figure 6. Density and extinction profile in rapid density gradient centrifugation of a graft copolymer latex of butadiene and styrene acrylonitrile (c ≈ *5 mg* L^{-1}, N = *40.000* min^{-1}, t − t_o = *60 min*, λ_v = *546 nm)*

The particles have a relatively uniform density of about 0.95 g cm^{-3} and hence a content of poly(butadiene) of 59 percent (ratio of the weight styrene/acrylonitrile = 88/12). With this gradient system the investigation of the chemical composition and heterogeneity of nearly all butadiene copolymer latices is possible. The experiment needs about 1 hour.

For particles with densities higher than 1.1 g cm^{-3}, no similar gradient system could be found. Therefore in this case mixtures of H_2O and D_2O with equal amounts of glycerol are used. Figure 7 shows the densities of these mixtures plotted against the volume fraction of glycerol (upper curves). Of course the width of the density range decreases as the glycerol content increases, but the density rises up to 1.26 g cm^{-3}. As a corresponding system for lower densities down to 0.8 g cm^{-3} mixtures of H_2O and D_2O with equal amounts of methanol are used (lower curves).

Figure 8 shows an example for application of this technique in the higher density range around 1.2 g cm^{-3}. The gradient system consists of mixtures of H_2O and D_2O with glycerol and the latex under investigation is a latex of poly(vinyl acetate). The particle density and consequently the chemical composition is uniform, as it is expected for homopolymers. The experiment needs 40 min.

Figure 9 shows an example for application in the lower density range. The gradient system consists of mixtures of H_2O and D_2O with methanol. The sample under investigation is a mixture of a normal and an additionally cross-linked poly(butadiene) latex. The particle density is not uniform. The normal poly(butadiene) has particle densities as expected around 0.89 g cm^{-3}, but for the additionally cross-linked poly(butadiene) density values between 0.9 and 0.92 g cm^{-3} are obtained. Additional cross-linking yields obviously higher particle densities. In this case the experiment needs 30 min.

These are some examples for application of the rapid density gradient centrifugation method for characterizing polymer latex particles. The main condition of this method is that the migration time of the particles to the right positions in the ultracentrifugation cell is negligible compared with the temporal change of the density profile. That this condition is fulfilled can be seen from Figure 10. Here the density of some chemically uniform latex particles, determined by rapid density gradient centrifugation, is plotted against the centrifugation time. In all cases the right density values are obtained

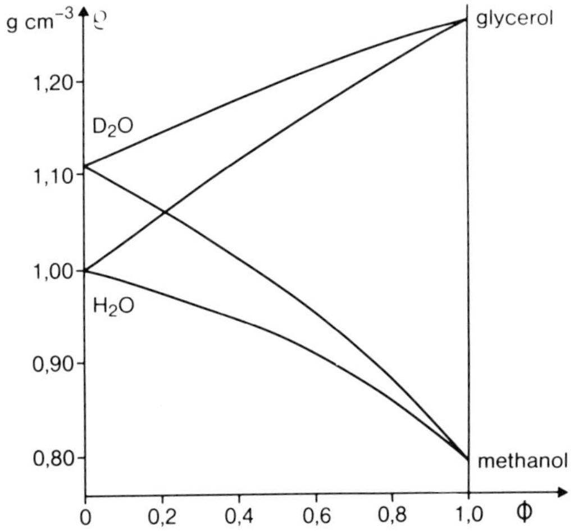

Figure 7. Density of mixtures of H_2O resp. D_2O with methanol and glycerol (Φ = volume fraction)

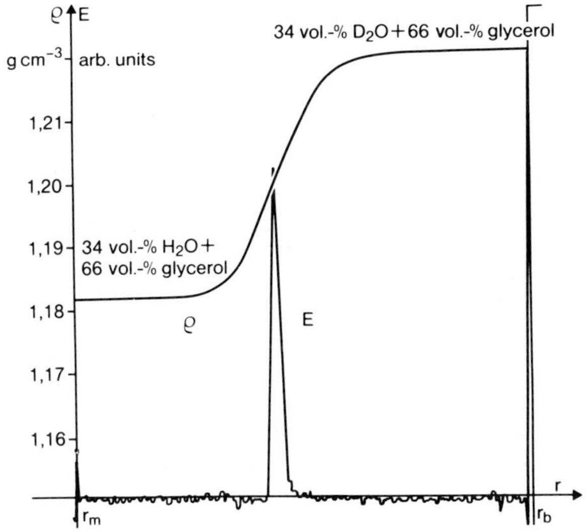

Figure 8. Density and extinction profile in rapid density gradient centrifugation of a latex of polyvinyl acetate (c $\approx$ 5 mg L^{-1}, N = 40.000 min^{-1}, t − t_o = 40 min, λ_v = 298 nm)

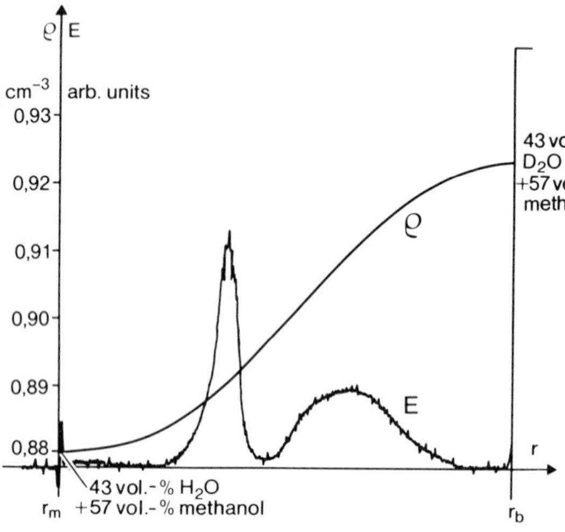

Figure 9. Density and extinction profile in rapid density gradient centrifugation of a mixture of two latices of polybutadiene ($c \approx 20$ *mg* L^{-1}, N = *40.000 min*$^{-1}$, $t - t_o$ = *30 min,* λ_v = *546 nm)*

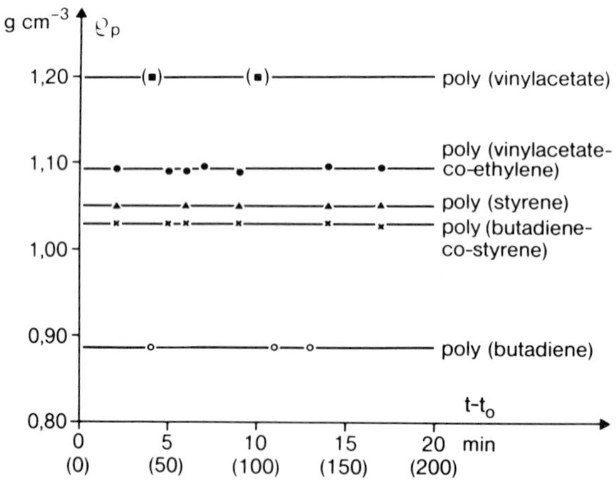

Figure 10. Density of different polymer latex particles as a function of centrifugation time (N = *40.000 min*$^{-1}$) *((■) polyvinyl acetate; (●) poly(vinylacetate-co-ethylene); (▲) polystyrene; (×) poly(butadiene-co-styrene); (○) polybutadiene)*

after only a few minutes and there are no changes during the further centrifugation time.

Thus the density and the density distribution and consequently the chemical composition and heterogeneity and also the cross-linking of polymer latex particles can be determined in a few minutes by rapid density gradient centrifugation.

Acknowledgement

I am very grateful to Mr. H. Hartung, who carried out and evaluated the experiments.

Abstract

For the investigation of the chemical composition of dispersed particles, e. g. in polymer latices, a rapid method of density gradient centrifugation is developed. The method yields in a few minutes the density distribution and by this the distribution of the chemical composition of the particles. The standard density gradient centrifugation needs for the same results because of the time-consuming equilibrium gradient formation more than 15 hours.

The rapid method is based on adding a layer of a dispersion medium of low density to a chemically similar medium of high density and forming the density gradient by diffusion (e.g. H_2O to D_2O or 3-butene-2-ol to ethylene glycol). The main field of application of this method is the investigation of the chemical composition and heterogeneity of copolymer latices and latex mixtures and of the particle cross-linking.

Literature Cited

1. Scholtan, W.; Lange, H. Kolloid-Z. u. Z. Polymere 1972, 250, 782.
2. Menold, R.; Kaiser, W.; Lüttge, B. Chem. Ing. Technik 1971, 43, 1030.
3. Menold, R.; Kaiser, W.; Schmidt, A. Chem. Ing. Technik 1972, 44, 1226.
4. Roth, C.; Gebhart, J.; Heigwer, G. J. Colloid and Interface Sci. 1976, 54, 265.

RECEIVED April 6, 1981.

15

The Relationship Between the Electrophoretic Mobility and the Adsorption of Ions on Polystyrene Latex

C. M. MA[1], F. J. MICALE, M. S. EL-AASSER, and J. W. VANDERHOFF

Department of Chemistry, Center for Surface and Coatings Research, Sinclair Laboratory #7, Lehigh University, Bethlehem, PA 18015

The stability of latexes is important for many industrial applications and in theoretical research. One of the measures of the stability is the electrophoretic mobility or zeta potential of the latex particle. Thus far the chemisorbed surface charge groups have been considered as the dominant factor determine the ζ potential. In a previous paper (1), we concluded that the origin of surface charge on polystyrene latex was not simply the charged ionizable groups on the surface of the latex particle, and that the hydrophobic surface of the latex might be playing a more significant role in the origination of the surface charge. In this work, the relationship between electrophoretic mobility and surface charge of latexes, especially the effect of the adsorption of ions from solution on the electrophoretic mobility of polystyrene latex particles, was investigated in more detail. The possible mechanisms of the origination of the surface charge on latex particles are the specific objectives of this investigation.

Experimental

The latexes investigated were the 357 nm Dow monodisperse polystyrene (LS-1010) and two polydisperse polystyrene latexes prepared in our laboratory (2) where the concentration of functional monomer, Cops II (Alcolac-ammonium salt of a short chain vinyl sulfonate), added to the recipe was 10^{-3} and 10^{-1}M for C_1 and C_5, respectively.

The latexes were cleaned by using a mixed ion-exchange resin bed (3). The number of acid surface groups were determined by using a conductometric titration technique (4). Initially the latexes contained only sulfate groups and a small concentration of hydroxyl groups. One of the latexes was hydrolized at 90°C for 120 hrs in order to convert the surface sulfate groups to hydroxyl groups. The electrophoretic mobility of the latexes was measured at 25°C by using a Rank-microelectrophoresis.

[1]Visiting Research Scientist, Peking University, Peking, Peoples Republic of China.

0097-6156/81/0165-0251$05.00/0

The adsorption of sodium laurylsulphate on PS latex was determined by using a differential refractometer (5). The sodium lauryl-sulfate was a product of Eastman Kodak Company and did not undergo further purification.

Results and Discussion

1. Characterization of the Polystyrene Latex Samples. The polystyrene, PS, latex samples under investigation were characterized according to particle size, concentration of surface sulphates and electrophoretic mobility, em, in deionized water. The results, Table I, show that all the samples were of similar size with the exception that the Dow latexes were monodisperse while C_1 and C_5 had a broad particle size distribution. The Dow and C_1 latexes had relatively low surface charge densities, as determined from conductometric titration, of 3.1 and 5.9 μc/cm^2, respectively, while the C_5 latex exhibited a high surface charge density of 74 μc/cm^2. The decrease in surface charge density for C_1 and C_5, indicated in Table I, is attributed to partial hydrolysis of the samples which were stored under room temperature conditions for about two years. The Dow latex, however, was intentionally hydrolyzed at 90°C with the result that no acid groups were detectable. A striking feature of the electrophoretic mobility results in deionized water is the total lack of dependence on the concentration of surface acid, sulphate, groups. The indication is that negative ions, the only possibility being hydroxyl ions, adsorb on the hydrophobic part of the PS surface.

2. The Effect of NaCl on the Electrophoretic Mobility of PS Latex Particle. The em of the Dow 357 nm latex in the H-form and Na-form, along with two other Dow monodisperse latexes in the H-form with diameters of 795 and 1100 nm, was measured as a function of NaCl concentration. The results in Figure 1 show that the em for all three latexes increased with increasing concentration of NaCl to a maximum at about 1 x 10^{-2} M NaCl followed by a rapid decrease. Converting the electrophoretic mobility to zeta potential, using tables derived by Ottewill and Shaw (6) from the results of Wiersma et al. in order to account for relaxation and retardation effects, led to the same dependency as shown in Figure 2.

The H-form latex was converted to Na-form latex by adding either an excess or an exact amount of NaOH as determined by conductometric titration in order to eliminate the possible effect of the ion-exchange between the H^+ ion on the latex particle and Na^+ ion in the solution. The results showed that the Na-form latex had the same em (3.2 μ.cm/sec. volt) with H-form latex (3.1 μ.cm/sec. volt) in deionized water and same increasing dependency of mobility with increasing NaCl concentration, Figure 1. A reasonable explanation for the increase in zeta potential is the adsorption of negative chloride ion from solution to the surface of latex particle. The decrease in em above 10^{-2}M NaCl is associated with compression of the electrical double layer.

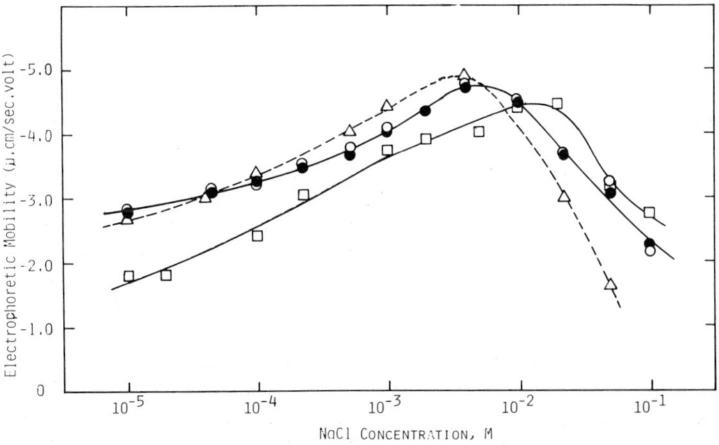

Figure 1. Electrophoretic mobility of PS latexes ((●) Na-form 357 nm; (○) H-Form 357 nm; (△) 795 nm; (□) 1100 nm) as a function of NaCl concentration

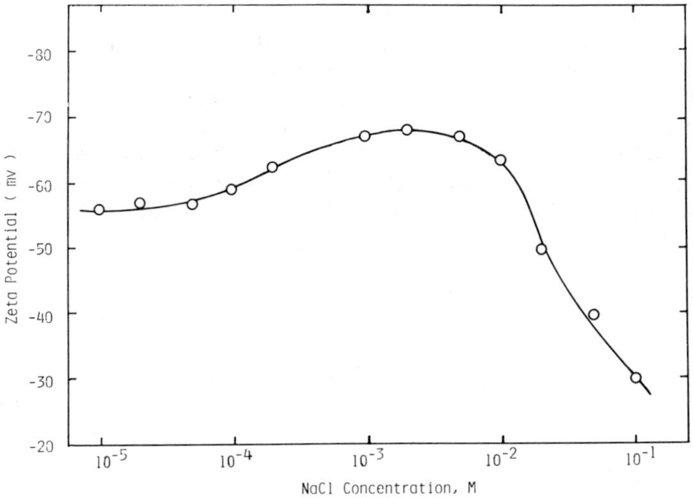

Figure 2. Zeta potential of 357-nm PS latex as a function of NaCl concentration

3. The Electrophoretic Mobility of the Hydroxyl-form 357 nm PS Latex. The hydroxylated 357 nm PS latex was shown in Table I to have the same em in deionized water as the original latex even though no acid groups were detectable by conductometric titration. Since the proton on the hydroxyl was assumed not to be dissociable, the origin of the negative charge was attributed to the adsorption of anions, in this case hydroxyl ions. The em of this latex in both the hydroxyl-form and the original sulphate-form was investigated in Figure 3 as a function of pH. The results for both samples are very similar above a pH value of 6. When the latexes are exposed to more acidic environments, however, the negative potential of the hydroxyl-form latex decreases more rapidly with decreasing pH than the original latex. Figure 4 shows the results for the same samples as a function of pH at a constant NaCl concentration of 2 x 10^{-3}M. No difference in results was observed at high pH except that the em for both samples was higher below pH values of 10. Furthermore, when the pH was decreased below 5, the em decreased faster with NaCl present. Also, the maximum at pH 2.5 was eliminated in the presence of NaCl. The em for the hydroxyl-form of the latex was also observed to decrease faster in the low pH range possibly due to the adsorption of protons on the surface hydroxyls.

The em of the hydroxyl-form latex was also investigated as a function of the concentration of the different electrolytes NaOH, NaCl and HCl. The results in Figure 5 show that the charge on this latex is independent of the type of electrolyte at concentration below 10^{-5}M. The results at higher concentrations show that hydroxyl ions adsorb more strongly than chloride ions, and that hydrogen ions adsorb more strongly than sodium ions. The same trends were observed for this latex in the sulphate form. These results are consistent with the concept that the origin of charge on PS latexes both with and without surface acid groups, is due primarily to the adsorption of ions onto the hydrophobic parts of the surface.

Table I
Particle Size and Surface Charge Density of PS Latexes

Latex	Particle Size, nm	Surface Charge Density (sulfate group), μc/cm^2	Electrophoretic Mobility in Deionized Water, μm cm/sec volt
Dow LS 1010 (sulfate form)	357	3.1	3.2
Dow LS 1010 (hydroxyl form)	357	0	3.1
C_1	416	5.9 (11)*	3.7
C_5	448	74 (174)*	4.3

*Original results obtained by S.M. Ahmed soon after sample preparation (2).

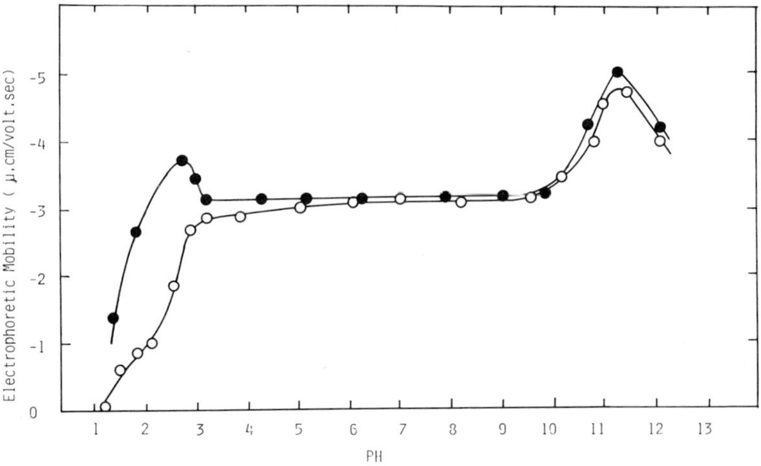

Figure 3. Electrophoretic mobility of 357-nm PS latex as a function of PH ((●) sulfate form; (○) hydroxyl form)

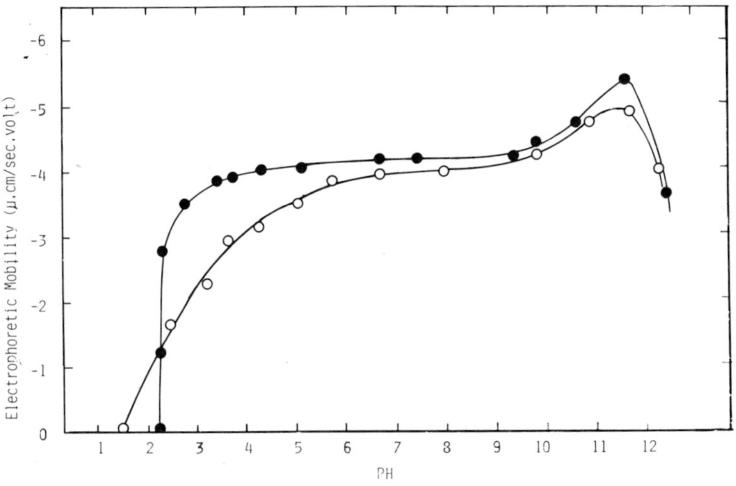

Figure 4. Electrophoretic mobility of 357-nm PS latex as a function of PH in 2×10^{-3}M NaCl ((●) sulfate form; (○) hydroxyl form)

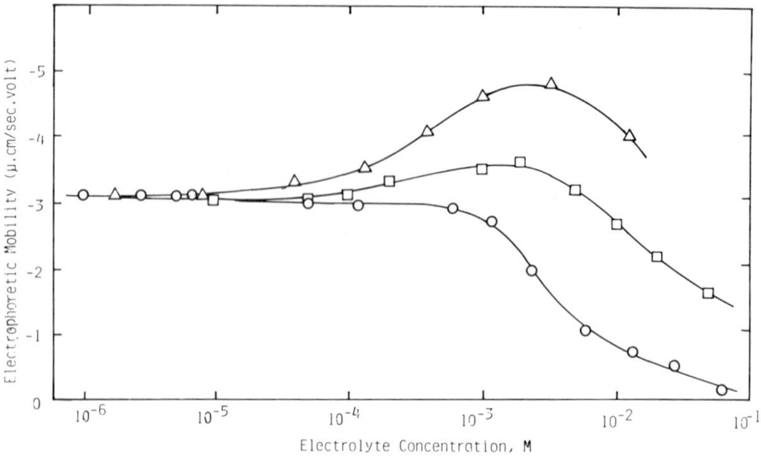

Figure 5. Electrophoretic mobility of the hydroxyl form 357-nm PS latex as a function of the concentration of electrolytes ((△) in NaOH; (□) in NaCl; (○) in HCl)

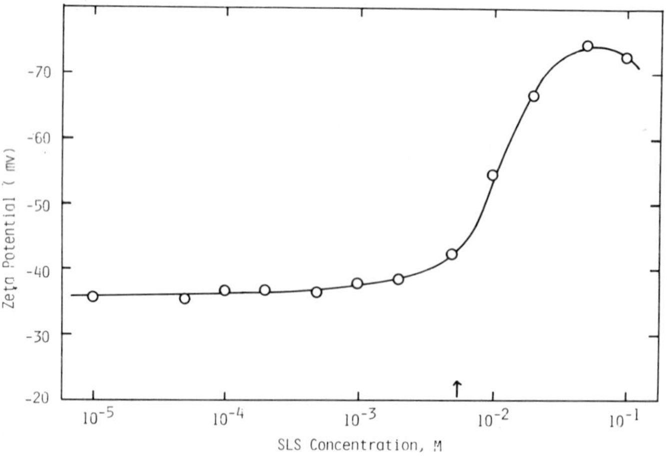

Figure 6. Zeta potential of 357-nm PS latex as a function of SLS concentration

4. The Effect of Adsorbed SLS on the Electrophoretic Mobility of PS Latex. The zeta potential, corrected for retardation and relaxation effects (6), of the original 357 nm PS latex was determined as a function of SLS concentration. The results in Figure 6 show that there is essentially no change in the zeta potential up to the CMC after which the zeta potential increases, which is consistent with results obtained by Kayes (7). Solution adsorption isotherms on the same latex at different NaCl concentration, Figure 7, show that the adsorption occurs up to the CMC, after which adsorption tends to level off. The fact is that the zeta potential measured under these conditions is not a function of the concentration of adsorbed SLS. The interpretation of these results, which is consistent with the results presented above, is that the negative charge on the PS latex surface in deionized water is controlled primarily by the adsorption of hydroxyl ions. The increase in SLS concentration results in adsorption LS ions, possibly through displacement of hydroxyl ions, and a decrease in the degree of dissociation of the surface sulphate groups. The increase in the zeta potential above the CMC is not well understood at this time, but could be due to the adsorption of micelles, which concentration range has been beyond the scope of solution adsorption measurements.

The em of the 357 nm PS latex was measured as a function of SLS concentration in the presence of different concentrations of NaCl. The results in Figure 8 show that NaCl affects the negative charge of this latex in a competitive fashion with SLS. As the concentration of NaCl increases the negative charge on the latex surface increases, and subsequent adsorption of SLS has a decreasing effect on the existing surface or zeta potential. Although Piirma (8) has shown that the saturated adsorption of SLS on PS latex increases with increasing NaCl concentration, our results, Figure 7, show only a small increase in the adsorption of SLS. The interpretation of the results in Figure 8 is that the Cl ion adsorbs on the hydrophobic part of the PS latex surface, and that LS ion displaces the Cl ion by preferential adsorption. As the adsorbed Cl ion concentration increases, subsequent adsorption of the LS ion by displacement of the Cl ion has a decreasing effect on the surface potential. These results also suggest that the decreasing order of adsorption of anions are LS, OH, and Cl respectively.

5. Electrophoretic Mobility of Latexes with High Concentration of Surface Sulphate Groups. The two PS latexes, C_1 and C_5, were prepared with increasing concentrations of sulphate functional monomers (2). Conductometric titration results for acid groups, Table I, showed that the surface charge density was 5.9 and 74 $\mu c/cm^2$ for C_1 and C_5, respectively, while the em in deionized water was essentially insensitive to surface charge density. The em for these two latexes was measured as a function of NaCl concentration. The zeta potential calculated from these results, Figure 9, show that for C_1 there is no change up to a con-

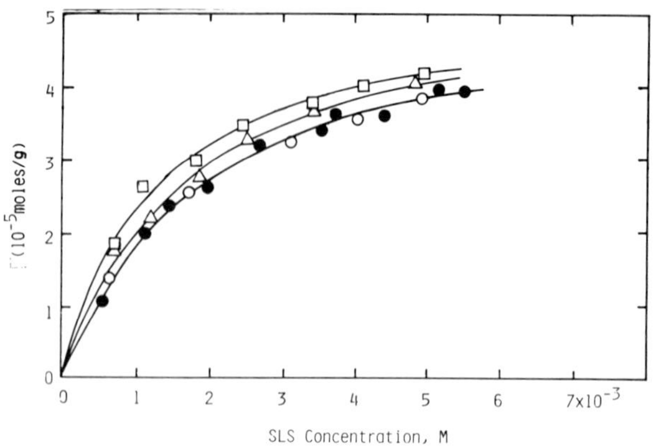

Figure 7. Adsorption of SLS on 357-nm PS latex ((●) without NaCl; (○) 10^{-4}M NaCl; (△) 10^{-3}M NaCl; (□) 10^{-2}M NaCl)

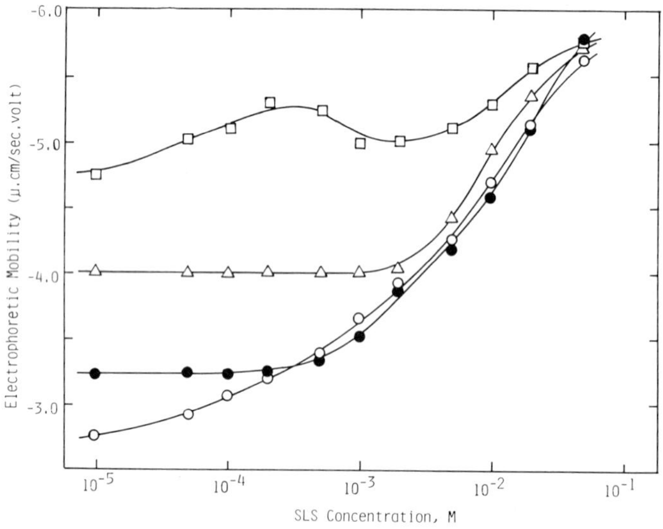

Figure 8. Electrophoretic mobility of 357-nm PS latex as a function of concentration of SLS in different concentrations of NaCl ((○) without NaCl; (●) 10^{-4}M NaCl; (△) 10^{-3}M NaCl; (□) 10^{-2}M NaCl)

centration of 10^{-3}M, after which the zeta potential increases to a maximum followed by a decrease. The C_5 sample, which has an order of magnitude higher concentration of surface acid groups, however, decreases linearly with increasing NaCl concentration up to 10^{-3}M, after which the zeta potential remains constant. The interpretation of these results is that the zeta potential of C_1 is controlled primarily by the adsorption of Cl ions replacing OH ions on the hydrophobic part of the surface, while the zeta potential of C_5 is controlled by the decreasing ionization potential of the surface sulphates with increasing concentration of Na ions.

The em was also measured for C_1 and C_5 as a function of SLS concentration. The results, Figure 10, show that the zeta potential increases with increasing SLS in a manner similar to what was obtained for the 357 nm PS latex in Figure 8. The C_5 sample, however, decreased linearly with increasing SLS concentration up to 10^{-3}M, followed by an increase above the CMC concentration. The concentration effect of SLS on the zeta potential of the C_5 sample is in quantitative agreement with the concentration effect of NaCl on this sample up to 10^{-3}M. These results suggest that SLS does not adsorb strongly on a latex sample containing a high concentration of surface sulphates, and that the decrease in zeta potential with increasing concentration of either NaCl or SLS is due to the decrease in the degree of dissociation of the surface sulphates with increasing Na ion concentration. Solution adsorption isotherms of SLS was measured on C_1 and C_5. The results on C_5 were limited because of the limited amount of sample available. The results, Figure 11, show that SLS adsorbs to a much greater extent on C_1 as compared to C_5 where the calculated cross sectional area were 57 and 300 Å^2/molecule for C_1 and C_5, respectively. These results are in agreement with the interpretation of the em results proposed above.

Conclusions

PS latexes are generally prepared by emulsion polymerization using potassium persulphate as initiator. The concentration of surface sulphates in these latexes are usually in the range of 1 to 5 sites/1000 Å^2. The results presented in this paper show that the zeta potential is independent of the concentration of surface sulphates when the concentration of these surface sulphates is relatively low. The controlling mechanism for the generation of surface potential in these cases is the tendency for the hydrophobic part of the surface, which represents a much greater fraction of the surface, to adsorb negative ions. The anions which were investigated were LS, OH, and Cl in that order with respect to preferential adsorption. When the concentration of surface sulphates is high, i.e. greater than 15 sites/1000 Å^2, the surface is hydrophilic and the zeta potential is predominantly a function of the degree of dissociation of the surface sulphates.

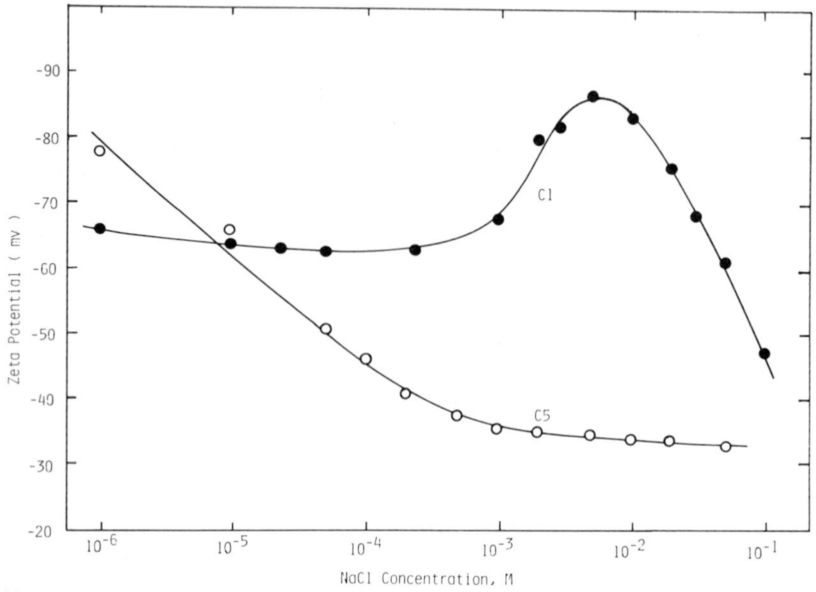

Figure 9. Zeta potential of PS latex C1 and C5 as a function of concentration of NaCl

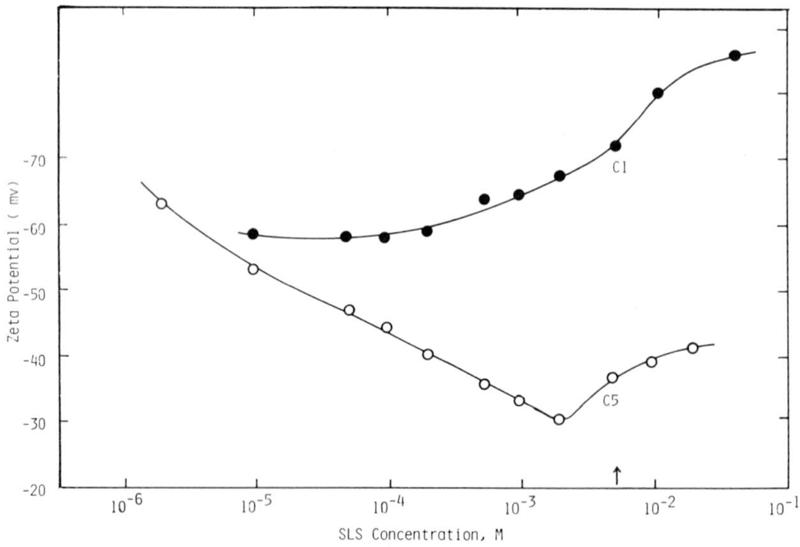

Figure 10. Zeta potential of the PS latex C1 and C5 as a function of the concentration of SLS

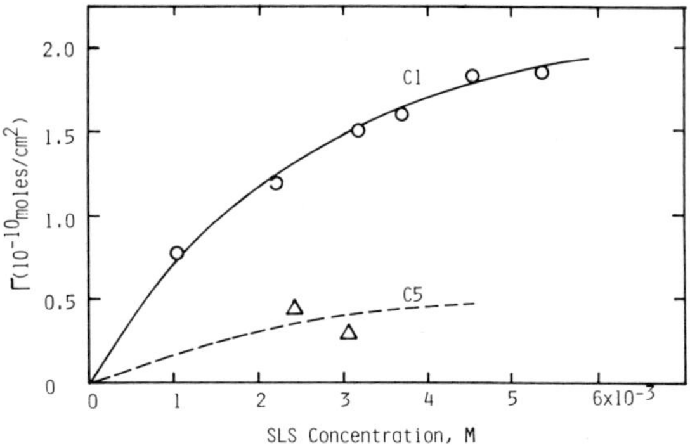

Figure 11. Adsorption of SLS on PS latex C1 and C5

Literature Cited

1. Kamel, A., Ma, C. M., El-Aasser, M. S., Micale, F. J., Vanderhoff, J. W., presented at 54th Colloid and Surface Science Symposium, June 1980.
2. Ahmed, S. M., Ph.D. thesis, Lehigh University, 1979.
3. Vanderhoff, J. W., Van den Hul, H. J., J. Colloid Interface Sci., 1968, 28, 336.
4. Van den Hul, H. J., and Vanderhoff, J. W., Brit. Polymer J. 1970, 2, 121.
5. Saunders, F. L., J. Colloid and Interface Sci., 1968, 28, 475-80.
6. Ottewill, R. H., and Shaw, J. N., J. Electroanal. Chem., 1972, 37, 133-42.
7. Kayes, J. B., J. Colloid and Interface Sci., 1976, 56, 426-442.
8. Piirma, I., Chen, S. R., J. Colloid and Interface Sci., 1980, 74, 90-102.

RECEIVED April 6, 1981.

16

Techniques for Measuring Particle Swelling of Carboxylic Emulsion Polymers

D. R. BASSETT, E. J. DERDERIAN, J. E. JOHNSTON[1], and T. B. MacRURY[2]

Technical Center, Union Carbide Corporation, South Charleston, WV 25303

Carboxylic monomers, such as acrylic or methacrylic acid, are included in emulsion polymerization recipes for a variety of reasons: to increase the stability of the latex particles, to improve the adhesion of the resultant films to various substrates, to provide functional crosslinking sites for interparticle thermosetting reactions and to control the viscosity of latex formulations via particle swelling upon neutralization. This swelling or partial solubilization, has been qualitatively investigated by titration (1) and turbidity measurements (2) as well as by viscometry and microscopy (3). More recently Nishida, et al. (4), characterized the swelling behavior of methyl methacrylate/methacrylic acid emulsion polymers using viscometry and light scattering. It is clear that particle swelling is influenced by the type and concentration of carboxylic monomer incorporated into the particle, the relative hydrophilicity of the comonomers employed in the polymerization, the stiffness (T_g) of the polymer chains, and the molecular weight of the polymer. While most studies of particle swelling have dealt with emulsion polymers containing high acid levels, we have restricted our attention to latexes containing relatively low incorporated acid (ca.2-3%) to avoid substantial solubilization of the particles. Questions under current consideration are related to the location of incorporated carboxyl groups within the latex particles, the morphology of expanded particles, and the detailed mechanism of expansion. Information of this type is valuable not only from a fundamental standpoint but is essential in the investigation of very practical problems such as the viscosity stability of latex formulations.

Central to any investigation of particle swelling is the use of a reliable method of measuring the size of the particles undergoing expansion. A sedimentation method was described previously (5) which can be used to explore the expansion characteristics of carboxylic emulsion polymers. In the present report, we present a comparison of sedimentation results with those obtained with two

1. Current address: Exxon Chemicals, Linden, New Jersey
2. Current address: IMC Corporation, Terre Haute, Indiana

0097-6156/81/0165-0263$05.00/0

other quantitative methods, precision viscometry and photon correlation spectroscopy. Of particular concern in this comparison is the effect of interparticle interactions. The special problems associated with the use of light scattering techniques to probe the internal structure of expanded particles is described elsewhere in this Monograph (6).

EXPERIMENTAL

Latex Preparation. Model latexes were prepared by a semi-batch technique in which a monomer mix was continuously fed into a stirred reactor at 80°C in the presence of ammonium persulfate and a single anionic surfactant, Aerosol OT (American Cyanamid). The initial surfactant concentration in the reactor was adjusted to produce a particle size of approximately 0.1 μm. The standard acrylic polymer composition was 40 parts methyl methacrylate, 54 parts ethyl acrylate and 6 parts butyl acrylate. The polymerization was arbitrarily divided into two stages, Stage I being the first half of the monomer feed and Stage II being the last half of the feed. Since previous expansion studies (5) revealed that particle swelling is accentuated by late addition of the carboxylic monomer, the Stage II addition was chosen for the present comparison. Acrylic acid, amounting to 2 percent of the total monomer mix, was added to the last half of the monomer feed used to prepare the model carboxylic latex. A second acrylic latex was prepared under the same conditions without the monomeric acid. Several much harder latexes were also prepared using methyl methacrylate with 2 or 3 percent acrylic acid added in Stage II.

Sedimentation. The sedimentation method, described previously (5), is based on the relative sedimentation rates of swollen and unswollen latex particles. Starting with the Stokes expression for centrifugal sedimentation, an equation can be developed for the ratio of the sedimentation coefficient of a particle, S_o, to the sedimentation coefficient, S, of the same particle having a surface layer:

$$\frac{S_o}{S} = \frac{r^2(\rho_c - \rho_o)(r+x)}{r^3(\rho_c - \rho_o) + (3r^2x+3rx^2+x^3)(\rho_e - \rho_o)} \tag{1}$$

where r is the radius of the particle, ρ_c is its density, ρ_o is the density of the suspending medium, ρ_e is the density of the layer and x is the thickness of the layer.

Surface layers (adsorbed, solvated, ionic) are of considerable importance in controlling the stability and rheological properties of colloidal systems. Sedimentation methods have proven effective in the measurement of adsorbed layer thickness using equations similar to Equation 1 when the density of the layer could be estimated (7,8). The equation can be considerably simplified if the density of the adsorbed layer is equal to that of the surrounding fluid. This simplification was used by Ottewill and Walker (7) in their study of the adsorption of a nonionic surfactant onto polystyrene latex in aqueous sodium chloride. In the case of carboxylated emulsion polymers, evidence from conductometric titrations suggests that the carboxyl groups are generally concentrated near the particle surface. The resultant model of an expanded particle is that of a hydrated acid-rich shell surrounding a compact polymer core. The hydrated shell may be viewed as a dilute polymer solution where the density is close to that of water, i.e., $\rho_e = \rho_o$. With this assumption, Equation 1 reduces to the form:

$$\frac{S_o}{S} = \frac{r+x}{r} \tag{2}$$

Actually, information regarding the internal structure of the swollen particle is not necessary since the change in particle volume is equal to the volume of water absorbed, and the expanded particle settles slower, as its average density decreases, according to Equation 2 in either case. As with other hydrodynamic methods, sedimentation does not offer easy access to information regarding particle morphology.

In this study, model latexes were diluted with distilled water to 1 percent solids by weight. Individual samples were adjusted with sodium hydroxide to various pH values from 3 to 13 and allowed to equilibrate 24 to 48 hours depending on the pH drift observed. Sedimentation coefficients were determined at 30°C with a Beckman Model E analytical ultracentrifuge using a single sector cell. Since the latexes had fairly narrow particle size distributions, precise measurements of the sedimentation coefficients were obtained from the change in boundary position with time. The operating speed of the rotor was selected to give a sedimentation time of at least 20 minutes.

Viscometry. At low concentrations of suspended solids, the viscosity, η, of a dispersion increases with the volume of the suspended phase. Einstein (9) proposed a relationship for spherical particles:

$$\eta = \eta_o\ (1+2.5\ \phi) \tag{3}$$

where η_o is the viscosity of the suspending liquid and ϕ is the volume fraction of the suspended phase. The Einstein equation assumes no interaction between particles and is strictly applicable only at very low values of ϕ (ca. 0.002 or below). This restriction necessitates extreme experimental accuracy due to the small difference in viscosity between the suspension and the liquid medium.

Many attempts have been made to extend Equation 3 to higher concentrations. Of these, we have chosen the equation of Guth, Simha and Gold (10,11).

$$\eta = \eta_o\ (1+2.5\ \phi + 14.1\ \phi^2) \tag{4}$$

which has been verified for polystyrene latexes at volume fractions up to 0.075 (12). Similar extensions have been utilized in studies of polymer adsorption on irregular particles (13,14). The particle volume fraction, ϕ, is the hydrodynamic volume of the suspended phase. In the case of an expanded latex particle, this value will be greater than that of the dry polymer such that

$$\frac{\phi}{\phi_0} = \frac{\rho\phi}{c} \tag{5}$$

where ϕ_0 is the volume fraction of the dry polymer, ρ is the density of the dry polymer and c is the concentration of the polymer (g/ml). This equation leads to a swelling ratio similar to Equation 2:

$$\frac{r+x}{r} = \left(\frac{\phi}{\phi_0}\right)^{1/3} = \left(\frac{\rho\phi}{c}\right)^{1/3} \tag{6}$$

with ϕ given by Equation 4.

Viscometric particle swelling experiments were carried out with a Cannon-Ubbelohde shear-dilution viscometer thermostated at 30 $\pm$ 0.05°C. The shear rate was approximately 2000 sec^{-1} with flow times determined to $\pm$ 0.1 sec. As in the sedimentation method, the model latexes were diluted with distilled water to a concentration of approximately 1 percent. Individual samples were adjusted with sodium hydroxide to various pH values and allowed to equilibrate at least 24 hours before measuring the viscosity.

Photon Correlation Spectroscopy. In the past several years, photon correlation spectroscopy (PCS), also known as quasielastic light scattering, has become an increasingly popular technique for measuring the particle size of colloidal dispersions and macromolecules in solution. A detailed account of the theory and applications has been given by Berne and Pecora (15). Briefly, when a colloidal dispersion is illuminated by a light source, the intensity of the scattered light fluctuates around its mean value because of the Brownian movement of the particles. In PCS, the time dependence of these intensity fluctuations is related to the translational diffusion coefficient of the particles. In a typical experiment, the autocorrelation function, $C(\tau)$, of the time-dependent intensity is determined. The autocorrelation function is a measure of the correlation in the scattering intensity and, for monodisperse spheres, is characterized by a single exponential decay with time:

$$C(\tau) = A\,[1 + B \exp(-\tau/\tau_c)] \tag{7}$$

where A is the average scattering intensity, B is a spatial coherence factor, and τ is the correlation delay time. The exponential time decay is determined by a single characteristic correlation time,

τ_c, of the intensity fluctuations. In a homodyne light scattering experiment, τ_c is related to the translational diffusion coefficient, D_T, of the particles by the relationship

$$\tau_c = \frac{1}{2K^2 D_T} \tag{8}$$

where K is the scattering vector which is a function of the scattering angle, the wavelength of the light source and the refractive index of the medium.

For dilute dispersions of spherical particles, the diffusion coefficient can be related to the hydrodynamic diameter of the particles by the Stokes-Einstein equation

$$d = \frac{kT}{3\pi\eta D_T} \tag{9}$$

where k is the Boltzmann constant, T the absolute temperature, and η the viscosity of the medium.

For this description of PCS, it is evident that, for monodisperse systems, the technique can provide an absolute measurement of hydrodynamic size; knowledge of the density or refractive index of the particles is not required, and no calibration or correction is needed. With the advent of digital correlators and microprocessors, PCS has also become a very fast and precise technique. Recent studies of latex using PCS include adsorbed layers (8), particle sizes (16), surface characterization (17) and aggregation (18).

The present study was carried out using a Chromatix KMX-6DC low-angle light scattering photometer (Chromatix, Inc., Sunnyvale, California) interfaced with a 64-channel digital correlator (Chromatix Model 64) which is linked to a DEC LSI-11 data processing system. The excitation source is a 2mW He-Ne laser (λ =6328 Å). A Tektronix T921 oscilloscope is used to display the temporal decay of the autocorrelation function.

The design of the photometer, shown in Figure 1, allows light scattering to be done in the backward as well as the forward direction. At angles near 180°, the correlation times of latex particles and dust are well separated; furthermore, at 180° there is a large attenuation in the scattering intensity from dust due to destructive interference. As a result, the autocorrelation function due to scattering from the latex particles is not affected by the presence of dust; consequently, the latex samples did not require filtration. All measurements were carried out at 175.5° in the backward direction at 20°C using the homodyne mode. The full (unclipped) autocorrelation function was used to obtain D_T. An optimum correlator sample time was determined for each latex. Details of this procedure are given in Reference 16.

The model latexes were diluted with doubly distilled water to a concentration of 5 x 10^{4}% by weight. Each sample was adjusted

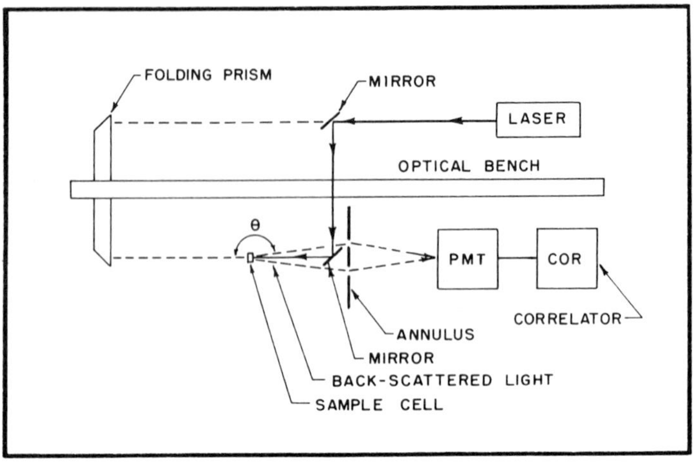

Figure 1. Diagram of the optical arrangement for back-scattering at $\theta = 175.5°$ used in photon correlation spectroscopy

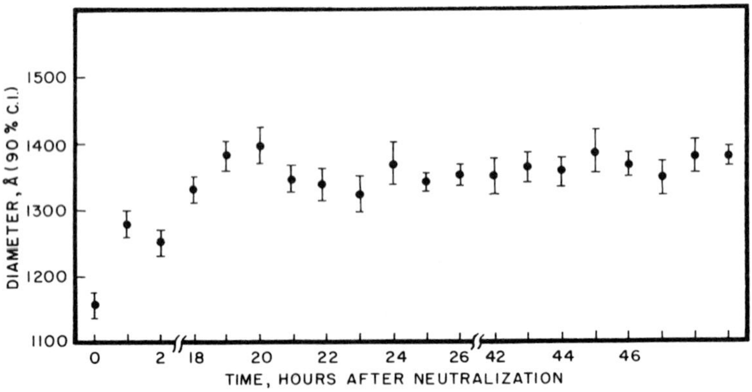

Figure 2. Effect of equilibration time on carboxylic particle expansion at pH = 12.5 as determined by photon correlation spectroscopy. The confidence interval on $\bar{d}$ *is at the 90% probability level. The initial particle size is given by* $\bar{d}_o$ *(acrylic latex, 2% AA(II),* $\bar{d}_o = 1120$ *Å).*

to the desired pH range with sodium hydroxide and equilibrated as in the sedimentation method; the final pH was measured at the time the light scattering measurement was made. The need for adequate equilibration time is shown in Figure 2 for a high pH sample of the model acrylic latex. Periodic measurement of the same sample after addition of sodium hydroxide showed that approximately 24 hours were required to reach the final equilibrated degree of expansion at this pH. Equilibration was much faster at the lower pH's where little expansion was observed. This exercise also illustrates one of the benefits of PCS in that time effects can be conveniently studied. Twenty measurements were carried out at each pH; error limits on each point represent the 90% probability level.

RESULTS AND DISCUSSION

The sedimentation ratio, S_0/S, is presented in Figure 3 as a function of pH for two model latexes. The upper curve illustrates the general characteristics of carboxylic particle expansion (5): an initial region in which the particle size is independent of pH, an abrupt increase in size over a fairly small range in pH, a narrow maximum, and finally an abrupt contraction. The behavior of the latex containing no polymerized acid is quite different, exhibiting no expansion over the entire pH range. In fact, the particles undergo a slight contraction at high pH.

A general explanation of the expansion behavior illustrated in Figure 3 is as follows. At low pH the incorporated acid groups possess little charge, but as the pH is increased, electrostatic charge is generated to develop sufficient repulsion to separate the acid-rich polymer chains. The ionized chains expand, absorbing water in the process, thereby reducing the density of the particles and hence causing a decrease in the sedimentation rate. The pH at maximum expansion corresponds to the pH (from potentiometric titration) at which most of the carboxyls are neutralized. Excess sodium hydroxide acts like a simple electrolyte, and the hydrated particle shrinks as the repulsive centers are shielded. The shrinkage of the latex having no incorporated acid is due to the same excess electrolyte effect and illustrates the presence of a hydration layer around most colloidal particles suspended in water. At present we lean toward a particle structure in which the acid groups are preferentially located near the particle surface, but hydrodynamic measurements cannot differentiate between homogeneous and core-shell particle structures.

The viscometric expansion results for the same model latexes are shown in Figure 4. Since sedimentation and PCS measurements include whatever hydration layer is already present on the particles before expansion (low pH), the initial point on the expansion curves in Figure 4 was set equal to 1.0 for comparison with the other two methods. Actually, the initial value of $(\emptyset/\emptyset_0)^{1/3}$ for the carboxylic latex was 1.08 indicating a layer thickness of about 45 Å present at pH = 3.8. The general characteristics of the

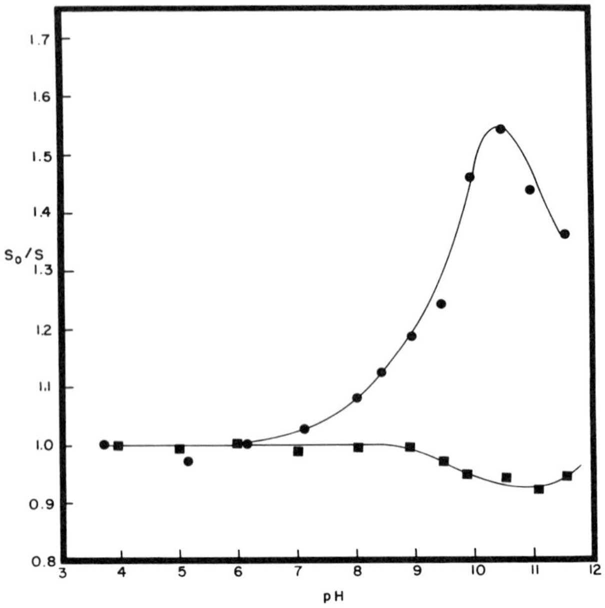

Figure 3. Expansion characteristics of model acrylic latexes as determined by sedimentation ((●) 2% acrylic acid—Stage II; (■) no acid)

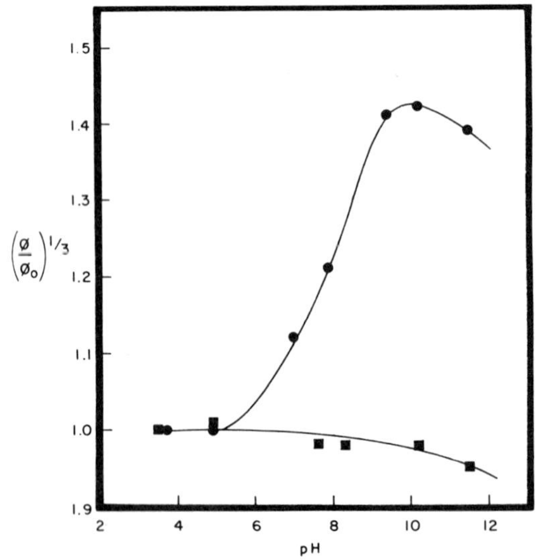

Figure 4. Expansion characteristics of model acrylic latexes as determined by viscometry ((●) 2% AA(II); (■) no acid)

curves in Figure 4 are quite similar to those obtained by sedimentation. Particle shrinkage of the noncarboxylic sample, evident in Figure 3, was also detected by viscometry at the higher pH values due to the excess sodium hydroxide.

Figure 5 shows the swelling behavior of the model acrylic latexes as determined by photon correlation spectroscopy. Changes in particle diameter are shown instead of radius as in Figures 3 and 4. The diameter, $\overline{d}$, at each pH was divided by the initial diameter, $\overline{d}_0$, determined at pH = 5 (unexpanded). In general shape, the curves are similar to those determined by sedimentation and viscometry. The initial points occurred at a higher pH due to the much lower particle concentration. No effort was made to lower the initial pH with acid; consequently, the undesirable salt effect was avoided.

A quantitative comparison of particle expansion determined by the three methods is given in Table I. The particle diameter of the standard acrylic latex was determined by PCS to be 1120 Å. This value was used in the calculation of the increase in particle radius at maximum expansion in each case. The sedimentation method yielded the largest increase in radius, 302 Å, followed by the viscometric value of 240 Å. Possibly the shear involved in the latter method resulted in a partial collapse of the surface layer. The value determined by PCS was found to be approximately half that determined by sedimentation. Since the PCS determination is presumed to be free of particle interactions at a concentration of 5×10^{-4}%, we must conclude that the other two methods (at 1% solids) exhibit such interactions. As a result, the charged particles settle slower (19) and yield a higher viscosity than in the absence of these (repulsive) interactions.

TABLE I

PARTICLE EXPANSION COMPARISON

Method	Particle Radius, Å	$\left(\frac{r+x}{r}\right)_{max}$	Increase in Radius, Å
Sedimentation	560	1.54	302
Viscometry	560	1.43	240
PCS	560	1.28	157

In an effort to estimate the magnitude of the decrease in sedimentation rate due to interparticle interactions, several poly (methyl methacrylate) latexes (PMMA) were prepared since PMMA particles are too hard for expansion to occur at low acid levels. Thus surface charge can be adjusted in the absence of expansion. The lower curve in Figure 6 shows the sedimentation behavior of a PMMA latex containing 2% acrylic acid added in Stage II in the same manner employed to make the model carboxylic acrylic latex. A contraction of approximately 75 Å was observed at pH = 11.

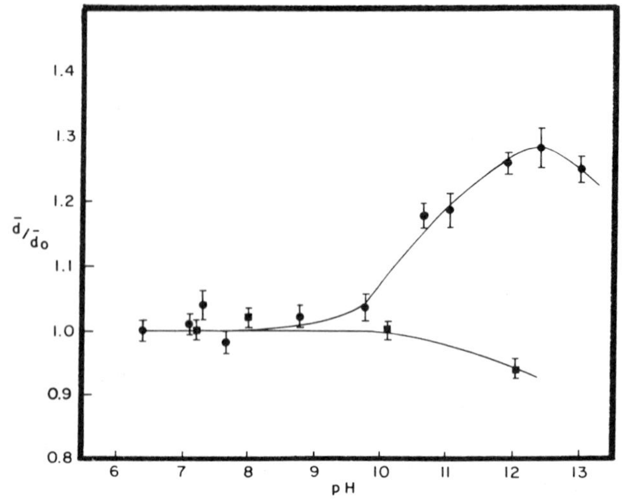

Figure 5. Expansion characteristics of model acrylic latexes as determined by photon correlation spectroscopy ((●) 2% AA(II), d_o = *1120 Å; (■) no acid,* $\bar{d}_o$ = *855 Å)*

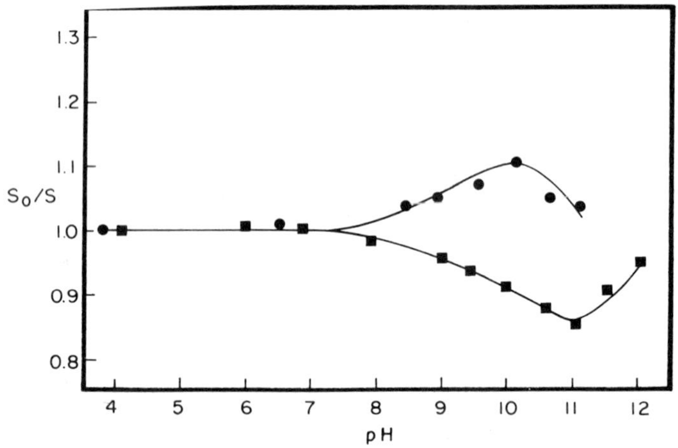

Figure 6. Expansion characteristics of PMMA latexes containing 2% ((●), $\bar{d}_o$ = *0.10 μm) and 3% ((■),* $\bar{d}_o$ = *0.11 μm) acrylic acid (Stage II) as determined by sedimentation*

In order to compare the PMMA results with those obtained with the carboxylic acrylic latex, the concentration of surface carboxyls must be determined. Acid location analysis (5) was carried out for this purpose. Briefly, the latexes were titrated conductometrically with 0.1N NaOH followed by a titration of the aqueous phase from which the particles had been removed by centrifugation. The difference in the two titrations provided the distribution between surface and soluble acid. The deficit between the total acid thus determined and the concentration of acrylic acid used in the polymerization was termed "buried". Although some drift occurred in the conductance with time, an equilibration time of approximately 10 minutes per addition of sodium hydroxide was generally sufficient to yield stable readings.

The results of the acid location analyses are given in Table II. In the case of the standard carboxylic latex used in this study, most of the acid (92%) was found to be incorporated in the surface region of the particles (easily titrated) with a small amount of soluble acid and relatively little unavailable. This analysis illustrates the basis for viewing the particle morphology of soft carboxylic latexes in terms of an acid rich surface region which leads to an expanded hydrated surface layer, at high pH, surrounding a rather compact polymer core. In contrast, the similarly prepared PMMA latex containing 2% acrylic acid (Stage II) yielded substantially less surface acid, 0.90% (only 45% of the total acid), with more than half the polymerized acid unavailable. This result is a good illustration of the effect of polymer T_g on the location of incorporated acid within the particles of carboxylic emulsion polymers. A comparison of the sedimentation properties of this latex with those of the standard acrylic latex is thus inappropriate with regard to surface charge.

TABLE II
ACID LOCATION ANALYSIS OF MODEL LATEXES

Latex (Acrylic Acid in Stage II)	% Acrylic Acid Based on Polymer		
	Surface	Soluble	Buried
STD Acrylic, 2% AA	1.84 (0.92)	0.12 (0.06)	0.04 (0.02)
PMMA, 2% AA	0.90 (0.45)	0.04 (0.02)	1.06 (0.53)
PMMA, 3% AA	1.89 (0.63)	0.21 (0.07)	0.90 (0.30)

Parentheses indicate fraction of total acid in respective location.

A second PMMA latex was prepared, with 3% acrylic acid (Stage II), in an effort to match the surface acid concentration of the standard acrylic latex. The acid location results in Table II show that the surface acid of this latex was indeed close to that of the standard acrylic. The sedimentation curve for this latex is shown in Figure 6 in which an apparent expansion maximum of about 60 Å occurred at pH = 10.2. PCS examination of the same latex, Figure 7, yielded an entirely different curve. Contraction of about 30 Å in radius was observed in the high pH range. The difference in the two results indicates that the sedimentation value at maximum expansion is too high by at least 90 Å due to particle-particle interaction. Figure 7 also shows conclusively that the particle expansion observed in Figure 5 is not an artifact due to particle-particle, electroviscous, double layer or charge effects.

A study of particle shrinkage due to added electrolyte is of interest because it provides an estimate of the size of the hydration layer around particles suspended in water. These values range from 26 Å (PCS) to 45 Å (SED) for the noncarboxylated acrylic latex, 75 Å (SED) for the PMMA + 2% AA and 31 Å (PCS) for the PMMA + 3% AA latex.

The unexpected difference in the PCS expansion results, when compared with those obtained by the other two methods, is that the pH at which maximum expansion occurs is shifted from approximately 10.5 in the case of sedimentation and viscometry to about 12.5 in the case of photon correlation spectroscopy. Since much lower particle concentrations were employed in the PSC experiments, it was of interest to explore the effect of ionic strength on particle expansion as determined by PCS. Figure 8 shows a comparison of expansion for the standard acrylic latex in distilled water and in 0.01 M NaCl. As expected, the size at maximum expansion decreased due to the shielding effect of the electrolyte on the electrostatic repulsion between charged groups. In addition, the pH at which the maximum occurred shifted downward to about pH = 11.2. This shift indicates that the carboxyl groups are more easily ionized the higher the ionic strength of the medium. These results suggest that at very low particle concentration, carboxyl groups on latex particle surfaces are much weaker acids than hitherto suspected (20). A further inference is that even when the ionic strength of the medium is low, the interactions between charged particles facilitates the ionization of weak acid groups in a manner similar to that of discrete ions in solution. The absence of such interactions, as in PCS at low particle concentration, makes the removal of a proton from the particle surface more difficult. Potentiometric titrations on carboxylic latexes at concentrations used in the PCS experiments (5×10^{-4} weight percent) are necessary to verify this hypothesis. Unfortunately, meaningful results are very difficult to obtain with such dilute systems.

CONCLUSIONS

The expansion characteristics of carboxylic latex particles have been measured using three independent techniques: sedimentation, which uses the change in particle density due to swelling to determine the change in particle size; viscometry, which measures volume changes; and photon correlation spectroscopy, which measures the diffusion coefficient of the particles. The sedimentation technique offers precise measurements at low shear but requires relatively

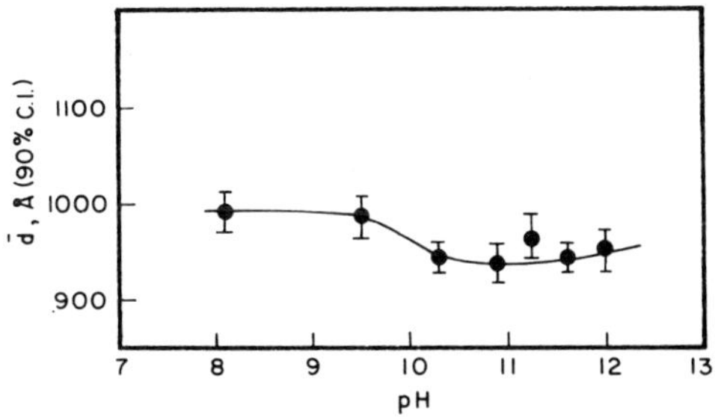

Figure 7. Expansion characteristics of PMMA latex containing 3% acrylic acid (Stage II) as determined by photon correlation spectroscopy

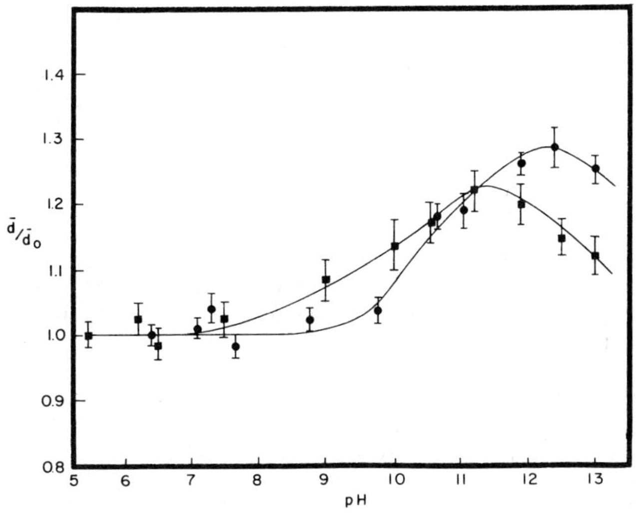

*Figure 8. Effect of added electrolyte (0.01*M *NaCl) on the expansion characteristics of a model carboxylic latex as determined by photon correlation spectroscopy ((●) acrylic latex, 2% AA(II),* d_o *= 1120 Å; (■) same latex in 0.01*M *NaCl,* d_o *= 1010 Å)*

narrow particle size distributions at fairly high solids (ca. 1%) for sharp sedimentation boundaries. The viscometric technique is simple in concept and requires a minimum expenditure for equipment although good temperature control is important. The density of the polymer is required for quantitative results, and shear effects can be important. Again, relatively high solids are required. Photon correlation spectroscopy is carried out at very low particle concentrations so that interparticle interactions are minimized. PCS offers fast, absolute measurements of hydrodynamic size at very low shear rate with excellent precision. The back-scattering mode described here minimizes the effect of dust particles and thus eliminates the need for sample filtration.

Photon correlation spectroscopy, carried out under very dilute conditions, has unambiguously demonstrated the expansion of carboxylic emulsion polymers at high pH, but it may not always be useful in predicting properties of practical interest. Of special concern is the apparent decrease in the intrinsic ionization constant of surface carboxyls at very low concentration. Since most uses of emulsion polymer occur at high concentrations, the measurement of particle-particle interactions is of great practical importance (21). It has been found that the sedimentation and viscometric techniques closely reflect viscosity changes in latexes at much higher solids. Extension of the PCS approach to more concentrated systems is underway but not without problems (22).

The existence of surface layers on aqueous colloids is an important consideration in the interpretation of hydrodynamic effects in viscosity, stability, electrophoresis, diffusion, sedimentation and adhesion measurements. In the case of charged systems, represented by expanded carboxylic latexes, these layers possess both steric and electrostatic properties and might properly be called "electrosteric" in nature. The expansion properties of carboxylic, as well as other ionogenic, latex particles are important in the interpretation of conductometric titration curves, acid location results and in the choice of model colloids for fundamental studies. The existence of a constant geometric surface over a wide range of pH is generally desirable in such studies. A variable surface layer obviously complicates the use of models designed to describe ionizable surface groups at interfaces (23). Polystyrene and poly(methyl methacrylate) latexes at reasonably low (< 5%) carboxyl contents seem appropriate choices in this regard.

ABSTRACT

Carboxylic emulsion polymers exhibit particle swelling when the incorporated acid groups are neutralized. Active areas of investigation include the location of the polymerized carboxylic monomers, the detailed mechanism of particle expansion and the morphology of the expanded particles. Several methods have been used to study particle expansion including light scattering, centrifugal sedimentation, precision viscometry and photon correlation spectroscopy (PCS). Although none of these techniques can distinguish homogeneous and non-uniform particle expansion, the hydrodynamic expansion curves were similar in shape. Particle expansions obtained by viscometry were slightly smaller than the values obtained by sedimentation. The PCS method however, yielded particle expansion values which were smaller by almost 50 percent. In addition, maximum expansion occurred at pH = 12.5 in the PCS experiments compared with approximately 10.5 observed in the sedimentation and viscometry experiments. Since PCS is carried out at much lower particle concentrations, interactions between the charged particles at the higher concentrations are probably involved. Similar comparisons with non-expanding carboxylic latex particles were carried out in an effort to separate interparticle charge effects from true particle expansion in interpreting apparent particle sizes determined by hydrodynamic methods.

ACKNOWLEDGMENTS

The authors thank the following people for helpful discussions and experimental assistance: K. L. Hoy, R. H. Peterson, L. C. Cantley, M. R. Doerflein and R. L. Pack.

LITERATURE CITED

1. Muroi, S., J. Appl. Polymer Sci., 1966, 10, 713.
2. Muroi, S.; Hosoi, K.; Ishikawa, T., J. Appl. Polymer Sci., 1967, 11, 1963.
3. Verbrugge, C. J., J. Appl. Polymer Sci., 1970, 14, 897,911.
4. Nishida, S.; El-Aasser, M. S.; Klein, A.; Vanderhoff, J. W.; This volume,
5. Bassett, D. R.; Hoy, K. L.; "Polymer Colloids II", R. M. Fitch, ed., Plenum Press, New York, 1980, p.1.
6. Ford, J. R.; Rowell, R. L.; Bassett, D. R.; This volume,
7. Ottewill, R. H.; Walker, T., Kolloid-Z., Z. Polymere, 1968, 227, 108.
8. Garvey, M. J.; Tadros, Th. F.; Vincent, B., J. Colloid Interface Sci., 1976, 55, 440.
9. Einstein, A., Ann. Physik, 1906, 19, 289.
10. Guth, E.; Simha, R., Kolloid Z., 1963, 74, 266.
11. Guth, E.; Gold, O., Phys. Rev., 1938, 53, 322.
12. Cheng, P. Y.; Schachman, H. K., J. Polymer Sci, 1955, 16, 19.
13. Doroszkowski, A.; Lambourne, R., J. Colloid Interface Sci., 1968, 26, 214.
14. Fleer, G. J.; Koopal, L. K.; Lyklema, J., Kolloid-Z., Z. Polymere, 1972, 250, 689.
15. Berne, B. J.; Pecora, R., "Dynamic Light Scattering," John Wiley and Sons, Inc., New York, 1976.
16. Derderian, E. J.; MacRury, T. B., Paper presented at the 54th Colloid and Surface Science Symposium, Lehigh University, June, 1980. To be published in J. Dispersion Sci. Tech.

17. Goosens, J. W. S.; Zembrod, A., Colloid and Polymer Sci., 1979, 257, 437.
18. Bauer, D. R., J. Phys. Chem., 1980, 84, 1592.
19. Dickinson, E., J. Colloid Interface Sci., 1980, 73, 578.
20. James, R. O.; Davis, J. A.; Leckie, J. O., J. Colloid Interface Sci., 1978, 65, 331.
21. Ottewill, R. H., Prog. Colloid & Polymer Sci., 1980, 67, 71.
22. Bauer, D. R., "Polymer Colloids II", R. M. Fitch, ed., Plenum Press, New York, 1980, p. 51.
23. Healy, T. W.; White, L. R., Adv. Colloid Interface Sci., 1978, 9, 303.

RECEIVED April 6, 1981.

17

Light Scattering Studies of the Internal Structure of Emulsion Polymer Particles

J. R. FORD and R. L. ROWELL
Department of Chemistry, University of Massachusetts, Amherst, MA 01003
D. R. BASSETT
Technical Center, Union Carbide Corporation, South Charleston, WV 25303

Wide angle light scattering is used as the principal probe to examine the core-shell structure proposed for certain acrylic acid acrylate ester copolymer latexes. Additional techniques were sedimentation and photon correlation spectroscopy.

The work represents an application of core-shell light scattering theory to polymer latex suspensions and addresses the separate identification of light scattering by dust, latex particles and low molecular weight solutes.

Core-Shell Theory and Model

The exact electromagnetic scattering theory of the concentric shell model was first solved by Aden and Kerker (1) and shortly thereafter by Güttler (2). The problem has been extensively studied both theoretically and experimentally for aerosols by Kerker and co-workers and is reviewed in Kerker's book (3). The aerosol system had a core of relative refractive index m_1=2.105 and a shell of m_2=1.482 corresponding to silver chloride coated with linolenic acid. The results indicated that for a smooth variation in the refractive index of the shell, the refractive index might not be sensitive to the form of the variation.

Olaofe and Levine (4) investigated the scattering for concentric spheres and spherically symmetric inhomogeneous spheres with Cauchy and parabolic distributions of the radial profile of refractive index. Comparisons were with the volume-weighted refractive index

$$\bar{m} = (1/V) \int m dV$$

They found, that for a concentric sphere with a variable coating, the scattering was not sensitive to the form of the variation in the shell whenever there was a constant amount of refractive material ($\bar{m}$ = 1.0738). For inhomogeneous spheres with either a Cauchy or parabolic distribution, the forward scattering

0097-6156/81/0165-0279$05.00/0

differed only slightly from the variable shell concentric sphere model. The backscattering, however, was considerably different from that of concentric spheres and was significantly different for different internal profiles.

Previous Work

Our first paper (5) appears to be the first wide-angle light scattering application of concentric-sphere theory to aqueous suspensions. In the investigation a laser light scattering apparatus with photon counting detection was used to obtain wide-angle light scattering data on a small particle size latex, 34BRD45, that conclusively established the existence of a concentric-shell particle with a core diameter of 0.20 μm, a shell thickness of 0.09 μm and a shell relative refractive index of 1.04. Some samples showed a slow, consistent decrease in pH after dilution that continued for at least 8 hours so that the details of the pH-dependent changes in the core-shell structure were not completely clear within the scope of the analysis. Since the accuracy of the light scattering apparatus and the reliability of the inversion procedure had been proven by comparison of four methods on Dow polystyrene latex LS-1028-E (6), it was decided to examine a larger latex in more detail as a further study of the method of determining the internal structure of a latex by wide-angle light scattering.

Present Work

Not only was a larger latex examined in more detail but also a greater study was made of the concentration dependence of the light scattering. The concentration dependence of lateral scattering was first examined by Heller and Tabibian (7). They measured the concentration dependence of the specific turbidity $I_{90}/I_o c$ and carried out an extrapolation to zero concentration defined by

$$(I_{90}/I_o c)_o = \lim_{c \to o} (I_{90}/I_o c)$$

Practical difficulties in obtaining I_o led to the use of a reference intensity I_R so that $I'_{90}/I_R c$ was plotted versus c and extrapolated to zero concentration. The method required a calibration constant which was obtained by the use of polyvinyltoluene latex spheres of known size from electron microscopy. For small particles the specific turbidity increased with dilution which was explained by decreased intensity loss from the primary and scattered beams due to particle turbidity. For certain particles of a size around the first turbidity maximum where the scattering is strongest, the specific turbidity decreased with dilution as the stray light from multiple scattering decreased.

In the present work with a laser source and photon counting detection we were able to make measurements at concentrations lower than the work of Heller and Tabibian where multiple scattering and particle turbidity were negligible and single scattering could be directly observed. It was possible to make direct measurements of I_o and utilize the Rayleigh ratio V_v and Mie theory coefficient $(i_1)_\theta$ by

$$V_v = (I_{V_v}/I_o)(R^2/V) = N\,(\lambda/2\pi)^2 \int_0^\infty p(\alpha)(i_1)_\theta \, d\alpha$$

where R is the observation distance from a scattering volume V, N is the number concentration of particles, λ the wavelength in the medium and $p(\alpha)$ is a distribution function over the size parameter $\alpha = 2\pi r/\lambda$. It is clear that V_v is directly proportional to N so that one needs only to find the range where the excess scattering is directly proportional to concentration in order to select a weight concentration suitable for measurement.

Latex Preparation

The latexes were prepared as in previous work (8) by one of us (DRB) using a semi-batch technique in which a monomer mix was fed into a reactor at 80^oC in the presence of ammonium persulfate and AEROSOL OT. The initial AEROSOL OT concentration was adjusted to produce a particle size in the range 0.10 to 0.50 μm. The monomer mix consisted of 40 parts methyl methacrylate, 54 parts ethyl acrylate and 6 parts butyl acrylate. For the "core" latexes the standard monomer feed was used; for the "core-shell" latexes, acrylic acid amounting to 2 percent of the total monomer feed was added. In earlier work (8) the manner of addition of the acrylic acid was studied: all in Stage I (the first half of the monomer feed), all in Stage II (the second half of the monomer feed) and equally divided between Stages I and II. Conductometric titrations indicated that most of the carboxyl groups were located in the surface region of the particles. Stage I gave 98% surface carboxyls; Stages I & II gave 99% surface carboxyls but Stage II gave 91% surface carboxyls. The question of the location of carboxyl groups within the latex particles remains a subject of current investigation. For the "core-shell" latex 42BRD47, measured by wide-angle light scattering, the acid was added in Stage II so that one might expect to find 91% surface carboxyls, 8% soluble carboxyls and 1% carboxyls buried within the latex particle on the assumption that the earlier conductometric titration results (8) were general.

Results and Discussion

Theoretical Feasibility. Earlier work (5) had established the general domain of size parameters and refractive index expected. This allowed a number of model calculations to be carried out to clearly establish the theoretical feasibility of the method. Computations were carried out for homogeneous spheres and for concentric core-shell spheres.

The theoretical feasibility is illustrated in Figure 1 which shows a homogeneous core-sphere A, a homogeneous expanded sphere B and a core-shell structure C. The core-shell structure has the same core diameter as A and the same shell diameter as B. The refractive index of the homogeneous expanded sphere B is the volume-weighted refractive index of the core-shell structure C. The two homogeneous spheres are clearly different in angular pattern from the core-shell structure. The intensity scale i_1 is logarithmic and has been arbitrarily shifted to facilitate comparison of the angular patterns. In the comparison of experiment with theory to determine the best-fit size parameters, it is the angular pattern or template that is important and not the magnitude of the intensity.

Dust and Extraordinary Particles. In earlier work (5) it was found that filtration of the latex with either a 0.4 μm or 0.65 μm Nuclepore filter to remove dust, introduced poor reproducibility due to the film-forming nature of the latex. Since the number of dust particles in the latex was small, the problem could be circumvented by using focused incident illumination. Focusing of all the illumination within the observation volume did not change the molecular scattering or the latex particle scattering since the increased irradiance was exactly compensated by a decrease in the number of scatterers. Dust particles, however, were very large and were present in a much lower concentration than latex particles and were observed as sharp spikes superimposed on a continuous signal as they diffused through the beam. It was a simple but time consuming matter to ignore the dust spikes and record the average continuous signal. The method allowed then in situ discrimination of extraordinary particles such as dust.

A second method was also used that did not require focused incident illumination. Since the dust particles were in low concentration compared to the latex particles it was possible to collect a time-averaged photon count over a time interval that was small compared to the average time between the dust "blips." The lowest of a series of time-averaged photon counts was taken as most representative of a dust-free measurement. Since the latter method did not require focused illumination it was preferred over the first method.

The focused-illumination method of signal discrimination depends on both the magnitude and frequency of the extraordinary signal whereas the time-averaged count depends only on the frequency of the extraordinary signal. In general concept the focused-illumination method is spatial discrimination whereas the time-average count method is time discrimination.

Since a vast number of photon counts may be rapidly accumulated, i.e. a full memory of 4096 measurements at 20 μsec per measurement is obtained in 82 msec, a statistical analysis of such data can, in principle, lead to separate characterization of the ordinary signal from latex particles and an extraordinary signal which may arise from dust particles as in the present instance or in general from a low population of any extraordinary particles.

Time-Dependence of pH. Previous work indicated that significant time-dependent changes in the sample pH could occur even after 24 hours of equilibration. An extensive study of the time dependence of pH was carried out.

In the first experiments, five samples at 1% solids concentration were adjusted to approximately pH 10 and the pH was monitored over the next 100 hours. Samples were stored in beakers with parafilm covers. In the first 5 hours the pH dropped sharply about 0.5 pH unit and then continued a steady but slower decline over the next 95 hours for a total drop of 1 or 2 pH units depending on the particular latex sample.

In a second procedure, the pH was adjusted to 10, in the morning, readjusted twice more during the course of the first day, and a final adjustment the following morning. After the last adjustment the sample pH was monitored for 7 hours. Again a two-stage decline was observed which was rapid in the first two hours and slower thereafter and dropped around 0.6 pH unit over a 7 hour period. The general trends were similar for 7 samples with detailed absolute pH values dependent on the particular sample.

It was concluded that the time-dependence of the pH represented interesting changes that should be examined more closely in a study of the kinetics of the system. For the present work the pH was stable over the time required for measurement so that an angular light-scattering scan at constant pH was obtained.

Linear Scattering Range. As discussed above, in a very dilute solution, the excess intensity becomes directly proportional to the concentration so that it is sufficient to determine the linear scattering range in order to carry out measurements. The concentration may then be treated as an unknown which may be determined using absolute Rayleigh ratios or, as in this work, it may be held constant in order to study structural changes. In the present investigation we have experimentally determined two distinct ranges in which the light scattering decreases linearly with dilution.

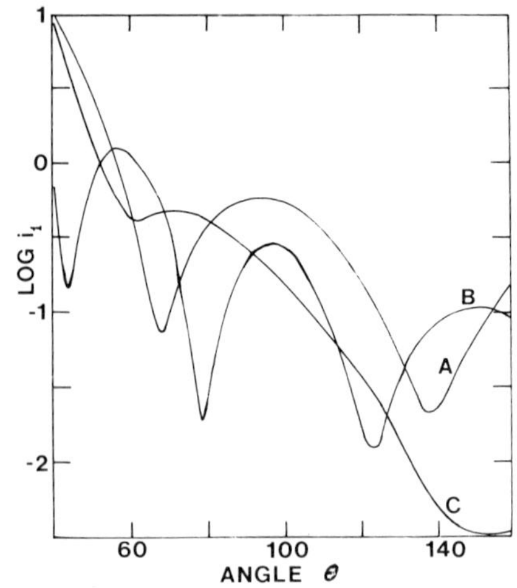

Figure 1. Theoretical calculations for (A) homogeneous sphere with $\alpha = 3.71$, m = 1.20 and (B) homogeneous sphere with $\alpha = 5.74$, m = 1.09, which is the volume-weighted refractive index for the core-shell structure of (C) with $\alpha = 3.71$, $\nu = 5.75$, $m_1 = 1.20$, $m_2 = 1.05$.

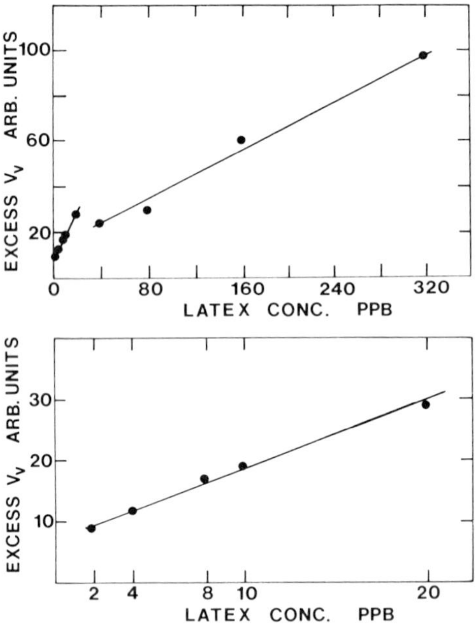

Figure 2. Concentration dependence of V_v scattering from Latex 42BRD47 at pH 3.64. The dilute range, expanded in the lower graph corresponds to residual molecular scattering in excess of the solvent.

Both ranges of linear excess scattering for V_v polarization are shown in the upper graph of Figure 2 and the most dilute range is plotted with an expanded scale in the lower graph of Figure 2. In the absence of multiple scattering the excess normalized intensity is proportional to the Rayleigh factor and was computed from

$$V_v' = I(\theta)/I_o - B(\theta)/B_o$$

where $I(\theta)$ and $B(\theta)$ are proportional to the intensities of sample and background at angle θ and I_o and B_o are proportional to the intensities incident upon the scattering volume.

Typical angular scans corresponding to the two distinct linear ranges of Figure 2 are shown in Figure 3. The upper curve is characteristic of particles and the lower curve is more Rayleigh like.

Calculations have been carried out to estimate the number concentration of latex particles in each of the linear ranges shown in Figures 2 and 3. In the upper range corresponding to latex mass concentrations of 40-320 ppb the number of particles in the scattering volume varies from 1,600 to 13,000 so that the scattering is clearly dominated by scattering from latex particles. In the lower range corresponding to latex mass concentrations of 2 to 20 ppb the number of particles in the scattering volume is estimated to vary from 80 to 800. The scattering pattern in the dilute range is clearly dominated by Rayleigh scatterers which is very likely molecular scattering from oligomers, monomer and emulsifier.

The upward shift in the excess scattering at the discontinuity between ranges as the dilution goes from 40 to 20 ppb is an interesting effect and may arise from a redistribution of scattering material at a critical concentration.

Light Scattering Results. The excess V_v scattering intensity was taken as the difference between the particle scattering range at 160 ppb and the background scattering range at 10 ppb. Typically, an angular scan was characterized by measuring at 56 angles from $40^o(2^o)150^o$, and from these, 46 angles $60^o(2^o)150^o$ were used in the data inversion program. The data were analyzed using both the theory for homogeneous spheres and core-shell structures.

The typical results of the wide-angle light-scattering (WALS) analysis on the latex particles that were presumed to be cores are given in Table I where the theoretical model to be fitted was either S = sphere or CS = core-shell, the variance measured the goodness of fit, the modal size parameter is given by $\alpha_M = 2\pi r/\lambda$, σ_o is the log normal breadth parameter, D_M is the core diameter in μm, m_1 the relative refractive index of the core and m_2 the relative refractive index of the shell of indicated thickness.

Table I. Wide-Angle Light Scattering Results on Latex 41BRD30 (No-acid cores with turbidimetric diameter of 0.41 μm)

Fit	pH	Variance	α_M	σ_o	D_M(μm)	m_1	m_2	Shell (μm)
S	3.65	0.742	3.1	0.06	0.36	1.20		
CS	3.65	0.566	3.2	0.01	0.37	1.18	1.03	0.14
S	9.96	0.768	3.2	0.005	0.37	1.18		
CS	9.96	0.622	3.3	0.01	0.38	1.18	1.03	0.14

The constant value of core diameter and low polydispersity σ_o in core diameter is consistent with the known uniformity of the latexes. Either the S or the CS analysis gives a core diameter of 0.37 μm and refractive index of 1.18-1.20. The core diameter is slightly smaller than the value of 0.41 μm which came from an independent analysis using a turbidimetric technique. The turbidimetric technique was an approximate method that required the assumption of refractive index and density. At both pH's, the main result is a constant core diameter and composition, independent of the theoretical model.

Since the core-shell fits were better than the homogeneous sphere fits there might be some material in a layer around the core other than pure solvent. The data for the shell, however, must be close to the uncertainty of the fit obtained with the crude step-function model so that at best the arguments for a shell are tenuous. Clearly, a further investigation is required in order to test the statistical significance of the fits and to consider other possible models.

In Table II are summarized representative results for the light scattering on a latex that was presumed to be capable of expansion at high pH. An independent estimate by the turbidimetric method gave a core diameter of 0.43 μm.

Table II. Wide-Angle Light Scattering Results on Latex 42BRD47

Fit	pH	Variance	α_M	σ_o	D_M(μm)	m_1	m_2	Shell (μm)
S	3.64	0.839	3.80	0.015	0.44	1.24		
CS	3.64	0.303	3.90	0.02	0.45	1.20	1.05	0.20
S	9.90	0.487	3.90	0.03	0.45	1.22		
CS	9.90	0.316	3.90	0.02	0.45	1.22	1.03	0.18

Independent of the model used, it is quite clear that the core diameter is 0.44 to 0.45 μm and the relative refractive index of the core is 1.22 $\pm$ 0.02. The evidence for a shell structure is again very tenuous. The "shell" thickness is unrealistically large, constant in extent, and little different from the results for the no-acid latex. If a shell structure was present it was at best marginally observable in the wide-angle light scattering.

Comparison of Sedimentation and Light Scattering. The sedimentation method has been presented elsewhere (8) where it was shown that

$$S_o/S = (r+x)/r$$

where S is the measured sedimentation coefficient, S_o refers to the unexpanded particle of radius r and x is the shell thickness. The model is controlled by the average density of the particle relative to the density of the surrounding liquid. In the derivation of the model it was assumed that the density of the shell was very close to that of the solvent which is equivalent to the statement that the decrease in density due to particle swelling is related to the volume of water adsorbed. The sedimentation results alone, then, could not differentiate between homogeneous and heterogeneous particle structures.

In essence, the sedimentation method gives a ratio of the hydrodynamic radius at a given pH to the hydrodynamic radius of the same particles at a reference pH. It is not an absolute method and requires an independent absolute measurement of particle radius.

The expansion of carboxylic latex 42BRD47 as a function of pH, determined by the sedimentation method, is given in Figure 4. More than a 50% increase in hydrodynamic diameter has been reached at high pH.

Taken together, the light scattering measurement of a constant core size of 0.45 μm and the sedimentation curve showing a pronounced increase in hydrodynamic size indicate that the particle expansion occurs in a layer of very low density. It is not surprising, therefore, that attempts to quantitatively measure the thickness and refractive index of the expanded layer by wide-angle light scattering met with some difficulty.

Conclusions

1. Theoretical calculations have demonstrated the feasibility of distinguishing a core-shell latex from a homogeneous latex.

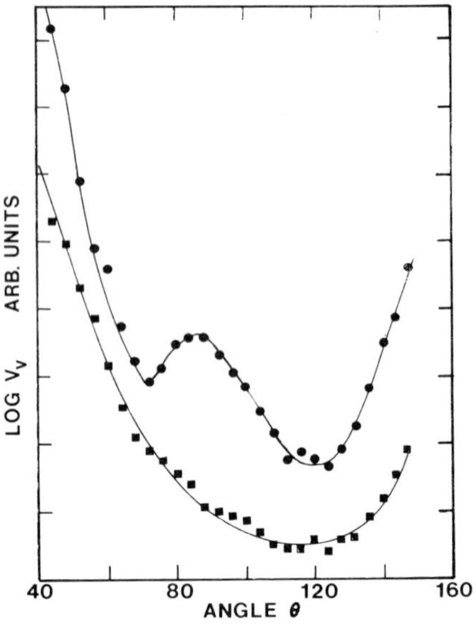

Figure 3. Angular scattering for Latex 42BRD47 at pH 3.64 for excess V_v scattering. Upper curve is from latex particles and lower curve is from residual molecular scattering in excess of the solvent.

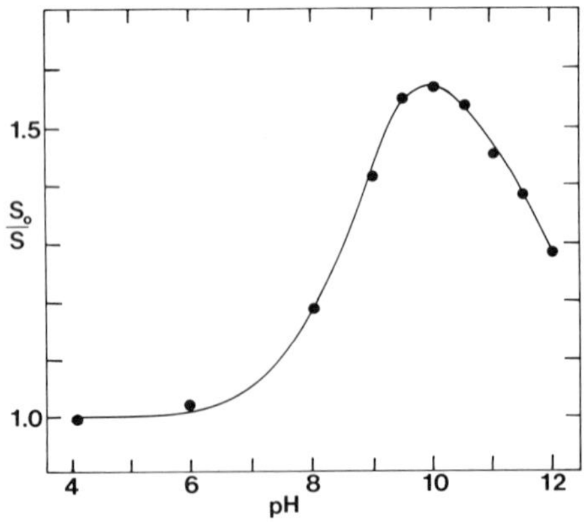

Figure 4. Expansion curve of carboxylic Latex 42BRD47 as determined by the sedimentation method. The estimate of particle diameter by turbidity was 0.43 μm.

2. Wide angle light scattering experiments have been carried out with sufficient accuracy to separately identify dust or extraordinary particles, latex particles and small-molecule scattering.

3. WALS has been used to establish conditions of stable particle size and to illustrate the possibility of examining the kinetics of the system by light scattering.

4. Inversion of WALS data gave evidence for a uniform sphere, independent of pH. At best, only marginal evidence for a shell of low relative refractive index and ill-defined extent was obtained.

5. Sedimentation studies of the same latex showed a pronounced increase in hydrodynamic size with pH so that particle expansion must occur by a layer of very low density.

The results are generally consistent with a broader treatment of the techniques for measuring particle swelling of carboxylic emulsion polymer latexes reported elsewhere in this Monograph (9). The broader study, which was carried out independently but concurrently, has shown that the magnitude and pH of maximum expansion depends on dilution and ionic strength. Studies of the concentration dependence in the dilute regime and more concentrated systems are underway.

Acknowledgements

The principal sponsor of the work was the Union Carbide Corporation. The light scattering apparatus and laboratory has been assembled with support from the National Science Foundation and the computations were carried out with a grant of computer time from the University Computing Center.

Literature Cited

1. Aden, A. L. and Kerker, M. J. Appl. Physics 1951, 22, 1242.
2. Güttler, A. Ann. Physik 1952, (6) 11, 65.
3. Kerker, M. "The Scattering of Light and Other Electromagnetic Radiation"; Academic Press: New York, 1969.
4. Olaofe, G. O. and Levine, S. in "Electromagnetic Scattering"; Rowell, R. L. and Stein, R. S., Eds., Gordon and Breach: New York, 1967; pp. 237-292.
5. Rowell, R. L., Ford, J. R., Parsons, J. W. and Bassett, D. R. in "Polymer Colloids II", Fitch, R. M., Ed., Plenum Press: New York, 1980; pp. 27-35.
6. Rowell, R. L., Farinato, R. S., Parsons, J. W., Ford, J. R., Langley, K. H., Stone, J. R., Marshall, T. R., Parmenter, C. S., Seaver, M. and Bradford, E. B. J. Colloid Interface Sci. 1979, 69, 590.
7. Heller, W. and Tabibian, R. M. J. Colloid Interface Sci. 1958, 13, 6.

8. Bassett, D. R. and Hoy, K. L. in "Polymer Colloids II", Fitch, R. M., Ed., Plenum Press: New York; 1980; pp. 1-26.
9. Bassett, D. R., Derderian, E. J., Johnston, J. E. and MacRury, T. B., this volume, p.

RECEIVED April 6, 1981.

18

Preparation and Characterization of Alkali-Swellable Carboxylated Latexes

S. NISHIDA[1], M. S. EL-AASSER, A. KLEIN, and J. W. VANDERHOFF

Emulsion Polymers Institute, Lehigh University, Bethlehem, PA 18015

Latexes prepared with ionogenic monomers containing carboxylic acid groups are a special class of latexes which have wide industrial application in carpet backing, adhesives, surface coatings, and paper coating. These are usually prepared using monomers such as itaconic, acrylic, methacrylic, maleic, fumaric and crotonic acids. These carboxyl-containing monomers are copolymerized with the main monomer and contribute to the stability of the latex, and exhibit alkali-swelling and/or dissolving behavior with increasing pH.

The alkali-swelling and dissolving behavior of carboxylated latexes has been qualitatively correlated to the glass transition temperature (Tg), hydrophilicity and molecular weight of latex polymers, and the carboxyl content and their distribution within the latex particles (1-5). These investigations have been useful in understanding the swelling and dissolving behavior of carboxylated latexes. However, no consideration was given to the effect of polymerization procedure and method of monomer addition on the properties of carboxylated latex systems used in these studies. Moreover, the alkali-swelling behavior of these latexes was studied on not-well-characterized latex systems; for example, no consideration was given to the location and distribution of the carboxyl groups within the particles. It is well known that in the copolymerization process, in which two monomers have different reactivity ratios, the instantaneous copolymer composition at the initial stage of a batch polymerization would be quite different from that at the final stage. On the other hand, in a semi-continuous process, if the feed rate (R_f) is adjusted to be much lower than the maximum rate of polymerization (R_{pmax}), [R_{pmax} is referred to the rate of polymerization in a system in which enough monomer is available at reaction sites, just as in the batch type polymerization]; the instantaneous copolymer composition could be substantially equal to feed monomer composition during the course of polymerization. Snuparek and Krska (6) have recently experimentally proven this finding with semi-continuous emulsion copolymerization of styrene and butyl acrylate.

[1] Japan Synthetic Rubber Co., Yokkaichi Mie, Japan

0097-6156/81/0165-0291$06.00/0

This microscopic difference in the copolymer composition could influence the particle morphology, especially the distribution of the carboxyl groups within the latex particle, which in turn could be expected to influence the alkali-swelling behavior.

Bassett and Hoy (7,8) investigated the influence of the method of monomer addition on the alkali-swelling behavior. In this case, carboxyl-containing monomers were fed in three different stages of the semi-continuous polymerization: (i) in the first half of the monomer feed, (ii) in the second half of the monomer feed, and (iii) in both segments. The results of conductometric titration with these latexes indicated that most of the carboxyl groups were located at the particle surfaces. However, the particle expansion, estimated by the rate of sedimentation, indicated that the carboxyl monomer feed in the second half was most effective in the particle swelling. This suggests that the method of monomer feed affected the location of carboxyl groups, and in turn the particle expansion, although the difference could not be detected by conductometric titration.

The purpose of the current work is to investigate quantitatively the alkali-swelling behavior of carboxylated latex systems. During this investigation a model methyl methacrylate (MMA) - methacrylic acid (MAA) copolymer latex system was prepared using both a batch and a semi-continuous polymerization process. The MAA content was varied over the range of 5 - 25% by weight. A kinetic study was conducted to establish the reactivity ratios of the comonomers and the permissible rates of monomer addition in the semi-continuous polymerization. All model latexes were characterized with respect to particle size, monomer sequence in the copolymer, and molecular weight. A new method based on conductometric titration was developed in order to determine the distribution of the carboxyl groups within the latex particles. Finally, the alkali-swelling behavior of these latexes was determined by investigating the light scattering and the viscosity behaviors of the latexes as a function of pH.

Experimental

1. Materials. Methyl methacrylate (MMA) and methacrylic acid (MAA) monomers (Rohm and Haas), and n-dodecyl mercapten (Eastman Kodak Co.) were purified by vacuum distillation under nitrogen atmosphere to remove the inhibitor. Distilled-deionized and deoxygenated water was used in all polymerizations. Sodium lauryl sulfate (99% Onyx Maprofix 563) and Triton X-100 nonionic surfactant (Rohm and Haas), and potassium persulfate initiator (Fisher Scientific Co.) were used as received without further purification.

2. Preparation of Carboxylated Latexes. Latexes were prepared by emulsion polymerization using recipes given in Table I. Several MMA-MAA copolymer latex samples were prepared with differ-

ent MAA content of 5, 10, 15, 20 and 25 weight %. The batch polymerization was carried out using the bottle polymerization method. All ingredients, except initiator, were charged into the bottles and purged with nitrogen. The bottles were tumbled end-over-end at 35 rpm, in a constant-temperature bath at 60°C for 10 minutes prior to initiator injection, and for 6 hours thereafter.

Table I
Polymerization Recipe for Preparation of Model Carboxylated Latex

	Batch	Semicontinuous
Water	270.00	360.00
Methyl methylate (MMA) Methacrylic acid (MAA)	30.00	40.00
N-Dodecylmercaptan	0.06	0.08
Potassium persulfate	0.09	0.12
Sodium Lauryl sulfate	0.15	0.20
Triton X-100	0.60	0.80

The semicontinuous polymerization was carried out in a 500-ml four-neck flask immersed in a constant temperature bath at 60° C, and equipped with a reflux condenser, a two-bladed stainless steel stirrer and a graduated dropping funnel. The water and surfactant were introduced into the flask, then nitrogen gas was bubbled with agitation for at least 20 minutes. The initiator solution was added, and after 5 minutes the mixture of monomer addition started and continued for a total of three hours. The polymerization was continued for at least one hour past this point. Under these conditions the rate of monomers addition was less than 1/10 of their maximum rate of polymerization (Rpmax).

During the semi-continuous polymerization, 4-5 small samples were withdrawn from the polymerization for the determination of the comonomer and copolymer composition. A few drops of the sample latex were mixed with hydroquinone, cooled in ice, and subjected to GC analysis to determine the amounts of unreacted monomer. The rest of the sample (5-8 ml) was poured into mixed solvent of ispropanol/hexane (45/55) containing hydroquinone, and the precipitated polymer, after it was washed with hexane, was dried in a vacuum oven at 45°C for more than 5 hours. A certain amount of the dried polymer was dissolved in dimethyl formamide (DMF), and titrated for the carboxyl content with NaOH solution using phenolphthalein as the indicator.

3. Determination of R_{pmax}, r_1 and r_2. A batch polymerization of MMA-MAA comonomers was carried out for the determination of R_{pmax} and reactivity ratios. The same procedure described in the batch process was used for this purpose except that the reaction vessel was a round bottom flask equipped with agitation, a nitrogen gas inlet, an outlet for taking samples, and a condenser.

A small amount of the reaction mixture was withdrawn by a syringe, then put in a vial which contained hydroquinone as a polymerization stopper, and cooled in ice. One or 2µl of the withdrawn mixture was subjected to gas chromatography (GC) analysis for determination of the amounts of unreacted monomers. The GC column (Supalco, Inc.), which was 6 ft. long and 1/8 inch in diameter and packed with Chromosorb W AW 10% SP-1200/1% H_3PO_3, is suitable for the separation of carboxylic acids. The GC (Chromalytic, Inc.) was equipped with a flame ionization (FI) detector which is convenient because it is insensitive to inorganic materials or water.

4. Molecular Weight Determination. The molecular weight and distribution of each copolymer latex sample was determined by gel permeation chromatography (GPC) using Waters Associate Unit equipped with a differential refractometer detector. Five columns packed with microstyragel (fully porous, highly crosslinked styrene-divinylbenzene copolymer) with pore size in the range 10^2-10^6. The latex samples were dried in vacuum oven at 45-50°C and dissolved in tetrahydrofuran THF with 0.5% by weight solids.

5. Monomer Sequence Distribution in the Copolymer by ^{13}C NMR. The latex samples which were precipitated in the mixed solvent (ispropanol/hexane = 45/55 and dried in a vacuum oven at 45-50°C, were used for the ^{13}C NMR study. The polymer samples were dissolved (5-7% weight volume) in deuterated pyridine (PYR d%) with tetramethylsilane (TMS) as a zero external reference. The NMR measurement and the theoretical calculation were carried out in CNRS, Laboratoire des Matriaux Organiques, in Lyon, France using ^{13}C NMR 30 MH_z (Bruker).

6. Determination of Particle Size. A Brice Pheonix light scattering apparatus was used for the determination of particle sizes using the dissymmetry method. The intensities of scattered light at 45° and 135° were measured for the diluted latex samples with 0.02, 0.002 and 0.0002 wt. % solids. Since ratios of intensity at 45° to that at 135° were not significantly changed for the concentration range of 0.02-0.0002 wt. %, the average value of them was taken to calculate particle diameter using the table computed by Pangonis and Heller (9). No attempts were made to correct for the changes in refractive index of polymer during neutralization with NaOH. Consequently, the particle size determined as a function of pH are only apparent and not absolute.

7. Characterization of Latex Particles with Respect to Carboxyl Distribution. All latex samples were cleaned by the serum replacement technique (10). Diluted latex sample (3% by weight) was placed in the cell confined with a Nuclepore membrane and distilled deionized water was fed into the cell from a reservoir placed at 1.5 meters above the cell. The serum from the cell exit was collected and monitored with conductance measurement. The cleaning of the latex was considered completed when the conductance of the serum was reduced to that of the pure water. Since the carboxyl groups were introduced in the acid form, and since no buffer was used during the polymerization, therefore, the carboxyl groups are expected to be kept in the H^+ form after the serum replacement. Consequently, no additional cleaning

steps were required prior to the titration of the carboxyl acid. The "cleaned" latexes and the corresponding serums were characterized with respect the carboxyl content using the conductometric titration technique (11).

The conventional conductometric titration was not suitable for the characterization of these carboxylated latexes because the surface of the latex particles was varied during the titration. In carboxylated latexes, the carboxyl groups located at the particle surface are neutralized and hydrated first, followed by the neutralization and hydration of the carboxyl groups located in the inner layers adjacent to the surface, and so on. These sequential reactions seemed to take longer than the experimental time (say, 30 minutes). Verbrugge (4) reported that it took almost two days to get to equilibrium completely. The overall rate of reaction should be controlled by a diffusion process in the neutralization reaction of carboxylated latexes. If this is the case, the rate of reaction must be a function of the distribution of carboxyl groups within the latex particles.

Consequently, modified conductrimetric titration, which takes into account the time dependence of the neutralization reaction was devised and used in this work, as follows (12).

a) A certain amount of latex sample is weighed in a 250 ml. beaker.

b) The electrodes of a conductance meter is set to measure conductance of the latex. The beaker is covered and nitrogen is introduced into the latex to eliminate carbon dioxide.

c) A certain amount of NaOH solution, which is 30-50% in excess of the amount required to complete neutralization, is quickly injected with a syringe into the beaker.

d) Immediately after the injection of NaOH, conductance of the latex is recorded and is continued for a period of time during which the initial rapid change in conductance settles down to a steady state.

e) The latex-NaOH mixture is sealed up in a glass bottle and left at room temperature for 24 hours.

f) After 24 hours, the latex is subjected to conductometric back-titration with HCl solution, which gives the total amount of carboxyl groups neutralized within 24 hours.

g) Exactly the same procedure is carried out for pure water as a blank.

8. Alkali-Swelling Behavior. The alkali-swelling behavior of the "cleaned" carboxylated latexes as a function of pH was determined by the change in particle size, and the viscosity of the latex system. The change in particle size of the carboxylated latex as a function of pH was measured by dissymmetry method of light scattering. Latex samples with 0.01 wt. % solids were used. Sodium hydroxide solution was used to adjust the pH, 24 hours before dissymmetry measurements.

The viscosity measurements were carried out using a cone-plate viscometer (Model R-16 Weissenberg Rheogoniometer-Sangamo

Controls, Ltd.). The apparatus was calibrated using certified viscosity standards No. S60 and S600 (Cannon Instrument Co.). The solid contents of latex samples were 2.0 wt. %. The pH adjustment was carried out using NaOH solution 24 hours prior to viscosity measurement. The viscosity measurements were carried out over a range of shear rate of 10^2-10^4 sec^{-1}.

Results and Discussion

1. Polymerization Results. A batch polymerization of MMA-MAA comonomer was analyzed for the determination of the reactivity ratios of the two monomers. The change in the ratio of the copolymer composition determined by GC was plotted against conversion as shown in Figure 1. Similarly, the calculated curves for some assumed reactivity ratios are also shown in the same Figure. The optimum values of the reactivity ratio for the emulsion polymerization at 50°C of the two monomers were obtained from the curve fitting by trial-and error, r_1 (MMA) = 0.60 & r_2 (MAA) = 1.55. Consequently, in a batch emulsion polymerization process, the copolymer composition will be rich in MAA till it is consumed.

The conversion of each monomer, as determined by GC, was plotted against the polymerization time, and the maximum rates of polymerization (R_{pmax}) were determined from the initial slopes. The results showed that R_{pmax} (MMA) = 6.28 X 10^{-2} (mol/l-min) and R_{pmax} (MAA) = 3.93 X 10^{-2} (mol/l-min). Based on these results, the feeding rate during the semicontinuous polymerization were adjusted to R_f (MMA) = 0.41 X 10^{-2} mol/l-min and R_f (MAA) = 0.16 X 10^{-2} mol/l-min.

Tables II & III show the particle size, percent conversion and percent solid of the MMA-MAA copolymer latexes prepared by batch and semicontinuous processes respectively. The amount of coagulum which was formed during the polymerization was negligible in all latexes. The particle size of all batch latexes are less than the corresponding semicontinuous latexes; with one exception for the copolymer with 5 weight percent MAA, where the reverse is true. Also, the particle size in the batch latexes seems to depend on the MAA content, where it is found to decrease with increasing the MAA content. The particle size of the semi-continuous copolymer latexes is found to be independent of the copolymer composition in the range of 10-25 weight percent MAA. This difference in particle size suggests that MAA, which is the monomer with the higher reactivity ratio and higher water solubility compared to the MMA, is active in the formation of stable, higher number of initial particles during the early stages of batch polymerization.

The instantaneous conversion of each monomer was calculated from the amount of unreacted monomers determined by GC, which was found to be higher than 90% during the course of semi-continuous polymerization and to increase slightly as the polymerization proceeded. The ratio of instantaneous conversion between the two monomers was found to be constant and very close to unity, (Figure 2), which indicates that the copolymer composition is homogeneous during the course of polymerization.

Table II

Results of the Batch Polymerization of MMA-MAA System

Latex No.	MMA/MAA Weight Ratio	Coagulum (%)	pH	Solid Content (%)	Conversion (%)	P.S. (by Light Scattering)(nm)
SPB-2-(1)	95/5	0	3.6	10.21	99.4	90
SPB-2-(2)	90/10	0	3.6	10.20	99.3	88
SPB-2-(3)	85/15	0	3.5	10.23	99.6	81
SPB-2-(4)	80/20	0	3.5	10.21	99.4	77
SPB-2-(5)	75/25	0	3.6	10.19	99.2	78

Table III

Results of the Semi-Continuous Polymerization of MMA-MAA System

Latex No.	MMA/MAA Weight Ratio	Coagulum (%)	pH	Solid Content (%)	Conversion %		P.S. (by Light Scattering)(nm)
					S.C.	G.C.	
SP-29	95/5	0	3.8	10.42	96.5	>99	70
SP-28	90/10	0	3.8	10.39	99.8	>99	127
SP-26	85/15	0	3.8	10.09	98.0	>99	144
SP-27	80/20	0	3.4	10.09	97.9	>99	144
SP-30	75/25	0	3.5	10.10	96.5	--	144

2. Molecular Characteristics of the MMA/MAA Copolymer Latexes.

a. Molecular Weight Distribution. The average molecular weights calculated from GPC data are shown in Table IV. It should be noted that the number average molecular weight is independent of the MAA content and the polymerization process. However, the batch latexes seem to have slightly wider molecular weight distribution than those of the semi-continuous latexes.

Table IV

Average Molecular Weights Obtained by GPC

Latex No.	MMA/MAA Weight Ratio	Number Average $Mn \times 10^5$	Weight Average $Mw \times 10^5$	Mw/Mn
SPB-2-(1)	95/5	1.60	6.67	4.17
SPB-2-(2)	90/10	2.03	8.27	4.08
SPB-2-(3)	85/15	1.66	5.22	3.15
SPB-2-(4)	80/20	1.28	4.05	3.18
SPB-2-(5)	75/25	1.75	5.01	2.92
SP-29	95/5	1.47	3.28	2.23
SP-28	90/10	1.70	4.30	2.47
SP-26	85/15	1.75	4.29	2.45
SP-27	80/20	1.74	4.26	2.45
SP-30	75/25	1.58	3.62	2.29

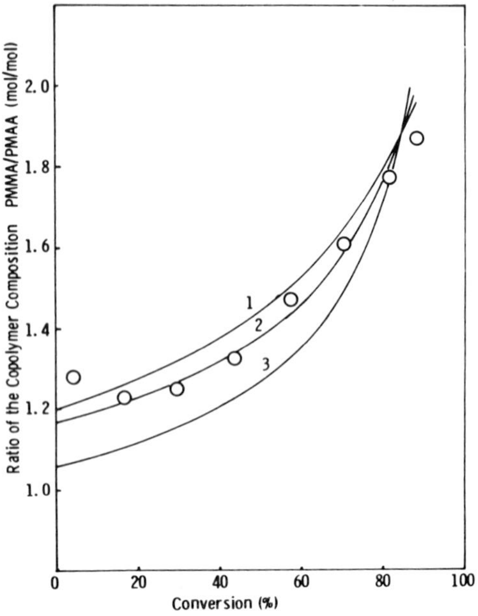

Figure 1. Changes in copolymer composition (PMMA/PMAA ratio) during batch polymerization of MMA–MAA copolymer system: comparison between theoretical and experimental values ((○) experimental point; (1) $r_1 = 0.60$, $r_2 = 1.50$; (2) $r_1 = 0.60$, $r_2 = 1.55$; (3) $r_1 = 0.45$, $r_2 = 1.60$)

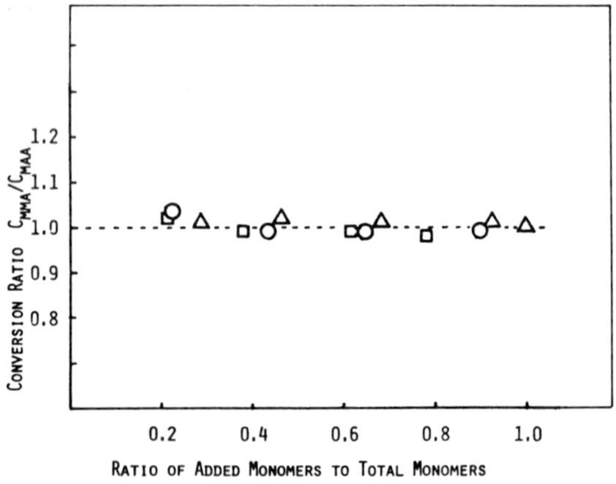

Figure 2. Changes in conversion ratio C_{MMA}/C_{MAA} during semicontinuous polymerization of MMA–MAA comonomer system: (C_{MMA}) instantaneous conversion of MMA; (C_{MMA}) instantaneous conversion of MAA ((△) SP-28 (90/10); (□) SP-26 (85/15); (○) SP-27 (80/20))

b. Monomer Sequence Distribution. The ^{13}C NMR studies were done on the final latex polymers prepared by either batch or semicontinuous process. Consequently, only average values of the sequence from beginning to end of the polymerization can be determined. The ^{13}C NMR spectra of the carboxylated latex was found to consist of many peaks as shown in Figure (3). Klesper et al (13-15) have assigned 18 NMR peaks to all possible pentad-triad sequences in the MMA-MAA. For the purpose of this work, only two peaks No. 12 and 13 (according to Klesper assignments), were considered, which depend mainly on the MMA-triad and all syndiotactic sequences (i.e. $_sA_sA_sA_s$) and represent the MMA sequence length in the copolymer.

For the semi-continuous MMA-MAA copolymer system, the MAA segments should be uniformly distributed along the copolymer chain, which means that the MMA should be interrupted more regularly. On the other hand, for the batch MMA-MAA copolymer system with the same overall composition, the heterogeneity of the copolymer composition would make the MMA sequence length much longer compared to the semi-continuous latexes. Consequently, it is expected that the peak fractions for Peaks No. 12 and 13 are less in the semi-continuous latex polymers compared to the batch latex polymers. This was shown to be the case from 13_C NMR data given in Columns 3 and 5 in Table V; which confirms the above expectation that the average sequence length for MMA in the semi-continuous MMA-MAA copolymer latex should be less than the sequence length in the batch latex.

The theoretical peak fractions listed in Column 4 in Table V were calculated by assuming r_1 (MMA) = 0.60 and r_2 (MAA) = 1.55

Table V
The Results of ^{13}C NMR Comparison
Between Theoretical and Experimental Peak Fractions

Copolymer Composition (MMA/MAA Wt. Ratio)	Peak No.	Batch Latexes Peak Fraction Exper.	Batch Latexes Peak Fraction Theor.	Semi-Continuous Latexes Peak Fraction Exper.
95/5	12	0.183	0.211	0.184
	13	0.360	0.343	0.320
90/10	12	0.179	0.211	0.175
	13	0.262	0.285	0.202
95/15	12	0.188	0.206	0.180
	13	0.216	0.237	0.156
80/20	12	0.162	0.198	0.160
	13	0.114	0.196	0.211
75/25	12	0.158	0.187	0.140
	13	0.160	0.160	0.140

according to the method outlined by Klesper et al (13,14). The theoretical values are in good agreement with the experimental values for the batch latex polymers, which means that the re-

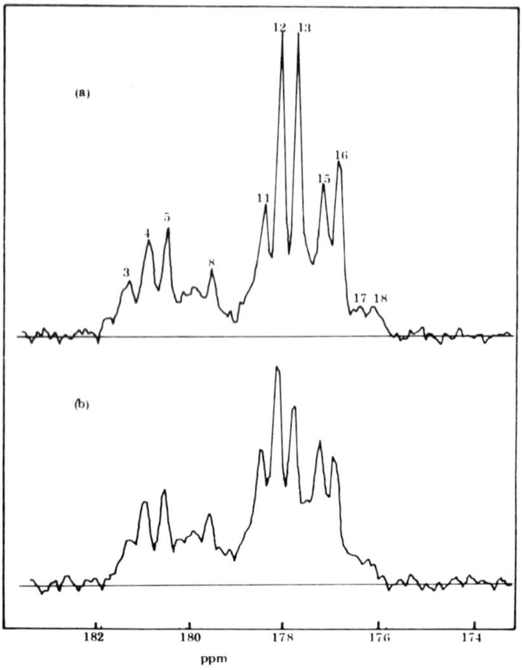

Figure 3. The ^{13}C NMR spectra of carboxylated latexes: (a) MMA–MAA (75:25) batch latex (SPB-2-(5)); (b) MMA–MAA (75:25) semicontinuous latex (SP-30)

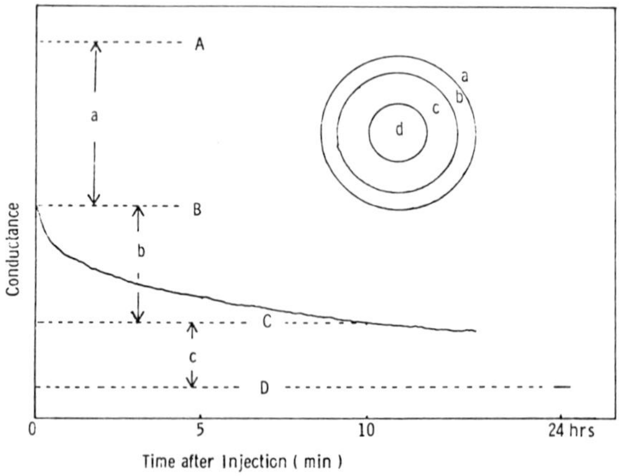

Figure 4. Changes in conductance of a "cleaned" MMA–MAA copolymer latex sample after injection of excess NaOH and assignment of location of carboxyl groups

activity ratios, r_1 (MMA) = 0.60 and r_2 (MAA) = 1.55, are quite correct in the emulsion copolymerization of the MMA-MAA system.

3. Characterization of Carboxyl Content and Distribution in MMA-MAA Copolymer Latex Particles. The results of the conductometric titration of serums and cleaned latexes are shown in Table VI, where the charged amount of MAA in wt. % and equivalent/g-polymer are shown together for comparison. It should be noted that the amount of carboxyl groups in the serum is very small in all latexes, regardless of the MAA content and polymerization process. The total amounts of carboxyl groups neutralized within 24 hours are greater in semi-continuous latexes than in batch latexes, for samples containing more than 15% MAA. However, in these results, the difference between the two types of latexes is not obvious.

The difference in the carboxyl distribution between the two types of latexes is more clearly shown in the time dependence measured by the change in conductance after injection of NaOH according to the method outlined in the experimental section. A typical result of the time effect on the conductance of the latex -NaOH mixture is shown in Figure 4. The conductance at level "A" in Figure 4 is the conductance obtained when the same amount of NaOH solution is added to water only, i.e., in absence of latex particles. The conductance at level "B" represents the conductance of the latex measured immediately after injection of NaOH solution, i.e. at time zero. The conductances at levels "C" and "D" represent the conductances measured at 10 minutes and 24 hours, respectively, after injection of NaOH.

The change in the conductance of the latex-NaOH mixture at various time intervals is then used to determine the amount of carboxyl groups titrated during that time period, which in turn is related to a certain location inside the latex particle. Thus, in Figure 4, value "a", calculated from the difference in conductances "B" and "A" (A-B), corresponds to the amount of carboxyl groups neutralized immediately after injection of the NaOH solution, which is assumed to represent the amount of carboxyl groups located on the particle surface. Value "b", calculated from B-C, is taken to represent the neutralization of carboxyl groups located inside the particle, but close to the particle surface. Value "c", calculated from C-D, corresponds to the amount of carboxyl groups buried deep inside the particle. The summation, a+b+c, is equivalent to the total amount of carboxyl groups neutralized within 24 hours. The difference between the total amount of MAA charged and the summation of a+b+c gives the amount of carboxylic acid which could not be neutralized after 24 hours by injecting the alkali, and is designated as "d".

In Figure 5, the distributions of carboxyl groups given as weight percent based on the amount of MAA charged are presented for the latexes prepared by batch and semi-continuous processes. In this figure, Column 1 represents the amount of carboxyl groups in the serum (see also Table VI), Column 2 represents the value

Table VI
Results of Conductometric Titration of MMA/MAA Copolymer Latexes

Latex No.	MMA/MAA Weight Ratio in Recipe	Amount of Methacrylic Acid (x 10^4 Eq./g. Polymer): Charged According to Recipe	In the Serum	Determined by Back-Titration with HCl 24 hrs after Addition of NaOH Solution to the "Cleaned" Latex	MAA Content in the Copolymer Based on the Back-Titration Results (wt. %)
Batch					
SPB-2-(1)	95/5	5.81	0.122	1.87	1.6
SPB-2-(2)	90/10	11.63	0.078	3.22	2.8
SPB-2-(3)	85/15	17.44	0.048	9.08	7.8
SPB-2-(4)	80/20	23.26	0.125	14.88	12.8
SPB-2-(5)	75/25	29.07	0.236	23.60	20.3
Semi-Continuous					
SP-29	95/5	5.81	0.0	1.30	1.1
SP-28	90/10	11.63	0.026	2.46	2.1
SP-26	85/15	17.44	0.077	10.82	9.3
SP-27	80/20	23.26	0.018	21.60	18.6
SP-30	75/25	29.07	0.156	26.50	22.8

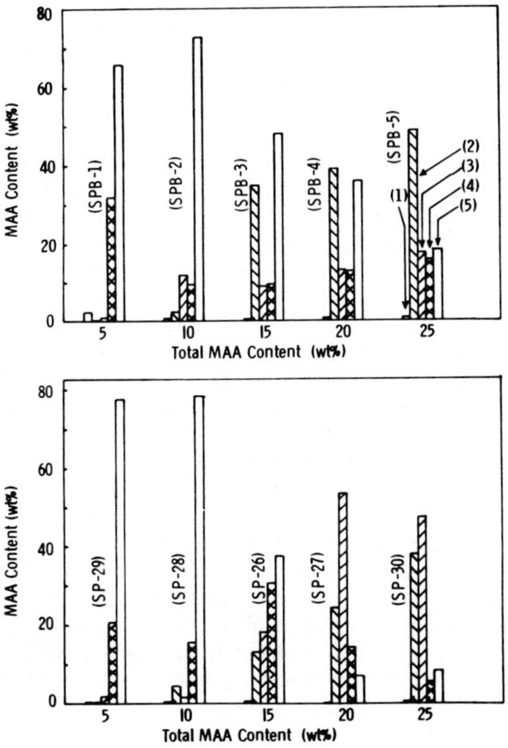

Figure 5. Distribution of carboxyl groups in MMA–MAA copolymer latexes as weight percent based on the MAA content used in the polymerization: (1) carboxyl groups in the serum; (2) surface carboxyl groups; (3) carboxyl groups next to the surface; (4) carboxyl groups buried deeply; (5) unneutralized carboxyl groups; (A) batch-type latexes; (B) semicontinuous-type latexes

"a" (surface carboxyl groups), Column 3 represents the value "b" (carboxyl groups next to the surface), Column 4 represents the value "c" (carboxyl groups buried deeply), and Column 5 represents the value "d" (unneutralized carboxyl groups). The corresponding carboxyl groups distribution, given as absolute value of weight percent MAA based on the polymer, is shown in Figure 6.

In the latexes of low MAA contents of 5% and 10%, there is no significant difference in the neutralization behavior between batch and semi-continuous latexes. On the other hand, for latexes of high MAA contents of 15%, 20% and 25%, it can be seen that batch latexes contain relatively more carboxyl groups in the region closer to the particle surface; whereas in semi-continuous latexes, carboxyl groups are more uniformly distributed inside the particle.

In Figure 7, the curves showing neutralization reaction-versus time after injection of alkali are given for batch and semi-continuous latexes with MAA content of 15%. These curves show that more carboxyl groups are neutralized initially in the batch latex than in the semi-continuous latex, which indicates that more carboxyl groups are located at the surface in the batch latex. But the extent of neutralization is reversed between the two latexes after 24 hours, which means that more carboxyl groups are located inside the particle in the semi-continuous latex. This reversal occurred at a much earlier stage in latexes with higher MAA contents of 20% and 25%.

Although MAA monomer possesses a larger reactivity ratio than MMA monomer, more MAA was found to exist in the outer side of the particle in the batch latex, as shown in Figures 5 and 6. This behavior could be explained if one can accept the fact that the MAA-rich polymers, which are formed early on during the polymerization, can migrate to the surface of the particle due to their higher hydrophilicity and plasticization of the polymer with the monomer. In the semi-continuous process, it could be expected that copolymer with the same composition as the comonomer feed is formed, and the particle contains a uniform distribution of carboxyl groups.

4. Swelling Behavior of the MMA-MAA Copolymer Latexes. When some alkali is added to the cleaned latexes of the MMA-MAA copolymer system, various changes can occur in the latexes. The carboxyl groups are neutralized and charged with negative ions, the charged groups are hydrated and exert electrostatic repulsive forces on each other. As a result, the latex particles are swelled and some segments of the polymer and/or whole polymer chains are dissolved in the aqueous phase. These dissolved polymer segments can interact with each other by hydrogen bond formation and/ or chain entanglements, which cause an interaction between the latex particles. When one examines the latex in these situations, macroscopically the above phenomena should be reflected in changes in particle size and viscosity. The change in particle size of the latex has often been measured as a change in the op-

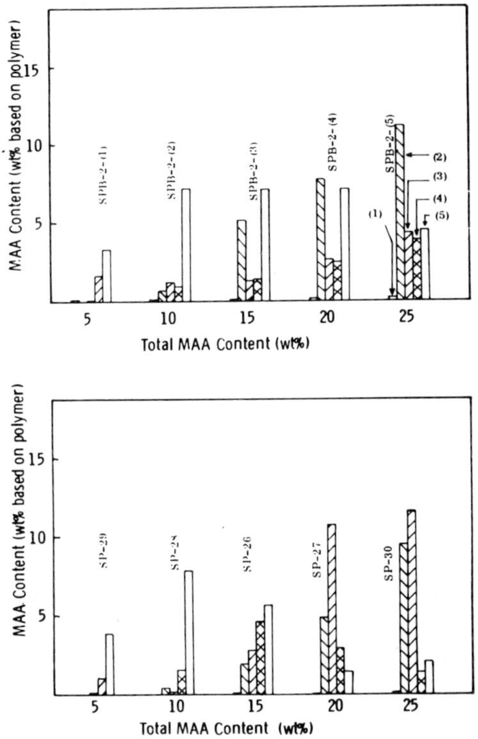

Figure 6. Distribution of carboxyl groups in MMA–MAA copolymer latexes given as absolute MAA amount based on polymer (the columns represent the same as in Figure 5; (A) batch latex; (B) semicontinuous latex)

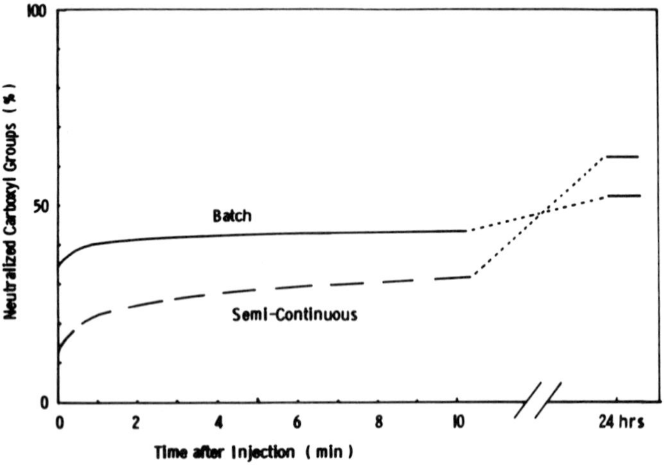

Figure 7. Percent conversion of neutralization reaction with time of latexes with the MAA content of 15 wt %

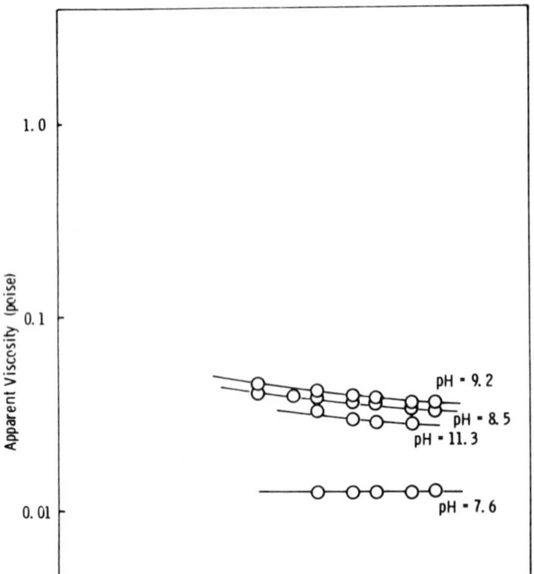

Figure 8. Dependence of viscosity on shear rate at various pH's for 85:15 MMA–MAA batch latex (SPB-2-(3))

tical properties (2,3,5). However, in the optical measurements of carboxylated latexes, there are some complicating factors, such as changes in the density and the refractive index of the latex particle due to swelling and dissolution of latex polymer. For this reason, Verbrugge (4) insisted that the viscosity measurement was better to investigate the behavior of carboxylated latexes with varying pH.

Viscosity of dispersion is known to follow Einstein's equation, in the extreme case where the particle concentration is so low that the particles do not interact with each other.

$$\eta_r = 1 + 2.5\phi , \qquad (1)$$

where η_r is the relative viscosity, and

ϕ is the volume fraction of the dispersed phase,

If the swelling behavior of carboxylated latexes could be characterized only by the increase in the volume fraction, i.e., in particle size, Eq. 1 would still hold; but this is not the case in carboxylated latexes. The latex particles are swelled to a great extent, and the polymer segments are dissolved into the aqueous phase and interact between the particles by the hydrogen bonds and chain entanglements. Thus, Eq. 1 is not expected to hold at all and the interparticle interactions are expected to be dominant in the viscosity development of the latex.

In the light scattering behavior of the carboxylated latex, there are some complicating factors, such as a change in refractive index of latex particles upon the swelling and dissolution of latex polymers. However, it can be expected that differences in the alkali-swelling behavior as measured by changes in viscosity and by changes in light scattering should give valuable information about the true scheme of alkali-swelling.

a. Viscosity

The apparent viscosity shows some dependence on shear rate; the magnitude of which is a function of pH as shown in Figures 8 and 9 for batch and semicontinuous carboxylated latexes respectively. It is difficult to compare the viscosities of non-Newtonian fluids. One possible method for comparison is to take the apparent viscosity value at a certain shear rate for all samples. In this work, the viscosity values at a shear rate of 215 (sec^{-1}) were taken to compare the behavior of the samples. The results of viscosity as a function of pH obtained in this manner are shown in Figure 10 for batch latexes, and in Figure 11 for semi-continuous latexes.

The semi-continuous latexes with relatively high MAA content were found to exhibit 5 orders of magnitude increase in viscosity at pH 8-9. Furthermore, visual observation of these latexes showed that the milky white color disappeared and the dispersion became slightly hazy at the pH of the peak viscosity. For the semi-continuous latexes with low MAA contents of 5 and 10%, no significant viscosity change with pH was observed. On the other hand, batch latexes, even with the highest MAA content, main-

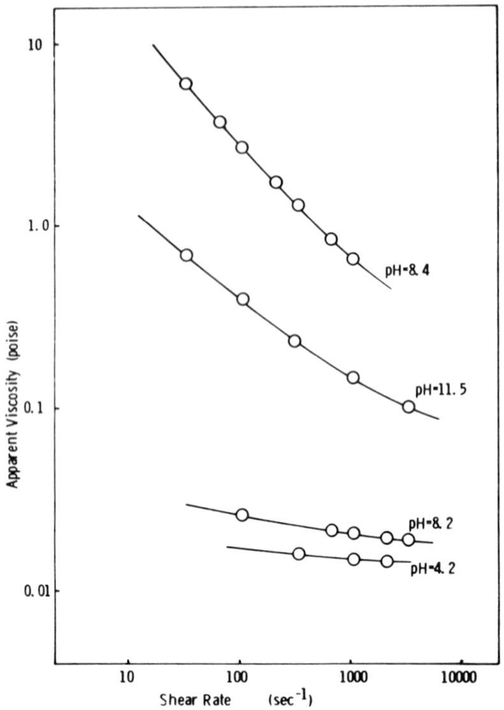

Figure 9. Dependence of viscosity on shear rate at various pH's for 85:15 MMA–MAA semicontinuous latex (SP-26)

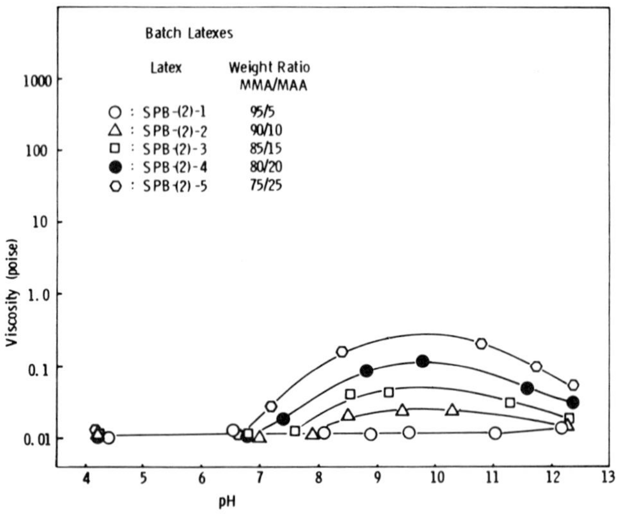

Figure 10. Changes in viscosity with pH of batch MMA–MAA copolymer latexes determined at shear rate of 215 s^{-1})

tained the milky white color at high pH, and the change in viscosity was not more than one order of magnitude.

The difference in viscosity behavior between the two types of latexes appeared in the dependence on the shear rate. A careful comparison between Figures 8 and 9 shows that the semi-continuous latex exhibits a stronger dependence on the shear rate than the batch latex. The latexes showing the large increases in viscosity, i.e. the semi-continuous latexes with the MAA contents of 15, 20 and 25%, exhibited such a strong dependence on the shear rate. All the batch latexes with varied MAA contents of 5-25% and the semi-continuous latexes with MAA contents of 5 and 10% which exhibited moderate or small increases in viscosity exhibited less significant dependence on the shear rate. Generally the high dependence of viscosity on the shear rate means an extensively spread structure formation in the dispersion due to a strong interaction between the dispersed materials. In the semi-continuous latexes, the hydrogen bonds constructed in the whole dispersion by the fully swelled latex particles could be a reasonable explanation for the extraordinary increases in viscosity and the higher dependency of viscosity on the shear rate. In the batch latexes, the swelling of the latex particles can be considered to extend to the region close to the particle surface, not too deep inside the particle, and hence insufficient to interact strongly with each other and form the hydrogen bonds in the whole dispersion.

These results on viscosity behavior are in good agreement with the results of the latex characterization by conductometric titration (see Figures 5 & 6), which showed that the carboxyl groups are uniformly distributed within the particle for the semi-continuous latex, whereas in the batch latex the carboxyl groups are concentrated at the water-particle interface.

b. Light Scattering Behavior

The change in apparent particle size measured by the dissymmetry method of light scattering are shown in Figure 12 for batch latexes, and in Figure 13 for semi-continuous latexes as a function of pH. The dependence of the apparent particle size on pH is found to depend on both polymerization process and MAA content. The batch latexes with MAA contents of 5, 10 and 15% exhibit a slow increase in particle size with increasing pH; whereas latexes with higher MAA contents exhibit peaks in particle size at pH = 10.5 followed by a slow decrease. These results can be explained in terms of swelling upon neutralization followed by dissolution and/or shrinkage of particle due to addition of excess alkali.

On the other hand, the changes in apparent particle size of the semi-continuous latexes with pH is found to be insignificant. The particle size decreases slightly to a minimum at a certain pH followed by a slight increase with increasing pH. These results can be explained on the basis of the homogeneous distribu-

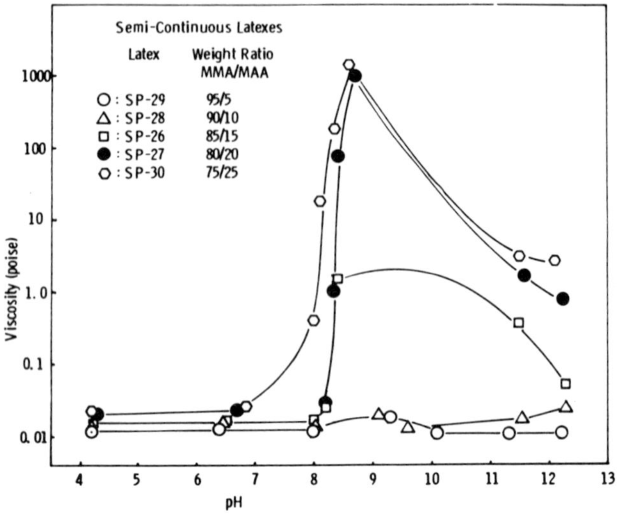

Figure 11. Changes in viscosity with pH of semicontinuous MMA–MAA copolymer latexes (determined at shear rate of 215 s^{-1})

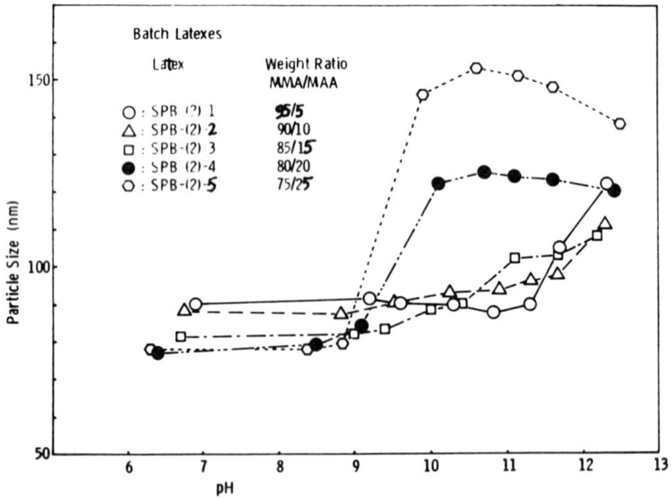

Figure 12. Changes in apparent particle size with pH of batch MMA–MAA copolymer latexes (determined by dissymmetry method of light scattering)

tion of carboxyl groups inside the particles. Upon neutralization of the semi-continuous latexes, swelling and simultaneous dissolution of the outer layer takes place, thus causing the particle size detected by light scattering to exhibit insignificant changes with pH. At pH higher than 11.5, the increase in particle size of latexes with lower MAA contents could be seen, which can result from flocculation of the latex particles due to the high ionic strength.

Yudelson and Mack (16) reported that certain carboxyl-containing polymer solutions showed extraordinary viscosity maxima in a narrow pH region at degrees of ionization that are less than 100% and explained these results by gel formation due to intermolecular interaction, particularly hydrogen bonding. They also showed that at the gelation pH, attractive force interactions due to hydrogen bonding should be equal to repulsive force effects due to the coulombic repulsion of charged carboxylate ions. In our present carboxylated latex systems, the latex particles should be swelled and polymer molecules or smaller segments should be dissolved in the aqueous phase upon neutralization so that the particle-particle and/or particle-dissolved polymer interaction may occur. At the pH of the viscosity peak, swelling would be quite high, and so particles should be close enough to interact strongly with each other. Some of the unneutralized carboxyl groups can form hydrogen bonds, both inter-and intra-particles, with other carboxyl groups and ester groups. Thus, a structure which is fully extended in the entire latex system could be formed. As the pH increases and neutralization proceeds, more carboxyl groups would be neutralized, and polymer molecules would be more expanded due to the increased electrostatic repulsion. However, the unneutralized carboxyl groups which play the main role in forming hydrogen bonds would disappear as the pH is increased further. Therefore, the viscosity decreases. In this special case, hydrogen bonding could be expected to be responsible for the viscosity peak, and the breakage of the hydrogen bond could be a reason for the viscosity decrease following the peak.

From the above Results and Discussion, one could speculate the following scheme given in Figure 14, of alkali-swelling and/or dissolving behaviors of the MMA-MAA carboxylated latexes. The semi-continuous latex is swollen extensively and forms structure due to hydrogen bonds which results in a large increase in viscosity and a high shear rate dependence of viscosity. At higher pH, carboxylated groups are neutralized completely and hence the hydrogen bonds disappear, which results in the decrease in viscosity. The batch latex is swollen only in the surface region and the swelling is not sufficient for the particles to interact strongly with each other, resulting in a moderate increase in viscosity. At higher pH, swollen latex would be shrunk due to the increased ionic strength.

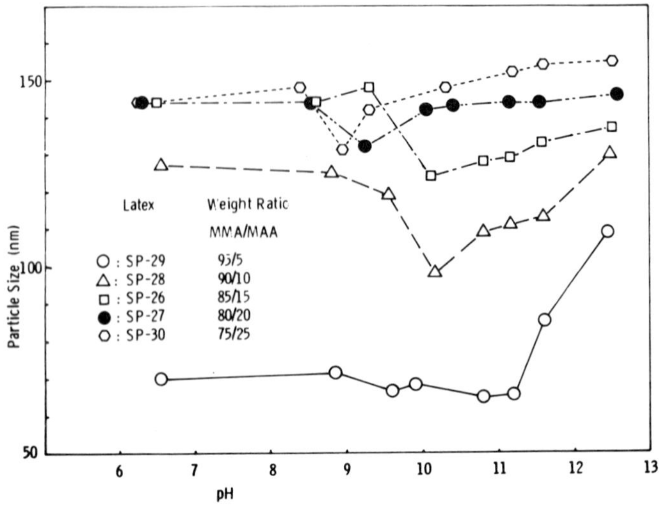

Figure 13. Changes in apparent particle size with pH of semicontinuous MMA–MAA copolymer latexes (determined by dissymmetry method of light scattering)

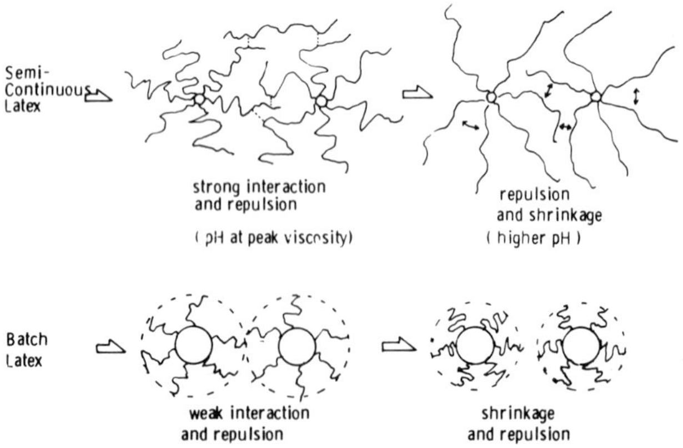

Figure 14. Schematic explanation of alkali-swelling and/or dissolving behaviors of carboxylated MMA–MAA copolymer latexes

Conclusions

In the emulsion polymerization of alkali-swellable carboxylated latexes, the polymerization process was found to influence the distribution of carboxyl groups in the final latex particles, which in turn, influence the alkali-swelling and/or dissolving behaviors of the latexes. MMA-MAA copolymer latexes prepared by batch process contained less uniform distribution of carboxyl groups inside the particle, and exhibited moderate increase in viscosity followed by slow decrease with increasing pH. The corresponding copolymer latexes prepared by semi-continuous process, in which the rate of monomer feed is much less than the maximum rate of polymerization, contained more uniform distribution of carboxyl groups inside the particle, and exhibited a great increase in viscosity followed by rapid decrease with increasing pH.

The distribution of carboxyl groups in alkali-swellable carboxylated latex particles could be detected by the modified conductometric titration after cleaning by serum replacement technique. The modified conductometric titration is composed of the following procedures: (1) following a change in conductance after injection of excess NaOH into a sample latex; (2) after 24 hours, conductometrically back-titrating the latex containing excess NaOH with HCl; and (3) assigning the amount of carboxyl groups neutralized during a certain period of time to a location of carboxyl groups within the particle.

The extraordinary increases in viscosity exhibited by the semi-continuous latexes with MAA contents of 15, 20 and 25% were attributed to the balance of repulsive forces due to dissociated carboxyl groups with attractive forces due to hydrogen bonds, which results in structure formation extending within the entire system. On the other hand the moderate increase in viscosity of the batch latex was attributed to limited swelling, and moderate interaction between the particles and restricted to the surface regions of the particles.

Literature Cited

1. Fordyce, D. B., Dupre, J., and Toy, W., Ind. Eng. Chem., 1959, 51, 115.
2. Muroi, S., J. Appl. Polym. Sci., 1966, 10, 713.
3. Muroi, S., Hosoi, K., and Ishikawa, T., J. Appl. Poly. Sci.; 1967, 11, 1963.
4. Verbrugge, C.J., J. Appl. Polym. Sci.; 1970, 14, 897.
5. Verbrugge, C.J., J. Appl. Polym. Sci.; 1970, 14, 911.
6. Snuparek, J. Jr., and Krska, F., J. Appl. Polym. Sci., 1976, 20, 1753.
7. Bassett, D. R., and Hoy, K. L., "Polymer Colloids, Vol. II", editor, R.M. Fitch, Plenum Press, New York, 1980, p.1.
8. Hoy, K.L., J. Coat. Tech., 1979, 51, No. 651, 27.

9. Pangoni, W.J. and Heller, W., "Angular Scattering Functions for Spherical Particles", Wayne State University Press, Detroit, 1960.
10. Ahmed, S.M., El-Aasser, M.S., Pauli, G.H., Peohlein, G.W., and Vanderhoff, J.W., J. Colloid Interface Science; 1980, 73, 388.
11. Van den Hul, H.J., and Vanderhoff, J.W., J. Electronal. Chem.; 1972, 37, 161.
12. Nishida, S., "Preparation, Characterization and Alkali-Swelling Behavior of Carboxylated Latexes", Ph.D. Dissertation, Lehigh University, 1980.
13. Klesper, E., J. Polymer Science; 1970, 8, A-1, 1191.
14. Klesper, E., Johnsen, A., Gronski, W., and Wehrli, F.W.; Makromol. Chem., 1975, 176, 1071.
15. Johnsen, A., Klesper, E., and Wirthlin, T., Makromol. Chem. 1976, 177, 2397.
16. Yudelson, J.S. and Mack, R.E., J. Polym. Sci., 1964, 2, Part A, 3683.

RECEIVED April 6, 1981.

19

Emulsion Polymerization Kinetics

Chain Entanglements and Glassy-State Transition

B. HARRIS
Dow Chemical Company, Midland, MI 48640

A. E. HAMIELEC and L. MARTEN[1]
Department of Chemical Engineering, McMaster University, Hamilton, Canada

Free radical polymerization in a latex particle occurs at polymer concentrations where chain entanglements and low free volume can significantly reduce the translational and segmental mobility of macroradicals. The termination reaction becomes diffusion-controlled and a reduction in the termination rate constant by several orders of magnitude can cause a large increase in the average number of radicals per particle and an acceleration in the rate of polymerization. At higher conversions a glassy-state transition may occur with the propagation reaction rate falling rapidly and approaching rates which are typical of solid state polymerizations. In the normal time scale a limiting conversion of less than 100% is reached. When molecular weight development is controlled mainly by termination reactions, a dramatic increase in the weight average molecular weight is observed at high conversions. In emulsion polymerizations chain transfer reactions are generally more important and thus molecular weight increases with conversion are not large. However, at very high conversions (X >80%), abnormal molecular structures have been observed and predicted in the emulsion polymerization of vinyl chloride (1-6). These include a large increase in the number of terminal double bonds associated with a sharp decrease in the number average molecular weight ($\bar{M}_N$) and an increase in the number of long branches (LCB) associated with an increase in the weight average molecular weight ($\bar{M}_W$). Abnormal changes in $\bar{M}_N$, $\bar{M}_W$ and LCB can be predicted on the basis of diffusion-controlled propagation (6).

A valid kinetic model of stage 3 emulsion polymerization must account for diffusion-controlled termination and propagation reactions. Marten and Hamielec (7) have proposed such a model based on a free-volume theory and have confirmed its validity for the bulk polymerization of methyl methacrylate (7). Herein is reported an evaluation of this model for the emulsion

[1] Present address: Air Products and Chemicals Inc., Allentown, PA 18105

0097-6156/81/0165-0315$05.00/0

polymerization of styrene. The experimental evaluation was done using monodispersed latex particles in the size range, 400-1650 Å with a measure of instantaneous rates using an adiabatic calorimeter as reactor. A knowledge of instantaneous rates permits one to determine the variation of k_t and k_p with conversion. In particular for the 400 Å latex particles, one can set the average number of radicals per particle ($\bar{n}$) equal to 0.5 and this permits a direct measure of the decrease of the propagation constant with conversion.

Kinetic Model Development

When the polymerization temperature is less than the glass-transition temperature (T_{gp}) of the polymer being synthesized, a glassy-state transition occurs and a limiting conversion is approached with the unreacted monomer acting as a plasticizer. The correlation of limiting conversion using free-volume theory has been quite successful (7-12). The limiting conversion (x_ℓ) can be estimated using equations (1) and (2) neglecting volume changes on mixing monomer and polymer.

$$x_\ell = \frac{(\rho_p/\rho_m)\,\phi_p}{(1-(1-\rho_p/\rho_m)\phi_p)} \tag{1}$$

$$T = T_g = \frac{\alpha_p \phi_p T_{gp} + \alpha_m (1-\phi_p)\, T_{gm}}{\alpha_p \phi_p + \alpha_m (1-\phi_p)} \tag{2}$$

where T is the polymerization temperature and T_g is the glass-transition temperature of the monomer/polymer solution, T_{gp} and T_{gm} are the glass-transition temperatures of pure polymer and monomer. ϕ_p is the volume fraction of polymer in the glassy state, α_p is the difference between the coefficients of volumetric expansion of the polymer in the melt and glassy state, and α_m is the corresponding quantity for the monomer. ρ_p and ρ_m are densities of polymer and monomer.

A graph of equations (1) and (2) is shown in Figure 1 for the homopolymerization of styrene. The parameters used are given in the caption. In an independent study Sundberg and James (12) found the following set of parameters to fit limiting conversions for styrene/polystyrene seed latices: $T_{gp} = 92.5\,^{o}C$, $T_{gm} = -106\,^{o}C$, $\alpha_p = 0.48 \times 10^{-3}\ {}^{o}C^{-1}$ and $\alpha_m = 1.0 \times 10^{-3}\ {}^{o}C^{-1}$. These data are also shown in Figure 1 and support somewhat lower limiting conversions. The trends are the same however.

A comprehensive model which is based on the free-volume theory and which accounts for the effect of molecular weight and solvent on chain entanglements and glassy-state transition has been recently developed by Marten and Hamielec (7). This model accounts for diffusion-controlled termination and propagation

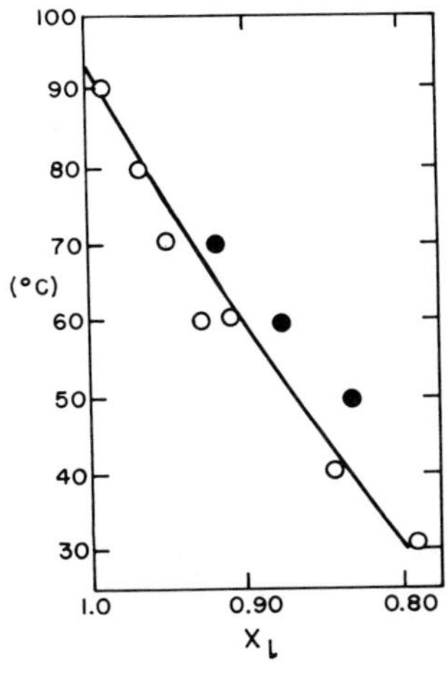

Figure 1. Styrene homopolymerization—limiting conversions (Equations 1 and 2 with parameters T_{gp} = 93.5°C; T_{gm} = −88.2°C; α_p = 0.48 × 10^{-3}°C^{-1}; α_m = 1.0 × 10^{-3}°C^{-1}). Experimental limiting conversions (bulk, suspension, and emulsion polymerization)—(○) present data; (●) from Ref. 12.

reactions with the following equations

$$\frac{k_t}{k_{t_o}} = \left(\frac{\overline{M}_{w_{cr1}}}{\overline{M}_w}\right)^{2\alpha} \exp[- A (\frac{1}{V_F} - \frac{1}{V_{F_{cr1}}})] \qquad (3)$$

$$\frac{k_p}{k_{p_o}} = \exp [- B (\frac{1}{V_F} - \frac{1}{V_{F_{cr2}}})] \qquad (4)$$

$$V_F = (0.025 + \alpha_p (T - T_{gp})) \phi_p + (0.025 + \alpha_m (T - T_{gm})) \phi_m \qquad (5)$$

where $\overline{M}_{w_{cr1}}$ is the cumulative weight average molecular weight at

$V_{F_{cr1}}$ the free volume fraction at which the termination reactions become diffusion-controlled. A, B and α are adjustable parameters which are obtained by fitting rate and molecular weight data. In solution polymerization the termination and propagation constants fall simultaneously and both affect the polymerization rate. This makes it difficult to observe the effect of say a diffusion-controlled propagation reaction on the polymerization rate alone without the confounding influence of the diffusion-controlled termination. Emulsion polymerization presents a unique opportunity to observe the effect of a diffusion-controlled propagation reaction on the polymerization rate at different levels of conversion or free volume. When very small monodispersed latex particles and low rates of radical initiation are used, the average number of radicals per particle should remain at 0.5 over the entire conversion range. The decrease in k_t will therefore have no effect on the rate and the fall in rate corrected for monomer concentration should be due to a fall in the propagation constant. These data permit one to make direct estimates of $V_{F_{cr2}}$ and B. This will be illustrated with data for 400 Å latex particles under Discussion and Results.

Experimental Details

An adiabatic batch reactor was constructed from a 2-liter Dewar flask. The reactor top was fabricated from cork, machined to fit and then coated with epoxy resin to seal its pores. Agitation was provided by a paddle agitator, Gast air motor, via a glass seal. Teflon stopcocks allowed venting, nitrogen purge and addition of reaction components. Temperature was monitored via a Hewlett Packard 2802 A platinum resistance thermometer and quick response probe. The 2 volt analog output was recorded on a Heath servo recorder, EU-20 B. A bucking voltage and voltage divider were used to modify the signal to allow recording on the millivolt scale.

A typical latex recipe follows:

Component	Weight (grams)
water	1300
polystyrene seed (monodispersed)*	104
styrene monomer (washed with caustic and water)	100
sodium persulfate	2-10
sodium bisulfite	0.1-2
$F_e(NO_3)_3\ (H_2O)_9$	0.01
caustic	to pH 6-7
anionic emulsifier (DOWFAX 2A1)**	0-5

* original 300 Å latex particles contained 1% divinyl benzene crosslinker

** Trademark of the Dow Chemical Co.

The polymer, water and monomer phases were heated externally and added to the reactor (previously purged with nitrogen). After thermal equilibrium (estabishment of steady state heat loss) and chemical equilibrium the reaction was initiated by a persulfate/bisulfite system, each component heated separately to reactor temperature. The total adiabatic temperature rise was approximately 10°C. This was used to calculate rates of polymerization, change in rate with temperature and limiting conversion.

The polystyrene seed latex was monodispersed. Even after several grow-ups (polymerizations) the final 1650 Å latex was monodispersed. Hydrodynamic chromatography on the 1650 Å latex gave a mean diameter of 1660 Å with a size variance as small as for normal polystyrene latex standards (typical standard of 1760 Å with a standard deviation of ± 23 Å). The final latex particle size could be accurately predicted from the initial particle size and the total amounts of monomer and polymer used.

With small latex particles (∿ 400 Å) the rate of free radical generation (varied by varying the persulfate level) did not affect the polymerization rate over the range of conversions investigated. In addition, small amounts of sodium bisulfite were added to increase radical generation at low temperatures (∿ 25°C) at various conversion levels. This also did not change the reaction rate. These experiments confirm the assumption that small latex particles contain on the average 1/2 radical/particle over the entire conversion range investigated.

Discussion of Results

Figure 2 illustrates some typical instantaneous rate data obtained using the adiabatic calorimeter reactor for monodispersed latices in the size range 395-1650 Å. The smallest latex particles (395 Å) experience a continuous decrease in rate with decrease in monomer concentration. At about a monomer concentration of 1 gmole/liter the rate falls very sharply. These observations are consistent with case II kinetics and $\bar{n} = 0.5$ and a rapidly falling propagation constant at higher conversions. This was further confirmed by showing that the rate data were independent of initiation rate over the entire conversion range investigated. The larger latex particles experience a rate maximum which is more pronounced for the larger particles. These observations are consistent with diffusion-controlled termination and latex particles which contain two or more radicals. For the larger latex particles the sharp drop in rate at very high conversions is due to diffusion-controlled propagation.

The rate data for the 395 Å latex particles will be treated first to examine the variation of the diffusion-controlled propagation constant during an adiabatic polymerization. Setting $\bar{n} = \frac{1}{2}$ permits one to calculate k_p at different temperature and conversion levels from the instantaneous rate data. These values are shown plotted in Fig. 3. The propagation constant obeys an Arrhenius type equation at intermediate conversions where propagation is kinetically-controlled. However, a sharp break occurs and then k_p falls rapidly in the diffusion-controlled region. Figure 4 shows these same falling k_p values plotted versus conversion with k_p set equal to zero at the limiting conversion $x_\ell = 0.948$ at 75°C. All the k_p values were normalized to 75°C. Figure 5 shows the k_p data for 395 Å latex particles plotted versus free volume fraction according to equation (4). A value for B = 0.1275 fits the higher temperature data quite well and suggests that equation (4) is suitable for correlating propagation constants in the diffusion-controlled region. The lower temperature data deviate somewhat from the higher temperature data at very high conversions. This may be due to experimental errors with the very low rates obtained at the lower temperatures. Figure 6 shows a correlation of X_{cr2}, the conversion at which the propagation reaction becomes diffusion-controlled, with polymerization temperature. The critical conversion increases with temperature and presumably will reach unity at some temperature above the glass-transition temperature of the polymer.

The rate data for the 1650 Å latex particles will now be treated. These rates are sensitive to the magnitude of the termination constant and therefore it should be possible to

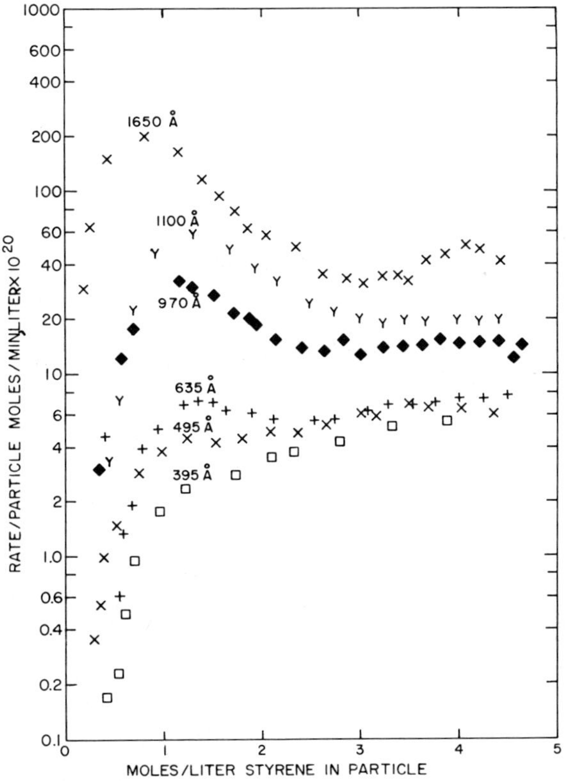

Figure 2. Styrene emulsion polymerization—instantaneous rate data for monodispersed latices showing the effect of particle size ($-\Delta Hp = 16.4$ kcal/gmol)

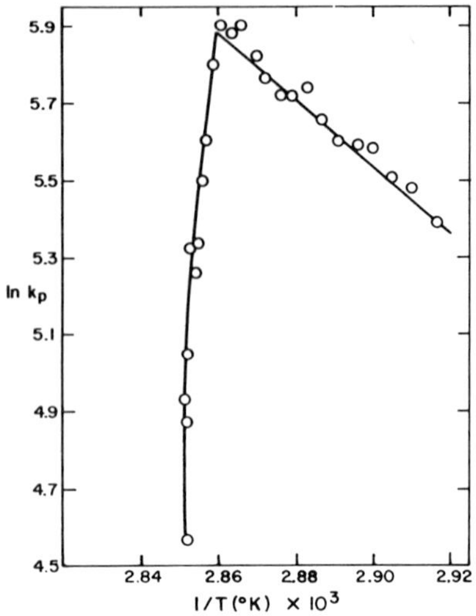

Figure 3. Styrene emulsion polymerization—variation of the propagation constant with temperature during adiabatic polymerization of 395-Å latex particles (k_p in L/mol · s)

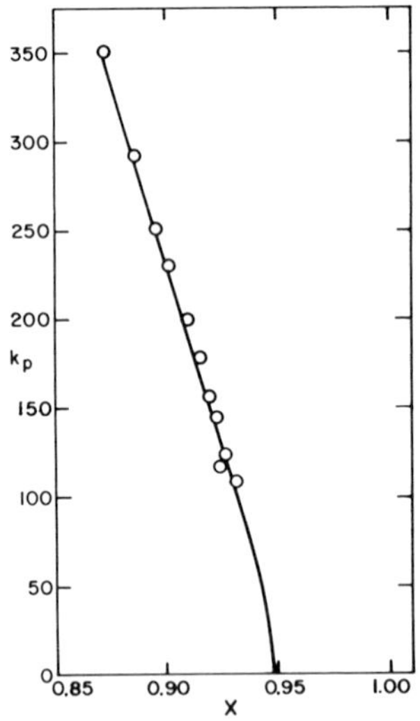

Figure 4. Styrene emulsion polymerization—variation of the propagation constant with conversion during adiabatic polymerization of 395-Å latex particles (k_p values normalized to 75°C with an activation energy of 7700 cal/gmol and in L/mol · s)

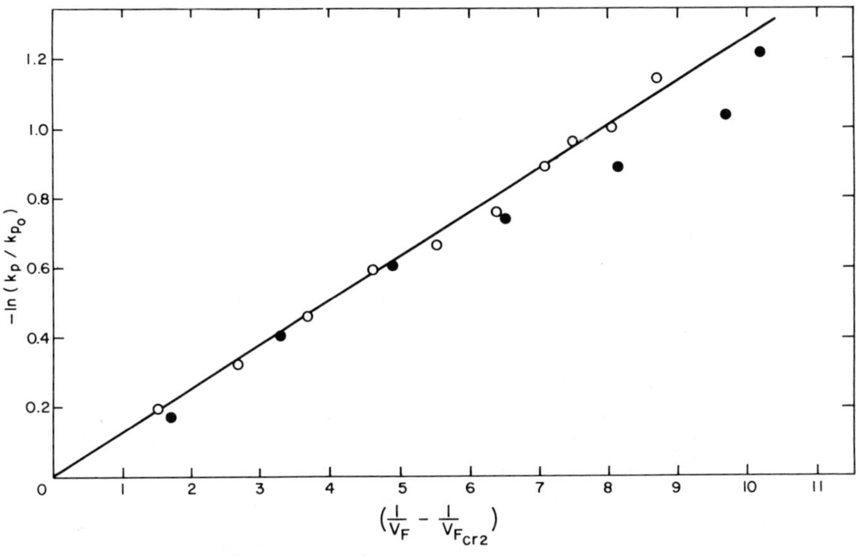

Figure 5. Styrene emulsion polymerization—variation of the propagation constant with free volume fraction according to Equation 4 during adiabatic polymerization of 395-Å latex particles at different temperature levels ((——) Equation 4, B = *0.1275; (○)* T = *76.4°C,* $V_{F_{cr2}}$ = *0.0426; (●)* T = *30.3°C,* $V_{F_{cr2}}$ = *0.0306)*

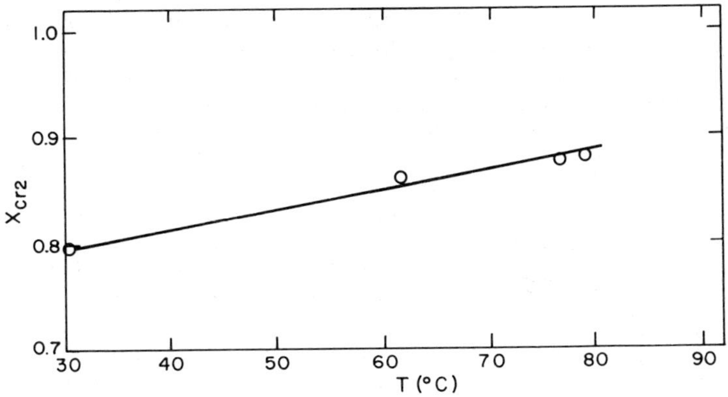

Figure 6. Styrene emulsion polymerization—critical conversion for diffusion-controlled propagation as a function of temperature (——) X_{cr2} = *0.740 + 1.846 ×* 10^{-3}T *(°C))*

determine the change in k_t at different levels of conversion once the decrease in k_p is accounted for. To increase the sensitivity of the rate to k_t, we will consider conversions greater than 0.7 where $\bar{n}$ is greater than about 4. To test the applicability of equation (3) in this conversion range one can rewrite the equation as follows

$$\frac{k_t}{k_t^*} = \exp\left[- A\left(\frac{1}{V_F} - \frac{1}{V_F^*}\right)\right] \qquad (3a)$$

where k_t^* is the termination constant when the free volume fraction equals V_F^*. The effect of molecular weight changes on k_t is neglected. Figure 7 shows the dependence of k_t on free volume fraction according to equation (3a) obtained with 1650 Å latex particles in the temperature range, 74.2 - 78.0°C. The data shown in Figure 7 were normalized to 75°C. The fit of the data to equation (3a) is excellent. This suggests that corrections for molecular weight change are either negligible or that changes in molecular weight with conversion are small.

At this point it is appropriate to point out an apparent inconsistency. In Figure 3 the propagation constant obeys an Arrhenius equation with an activation energy of about 16,000 cal/gmole before the onset of diffusion control. This value is about double the usually accepted values for styrene polymerization. A possible explanation for this discrepancy is that the termination constant may have fallen to the point where a small fraction of the 400 Å latex particles might contain 2 radicals. This of course would increase the rate of polymerization and if not accounted for would give k_p values that are too large. At the point of diffusion control of k_p it is estimated that an $\bar{n}$ = 0.6 could account for a doubling of the apparent activation energy. In fact, when k_p values were measured at different temperatures but for the same sized latex particles and at the same monomer concentration the activation energy for propagation was found to be about 8400 cal/gmole. A reasonable value. The small increase in $\bar{n}$ would introduce a relatively small error in the estimation of k_p values and thus the conclusions reached are basically unchanged.

In Figure 1, the parameters T_{gp} = 93.5°C and T_{gm} = - 88.2°C were estimated by finding the best fit to the limiting conversion data. The remaining parameters were arbitrarily set to reasonable values.

In conclusion it may be said that equations (3a) and (4) adequately represent the changes of termination and propagation constants in the diffusion-controlled regions for the emulsion polymerization of styrene.

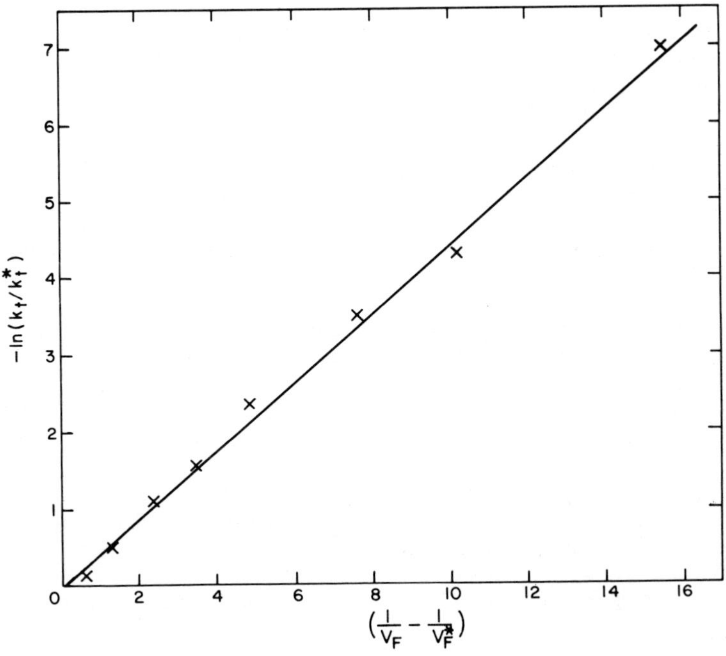

Figure 7. Styrene emulsion polymerization—variation of the termination constant with free volume according to Equation 3 with molecular weight changes neglected during adiabatic polymerization of 1650-Å latex particles over the conversion range, 0.7–0.911 ((A = 0.44; V_F = 0.0708)*

Literature Cited

1. Sorvik, E.M., Hjertberg, T., J. Macromol. Sci.-Chem. 1977, A11(7), 1349.

2. Hjertberg, T.; Sorvik, E.M., J. Polymer Sci. Chem. Ed. 1978, 16, 645.

3. Hjertberg, T.; Sorvik, E.M., "Anomalous Structures Influencing the Thermal Stability of PVC", Abstracts - Third International Symposium on PVC, Case Western Reserve University, Cleveland, August 10-15 (1980).

4. Hjertberg, T; Sorvik, E.M., "Polymerization of VCM at Subsaturation Pressure", Abstracts - Third International Symposium on PVC, Case Western Reserve University, Cleveland, Aug. 10-15 (1980).

5. Braun, D; Holzer, G., "Polymerization of Vinyl Chloride at Atmospheric Pressure - Structure and Thermal Stability of the Polymers", Abstracts - Third International Symposium on PVC, Case Western Reserve University, Cleveland, August 10-15 (1980).

6. Hamielec, A.E.; Gomez-Vaillard, R.; Marten, F.L., "Diffusion-Controlled Free Radical Polymerization - Effect on Polymerization Rate and Molecular Properties of PVC", Abstracts - Third International Symposium on PVC, Case Western Reserve University, Cleveland, August 10-15 (1980).

7. Marten, F.L.; Hamielec, A.E., in Polymerization Reactors and Processes, ACS Symp. Ser., 104, American Chemical Society, Washington, D.C., 1978, p. 43.

8. Ross, R.T.; Laurence, R.L., A.I.Ch.E. Symp. Ser. 160 1976, 72, 74.

9. Friis, N.; Hamielec, A.E., J. Poly. Sci. Chem. Ed. 1973, 11, 3321.

10. Friis, N.; Hamielec, A.E., J. Poly. Sci. Chem. Ed. 1974, 12, 251.

11. Friis, N; Hamielec, A.E., in Emulsion Polymerization, ACS Symp. Ser., 24, American Chemical Society, Washington, D.C., 1976, p. 82.

12. Sundberg, D.C.; James, D.R., J. Poly. Sci. Chem. Ed., 1978, 16, 523.

RECEIVED April 6, 1981.

20

Diffusion-Controlled Kinetics in the Emulsion Polymerization of Styrene and Methyl Methacrylate

D. C. SUNDBERG, J. Y. HSIEH, and S. K. SOH
Department of Chemical Engineering, University of New Hampshire, Durham, NH 03824

R. F. BALDUS[1]
Department of Chemical Engineering, University of Idaho, Moscow, ID 83843

The locus of reaction during an emulsion polymerization is nearly exclusively within particles in which the ratio of polymer to monomer is high enough so that the reacting fluid is quite viscous. Relative to bulk polymerization, this is beyond the start of the "gel effect" and one should expect the influence of restricted diffusion of the polymer to be felt during the entire emulsion polymerization reaction. The most common approach for treating this behavior in bulk polymerization is to treat the termination rate constant, k_t, as a function of conversion level and to predict or correlate its decrease with increasing conversion. The decrease in k_t causes a buildup in free radicals and increases the reaction rate. For latex systems in which radical desorption from the particles is not significant, this restricted diffusion will lead to particles having $\bar{n} > 1/2$. Some time ago van der Hoff (1) noted that a k_t value of two or three orders of magnitude below that at infinite dilution was required in order to explain rate data for various latex particle sizes. Similar results were obtained by James and Sundberg (2) in a more detailed study of this effect for styrene. From this work it is quite clear that the gel effect is prevalent in all styrene emulsion polymerizations, only masked at times when the initiator level or temperature is quite low or the particle is very small.

The first attempts to translate the gel effect work accomplished in bulk polymerization into the context of emulsion polymerization were offered by Friis and Hamielec (3,4). Here they used the empirical relationships developed to express k_t as a function of conversion and temperature for the mass polymerization of styrene (5) and methylmethacrylate (6) in order to modify the rate expressions for emulsion polymerization. The modification came about by adjusting Stockmayer's (7) dimensionless group $\alpha = \rho V/(Nk_t)$ with new values of k_t (from mass polymerization) as

[1]Current address: Crown-Zellerbach Corp., Camas, WA 98607

0097-6156/81/0165-0327$05.00/0

the conversion level increased. They were reasonably successful in fitting the conversion vs. time curve for one experimental condition for styrene, and one for methylmethacrylate. The purpose of the present communication is to remove the empirical nature of the k_t dependency discussed above and to use the free volume concept to interpret the same behavior. Experimental data obtained over a wide range of conditions will be presented to provide a broad base against which the theoretical results may be judged.

QUANTITATIVE DEVELOPMENT

Utilizing the seed latex polymerization method to avoid the occurance of new particle formation, the kinetic treatment of an emulsion polymerization is quite straight forward. Assuming that all the particles are the same size, the rate of polymerization, R_p, can be expressed as

$$R_p = -dM/dt = k_p \bar{n} M/(vN_{Av}) \quad (1)$$

where M is the monomer concentration in the particles (gmoles/lit.), k_p the propagation rate constant (liters/gmole, sec.), v the volume of the particle (liters), and N_{Av} is Avogadro's number. It is convenient to remove the monomer concentration from the right hand side of the expression and to write the relationship in terms of the fractional conversion (2), X,

$$d \ln (1-X)^{-1}/dt = k_p \bar{n} / (vN_{Av}) \quad (2)$$

As long as the distribution of radicals among the particles approximates a steady state situation (a condition that may be violated during periods of rapid acceleration in latices), one may use Stockmayer's result (7) to compute $\bar{n}$. When radical desorption is insignificant and water phase termination of radicals is neglected,

$$\bar{n} = (a/4) [I_0 (a) / I_1 (a)] \quad (3)$$

where $a^2 = 8\alpha$, and $\alpha = \rho v/(k_t N)$, where ρ is the rate of radical production in the aqueous phase. Assuming that one has a method for calculating k_t as a function of conversion and temperature for the particular polymer/monomer system at hand, one can make calculations in a stepwise fashion by computing $\bar{n}$ from equation (3) and then integrating equation (2) for one time step. The procedure is then repeated by updating $\bar{n}$ (via k_t and α) with the new conversion level. The procedure needs some modification when the steady state approximation does not hold and when deceleration is experienced as the conversion approaches its limiting value.

Diffusion controlled reactions have been the subject of considerable study and several models have been proposed to relate the rate constant to the appropriate diffusion coefficient. Allen and Patrick (8) summarize these models and show that they may be written as

$$k_{AB} = (\text{constant})\ D_{AB}\ R_{AB} \tag{4}$$

where k_{AB} is the rate constant for reaction between species A and B, D_{AB} is the mutual diffusion coefficient, and R_{AB} is the collision radius for an encounter. For the termination step in a polymerization reaction, R_{AB} is related to the size of a freely rotating polymer segment and to the molecular weight. D_{AB} is dependent upon whether the molecular motion required to cause an encounter is controlled by the segmental or translational diffusivity of the chain, and whether or not that diffusion is influenced by chain entanglements.

Since the entire emulsion polymerization takes place under diffusion controlled conditions, the variation in the termination rate constant may be conveniently expressed as a ratio of the values appropriate at different times during the reaction. Letting k_{to} refer to the conditions of the beginning of the reaction, eq'n (4) gives;

$$k_t/k_{to} = (D_p/D_{po})(R/R_o) \tag{5}$$

As a first approximation to this description, the variation of the collision radii with conversion will be neglected, yielding

$$k_t/k_{to} = D_p/D_{po} \tag{6}$$

Performing this approximation is essentially equivalent to ignoring changes of molecular weight with conversion. Because of the dominance of chain transfer reactions throughout most of the molecular weight development of an emulsion polymerization (resulting in little, if any, molecular weight change with conversion (9)), this approximation may be reasonable.

As noted earlier, the concept of free volume has been used to relate changes in k_t with conversion. The relationship between the diffusion coefficient and free volume has been discussed in detail by Bueche (10) and accounts for the separate effects of temperature and conversion level. This relationship is written as

$$D_p = (\phi_o \, \delta^2/6S) \exp(-v^*/v_f) \qquad (7)$$

where ϕ_o is the jump frequency of the polymer segments, δ is the jump distance for polymer segments, v^* is the critical free volume for a segment to jump, and v_f is the free volume of the polymer solution. S is related to the number of freely rotating segments in the polymer chain, but its value depends upon whether segmental or translational diffusion is being considered, and in the latter case whether or not the polymer chains are entangled. According to Bueche,

S = 1.0 (segmental diffusion)
= N (unentangled translational diffusion)
= N* (entangled translational diffusion)

where N is the number of segments in the polymer chain, and N* is greater than N due to entanglements. Although it is certain that the polymer chains are entangled in a latex particle, this effect has been neglected in the description that follows. Further, since the molecular weight does not change markedly with conversion, the factor N cancels out of the unentangled translational diffusion expression when the ratio of diffusion coefficients are written as in Equation (6). This form is mathematically equivalent to that obtained by simply employing the segmental diffusion relation. Further comments will be made on this point in the Discussion Section. With the above considerations, eq's (6) and (7) reduce to

$$k_t/k_{t_o} = \exp\,[v^*(1/v_{fo} - 1/v_f)] \qquad (8)$$

The free volume of the polymer solution may be written by assuming additivity of the free volume of polymer and monomer, and is given by

$$v_f = [0.025 + \alpha_p\,(T-T_{gp})]\,\phi_p + [0.025 + \alpha_m\,(T-T_{gm})]\,(1-\phi_p) \qquad (9)$$

where T is the reaction temperature (°C), T_{gp} and T_{gm} are the glass transition temperatures of the polymer and monomer, respectively, ϕ_p is the volume fraction of polymer, α_p is the difference between the coefficients of volumetric expansion of polymer in the melt and glassy states, and α_m is the corresponding difference for the monomer. The value of 0.025 is the fractional free volume of pure polymer at its glass point. In this work α_p was taken as 0.48×10^{-3}/°C, α_m as 1.0×10^{-3}/°C, and T_{gm} as - 106°C (11). The T_{gp} of polystyrene was taken to be 100°C and that for methyl methacrylate as 115°C. Given a value for v* (to be discussed below), eq's (8) and (9) allow computation of k_t/k_{to} at any point in the reaction.

During periods of rapid acceleration and at conversion levels approaching the limiting conversion, the above approach needs some modification. There is a physical limit to the rate at which free radicals may accumulate in the latex and that is equivalent to the dissociation rate of the initiator. Thus one needs to continually provide a check to limit the rate of change of free radicals in the latex, $\bar{n}N$ (N in this case being the total number of latex particles), as

$$d(\bar{n}N)/dt \leq \rho \tag{10}$$

Although k_t decreases rapidly as the conversion increases (free volume decreases), it does not approach a value of zero but reaches a limiting behavior. This was first discussed by Gardon (12) who pointed out that this limiting behavior should be related to the propagation step, since termination can take place even when all chain segments are immobile. Thus it is seen that propagation provides chain end mobility in the absense of overall chain mobility. A quantitative description of this behavior is quite involved and is left to future discussion (13). In the present work k_t is allowed to decrease to a certain point (actually defined as a critical free volume) and then held constant throughout the remainder of the reaction. This has the effect of creating an upper limit for the value $\bar{n}$ as calculated from eq'n (3).

At conversion levels near the glass point of the monomer/polymer solution, the free volume is low enough that the propagation reaction becomes diffusion controlled. In a development analogous to that for k_t, the k_p behavior is written as

$$k_p/k_{po} = D_m/D_{mo} = \exp[v_m{*}(1/(v_{fo})_m - 1/v_f)] \tag{11}$$

where k_{po} is the normal, or diffusion unaffected propagation rate constant, D_m is the diffusion coefficient for the monomer, and v_m* is the critical free volume for the monomer to jump. $(v_{fo})_m$ is the free volume of the monomer/polymer solution at the point at which k_p begins to decrease.

EXPERIMENTAL WORK

The approach taken in this study was to use the variables embodied in the dimensionless parameter α to change the experimental conditions over as wide a range as possible. The seed latex technique avoids the occurance of new particle formation and thus leaves the specific experimental variables as particle size (v), initiator level (ρ), and temperature (through ρ and k_t). Base latices were prepared having a particle size range between 500 and 2500 Å and were subsequently swollen with monomer to the extent of 150-175 parts of monomer per hundred parts of polymer. The surfactant level was adjusted to 45% of saturation on the swollen particle surface so that each experiment was equivalent in surface density of surfactant at the start of the reaction. That maintained a stable latex throughout the polymerization and avoided new particle formation.

Polymerizations were conducted in simple glass reactors with temperature control (± 1°C) provided by a circulating water jacket around the reactor. The inhibitor was removed from the monomers by washing them with a 5% sodium hydroxide solution followed by double vacuum distillation of the monomers. Reagent grade potassium persulfate was used as the initiator and sodium lauryl sulfate was used as the surfactant. Water was distilled and/or deionized prior to use.

The seed latices were charged with monomer and any necessary surfactant, and stirred under a nitrogen blanket for > 4 hours to assure solubilization of the monomer. The latices were then brought to reaction temperature and charged with initiator solution to begin the polymerization. No reaction took place during the monomer solubilization period as evidenced by samples withdrawn from the latices just after the initiator charge. The extent of conversion was determined periodically throughout the reaction by withdrawing ∿ 5 ml. samples of the latex and gravimetrically determining the conversions. Particle sizes were determined by standard turbidity techniques. Electron micrographs were obtained for selected samples to check the turbidity and to verify that no new polymer particles were formed during the seed polymerization.

RESULTS AND DISCUSSION

Experiments were carried out for both styrene and methyl methacrylate in the temperature range of 40-70°C. For each monomer the experiments were divided into two groups. Those in the first group were run at the same temperature but at varying conditions of particle size, particle number, and initiator level to obtain a wide range of α_o values. Those in the second group were run under identical formulation conditions, but at different temperatures. Since the seed latices were swollen with monomer to the extent of 150 parts/100 parts polymer, the effective conversion level starts at X = 0.40 at the beginning of the experiment.

There are several parameters that need to be quantified before computations can be made, the first of which is k_{to}. For each of the two monomers, a series of runs were made at a constant temperature of 50°C, the conversion profiles corrected for inhibition time, the data plotted as $\ln(1-X)^{-1}$ vs. time, and the slope determined at the beginning of the reaction (X = 0.40). Using eq'n (2) the corresponding value of $\bar{n}$ was calculated. These n values were then plotted against $\rho v_o/N = \alpha_o k_{to}$ as shown in Figure 1. For each set of data several trial values of k_{to} were assumed, corresponding values of $\bar{n}$ calculated from eq'n (3) for a range of arbitrary α_o's, and the results plotted as a solid curve to be compared with the data. The curves shown in Figure I are the results of this procedure, compare well with both sets of data, and correspond to values of k_{to} of 1.8×10^6 and 1.9×10^4 liter/mol, sec. for styrene and methyl methacrylate, respectively. These values assume that the radical efficiency factor is unity. A more complete description of this procedure is shown in reference (2) for the polystyrene data. The poly(methymethacrylate) data are from Hsieh (14). It should be noted here that the k_{to} values for the two systems are different by two orders of magnitude.

A number of parameters have to be determined in order to use eq's (8) and (11) to predict the conversion profiles. These are v* in eq'n (8) (since v_{fo} corresponds to v_f at X = 0.4 where k_{to} was computed), and v_m^* and $(v_{fo})_m$ in eq'n (11). Over the course of this work several independent sources were used to obtain estimates of v*. The first was to use eq's (8) and (9) to treat the published k_t/k_{to} data of Hui and Hamielec (5) for the mass polymerization of styrene. This treatment yielded a value of 0.6. Second, use was made of the "universal constants" for the WLF equation (15) and this yielded approximately the same value. Fjuita (16) studied the viscosity behavior of a polystyrene/ethylbenzene solution to obtain the variation of free volume with changing polymer concentration. This work suggested a value of v* = 0.68 at 50°C and 0.65 at 70°C. The present work has made use of v* = 0.6 for the entire 40-70°C temperature range and assumes that the same value is applicable to both styrene and methyl methacrylate. Fujita (16) also suggested that the critical free volume for the jump of a small molecule be 0.38. Thus v_m^* was taken as this value for both monomers.

The free volume at which k_t is assumed to have reached its limiting value (henceforth called v_{fc}), as well as the value of $(v_{fo})_m$, must be determined from the emulsion polymerization data

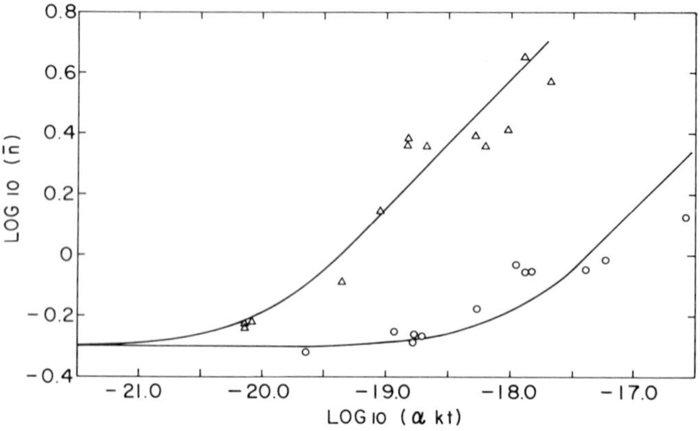

Figure 1. Plot of log_{10} ($\bar{n}_o$) vs. log_{10} ($\alpha_o k_{to}$) for styrene (○) and MMA (△) polymerizations at 50°C. Solid lines represent theory and the data points show experimental behavior.

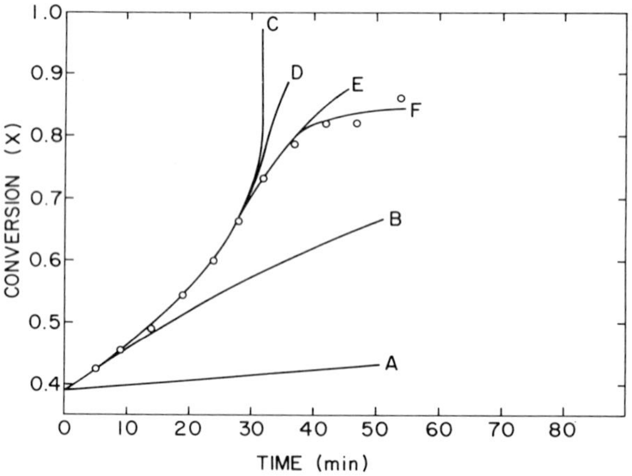

Figure 2. Conversion profile for MMA at 50°C. Curves show influence of each part of the reaction model ((○) Exp. 67).

and thus represent "fitted" constants. However, it is clear that k_t should reach its limiting value prior to the onset of diffusion control of the propagation step, and thus $v_{fc} > (v_{fo})_m$. For each of the two monomers used, one of the experimental runs at 50°C was used to determine the best values of these two parameters. The value of v_{fc} for styrene was found to be 0.049 and that for methyl methacrylate to be 0.074. The reason that k_t becomes limited for methyl methacrylate at a lower conversion (higher free volume) than for styrene is because the k_p for the former is several times larger than that of the latter. The computed results are not very sensitive to the value of $(v_{fo})_m$ and a value of 0.047 was used for both systems.

Figure 2 shows the effect of all of the above considerations, separately and collectively, upon the computed results for a methyl methacrylate run. Curve A neglects any diffusion controlled reactions and assumes that $\bar{n} = 1/2$ throughout. The curve does not come close to describing the conversion behavior even at the beginning of the reaction. Using the previously computed value of k_{to}, $\bar{n}_o$ should be 2.4 and Curve B shows the anticipated behavior if this remains constant for the remainder of the reaction. Curve C computes the gel effect through eq'n (8) but ignores anything else. Curve D further places the physical restriction of a finite accumulation rate of free radicals via eq'n (10). Adding the fact that k_t should have a lower limit, makes a significant change in the high conversion behavior as seen in Curve E. However, it takes the additional consideration of diffusion controlled propagation to show limiting conversion via Curve F. The actual data are shown as the open circles.

The goodness of the model presented here can be judged by comparison to a wide range of experiments. This is shown in Figures 3-6 which display isothermal and variable temperature conversion profiles for both monomers. The values obtained for v_{fc} and $(v_{fo})_m$ were derived from the $\alpha_o = 0.458$ curve in Figure 3 and the $\alpha_o = 30.3$ curve in Figure 4. Once these values were set for one curve, they were not changed when considering the other curves. As can be seen, there is excellent agreement between the predicted and experimental results over a broad range of conditions. Table 1 lists the actual experimental formulations for the data shown in Figures 3-6.

In order to compute the conversion profiles at various temperatures, the k_{to} and k_{po} values were adjusted by using the appropriate activation energies. The activation energy for termination

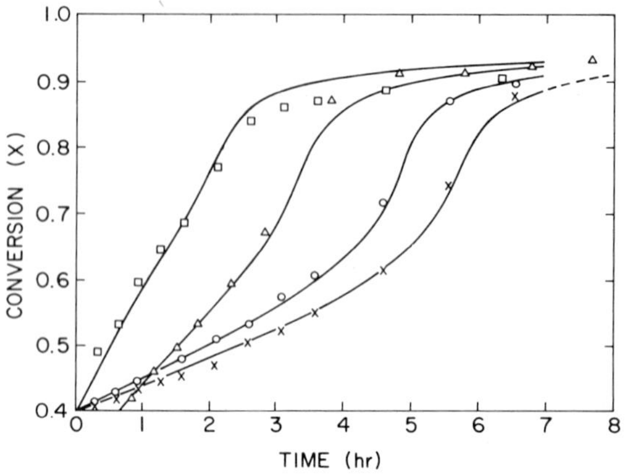

Figure 3. Conversion profiles for styrene seed latex polymerization at 50°C. Solid curves are theoretical predictions and data points are experimental results ((□) α_o = 0.0076; (△) α_o = 0.0393; (○) α_o = 0.458; (×) α_o = 1.37).

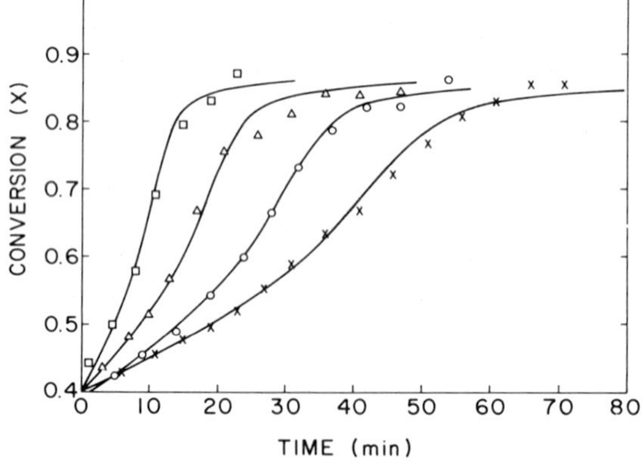

Figure 4. Conversion profiles for MMA seed latex polymerization at 50°C. Solid curves are theoretical predictions and data points are experimental results ((□) α_o = 1.40; (△) α_o = 16.2; (○) α_o = 30.3; (×) α_o = 65.9)

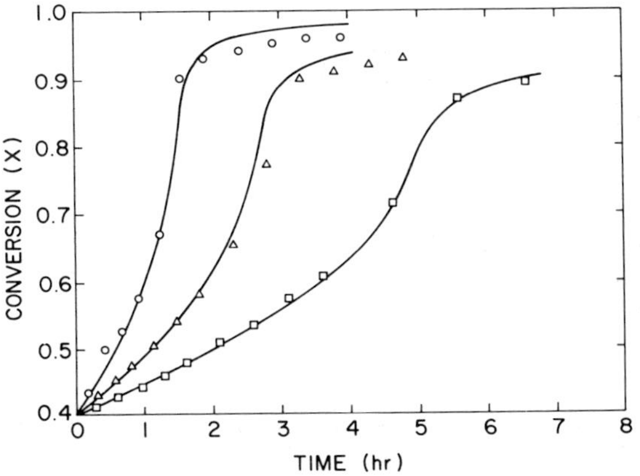

Figure 5. Conversion profiles for styrene seed latex polymerization at various temperatures. Solid curves are theoretical predictions and data points are experimental results ((□) $\alpha_0 = 0.458$; (△) $\alpha_0 = 1.82$; (○) $\alpha_0 = 7.71$)

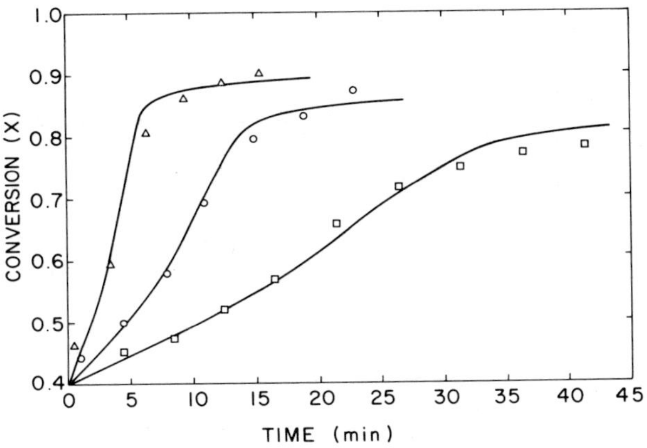

Figure 6. Conversion profiles for MMA seed latex polymerization at various temperatures. Solid curves are theoretical predictions and data points are experimental results ((△) $\alpha_0 = 5.24$; (○) $\alpha_0 = 1.40$; (□) $\alpha_0 = 0.35$)

Table I

Seed Latex Formulations

α_o	Seed particle size (Å)	Temperature (C°)	Recipe (by parts) Polymer	Monomer	Water	Initiator
Styrene:						
0.0076	920	50	100	150	476	1.00
0.0393	1110	50	100	206	578	1.49
0.458	1700	50	100	150	468	1.26
1.37	2090	50	100	150	470	1.26
1.82	1700	60	100	150	468	1.26
7.71	1700	70	100	150	468	1.26
Methyl Methacrylate:						
0.35	840	40	100	150	1465	2.50
1.40	840	50	100	150	1463	2.50
5.24	840	60	100	150	1463	2.50
16.2	1500	50	100	150	1435	0.95
30.3	1880	50	100	157	841	0.45
65.9	2450	50	100	150	449	0.20

was taken to be 2370 cal/mol for styrene and 2490 for methylmethacrylate. The Arrenhius expressions for the k_p's were taken from the literature as

k_{po} = 2.17 x 10^7 exp (-7760/RT), lit/mol, sec for styrene (17) and

k_{po} = 1.19 x 10^7 exp (-6840/RT), lit/mol, sec. for methyl methacrylate (18).

With no further change from the computations at 50°C, the results at variable temperatures are shown in Figures 5 and 6. Again the agreement between computations and experiment are very good. The only significant disagreement occurs for the computations for styrene at 60 and 70°C (Figure 5) which overestimate the limiting conversions actually measured by several percent. The T_g used for polystyrene was 100°C but there is some indication in the literature that it should be lower (4). If so, this would provide better agreement at these high conversion levels as well.

As a final illustration of the comparison between anticipated behavior and actual performance, it is interesting to consider the polymerization of two seed latices which differ markedly in their physical characteristics and which are reacted under significantly different formulation conditions. In particular it is of interest to note the conditions under which the conversion profiles of two such latices will be nearly identical. Eq'n (2) indicates that as long as the value of (k_p $\bar{n}/vN_{Av}$) is the same for the two latices, the conversion profiles should be the same. Among the numerous methyl methacrylate reactions carried out by Hsieh (14) were two which were not planned as such, but which are suitable for such a comparison. Table II lists the conditions of the experiments and the values necessary to compute (k_p $\bar{n}/vN_{Av}$). Although the above individual parameters are vastly different, the values of (k_p $\bar{n}/vN_{Av}$) at the beginning of the reactions are 2.31 x 10^{-4}/sec and 2.54 x 10^{-4}/sec, for experiments 62 and 73 respectively. Thus one should expect that the initial slopes of the X vs. t plot should be identical. Further, if the variations in $\bar{n}$ with conversion are the same, then the entire conversion profile should be identical, except close to the limiting conversions because of the different temperatures involved. $\bar{n}$ should change due to the variation in k_t via eq'n (8), that in turn depending upon changes in free volume via eq'n (9). Since the two temperatures involved are fairly close, there will be no significant difference in free volume of the two systems until the limiting conversions are approached, and therefore the conversion profiles should be identical. Near the limiting conversions, the curve for the higher temperature run should deviate above the other. This anticipated

Table II

Seed Latex Formulations and Rate Parameters

Exp. No.	Seed Particle Size (Å)	Temperature (°C)	Recipe (by parts)			
			Polymer	Monomer	Water	Initiator
62	1875	50	100	150	1563	0.63
73	840	40	100	150	1465	2.50

$\rho \times 10^9$ mols/lit, sec	$N \times 10^{-17}$ particles/lit	$v_o \times 10^{18}$ lit/particle	$k_{to} \times 10^{-4}$ lit/mol, sec	α_o	$\bar{n}_o$	$k_{po} \times 10^{-2}$ lit/mol, sec
3.95	0.17	9.20	1.94	39.5	4.57	2.80
3.18	1.94	0.85	1.45	0.35	0.66	2.00

behavior is shown as the solid curves in Figure 7. The data are shown to be in very strong agreement with these predictions, although the differing limiting conversions are not as distinctive as anticipated. When combined with the results shown in Figures 3-6, those of Figure 7 provide complementary evidence that the present interpretation of diffusion controlled emulsion polymerization behavior for the two monomers investigated is quite adequate to explain reaction rate data over very broad ranges of experimental conditions.

CONCLUDING REMARKS

In closing, it is interesting to consider that a model which is mathematically equivalent to assuming segmental diffusion control of the termination reaction during the gel effect works so well for emulsion polymerization data while it is clear that translational diffusion of entangled molecules controls the termination step in bulk polymerization (13,19,20,21). It may be that rate data alone, although displaying very characteristic behavior, are not sensitive enough to critically discern between two somewhat similar models (i.e. segmental and translational). On the other hand it may be that the polymer radical proximity within the latex particles reduces the need for significant translational motion of the radicals for termination to occur. Molecular weight data and predictions might clarify the situation somewhat, but typically there is little change in emulsion polymer molecular weight during the reaction (9) due to the predominance of chain transfer reactions. With this situation, our near term future efforts will be devoted to working with emulsion polymerization systems which differ significantly from styrene and methyl methacrylate in order to determine the generality of the reaction model presented here.

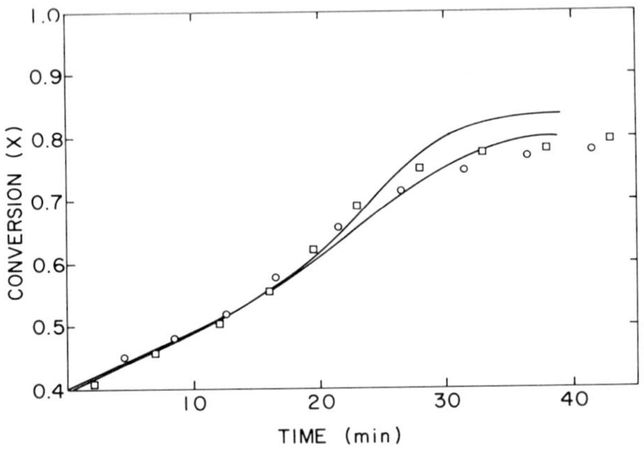

Figure 7. Conversion profiles for MMA seed latex polymerization. Solid curves are theoretical predictions and data points are experimental results ((○) $\alpha_0 = 0.35$; (□) $\alpha_0 = 39.49$)

ACKNOWLEDGEMENT

Several of the authors would like to acknowledge the Central University Research Fund of the University of New Hampshire for partial funding of this work.

LITERATURE CITED

1. B.M.E. van der Hoff, J. Polymer Sci., (1958), 33, 487.
2. D.R. James and D.C. Sundberg, J. Polymer Sci., Polymer Chem. Ed., (1980), 18, 903.
3. N. Friis and A.E. Hamielec, J. Polymer Sci., Polymer Chem. Ed., (1973), 11, 3321.
4. N. Friis and A.E. Hamielec, ACS Symp. Ser., (1976), 24, 82.
5. A.W. Hui and A.E. Hamielec, J. Appld. Polymer Sci., (1972), 16, 749.
6. S.T. Balke and A.E. Hamielec, J. Appld. Polymer Sci., (1973), 17, 905.
7. W.H. Stockmayer, J. Polymer Sci., (1957), 24, 314.
8. P.E.M. Allen and C.R. Patrick, "Kinetics and Mechanics of Polymerization Reactions", p. 94, Halsted Press, NY (1974).
9. H.L. James and I. Piirma, Polymer Preprints, (1975), 16, 198.
10. F. Bueche, "Physical Properties of Polymers", p. 61-124, Interscience, NY (1962).
11. K. Horie, I. Mita and H. Kambe, J. Polymer Sci. A-1, (1968), 6, 2663.
12. J.L. Gardon, J. Polymer Sci., A-1, (1968), 6, 2851.
13. S.K. Soh and D.C. Sundberg, in preparation.
14. J.Y. Hsieh, M.S. Thesis, University of New Hampshire, 1980.
15. M.L. Williams, R.F. Landel and J.D. Ferry, J. Am. Chem. Soc., (1955), 77, 3701.
16. H. Fujita and A. Kishimoto, J. Chem. Phys., (1961), 34, 393.
17. M.S. Matheson et al, J. Am. Chem. Soc., (1951), 73, 1700.
18. S.K. Soh, in press, J. Appld. Polymer Sci., (1981), 25.
19. S.W. Benson and A.M. North, J. Am. Chem. Soc., (1962), 84, 935.
20. J.N. Cardenas and K.F. O'Driscoll, J. Polymer Sci., Polymer Chem. Ed., (1976), 14 883.
21. F.L. Marten and A.E. Hamielec, ACS Symp. Ser., (1979), 104, 43.

RECEIVED April 6, 1981.

21

Polymerization Kinetics by Precision Densimetry

K. J. ABBEY

Glidden Coatings and Resins, Division of SCM Corporation,
16651 Sprague Road, Strongsville, OH 44136

Frequently the extent of conversion in commercial polymerization reactors can be monitored by measuring the total heat evolved. This is possible because of the large volume-to-surface area ratio of these reactors. The small scale laboratory preparations become much more difficult to follow by calorimetry. Meeks(1) has reported a suitable laboratory scale isothermal reactor utilizing an analog computer. Many other reaction calorimeter designs are known. Other methods, such as dilatometry and gravimetric analysis, have been used in the laboratory. A method for monitoring polymerizations was desired which would avoid the restrictions on reactor design, provide rapid response, and allow for automatic data collection and possibly control. The apparatus of Meeks appeared appropriate except that the reaction vessel was an integral part of the instrument.

The use of a precision digital density meter as supplied by Mettler Instruments (Anton Paar, Ag.) appeared attractive. Few references on using density measurements to follow polymerization or other reactions appear in the literature. Poehlein and Dougherty (2) mentioned, without elaboration, the occasional use of γ-ray density meters to measure conversion for control purposes in continuous emulsion polymerization. Braun and Disselhoff (3) utilized an instrument by Anton Paar, Ag. but only in a very limited fashion. More recently Rentsch and Schultz(4) also utilized an instrument by Anton Paar, Ag. for the continuous density measurement of the cationic polymerization of 1,3,6,9-tetraoxacycloundecane. Ray(5) has used a newer model Paar digital density meter to monitor emulsion polymerization in a continuous stirred tank reactor train. Trathnigg(6,7) quite recently considered the solution polymerization of styrene in tetrahydrofuran and discusses the effect of mixing on the reliability of the conversion data calculated. Two other references by Russian authors(8,9) are known citing kinetic measurements by the density method but their procedures do not fulfill the above stated requirements.

0097-6156/81/0165-0345$05.00/0

This paper includes critical comments on the instrumentation and methodology. Examples from solution and emulsion polymerization are given for illustration.

EXPERIMENTAL

For the work presented in this paper a DMA 60/DMA 601H high precision, high temperature, digital density meter (Mettler Instruments Corp.) was used. The continuously monitored emulsion polymerization reactions utilized either a Masterflex peristaltic pump (Cole Palmer Instrument Co.) equipped with various tubing discussed below of 1.66 mm I.D. or a PULSAfeeder Microflo metering pump (Interpace Inc.) with a remote, heat-traced, diaphram pump head. The remote, high temperature density cell was equipped with a continuous flow adapter. Teflon tubing of 1.5 mm I.D. was used for the interconnecting plumbing.

Failure of the pumps was a persistent problem. Viton tubing would typically fail in the peristaltic pump due to monomer swelling, acrylates in 10-30 minutes and after about 3 hours for styrene. Kalrez tubing (DuPont Co.) was chemically resistant but failed from mechanical fatigue after 1-2 hours. Silicon tubing is reported(5) to survive 8-10 hours with a solution of methyl methacrylate in ethyl acetate. This tubing, however, was observed to undergo swelling by methyl methacrylate in this study. The diaphram pump failed after a couple hours because a polymer plaque formed on the check valves. This occurred for both stainless steel and Teflon valves. Plaque formation was more pronounced with latices formulated for low T_g.

All syntheses discussed below used commonly available commercial materials without further purification. Manually sampled reactions were analyzed for their density at temperatures between 20-30°C. Continuously sampled reactions were analyzed at the polymerization temperature. Temperature control in the density cell was better than ±0.1°C for static samples. Temperature fluctuated more for continuously sampled reactions, sometimes rising .1-.3°C because of the exothermic reaction continuing in the density cell. Calibration of the density cell below 95°C was accomplished using values for densities for moist air(10) and degassed, distilled water(11). For measurements above 95°C a certified viscosity standard oil (Number S-200 oil from Cannon Instrument Co.) was used. Densities at six temperatures accurate to four significant figures over the range of 20° to 100°C for this viscosity standard was extrapolated to higher temperatures using an excellent linear fit (correlation coefficient of 0.999998).

The mass fraction polymer, F, at a given time, t, is given by equation 1. $W_p(t)$ and $W_m(t)$ are the instantaneous masses of polymer and monomer, respectively.

$$F(t) = \frac{W_p(t)}{W_p(t) + W_m(t)} \tag{1}$$

This equation can be put in a more usable form for semi-continuous solution or emulsion polymerization by algebraic manipulation, equation 2.

$$\frac{\dfrac{W_p(t) + W_m(t)}{\rho^m} + \dfrac{W_a(t)}{\rho^a} - \dfrac{W_m(t)}{\rho^m} - \dfrac{W_p(t)}{\rho^p} - \dfrac{W_a(t)}{\rho^a}}{\dfrac{W_p(t) + W_m(t)}{\rho^m} + \dfrac{W_a(t)}{\rho^a} - \dfrac{W_p(t) + W_m(t)}{\rho^p} - \dfrac{W_a(t)}{\rho^a}} \tag{2}$$

The new variables in equation 2, ρm, ρp, ρa and $W_a(t)$, are the densities of the monomer, polymer, and aqueous or solvent components and the mass of solvent or aqueous phase. Collecting terms with the assumption that no change in volume occurs on mixing yields equation 3.

$$\frac{V_m(t) - W_r(t)/\rho_r(t)}{V_m(t) - V_p(t)} \tag{3}$$

$V_m(t)$ and $V_p(t)$ are the theoretical volumes of the entire reaction mixture at any time assuming no polymerization and complete polymerization, respectively. $W_r(t)$ and $\rho_r(t)$ are the entire reaction mass at a given time and the measured density of this mixture. $V_m(t)$ is calculated from known masses and densities measured initially at the operating temperature of the densimeter. The quantity $V_p(t)$ requires the further knowledge of the density of the polymer being produced at any given instant.

RESULTS AND DISCUSSION

Several companies supply density equipment which was considered suitable for automatic, continuous operation with sufficient precision for calculation of polymerization conversion. These break down into three classes based on mode of operation: γ-ray absorption, oscillatory frequency of a sample filled tube, and mass measurement at fixed volume. Only one of these, an oscillator-based system distributed by Mettler Instrument Corp. (representing Anton Paar Ag.) has models with dead volumes small enough for laboratory scale experimentation. The other units generally also suffered from narrow density spans when the precision was sufficient for conversion studies. Table

I represents some calculations based on styrene models of the Mettler-Paar instrumentation.

TABLE I

SUITABILITY OF DENSIMETRY FOR MONITORING POLYMERIZATIONS

Monomer	$\Delta\rho$ 20°C (gcm^{-3})	Resolution (% conversion) DMA 40 ($\pm 1x10^{-4}$ gcm^{-3})	DMA 60/601 ($\pm 1x10^{-6}$ gcm^{-3})
Styrene (bulk)	0.159	6×10^{-2}	6×10^{-4}
Styrene (20 wt. % solution)*	0.033	3×10^{-1}	3×10^{-3}
Styrene (4 wt. % solution)*	0.006	1.5	1.5×10^{-2}

† Calculated from data in (12)
* Assuming ideal mixing with solvent of density 1.000 gcm^{-3}

Under the flow conditions of continuously sampling the maximum precision cannot be maintained for the DMA 60/DMA 601 instrument. This is shown by Table II where a small subset of data is presented from an emulsion polymerization of styrene for which the latex was pumped through the densimeter operating at 70.0°C. The latex passed through an internal heat exchanger immediately before entering the actual measuring compartment. If the rate of polymerization over this small range of overall conversion can be assumed to be constant, then the average $\Delta\rho$ is $1.16 \times 10^{-4} \pm 0.10 \times 10^{-4} g/cm^3$. The standard deviation leads to an error in the percentage conversion of ±0.02% for the 59% conversion datum. By comparing the estimate for the error in the percentage conversion within the calculated value in Table I shows that the actual pumped reaction mixture leads to an order to magnitude larger error.

Four polymerization examples are presented here to illustrate both available sensitivity, experimental difficulties, and hopefully some interesting aspects of the polymerization processes. The first two examples are the semi-continuous emulsion polymerization of methyl methacrylate (MMA) and styrene, respectively. The third example is a batch charged copolymerization of butyl acrylate (BA) with MMA. The fourth example is a semi-continuous solution polymerization of an acrylic system. In this last example aliquots were taken manually and analyzed at 29.7°C under static conditions. No further polymerization occurred after the samples were cooled to this temperature.

TABLE II

EMULSION POLYMERIZATION OF STYRENE

Emulsion ρ 70°C(gcm^{-3})*	Fraction Conversion	Δρ ($x10^6$) (gcm^{-3})
.973906	.5830	
.974030	.5860	124
.974133	.5885	103
.974251	.5914	118
.974377	.5944	126
.974487	.5971	110

*22% monomer plus polymer; density measurements made every 0.95 minutes.

Figure 1 shows the conversion history for the MAA latex, the recipe for which is given in Table III. The polymerization was conducted essentially isothermally at 82-84°C. The density cell was set at 83.0°C for a static sample. Curve A is the calculated response when the delay between reactor and density cell is ignored. Curve B was calculated considering a fixed delay of 2.7 minutes. The delay was in fact not constant and tended to increase as the reaction progressed. This is believed to be related to the problems mentioned above with the sampling pumps, but could also be related to a viscosity change in the reaction medium. Further, the responses measured in the density cell are not exactly those occurring in the reactor since reaction continues in the circulation lines and density cell. Thus, curve B is an upper limit to the true reactor behavior. These observations suggest the use of the lowest dead volume and lag time possible in the sampling loop.

TABLE III

MMA LATEX RECIPE FOR FIGURES 1 AND 2

Initial Charge	
Deionized water	400.0g
Surfactant	4.0g
Buffer	0.8g
$K_2S_2O_8$	0.3g
MMA + modifier	20.0g
Feed	
MMA + modifier	380.8g

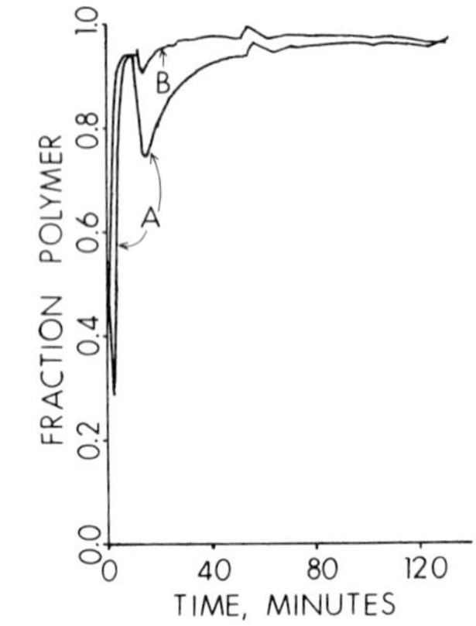

Figure 1. Conversion representation of MMA semicontinuous emulsion polymerization. Curve A results from neglecting lag in sampling while curve B is corrected for 2.7 min of lag.

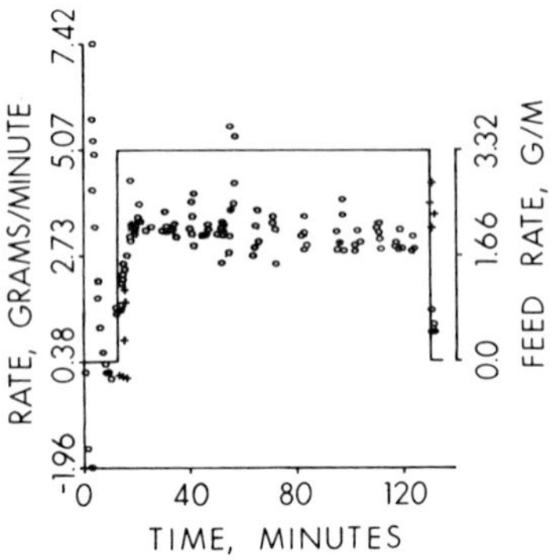

Figure 2. Rate representation for MMA semicontinuous emulsion polymerization ((——) the monomer feed rate (right ordinate); (○) the polymerization rate without correction for the lag in sampling (left ordinate); (+) the most significantly changed points when the 2.7-min lag is used)

Figure 2 is the rate of polymerization corresponding to curve A in Figure 1. The only significant alteration caused by including the 2.7 minute delay in response is a flattening of the minimum that occurs at about 10 minutes (represented by +'s in the figure). The rate representation has several advantages, one of which is greater sensitivity to process perturbations especially at high concentration and high conversion. The rapid rate is possible because of the fifth significant digit available when the DMA 60/DMA 601 instrument combination is used.

The styrene polymerization is shown in Figures 3 and 4. As with the MMA latex, the styrene latex began with a small portion of the monomer added initially. The remaining monomer was added as shown in Figure 4. Apparent "negative" conversion was caused by the collection of a separated monomer phase on the glass walls of the density cell. This behavior was only noted with styrene and was a function of the proportion of monomer to surfactant in the initial charge. The separated monomer was slowly absorbed into the circulating reaction medium. After about 70 minutes, or above 50% conversion, the responses are reliable. The sharp rate acceleration due to the Trommsdorff gel effect can be seen in both figures at about 100 minutes. The lag between density cell response and reactor events were considerably less for this example and the figures ignore any correction. After establishing a "steady state" response to the monomer feed (about 160 minutes into the reaction), the incremental increase of the feed rate is seen not to alter the overall fractional conversion since the rate of polymerization increases to parallel the monomer feed rate. At the end of this set of data the rate is 2-3 times that observed earlier before the feed.

The batch charged emulsion copolymerization of BA and MMA is shown in Figures 5 and 6. The reaction proceeded more slowly than the above MMA latex and the lag time was also less (ignored in these plots). The rate acceleration at about 40% conversion is believed to be the Trommsdorff effect. The much lower slope of the conversion curve above 90% conversion in Figure 5 is believed to be an artifact of the simple model used in the calculations. The model assumed that the composition of the polymer produced was the same as the initial monomer mix. This, of course, is not the case. Given an initial composition of 25 mole % BA, the polymer produced can be calculated using literature reactivity ratios (12) to vary in composition from about 15 mole % BA early in the reaction to about 50 mole % during the last 10% of the reaction. The entire plot is thus skewed, with the composition changing most rapidly late in the polymerization.

The use of densimetry provides a rapid and, with proper care, precise measure of the polymerization process. By using a second independent method of measuring the same process, the

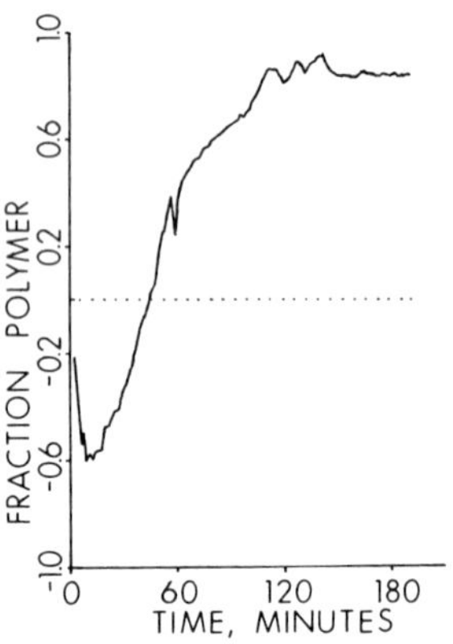

Figure 3. Conversion representation for styrene semicontinuous emulsion polymerization ((· · ·) zero conversion axis)

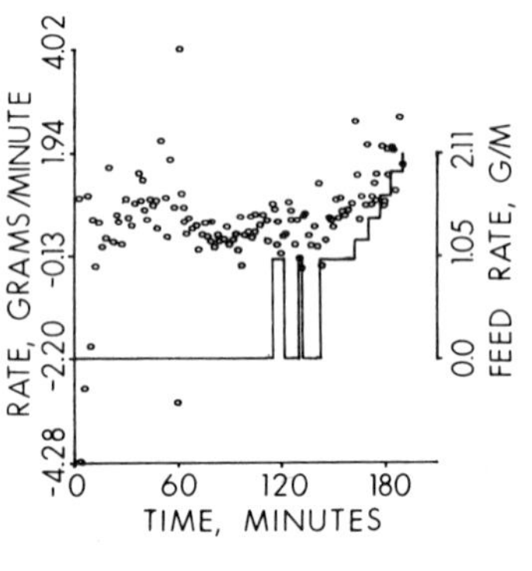

Figure 4. Rate representation for styrene semicontinuous emulsion polymerization ((——) the monomer feed rate (right ordinate); the scatter plot is the polymerization rate (left ordinate))

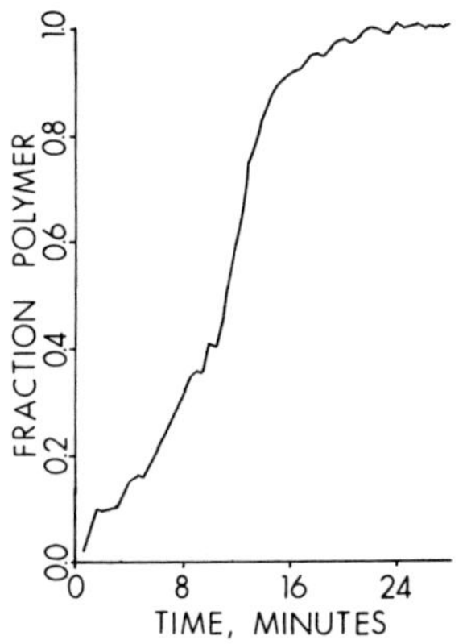

Figure 5. Conversion representation for a BA/MMA (30/70 by weight) batch emulsion polymerization

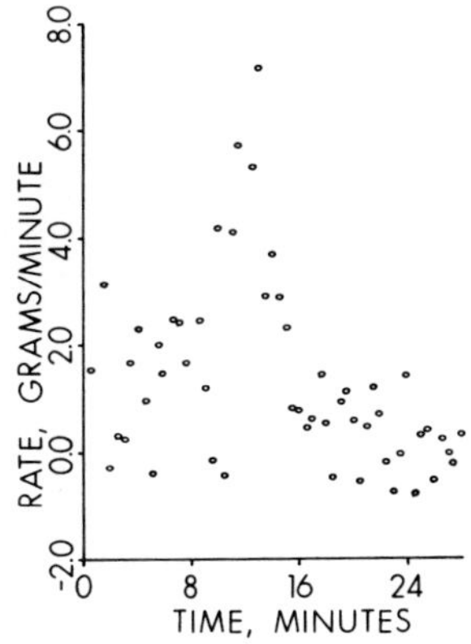

Figure 6. Rate representation for a BA/MMA (30/70 by weight) batch emulsion polymerization

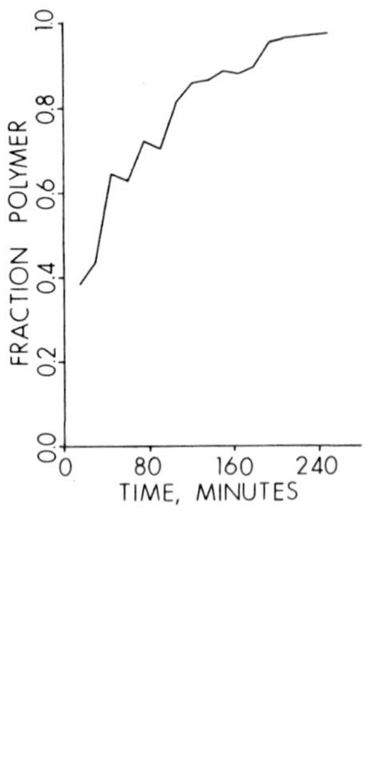

Figure 7. Conversion representation for a multicomponent acrylic, semicontinuous solution polymerization

copolymer reactivity ratios could be determined as a function of conversion. The study of the copolymerization of styrene and methyl methacrylate at high conversion (13,14) has shown that the reactivity ratios do deviate from the values reported from dilute solution measurements.

The last example, for which Figure 7 shows the running fractional conversion as a function of time, was conducted by feeding a catalyzed acrylic monomer mixture into a hot solvent mixture. The control on monomer feed rate was not as good as in the previous examples. This accounts for much of the scatter in the data as the calculations assume a constant feed rate. Unlike the semi-continuous emulsion polymerizations, the fractional conversion did not remain constant during the monomer feed, but increased throughout. Indeed, the actual weight of free monomer remained remarkably constant irrespective of the running total reaction mass or conversion, Table IV.

TABLE IV

WEIGH UNREACTED MONOMER IN SOLUTION ACRYLIC DURING FEED

Time (min.)	Monomer (g)	Time (min.)	Monomer (g)
15	5.2	105	11.2
30	9.6	120	9.7
45	9.1	135	10.3
60	12.7	150	9.6
75	11.8	165	11.1
90	15.1	180	12.7

CONCLUSIONS

The use of precision density measurements for monitoring polymerization reactions can be done rapidly and automatically using commercially available instrumentation. The method is independent of the reactor size and design but suffers from sampling difficulties. The examples of this paper show the rapidity of data collection and three distinct sampling problems; pump failure from either monomer attack or polymer scale formation, monomer phase separation in the density cell, and the lag time for rapid polymerizations. Techniques have or can be devised to avoid or reduce the influence of these problems.

LITERATURE CITED

1. Meeks, M. R., Polym. Eng. Sci. 1969, 9(2), 141.
2. Poehlein, G. W. and Dougherty, D. J., Rubber Chem. Techn. 1977, 50(3), 601.
3. Braun, D. and Disselhoff, G., Polymer 1977, 18(9), 963.

4. Rentsch, C. and Schultz, R. C., Makrom. Chem. 1978, 179, 1131.
5. Ray, W. H., private communication 1980.
6. Trathnigg, B., Makrom. Chem. 1978, 181, 1979-1986.
7. Trathnigg, B., Angew. Makrom. Chem. 1980, 88, 127-133.
8. Pugachevich, P. P. and Toknev, A. G., Zavod. Lab. 1970, 36(7), 817.
9. Zotov, L. I. and Sedov, L. N., Plast. Massy 1969(9), 68.
10. Weast, R. C. (ed.), "CRC Handbook of Chemistry and Physics, 57th Edition", p. F-9, CRC Press, Cleveland, 1976.
11. Weast, R. C. (ed.), "CRC Handbook of Chemistry and Physics, 53rd Edition", p. F-5, CRC Press, Cleveland, 1972.
12. Brandrup, J. and Immergut, E. H., "Polymer Handbook, 2nd Edition", John Wiley and Sons, New York, 1975.
13. Johnson, M., Karmo, T. S., and Smith, S. S., European Polym. J. 1978, 14, 409-14.
14. Dionisio, J. M. and O'Driscoll, K. F., J. Polymer Sci., Polymer Letters Ed. 1979, 17(11), 701-7.

RECEIVED April 6, 1981.

22

The Effect of Nonreactive Additives on the Kinetics of Emulsion Polymerization

A. R. M. AZAD[1] and R. M. FITCH

Institute of Materials Science and Department of Chemistry,
The University of Connecticut, U-136, Storrs, CT 06268

M. NOMURA

Department of Industrial Chemistry, Fukui University, Fukui, Japan

Various substances are often added to the monomers in an emulsion polymerization in order to achieve desirable end use properties or to investigate specific effects such as dilution or viscosity of the particles. Chain transfer agents (CTA) have long been used to control the molecular weight of synthetic rubbers and other emulsion polymers. Furthermore, plasticizers or other liquids (1 - 4) are often added in order to modify the polymer properties or polymerization conditions. In accordance with previous usage, we have considered the CTA as 'reactive' and the other additives as 'non-reactive' (3, 4). The effect of these additives on the polymerization has generally been to a) decrease the rate beyond that due to dilution alone; and b) to lead to the formation of low molecular weight polymer. However, the explanations given for the effect of the 'non-reactive' additives on the kinetics of the emulsion polymerization (3, 4) is open to some doubt.

When various non-reacting diluents were combined with the monomer in styrene emulsion polymerization, Owen and coworkers found both increases and decreases in the rates of polymerization (3). They postulated that the additives such as octane, nujol and dioctyl phthalate would enter the monomer-swollen polymer particles (even though highly water-insoluble) and exert two effects, depending upon solubility parameters and viscosity: a) phase separation (at least in part) leading to changed reactivity of propagating radicals and, b) changes in reaction rates due to differences in viscosity. Because they did not control nor measure particle sizes, differences in particle number concentration, which undoubtedly were obtained, were not accounted for. Blackley and Haynes followed up on these experiments using lower molecular weight additives which are all capable of transport through the water to the

[1] Current Address: Research and Development Dept.
Ortho Diagnostic Systems Inc.,
410 University Avenue, Westwood, Mass. 02090

0097-6156/81/0165-0357$05.00/0

particles (4). They found reductions in the rate of polymerization in every case which they showed was due to a decrease in the rate of polymerization per particle. They attributed the effects to a reduction of the Trommsdorff "gel" effect because of the lowering of the viscosity in the polymerizing particles by the non-reacting additives. However, the gel effect should not be in evidence until the latter stages of the polymerization, whereas the reductions in rate were observed from the earliest times of reaction. Furthermore, their particle sizes were on the order of 60 nm diameter, too small for Smith-Ewart Case 3 kinetics where the gel effect would be observed.

Azad and Fitch (5) investigated the effect of low molecular weight hydrocarbon additives on the formation of colloidal particles in suspension polymerization of methyl methacrylate and vinyl acetate. It was found that the additives n-octane, n-dodecane, n-octadecane, n-tetracosane and mineral oil exerted a thermodynamic affect depending upon water-solubility and molecular weight. Since these effects on emulsion polymerization have not been considered by the earlier investigators, we have chosen n-pentane and ethyl benzene as additives with limited water-solubility and n-octadecane, and n-tetracosane as water-insoluble ones. Seeded emulsion polymerization was chosen so that the number of particles could be kept constant throughout the experiments and only the effect of the other parameters on the rate could be determined.

We now report on some experiments using seeded emulsion polymerization of styrene in which conditions were carefully chosen to ensure that Smith-Ewart Case 2 kinetics (6) would obtain throughout, in the absence of chain transfer/radical desorption effects. Various hydrocarbons were investigated for their effects on kinetics of polymerization and equilibrium swelling of the latex particles.

Experimental

Materials. All chemicals were reagent grade purity and used as received. Styrene (Aldrich) was distilled twice under vacuum to remove inhibitor.

Apparatus and Procedure. The apparatus and procedures are the same as reported earlier (5) except for the following changes: Seed latex 200g, distilled water 220g, potassium persulfate (PPS) 0.675g, styrene monomer 60g, additive dissolved in monomer-variable. Polymerization temperature was 50°C. The PPS, dissolved in water, was added 5 min before the addition of monomer. All oxygen was removed from the system by purging with high purity nitrogen. Conversions were obtained by shortstopping samples with a solution of p-benzoquinone in methanol followed by drying and gravimetry.

The seed latex was prepared by emulsion polymerization at 50°C using distilled water, 4 dm^3; SDS, 50g; PPS, 5g; and distilled styrene, 440g. Two batches were prepared, one in U.S.A. and one in Japan with same recipe.

TABLE I

Effect of Additives on the Rate of Seeded Emulsion Polymerization of Styrene

Run No.	Additive	Mol additive /60g styrene $\times 10^3$	Initial volume fraction of additive in styrene, $\times 10^2$	R_p(d), mol min^{-1} dm^{-3} H_2O $\times 10^3$	R_p, mol min^{-1} dm^{-3} H_2O $\times 10^3$	Relative Rate, %	$[M]_p$, mol dm^{-3} from R_p	$[M]_p$, from swelling measurements*
AFN-7	None	-0-	-0-	20.2	20.2	100	5.6	5.6
AFN-10	$C_{24}H_{50}$	5.89	3.32	19.5	18.0	89	5.0	
AFN-8	$C_{18}H_{38}$	5.89	2.72	19.6	16.1	80	4.5	4.8
AFN-9	C_5H_{12}	5.89	0.97	20.0	19.2	95	5.3	
AFN-12	$C_{18}H_{38}$	23.6	10.1	18.2	13.8	68	3.8	4.1
AFN-13	C_5H_{12}	83.1	12.2	17.7	16.0	79	4.4	

R_p(d) theoretical rate due to dilution alone of the monomer

*Estimated at 15% conversion

Equilibrium swelling measurements with toluene were carried out by agitation with a mechanical shaker, a mixture of seed latex (10ml) and toluene-alkane solution (3 to 6ml) for 16 to 20h at 20-22°C, centrifugation to separate the oil phase, and measurement of the change in solvent and latex heights by means of a precision micrometer. Equilibrium swelling measurements with styrene/ethyl benzene were carried out by thoroughly mixing the initiator-free seed latex with styrene-ethyl benzene mixture, centrifugration and analysis by gas chromatography.

Results and Discussion

Figure 1 gives the conversion-time curves for the seeded emulsion polymerization of styrene in the absence and presence of various low molecular weight additives. Table I summarizes the results given in Figure 1. The rates of polymerization were determined from the straight line portion of the conversion-time curves (below 40% conversion) by least squares analysis of the experimental points. Table I also gives the calculated rates assuming a mere dilution of the monomer in the seed by the additive. It is clear that in every case the rate of polymerization is retarded much more than that due to dilution alone.

Electron microscopy of the final latex of the experiments given in Table I showed almost no new nucleation. The particle size distributions were narrow and indicated no noticeable coagulation as well. New nucleation would lead to increased rates whereas coagulation would have the opposite effect. Any decrease in the rate therefore must be due to a decrease in $[M]_p$, if we assume $\bar{n}$ to be constant. We therefore determined the toluene/polymer ratio in the seed latex in the absence and presence of the various additives. Toluene was chosen as the solvent, because it is similar to styrene and allows the measurement of equilibrium solubilities without the risk of polymerization. Table II gives the experimental values of the toluene solubility in the seed as a function of time. The results indicate that the swelling is nearly complete within 5 to 10 min.

TABLE II

Rate of Swelling of Polystyrene Seed Particles by Toluene

Diameter of polystyrene seed = 64 nm, 10% seed latex = 10 ml, 5% octadecane in toluene (v/v) = 4 ml.

Time	Toluene Absorbed, ml	Percent Saturation
5 min	1.55	95
10 min	1.58	96
19 h	1.64	100

Figure 2 gives the equilibrium swelling values of the seed by various octadecane-toluene mixtures. It is apparent that the concentration of toluene in the particles at all octadecane levels in the monomer phase is lower than that calculated on the basis of a similar dilution effect. This arises from the fact that the water-insoluble additive depresses the activity of the monomer (or toluene) and concentrates only in the monomer droplets so that at swelling equilibrium there will be less in the latex particles than expected from simple dilution. The activity of "solvent" (monomer or toluene) in the swollen particles is further affected by the Kelvin effect. These thermodynamic considerations have been quantitatively treated by Azad (7), in detail by Ugelstad and coworkers (8-10, 12-17) and later by Azad and Fitch (5). Ugelstad has also extended the two component swelling equation given by Morton et al. (19) to a three component system and has discussed the effect of chain length, interaction parameters and water-solubility on the degree of swelling of latexes and emulsions. Expressions for the three component competitive swelling has been given by Azad and Fitch (5) and Ugelstad and coworkers (13-16). Expressions have also been given by the above authors for the cases where one has in a system polymer in one set of particles and low-molecular weight compound in another.

The swelling of the seed latex in the presence of water-insoluble, low molecular weight compounds in the monomer phase may be quantitatively expressed by (5, 7-18),

$$\overline{\Delta G}_{1(1)}+2\gamma_1\overline{V}_{1(1)}/r_1=\overline{\Delta G}_{1(2)}+2\gamma_2\overline{V}_{1(2)}/r_2 \quad (1)$$

where

$$\overline{\Delta G}_{1(1)}=RT[\ln\phi_{1(1)}+(1-1/j_{(1)})\phi_{2(1)}+\chi_1\phi^2_{2(1)}] \quad (2)$$

and

$$\overline{\Delta G}_{1(2)}=RT[\ln\phi_{1(2)}+(1-1/j_{(2)})\phi_{2(2)}+\chi_2\phi^2_{2(2)}] \quad (3)$$

if $r_2 \gg r_1$

$$\overline{\Delta G}_{1(1)}=\overline{\Delta G}_{1(2)}-2\gamma\overline{V}_{1(1)}/r_1 \quad (4)$$

and $$[M]_p=\frac{d_1}{M_1}\phi_{1(1)} \quad (5)$$

where γ, r and χ refer to the interfacial tension, radius and Flory-Huggins interaction parameter, and subscripts 1 and 2 refer to the seed latex and "solvent" (monomer) droplet respectively. For the other

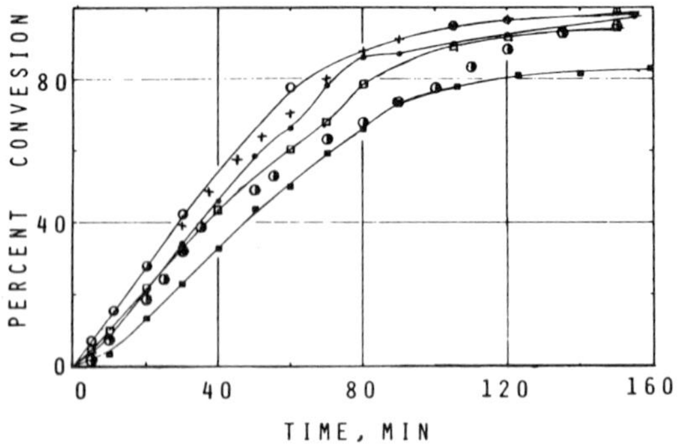

Figure 1. Effect of addition of low molecular weight hydrocarbons as additives upon the progress of the seeded emulsion polymerization of styrene at 50°C.

N = *3.5 × 10^{17} particle dm^{-3} water; diameter of seed = 64 nm; styrene monomer = 150 g dm^{-3} water; PPS = 1.69 g dm^{-3} water; amount of additive variable* (see *Table I); 60 g styrene/200 mL seed latex with: (○) no additive; (□) 5.89 × 10^{-3} mol* n-*octadecane; (■) 23.6 × 10^{-3} mol* n-*octadecane; (●) 5.89 × 10^{-3} mol* n-*tetracosane; (+) 5.89 × 10^{-3} mol* n-*pentane; (○) 83.2 × 10^{-3} mol* n-*pentane.*

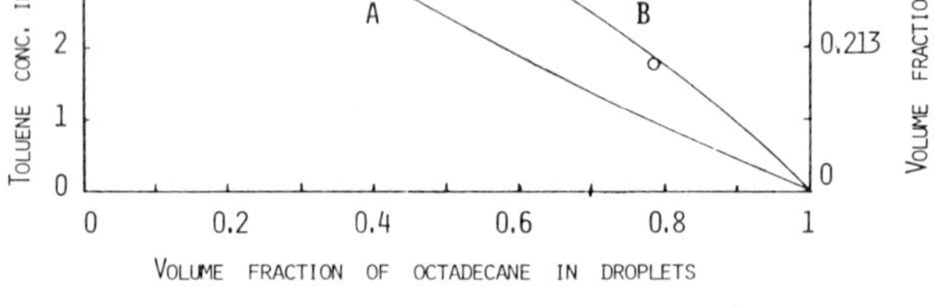

Figure 2. Toluene concentration and volume fraction in seed as a function of the volume fraction of octadecane in the toluene phase. Curves A (χ_2 = 0) and B (χ_2 = 0.8) are theoretical curves with $2\gamma_1 V_{1(1)}/r_1RT$ = 0.0236 ((○) experimental points).

parameters subscript 1 represents "solvent," and 2, polymer or additive; (1) represents the polymer particle and (2), the "solvent" droplet; $\overline{\Delta G}$ is the partial molar free energy of mixing; ϕ is the volume fraction; $\bar{V}$ is the partial molar volume; j is the ratio of molar volumes of polymer or additive to "solvent"; $[M]_p$ is monomer (solvent) concentration in polymer particles; d is the density and M_1, is monomer (solvent) molecular weight.

Figure 2 gives the calculated values of the concentration as a function of the volume fraction using Eq. 4 for various values of χ_2. The value of $2\gamma_1\bar{V}_{1(1)}/r_1RT$ was taken to be 0.0236, obtained from the experimental value of $\phi_{2(1)} = 0.33$ (swelling with pure toluene) and $\chi_1 = 0.43$ (19) and calculated using Eq. (4), with $\phi_{1(2)}=1$. We have assumed that $\gamma_1\bar{V}_{1(1)}/r_1RT$ is approximately constant, throughout the entire swelling process.

Figure 2 shows that with octadecane the theoretical curves fit the experimental values well if $\chi_2 = 0.8$. If $\chi_2 = 0$, the swelling of the seed would be lower than that for all cases where $\chi_2 = 0.8$. The interaction parameter, χ_2, was expressed by Huggins (20) using the van Laar - Scatchard - Hildebrand approximations as

$$\chi_2 = \chi_{\Delta s} + V_1 (\delta_1-\delta_2)^2/RT$$

where $\chi_{\Delta s} = (1-1/j)/z' \simeq 1/z'$ if $j>>1$

Here $\chi_{\Delta s}$ is the entropy contribution to χ_2, V_1 the molar volume of the solvent (monomer), δ_1 and δ_2 are the solubility parameters of the solvent and additive, and z' is approximately the coordination number, i.e. the number of nearest neighbors of the additive molecule or segment. Thus, χ_2 is determined by the solubility parameter of the additive and solvent (20). This may explain the observed dependence of the rate on the solubility parameter of the additive found by Owen and coworkers (3).

The value of $\bar{n}$ calculated from the experimental values of R_p, $[M]_p$ = 5.6 mol dm^{-3} and N = 3.5 x 10^{17}/dm^3 water with k_p = 206 dm^{3}mol^{-1}s^{-1} (21) is 0.5. The results indicate Smith-Ewart Case 2 kinetics in the absence of additives. Table I, columns 2 to 6 give the value of $[M]_p$ using $\bar{n}$ = 1/2. The calculations indicate that the decrease in R_p in these systems may be due primarily to decrease in $[M]_p$. This is especially so in the case of $C_{24}H_{50}$ and $C_{18}H_{38}$ where the additives would not be present in the seed particles. In the case of pentane (and ethyl benzene (see below)) it is expected that the hydrocarbon may rather rapidly diffuse through the aqueous phase and thus also be present in the swollen latex particles. Equation 2 should then be replaced by Flory's expression for a three component system (22,9)

$$\overline{\Delta G}_{1(1)} = RT\,[\ln\phi_{1(1)}+(1-1/j_2)\phi_{2(1)}+(1-1/j_3)\phi_{3(1)} + \chi_{12}\phi^2_{2(1)}+\chi_{13}\phi^2_{3(1)}+(\chi_{12}+\chi_{13}-\frac{\chi_{23}}{j_2})\phi_{2(1)}\phi_{3(1)}] \quad (6)$$

where the χ's are the interaction parameters and the subscripts 1, 2 and 3 now refer to the "solvent" (monomer) additive (diluent) and polymer respectively in the polymer particle. For ethyl benzene which is the saturated analog of styrene, the interaction parameter with polymer, the ratio of molar volumes and possibly the interfacial tension would be almost identical (i.e. $\chi_{13} = \chi_{23}$, $\chi_{12} = 0$, $j_2 = 1$) to that of styrene (23). Equation (6) would then reduce to

$$\overline{\Delta G}_{1(1)} = RT \left[\ln\phi_{1(1)} + (1-1/j_3)\phi_{3(1)} + \chi_{13}\phi^2_{3(1)} \right] \quad (7)$$

This may explain the fact that the degree of swelling with ethyl benzene, given in Figure 4, is identical to that for styrene alone. In the ease of pentane, since the interaction parameters and j_2 are not identical to that of the "solvent," the swelling would be lower than that due to simple dilution. A large difference between χ_{13} and χ_{23} favors selective absorption of the better solvent in the polymer particle. The effect of pentane on the rate of polymerization is most probably similar to that of ethyl benzene, but with a lower $[M]_p$ in the particles, and with less chain transfer, as discussed below.

Table III

Kinetics of the Seeded Emulsion Polymerization of Styrene in the Presence of Ethylbenzene, at 50°C

Data taken from Figures 3 and 4

St/(St+EtBz), Weight fraction	$[M]_p$, mol dm^{-3} particles	Rp x 10^4 mol s^{-1}dm^{-3}H_2O	$\bar{n}$
1	5.5	1.31	0.50
0.769	4.5	0.82	0.38
0.625	3.5	0.58	0.35
0.5	2.75	0.35	0.27
0.4	2.2	0.27	0.25

When ethyl benzene replaces styrene in seeded emulsion polymerizations, using the recipes and seed latex otherwise identical to those above, the rate of polymerization is also depressed below that expected for simple dilution (Figure 3) even though ethyl benzene is indistinguishable from styrene in its swelling behavior (Figure 4). In these experiments, then, reduction in rate must be attributable to chain transfer to the ethyl benzene which accumulates in the polymerizing particles. The small radicals formed are capable of diffusing out of the particles which has the effect of lowering the average number of radicals per particle, $\bar{n}$, as can be seen from the results given in Table III which were calculated from the data in Figures 3 and 4. This may be explained on the basis of

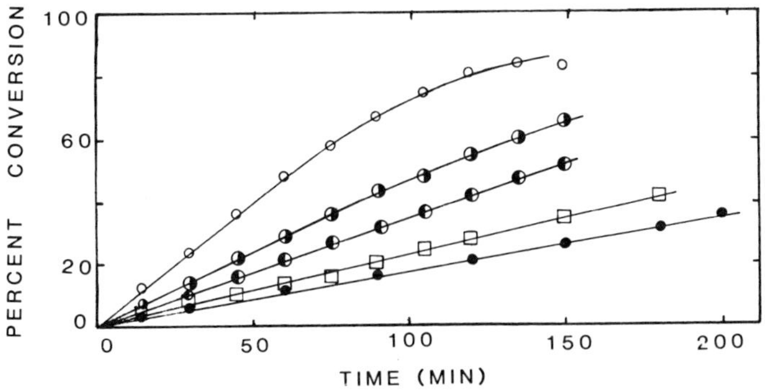

Figure 3. Effect of progressive addition of ethyl benzene upon the progress of the seeded emulsion polymerization of styrene at 50°C. N = 1.4×10^{17} particles dm^{-3} water; monomer = 102 g dm^{-3} water; PPS = 1.25 g dm^{-3} water; and ethyl benzene = 0–153 g dm^{-3} water. The monomer weight fractions St/(St + Et Bz) are: (○) 1.0; (◑), 0.769; (◐) 0.625; (□) 0.5; (●) 0.4.

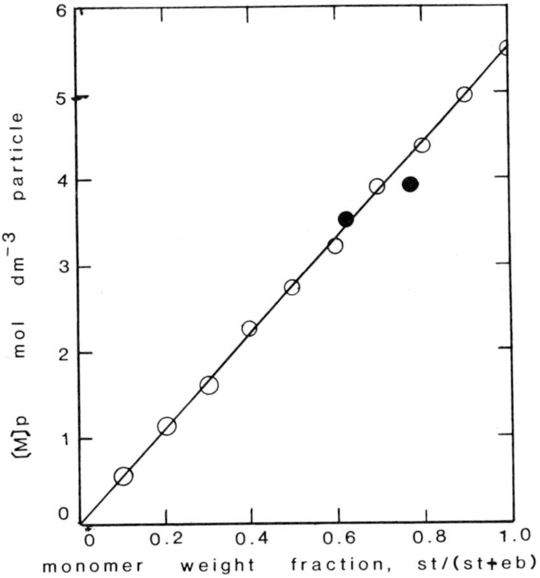

Figure 4. Effect of added ethyl benzene on the monomer concentration in seed particles ((○) data obtained from swelling experiments; (●) data obtained from seeded emulsion polymerization at start)

the general theory for emulsion polymerization (24) that because of the low value of the ratio of entry to termination frequencies per particle, α', where

$$\alpha' = \rho_w v / k_{tp} N$$

and of the increase in the ratio of desorption to termination frequencies, m, where

$$m = k_d v / k_{tp}$$

the system is sensitive to these processes and deviates from Smith-Ewart Case 2 in the direction of Case 1. Here ρ_w denotes the rate of radical production in the water phase, k_d and k_{tp} denote desorption rate constant and termination rate constant respectively and v is the average volume of a polymer particle.

These results explain the findings of Blackley and Haynes who also showed that the molecular weight of the polymer formed in the presence of ethyl benzene was lower than that in its absence. Calculation from their experimental data shows that their $\bar{n}$ varied from 0.005 to 0.039 radicals per particle, well into Case 1. Thus, their explanation on the basis of the Trommsdorff "gel" effect cannot be correct since this requires the mutual termination of two macroradicals in a particle, which obtains only under Case 3 kinetics. Similar experiments on the effect of the diluents on "insitu" (unseeded) and seeded emulsion polymerization indicates that $\bar{n}$ decreases due to desorption of free radicals from the particles (27).

Nomura (25) investigated the effect of carbon tetrabromide, carbon tetrachloride and long chain mercaptans on the kinetics of emulsion polymerization of styrene. In the case of CBr_4 and CCl_4 the effect on the polymerization was attributed to desorption of the small chain transferred radicals. Similar results were obtained by Napper et al (26). Nomura also observed that the long chain mercaptan (n- dodecyl mercaptan) did not affect the number of particles and the rate, presumably due to the water-insolubility of the chain transferred radicals.

It should be noted that in the case of water-insoluble long chain mercaptans, the thermodynamic effect as in the case with n-octadecane, may also depress the rate. Similar effects may be anticipated in emulsion copolymerizations, and polymerizations involving several monomers where one or more are highly water-insoluble. These effects would then be superimposed on the other variables of the polymerization.

Summary

When water-insoluble compounds are mixed with the monomer in styrene seeded emulsion polymerizations, the rate of polymerization is lowered below a simple dilution effect. Since the additive is not transported through the water, it remains in the droplets of monomer, reducing its activity, which in turn leads to a lower activity of monomer in the polymerizing particles. The results with octadecane and tetracosane agree well with theoretical predictions based on equations for equilibrium swelling. When the additive may diffuse rapidly through the aqueous phase, then simple dilution governs the monomer concentration, but chain transfer with the additive may have the effect of further depressing the rate. Experiments with ethyl benzene show that its effect on decreasing $\bar{n}$, the average number of radicals per particle, can be pronounced when the particles are small. In this sense, of course, ethyl benzene cannot be considered as "non-reactive."

Acknowledgement

This work was supported in part by a grant from Hercules Inc. The authors wish to thank Professor John Ugelstad, of the University of Trondheim, Norway for discussions which led to the proper choice of an interaction parameter in Figure 2, and in general to his seminal contributions to the theories involved.

Abstract

Various substances such as plasticizers and other hydrocarbons are sometimes added to the monomer in an emulsion polymerization. The effect on the kinetics of the seeded emulsion polymerization of styrene in the presence of various 'non-reactive' hydrocarbons as low molecular weight additives have been investigated. By the use of seed latex, the number of particles was kept constant throughout the experiments, so that the effect of other parameters on the rate could be determined. All the additives decreased the polymerization rate more than could be accounted for by dilution alone. In the case of the water-insoluble alkanes n-tetracosane and n-octadecane, the decrease in rate could be attributed to a reduction in the monomer concentration in the latex particles, $[M]_p$, resulting from a lowering of the activity of the monomer in the monomer droplets. No alkane is transported to the polymerizing particles. Theoretical equations have been developed for calculating $[M]_p$ at equilibrium swelling under these conditions. The values are in good agreement with those calculated from Smith-Ewart case 2 kinetics. Independent measurements confirm that the average number of radicals per particle, $\bar{n} = 1/2$ in these experiments. When the hydrocarbon additives were of lower molecular weight, e.g. n-pentane and ethylbenzene, they are capable of transport through the aqueous phase and of swelling the latex particles. Ethyl benzene has been shown to be indistinguishable from styrene in its swelling behavior. In these cases the depression in the rate of polymerization comes about from not only dilution of the monomer, but also from chain transfer with the diluent as it accumulates in the particles. Chain transfer leads to formation of small radicals which may diffuse out of the particles, lowering the value of $\bar{n}$. These conclusions differ from those of earlier workers on similar systems: Owen, McLemore, Liu, Seymour and Tinnerman suggested phase separation within the particles may be responsible, but they did not control nor measure their particle number concentrations. Blackley and Haynes invoked the Trommsdorff "gel" effect, but calculation shows that their kinetics follow Smith-Ewart Case 1 where this cannot occur.

Literature Cited

1. D.E. Williams, results obtained at University of Minnesota, 1944, given in "Emulsion Polymerization", F.A. Bovey, I.M. Kolthoff, A.I. Medalia and E.J. Meehan, p. 357, New York, Interscience (1955).
2. A.S. Dunn, private communcations (1980), K.Kirkman, MSc. Tech. Thesis, University of Manchester (1959). Cited in reference 2 .
3. D.R. Owen, D. McLemore, Wan-Li Liu, R.B. Seymour and W.N. Tinnerman, ACS Symp. Ser., "Emulsion Polymerization" No. 24, 299 (1976).
4. D.C. Blackley and A.C. Haynes, British Polym. J., 9,312 (1977).
5. A.R.M. Azad and R.M. Fitch, Sept. 1978. Polymer Preprints, Vol 70, p. 585 (1978). Polymer Colloids "II" Editor R.M. Fitch, Plenum Press, N.Y. (1980), p. 95.
6. W.V. Smith and R.W. Ewart, J. Chem. Phys., 16, 592 (1948).
7. A.R.M. Azad, thesis submitted to the University of Trondheim, The Norwegian Institute of Technology, Laboratory of Industrial Chemistry in partial fulfillment of the requirements for the Doktor Ingeniør Degree (June, 1975), pp. 71-73
8. J. Ugelstad, F.K. Hansen and K.H. Kaggerud, Faserforsch. und Textiltechn. 28 (7), 310 (1977).
9. J. Ugelstad, Makromol. Chem. 179, 815 (1978).
10. J. Ugelstad, T. Ellingsen and K. H. Kaggerud. "Adv. in Organic Coating Sci.", P. Techn. Athens 17-21 July 1978. Preprints.
11. J. Ugelstad, K.H. Kaggerud and R.M. Fitch, September 1978. Polymer Preprints. Vol. 70, p. 585 (1978). "Polymer Colloids", "II." Editor R.M. Fitch, Plenum Press, New York (1980), p. 83.
12. J. Ugelstad, K.H. Kaggerud, F.K. Hansen and A. Berge, Makromol. Chem. 180, 815 (1979).
13. J. Ugelstad, Polymer Colloid Group Newsletter, Vol. 9, No. 1, May 1980.
14. J. Ugelstad and P.C. Mørk, paper presented at Third International Conf. on Surface and Colloid Science, Stockholm 20-25 August 1979, Preprints.
15. J. Ugelstad, P.C. Mørk, K. H. Kaggerud, T. Ellingsen and A. Berge, "Adv. in Coll. and Interface Sci.", 13, 101-140 (1980).
16. J. Ugelstad Polymer Colloid Group Newlsetter Vol. 10, No. 1, June 1980.
17. J. Ugelstad, P.C. Mørk and A.A. Kahn, ACS Org. Coat. and Plast. Chem., Preprints. 43, 514 (1980).
18. A.R.M. Azad, R.M. Fitch and M. Nomura, Ibid, p. 537.
19. M. Morton, S. Kaizermann and M.W. Altier, J. Coll. Sci., 9, 300 (1954).
20. M.L. Huggins, Ann. N.Y. Acad. Sci. 43, 1 (1942).

21. H. Gerrens, Ber. Bunsengs. 67, 741 (1963). Cited in Polymer Handbook, Editor I. Immergut, Wiley Interscience (1975).
22. P.J. Flory, "Principles of Polymer Chemistry", Cornell University Press, Ithaca, N.Y., 1953, p. 549.
23. P.J. Flory, Ibid, p. 553.
24. J. Ugelstad, P.C. Mørk and A. Aasen, J. Polym. Sci., Part A1, 5, 2281 (1967).
25. M. Nomura. Polymer Colloid Group Newsletter, 1976.
26. D.H. Napper, Polymer Colloid Group Newsletter, Vol. 9, No. 1, May 1980.
27. A.R.M. Azad, M. Nomura R.M. Fitch, to be published.

RECEIVED April 6, 1981.

23

Nonuniform Emulsion Polymers

Process Description and Polymer Properties

D. R. BASSETT and K. L. HOY

Technical Center, Union Carbide Corporation, South Charleston, WV 25303

Because of its heterophase nature, emulsion polymerization is generally more complicated than simple solution polymerization in which monomers and polymers are soluble in a suitably chosen solvent. In emulsion polymerization the different relative solubilities of monomers in water and in the polymer particles lead to different reaction locales and to different particle structures. Another complicating factor is the need to achieve and maintain colloidal stability throughout the polymerization and subsequent handling of the dispersions. Emulsion polymers can properly be called products by process since the process details exert such a powerful effect on the properties of the particles and resultant films. Consequently, an emulsion polymer is far more than a product defined by a simple polymer composition.

One way of altering the properties of latex particles is to change the monomer feed composition during the polymerization. Much work has been carried out on multistage processes in which the composition of each stage differs from that of the preceding stage (1). Film properties as well as filming properties of latexes can be altered in this manner. Not only are multistage processes cumbersome to carry out in practice, but often incompatibility of the copolymers produced in the various stages leads to poor end use properties, especially in thin films. We report here a process for continuously changing the composition of the monomer mix fed into a reactor producing, thereby, copolymers whose instantaneous compositions vary as the polymerization proceeds. The exact nature of the variation in composition is controlled by suitable choices of simple variables. Characterization of several model emulsion polymers produced in this manner serves to highlight the polymer differences induced by the process.

PROCESS DESCRIPTION

A simple arrangement for gradually changing the monomer mix composition of the feed stream entering the reactor is shown in Figure 1. In this arrangement, the monomer mixture in the far tank is continuously added to the monomer mixture in the well-stirred

0097-6156/81/0165-0371$05.00/0

near tank. The continuously changing mixture in the near tank is simultaneously fed into the reaction vessel in the usual manner of emulsion polymerization. Assuming monomer starved conditions, i.e., no build up of unreacted monomers, the composition of the growing particles varies as the polymerization takes place, thereby avoiding the abrupt changes in composition encountered in multistage processes. In effect, each particle is an alloy of polymers of a nearly infinite compositional variety within set compositional limits (2).

An expression for the instantaneous composition of the mixture entering the reactor can be developed by assuming perfect mixing and writing the material balance for monomer A in the near tank:

$$\text{Input (A)} - \text{Output (A)} = \text{Accumulation (A)} \tag{1}$$

At constant flow rates, R_1 and R_2, this equation becomes

$$C_2 R_2\, dt - C_1 R_1\, dt = d(W_1 C_1) \tag{2}$$

where C_2 is the concentration (weight fraction) of monomer A in the far tank (constant), C_1 is the concentration of monomer A in the near tank, and W_1 is the weight of the monomer mixture in the near tank at any given time.

Rearrangement of Equation 2 and integration yields the following expression:

$$-\frac{R_2}{(R_2-R_1)} \ln\left[\frac{W_1^o + (R_2-R_1)t}{W_1^o}\right] = \ln \frac{C_2^o - C_1}{C_2^o - C_1^o} \tag{3}$$

where W_1^o is the initial weight of the monomer mix in the near tank, C_1^o is the initial concentration of monomer A in the near tank. If we require both monomer mix tanks to empty simultaneously, Equation 3 becomes

$$\left[1 + \frac{(R_2 - R_1)}{W_1^o} t\right]^x = \frac{C_2^o - C_1}{C_2^o - C_1^o} \tag{4}$$

where $x = -R_2/(R_2-R_1)$. It can be shown that x is also the ratio of the initial monomer weight in the far tank to that in the near tank. Since the quantity $-(R_2-R_1)t/W_1^o$ is really the fraction of the monomer fed, α, at time t, the final general equation for the process can be given:

$$C_1 = C_2^o - (C_2^o - C_1^o)(1-\alpha)^x \tag{5}$$

Equation 5 expresses the variation in the concentration of monomer A in the feed stream entering the reactor as a function of time. Since this variation is a power function of time, the process has been named "power feed."

Generally, the feed rate into the reactor, R_1, is fixed by the total feed time desired for a given polymerization. The feed rate from the far tank is then:

$$R_2 = R_1 \left(\frac{x}{1+x}\right) \tag{6}$$

Another useful power feed equation permits the calculation of the fraction of a given component, A, in the final polymer,

$$F_A = \int_0^1 C_1 d \quad = C_2^o - \frac{(C_2^o - C_1^o)}{(x+1)} \tag{7}$$

Normalized plots of several power feed profiles are shown in Figure 2. For the special case where $x = 1, (W_1^o = W_2^o, R_1 = 2R_2)$, the feed profile is linear with time. Curvature is introduced by suitable changes of the initial monomer weights in the two tanks: when $x > 1$, the curve is concave to the abscissa; when $x < 1$, the curve is convex. With proper mixing in the near tank, these feed profiles can be verified experimentally.

The addition of a third monomer tank to the basic power-feed arrangement expands the possible feed profiles available for investigation. As illustrated in Figure 3, one such arrangement involves a stirred middle tank which receives a monomer mix from the far tank and pumps a varying mixture to the near tank. The arrangement is essentially a power feed on top of a power feed and can be analyzed in the same manner as carried out with the two tank systems, except that C_2, the concentration of monomer A in the second (middle) tank is not constant but is given by

$$C_2 = C_3^o - (C_3^o - C_2^o)\,(1 - \alpha)^y \tag{8}$$

Where $y = W_3^o/W_2^o$. The easiest way to handle this problem is by computer integration, but an explicit equation can be developed (3) for C_1, the concentration of monomers A in the feed stream entering the reactor:

$$C_1 = C_3^o + \frac{x}{y-x}\,(C_2^o - C_3^o)\,\left[(1-\alpha)^x (1-\alpha)^y\right] + (C_1^o - C_3^o)(1-\alpha)^x \tag{9}$$

In Equation 9, C_3^o is the initial concentration of monomer A in the far tank of Figure 3, and

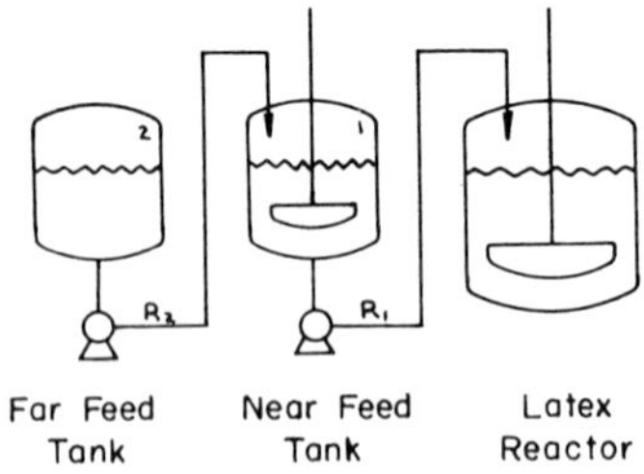

Figure 1. Monomer tank arrangement for continuously changing the monomer feed stream composition entering the reactor

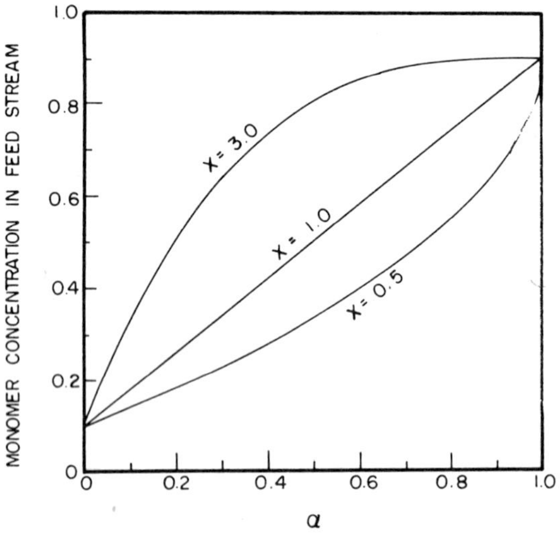

Figure 2. Power feed concentration profiles of a single monomer in the feed stream entering the reactor as a function of time. The curvature is controlled by the value of x*, the ratio of monomer weight in the far tank to that in the near tank.*

$$x = \frac{W_3^o + W_2^o}{W_1^o} \tag{10}$$

The monomer feed rates are related as follows,

$$R_2 = R_1 \left(\frac{x}{1+x}\right) \tag{11}$$

$$R_3 = R_2 \left(\frac{y}{1+y}\right) \tag{12}$$

Whereas the two-tank arrangement permits monomer feed profiles which vary smoothly in one direction, the three-tank scheme leads to inflections and concentration reversals as illustrated in Figure 4. Such reversals are useful in preparing hard-soft-hard, hydrophilic-hydrophobic-hydrophilic polymer variations and the like. In addition, three tank power feed has been useful as a means of calculating monomer inventory in copolymerization experiments (4).

While other monomer tank arrangements can be devised, the two-and three-tank configurations described here can generate most of the monomer feed profiles likely to be of interest. Variable feed rates can also be employed to generate similar profiles, but the use of constant feed rates simplifies laboratory and plant operations.

EXPERIMENTAL

Model latexes were prepared using a conventional semi-batch technique in which the reactor initially contained only water and an anionic surfactant, Aerosol OT (American Cyanamid), for particle generation. Generally the polymerizations were carried out at 80-85°C using ammonium persulfate as the initiator, although some redox polymerizations were also carried out. Uniform staged, and two-tank power feed techniques were used to introduce the monomers into the reactor. Samples were taken periodically during the three-hour feed time for residual monomer analysis. Generally, not more than 3-5% unreacted monomer was detected at any time during the reaction, thus assuring monomer starved conditions. The finished latexes were cast or molded into films for testing.

RESULTS AND DISCUSSION

The properties of polymers prepared by different processes can be studied in a variety of ways, the choice often depending on the intended end use of the polymer. For coatings applications, mechanical testing of films can lead to an understanding of polymer structure as well as to a prediction of end use performance (5). As an illustration of the influence of monomer feed profiles on

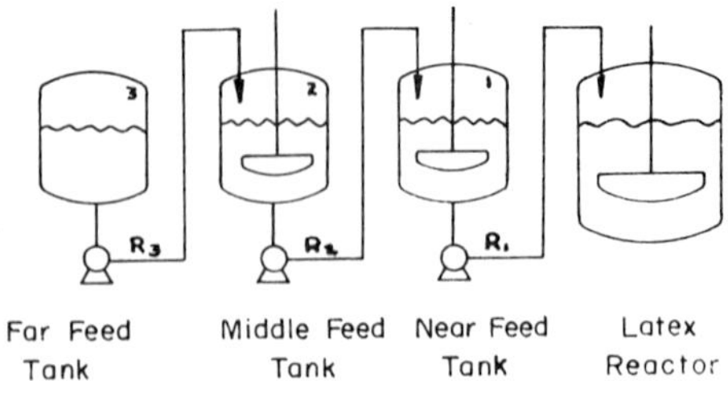

Figure 3. Three-tank monomer feed arrangement in which the near and middle tanks are stirred during operation

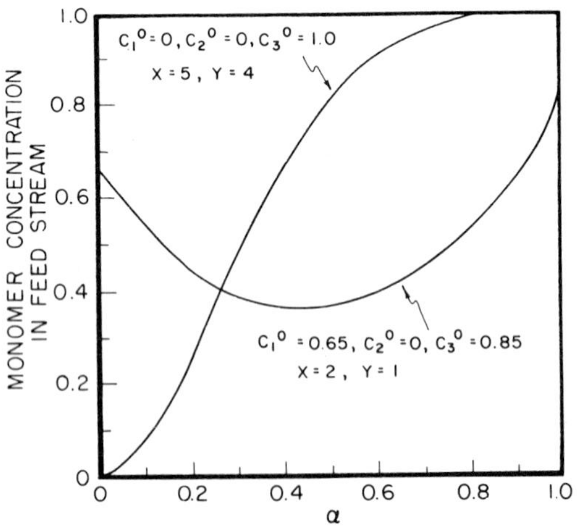

Figure 4. Monomer concentration profiles of a single monomer in the feed stream entering the reactor as a function of time for the three-tank arrangement

the properties of latex films, a series of emulsion polymers was prepared and examined in this manner. Examples of mechanical measurements included dynamic mechanical, stress-relaxation, and brittle-ductile experiments. Particle structure was investigated by particle expansion measurements utilizing a sedimentation technique.

Stiffness-Temperature - One simple tensile test of polymers involves the measurement of a modulus as a function of temperature. Figure 5 illustrates the characteristics of such a measurement. The modulus in this case is the secant modulus at 1% strain. Both polymers had the same composition: 39/59/2 - methyl methacrylate/ butyl acrylate/acrylic acid. The power feed example was prepared such that butyl acrylate varied 0.83 → 0.30 and methyl methacrylate varied 0.15 → 0.68 as the polymerization proceeded, with x = 0.83. While both examples show the usual trend of high modulus at low temperature to low modulus at higher temperature, the power feed polymer exhibits a much broader transition region than the uniform example.

Stress Relaxation - The ability to predict long term properties of polymers using short term tests is of obvious benefit in developing high performance coatings from latexes. Relaxation tests have been developed for this purpose and serve as a useful means of investigating the viscoelastic properties of polymers. In a stress relaxation experiment, a constant strain is applied to the specimen and the resulting stress is measured as a function of time. The time-dependent relaxation modulus can be extended to long times by making measurements over a wide temperature range and using the principle of time-temperature superposition (6) to shift the moduli to form a master curve as illustrated in Figure 6. In this case two polymers of the same composition (62/38-methyl methacrylate/ butyl acrylate) are compared at a reference temperature of 26°C. The power feed example utilized a feed profile with increasing butyl acrylate, 0 → 0.65, decreasing methyl methacrylate concentration, 1.0 → 0.35, and x = 1.3.

The similarity of the stress relaxation results in Figure 6 and the stiffness-temperature results in Figure 5 is quite apparent While the uniform (or random) copolymer exhibits the usual abrupt transition between brittle and rubbery regions with increasing time, the power feed example undergoes a much more gradual change. This broadened transition region may be an important factor in the ability of a film to relieve stress gradually to avoid cracking over long periods of time.

Dynamic Mechanical Testing - Film properties such as impact resistance and the cure response of thermosetting resins are conveniently investigated by dynamic measurements in which an oscillatory or torsional strain is applied to the sample with the stress and phase difference between the applied strain and measured stress being determined. In the present study, a Rheovibron Viscoelastometer was used which employed a sinusoidal strain at a

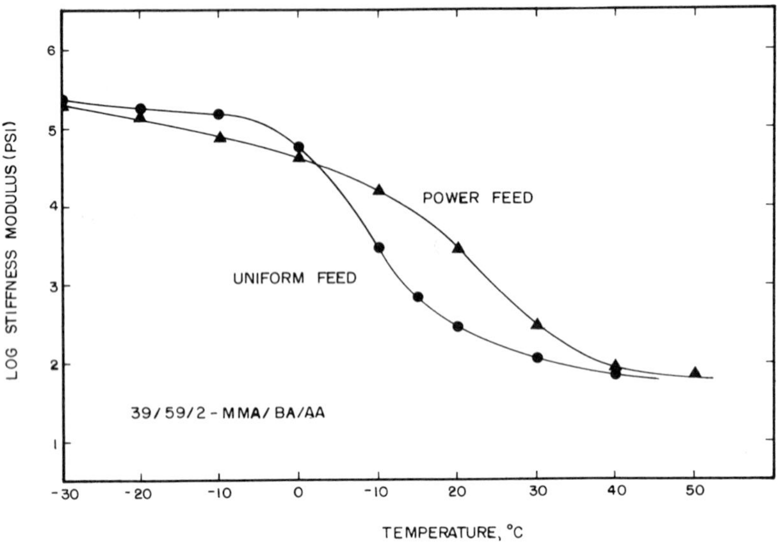

Figure 5. Stiffness–temperature comparison of two emulsion polymers having the same overall composition. In the power feed example, the methyl methacrylate concentration varied (0.15 → 0.68) and the butyl acrylate concentration varied (0.83 → 0.30), with x = *0.83.*

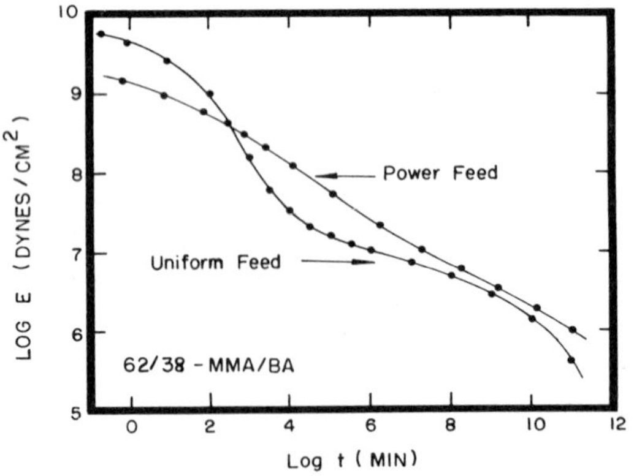

Figure 6. Stress relaxation comparison of two emulsion polymers having the same overall composition. The power feed example utilized a feed profile with increasing butyl acrylate (0 → 0.65), decreasing methyl methacrylate (1.0 → 0.35), and x = *1.3. The time axis has been shifted to a reference temperature of 26°C.*

frequency of 11 Hz and a heating rate of 1°C per minute. Using the standard treatment of data from this type of experiment (7), the storage modulus, E' (a measure of elastic response), and the loss modulus, E" (a measure of the viscous response), were calculated and displayed as a function of temperature in Figures 7-10 for a series of four latexes having the same composition: 50/50 - styrene/ethyl acrylate.

Figure 7 shows the dynamic mechanical response obtained with the latex prepared with a constant monomer feed composition. As expected, a single sharp transition is observed characteristic of a reasonably uniform copolymer. Figure 8 shows the dynamic mechanical spectrum for a two-stage process in which the first stage feed was ethyl acrylate and the second stage feed was styrene. This time, two well-defined transitions are observed characteristic of the hard and soft homopolymers.

The dynamic mechanical responses of the two power feed latexes are quite different from either of those discussed above. In Figure 9, the polymer was prepared via a linear power feed profile in which the near tank contained only ethyl acrylate and the far tank contained only styrene. In Figure 10, the polymer was prepared with the tanks reversed: the monomer feed began with styrene and ended with ethyl acrylate. In both cases, the transition regions are much broader than those observed with the uniform feed or staged feed examples.

The broadened transition response is a general characteristic of power feed polymers as evidenced by the results of stiffness-temperature, stress relaxation and dynamic mechanical measurements. The broadening phenomenon is probably a result of an alloying effect caused by the wide variety of sequencing induced by the continuously changing monomer feed composition. Indeed, sequence distribution analysis is an important aspect of the characterization of non-uniform emulsion polymers (4). The broadening effect may also result from compositional differences in polymer domains that form at different stages in the reaction. In terms of practical film properties, this broadened response suggests a wider use-temperature range which is of importance in many coatings applications.

Brittle-Ductile - Flexibility is a mechanical property of great interest in polymer design especially for polymers intended for use in films. Impact tests provide information about the ability of a film to withstand a high rate of deformation. Bending and drawing operations generally occur at lower strain rates and test the ability of a film to withstand severe elongations. The analysis of stress-strain behavior of polymers over a range of temperatures shows that a brittle specimen generally breaks at low elongations without exhibiting yield, whereas a ductile specimen exhibits a yield point which permits greater elongation before failure occurs. The temperature at which a sample changes from brittle to ductile can be called the brittle-ductile transition temperature (T_{BD}).

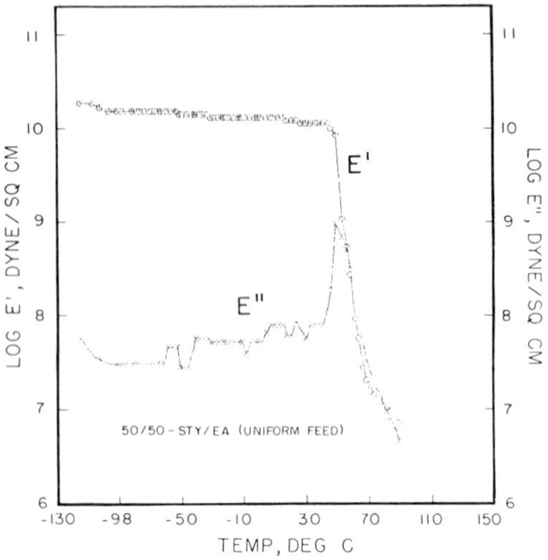

Figure 7. Dynamic mechanical properties of a model latex made with a uniform monomer feed composition

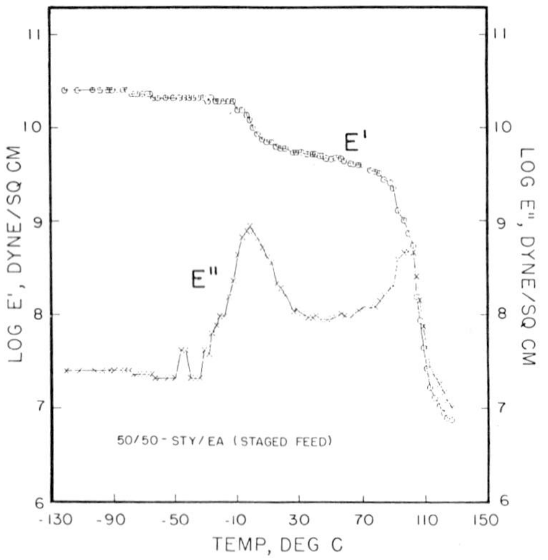

Figure 8. Dynamic mechanical properties of a model latex made with a two-stage monomer feed: Stage I contained ethyl acrylate, Stage II contained styrene.

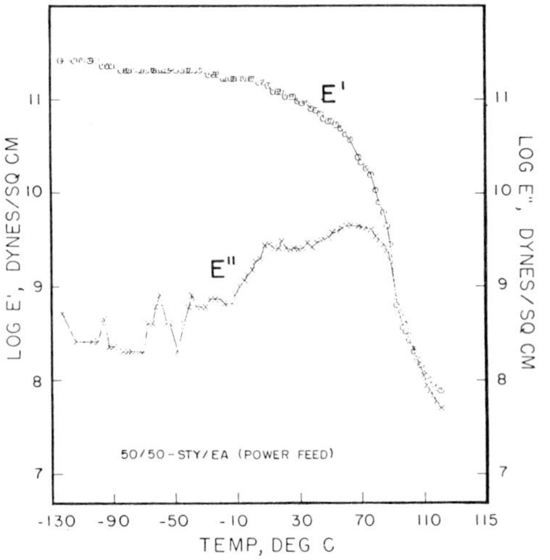

Figure 9. Dynamic mechanical properties of a model latex made with a linear power feed process; ethyl acrylate varied 1 → 0 and styrene varied 0 → 1.

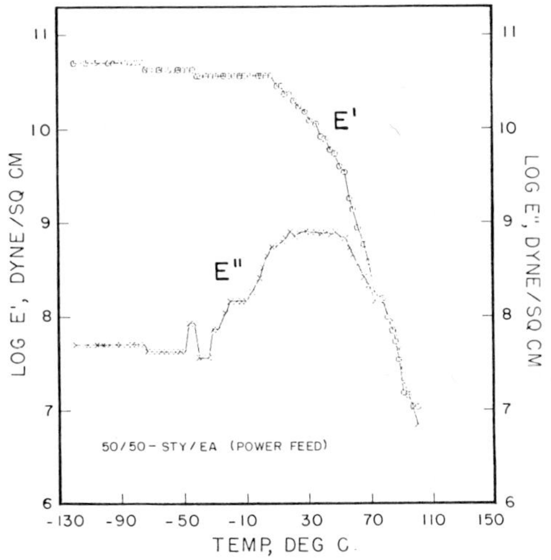

Figure 10. Dynamic mechanical properties of a model latex with a linear power feed process; ethyl acrylate varied 0 → 1 and styrene varied 1 → 0.

Although the exact nature of the B-D transition is not well understood, Wu (8) has shown that T_{BD} is related to the ability of plastics to be deep-drawn. Since the implications of such behavior to coatings are obvious, we desired to learn if non-uniform polymers offered any advantage in this respect.

A convenient way of measuring T_{BD} is to determine the ultimate strain (at break) and the yield of a specimen as a function of temperature. The intersection of the two curves defines T_{BD}. A constant strain rate of 10% per minute was employed. Polymers were isolated by air drying or freeze drying and molded at 150°C into dumbbell speciments for the tensile measurements. Glass transition temperatures (Tg) were measured by differential scanning calorimetry.

The relationship between flexibility and copolymer composition was explored by measuring Tg and T_{BD} for a series of copolymers consisting of styrene or methyl methacrylate paired with several "softer" comonomers. An example of the kind of results obtained is illustrated in Figure 11 with the methyl methacrylate/ethyl acrylate pair. Included in this plot is a ductility, or toughness, parameter, q, defined by Wu as,

$$q = \frac{T_g - T_{BD}}{T_g} \tag{13}$$

with temperatures in °K. In general, the objective is to minimize T_{BD} relative to T_g so that q is as large as possible, i.e., design a polymer which is hard but flexible over a wide range of temperatures.

As shown in Figure 11, the MMA/EA compositions yielded a fairly smooth variation in T_g but a much more erratic variation in T_{BD}. The two most flexible composition ranges (highest q values) occurred at the extremes of the series, homopolymers of methyl methacrylate or ethyl acrylate, with inconvenient T_g's for most practical uses. The composition styrene/ethyl acrylate series yielded q values of less than 0.05 indicating less flexibility over the entire composition range. This prediction was borne out by impact tests on the polymers of the series. Some of the conclusions of this study are: i) no simple relationship between T_g and T_{BD} was found; ii) no simple correlation was detected between T_{BD} and sub-T_g transitions in the dynamic mechanical spectra of the samples, and iii) copolymer pairs containing methyl methacrylate were generally more flexible (lower T_{BD}) than the corresponding copolymers containing styrene.

The effect of non-uniform polymer composition is shown in Figure 11 for the case of a 50/50-methyl methacrylate/ethyl acrylate copolymer made by a linear power feed process in which the near tank initially contained only ethyl acrylate and the far tank only methyl methacrylate. Compared with its uniform counterpart, the non-uniform polymer had a T_{BD} approximately 25° lower and a respectable

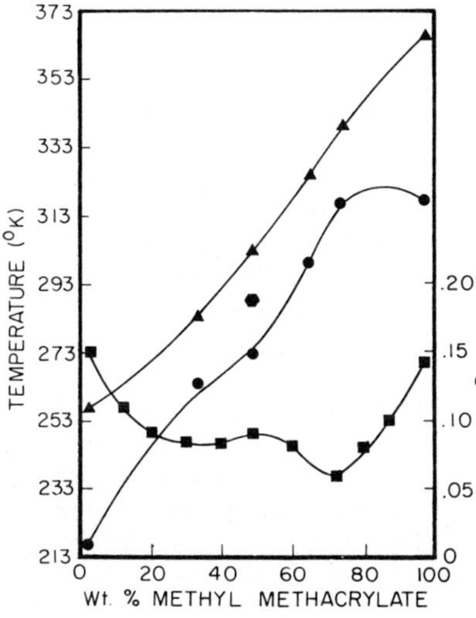

Figure 11. Brittle-ductile behavior of a series of methyl methacrylate/ethyl acrylate copolymers.

The "toughness" parameter q *is defined by Equation 13. The effect of nonuniform polymerization is shown for a 50/50 copolymer in which the power feed profile involved a decreasing ethyl acrylate concentration (0→ 1), with* x = *1 (linear). ((●) power feed,* q = *0.19)* q value of almost 0.20. Indeed, most of the power feed polymers, as well as the multi-staged polymers examined, were characterized by a significant lowering of T_{BD}. This behavior could be another manifestation of the alloying effect discussed earlier.

Particle Structure - So far, the discussion of non-uniform emulsion polymers has centered around mechanical properties, interpreted in terms of gradations of sequences between set compositional limits. But latexes are aqueous dispersions, and an exploration of particle morphology is equally interesting and important. For instance, can polymer particles be constructed such that particle sequence compositions are located in a desired region of the particle, and what evidence exists that such particle morphologies have been realized?

We have carried out extensive investigations of the swelling properties of carboxylic latex particles with increasing pH (9,10). Particles expand on neutralization due to electrostatic repulsion of the charged carboxylate groups and subsequent absorption of water. The major factors which control particle expansion are: the type and concentration of copolymerized acid, the stiffness (T_g) and hydrophilicity of the backbone, and the polymerization process employed. The process variable of most interest was the use of varying monomer compositions in stepwise changes. It was found that the expansion properties of carboxylic particles were controlled to a large extent by the order of addition of the different monomer compositions. Particle expansion, then, may be useful as a means of exploring the structures of emulsion polymers made in various ways.

The expansion behavior of carboxylic latex particles can be studied by several methods (10). The present comparison was made using a sedimentation method which involved the measurement of particle sedimentation rates in an ultracentrifuge at various degrees of neutralization. Assuming the change in particle volume is equal to the volume of water absorbed, an expanded particle settles slower, as its density decreases, according to the equation:

$$\frac{S_o}{S} = \frac{r+x}{r} \qquad (14)$$

Where S is the sedimentation coefficient of the particle at an adjusted pH, S_o is the sedimentation coefficient of the unswollen particle (low pH), r is the unswollen particle radius, and x is the increase in radius of the swollen particle. In this study, the model latexes were diluted with distilled water to 1 percent solids by weight. Individual samples were adjusted to various pH values with sodium hydroxide and allowed to equilibrate for at least 24 hours. Sedimentation rates were obtained at 30°C using a Beckman Model E analytical ultracentrifuge.

Two latexes were prepared with the composition: 47.5/47.5/5-styrene/ethyl acrylate/methacrylic acid. The power feed example had the monomer feed profile shown in Figure 12, with the styrene concentration increasing with time and ethyl acrylate decreasing as the reaction proceeded. The methacrylic acid concentration in the monomer feed was held constant in both cases. The particle expansion characteristics of the two latexes are shown in Figure 13. With 5 percent incorporated acid, both latexes have considerable potential for expansion upon neutralization. Not only did the power feed example fail to show the substantial expansion exhibited by the uniform feed example, but particle contraction was observed at the higher pH's characteristic of latex particles having no incorporated acid (9). Assuming carboxyl groups tend to locate in the surface regions of particles, the restriction in the expansion of the power feed example may be explained by the preferential polymerization of the styrene-rich segments near the sruface of the growing particles as opposed to polymerization in the particle interior.

The "onion skin" growth mechanism is supported by filming experiments in which film formation is greatly effected by the nature of the monomer composition added last in the polymerization. In power feed examples, as well as in staged feeds, hard and hydrophobic compositions hinder film formation while softer and more hydrophilic compositions aid film formation. Curiously, in this respect, it was found that the filming characteristics of all-acrylic latexes responded to non-uniform polymerization techniques much more dramatically than did their styrene-acrylic counterparts.

Molecular Weight Control - In addition to the control of monomer reaction sequences and particle morphology, the power feed process lends itself to molecular weight manipulation in ways not possible with uniform polymerizations. Molecular weight modifiers can be used separately or in combinations to produce unusually broad molecular weight distributions. One particularly useful technique involves the use of a multifunctional monomer in the near tank and a chain transfer agent in the far tank to produce latex particles with very high molecular weight capable of forming smooth, glossy films (11). The response of particle coalesence to small amounts of a chain transfer agent added late in a polymerization is another indication of the surface growth mechanism mentioned above and is consistent with other film studies of heterogeneous latexes (12).

CONCLUSIONS

A simple multi-tank arrangement for continuously changing the composition of a monomer feed stream into a latex reactor has been described and analyzed. The process is useful for preparing controlled non-uniform emulsion polymers. Such polymers exhibit broad transitions in dynamic mechanical and stress-relaxation measurements which imply a broad range of sequence distributions within the polymers and possibly polymer domains that form during the changing feed composition. The process can also be used to control the particle morphology of emulsion polymers to affect particle coalescence, viscosity control and functional group

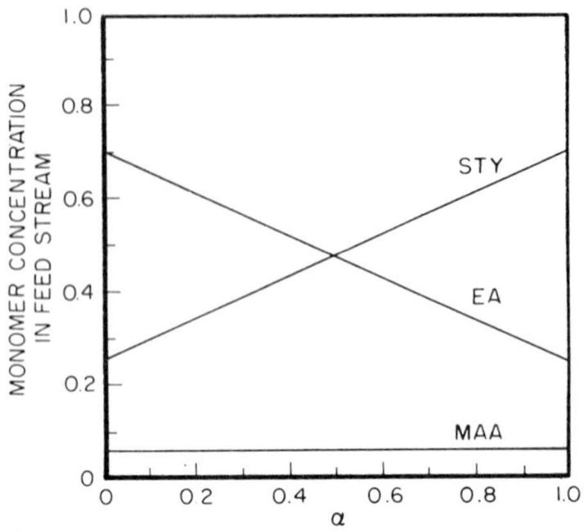

Figure 12. Monomer feed profile using a linear power feed. Overall polymer composition: 47.5/47.5/5—styrene/ethyl acrylate/methacrylic acid.

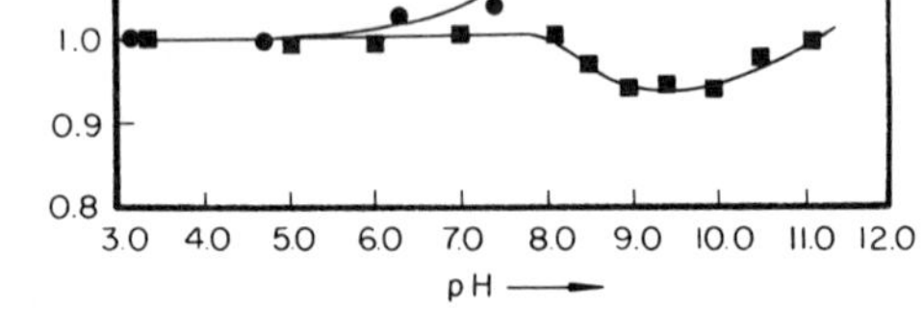

Figure 13. The effect of nonuniform polymerization on the expansion behavior of carboxylic emulsion polymers. The power feed example was prepared using the monomer feed profile illustrated in Figure 12 ((●) uniform feed; (■) power feed).

location. Finally, the multiple tank arrangement offers unique opportunities for molecular weight control through the use of molecular weight modifiers.

ABSTRACT

A novel polymerization technique has been developed by which non-uniform emulsion polymers can be produced in a controlled manner. The technique involves the continuous addition of one monomer mixture into a stirred tank containing another monomer mixture. This continuously changing mixture is then fed into a reaction vessel producing, thereby, polymers whose instantaneous copolymer composition varies as the polymerization proceeds. A general equation is developed which expresses the composition of the feed stream as a function of time. The process, call "power feed", is used to prepare emulsion polymers which exhibit dynamic mechanical spectra having broad transition ranges. Stress relaxation measurements also show a broadened transition range compared with analogs prepared using a uniform monomer feed profile. The characteristics of power feed latexes are interpreted in terms of an alloying of a wide range of sequence distributions and/or polymer domains of differing compositions superimposed on a particle morphology in which the surface properties of the particles reflect the characteristics of the polymer formed last. Practical applications include the control of particle expansion of carboxylic emulsion polymers on neutralization, the increased flexibility of relatively hard polymers and the control of molecular weight distributions for specific latex applications.

ACKNOWLEDGMENTS

The authors gratefully acknowledge the following people for their experimental contributions to this work: R. W. Callard, L. C. Cantley, Y. P. Chang, S. L. Hager and R. E. Kelchner.

LITERATURE CITED

1. See for example: Ryan, C. F., U.S. Patent 3,562,235; Dickie, R. A.; and Newman, S., U.S. Patent 3,787,522.
2. Bassett, D. R.; Hoy, K. L.; U.S. Patent 3,804,881.
3. Whelan, J. M.; Union Carbide Corporation.
4. Johnston, J. E.; Bassett, D. R.; MacRury, T. B.; This Volume, following paper.
5. Hill, L. W., Prog. Org. Coatings, 1977, 5, 277.
6. Aklonis, J. J.; MacKnight, W. J.; Sher., M.,"Introduction to Polymer Viscoelasticity," Wiley-Interscience, New York, 1972.
7. Nielsen, L. E., Mechanical Properties of Polymers and Composites, Marcel Dekker, New York, 1974, Volume 1.
8. Wu, S., J. Appl. Polymer Sci., 1976, 20, 327.
9. Bassett, D. R.; Hoy, K. L.; "Polymer Colloids II,", Fitch, R. M., ed., Plenum, New York, 1980, p.1.
10. Bassett, D. R.; Derderian, E. J.; Johnston, J. E., MacRury, T. B., This Volume.
11. Bassett, D. R.; Hoy, K. L.; U.S. Patent 4,039,500.
12. Eliseeva, V. I., Acta Polymerica, 1979, 30, 273.

RECEIVED April 6, 1981.

24

Nonuniform Emulsion Polymers

Carbon-13 NMR Spectroscopy

J. E. JOHNSTON[1], D. R. BASSETT, and T. B. MacRURY[2]

Technical Center, Union Carbide Corporation, South Charleston, WV 25303

With the advent of advanced characterization techniques, such as multiple detector liquid exclusion chromatography and ^{13}C Fourier transform nuclear magnetic resonance spectroscopy, the study of structure/property relationships in polymers has become technically feasible (1-6). Understanding the relationship between structure and properties alone does not always allow for the solution of problems encountered in commercial polymer synthesis. Certain processes, of which emulsion polymerization is one, are controlled by variables which exert a large influence on polymer infrastructure (sequence distribution, tacticity, branching, enchainment) and hence properties. In addition, because the emulsion polymerization takes place in an heterophase system and because the product is an aqueous dispersion, it is important to understand which performance characteristics are influended by the colloidal state, (i.e., particle size and size distribution) and which by the polymer infrastructure.

In order to design experiments to test the influence of process variables on polymer infrastructure, a simple but general process design is needed. For these studies a new sequential feed polymerization process called "power-feed" was chosen (7-8,9). The advantage of this technique is that almost any conventional monomer feed profile can be simulated and described by an equation containing only three independent variables. In addition, a number of novel monomer composition profiles can also be constructed with this approach. The composition of the monomer feed to the reactor can be described by:

$$C_1 = C_2^o - (C_2^o - C_1^o)(1 - \alpha)^x \qquad (1)$$

where C_1 is the instantaneous concentration of a given monomer A entering the reactor, C_2^o is the initial concentration of the

[1] Current address: Exxon Chemicals, Linden, New Jersey
[2] Current address: IMC Corporation, Terre Haute, Indiana

0097-6156/81/0165-0389$05.00/0

monomer in the second of two feed tanks, C_1^o is the initial concentration in the first feed tank, the one which feeds the reactor, α is a dimensionless quantity which expresses the percent of total monomer fed; $\alpha = R_2t/W_2 = t/t_f$, R_2 is the feed rate from the second feed tank to the first tank, t is the elapsed time, t_f is the total monomer feed time, W_1 is the initial weight of monomer mix in the first feed tank, x is the power exponent, i.e., W_2/W_1, which assumes that the feed tanks empty simultaneously.

The power exponent, x determines what the monomer feed profile will look like as a function of α. Some examples of more common feed profiles are shown in Figure 1. If the polymerization is carried out under monomer-starved conditions, the composition of the polymer being formed at any instant is the same as the feed composition, C_1. Therefore, the cumulative polymer composition at any time α may be obtained from:

$$\text{Cumulative polymer composition} = \int_0^{\alpha} C_1 d\alpha \qquad (2)$$

where C_1 is given by Equation 1. The total composition in component A can be determined by integrating from $\alpha = 0$ to $\alpha = 1$,

$$F_A = C_2^o - \frac{(C_2^o - C_1^o)}{(x + 1)} \qquad (3)$$

As can be seen by comparison of curve I and IV in Figure 1, there can be a significant difference in monomer composition distribution as a function of α in spite of the fact that the total content of monomer A in both cases is the same, as determined by Equation 3.

It is anticipated, and in fact is true, that the physical properties and performance characteristics of emulsion polymers prepared by routes I and IV are different. The question here is whether it is possible by characterization techniques to distinguish polymers made by different routes and, more subtly, by the same nominal route but with some process aberration such as a feed upset. In order to determine the feasibility of such an approach, an emulsion copolymer system was selected: the copolymer of styrene and ethyl acrylate.

We speculate that process changes could cause changes in sequence distribution. ^{13}C NMR spectroscopy has been used to study similar copolymers (10,11,12) and hence should be of value in correlating process with sequence distribution. ^{13}C NMR is a means of looking at changes in the electronic environment of nucleii of the isotope of carbon, ^{13}C. There are a number of texts on ^{13}C NMR spectroscopy and its application to organic molecules, and polymers (13,14,15). The chemical shifts which one observes as characteristic of different carbons can be caused by substitution, that is, oxygen, halogen, etc. substituted, primary, secondary, tertiary carbon. They may be caused by changes in bond hybridization, i.e., sp^3,sp^2. They may be caused by changes in tactic placement within a sequence, i.e., isotactic, heterotactic, syndiotactic, or they may be related to nearest neighbors on a chain, as for instance, whether a given carbon of an A monomer unit is in a sequence surrounded by A monomer units, B monomer units or both. The changes caused by this last effect, that is, sequence distribution, are the ones of predominant interest in this study. It is by correlation of sequence distribution data with process variables and performance characteristics that we hope to establish some generally valid guidelines for predicting structure/process/property relationships in emulsion polymers.

EXPERIMENTAL

The latexes were prepared using a conventional semi-batch emulsion polymerization system modified for power-feed by the addition of a second monomer tank. Polymerization temperatures ranged from 30-85°C using either redox or thermal initiators. Samples were taken periodically during the polymerization and analyzed to determine residual monomer in order to assure a "starved-feed" condition. As used in this study this is a condition in which monomer feed rate and polymerization rate are identical and residual monomer levels are less than 5%.

The ^{13}C NMR spectra were obtained using a Varian FT80A $^1H/^{13}C$ spectrometer (20 MHz ^{13}C) or a JEOL FX 90Q spectrometer (22.5 MHz ^{13}C). Unless otherwise indicated, the samples were analyzed at room temperature using deuterated chloroform, $CDCl_3$, as solvent and hexamethyldisiloxane, HMDS, as reference. Sample concentrations were 10-20% polymer by weight.

In order to obtain quantitative data, there must be complete relaxation between pulses. Thus a pulse width of 37.5° was employed rather than the full excitation pulse width of 90°. Using the shorter pulse width permits immediate repulsing but requires somewhat higher numbers of pulses to obtain the same signal-to-noise ratio (2). However, our studies indicated that adequate signal-to-noise could be obtained in overall shorter times with this method as compared to using a 90° pulse width.

All the spectra discussed here were pseudo-white noise decoupled to eliminate splitting due to spin-spin interactions between protons and ^{13}C nucleii.

RESULTS AND DISCUSSION

Styrene/ethyl acrylate random copolymers were prepared with ethyl acrylate contents of 10,25,50,75 and 90% by weight. Examination of the ^{13}C NMR spectra of the copolymers reveals that the carbonyl carbon of the acrylate ester is a sensitive probe of the sequence distribution in these copolymers. Figure 2 shows the carbonyl carbon resonance region for a sample containing 50%

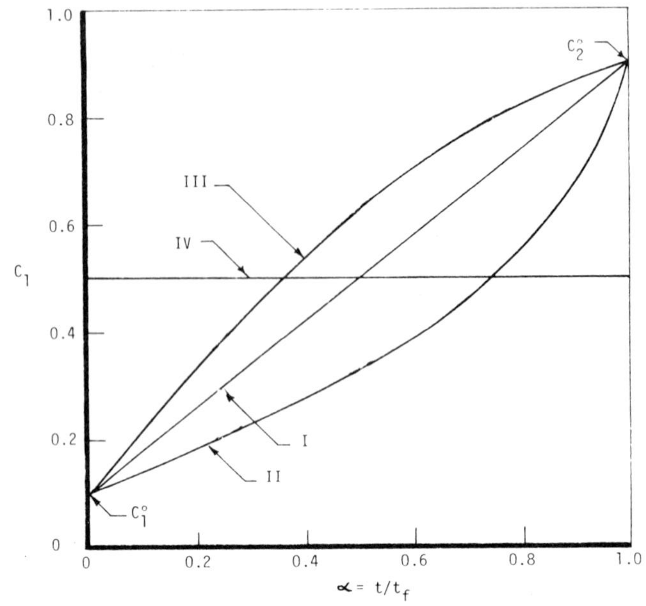

Figure 1. Monomer feed profiles for the power feed process ((I) $C_2°$ = 0.90, $C_1°$ = 0.10, X = 1.0; (II) $C_2°$ = 0.90, $C_1°$ = 0.10, X = 0.5; (III) $C_2°$ = 0.90, $C_1°$ = 0.10, X = 1.5; (IV) $C_2°$ = $C_1°$ = 0.5)

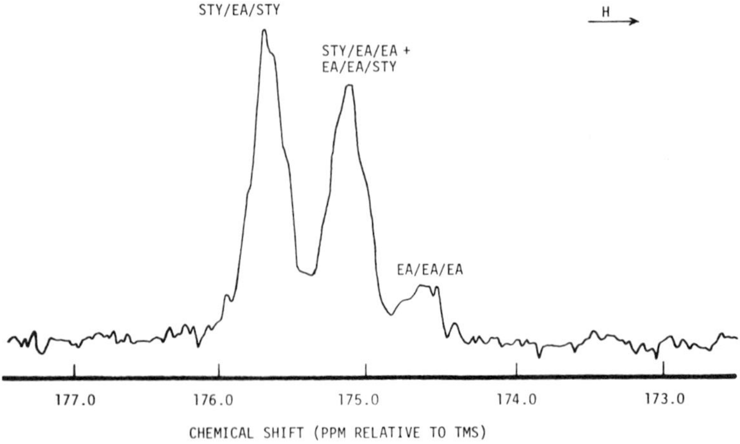

Figure 2. The ^{13}C NMR spectrum of a 50/50—styrene/ethyl acrylate copolymer prepared using a conventional uniform monomer feed process

ethyl acrylate. Based on the five random copolymers, assignments have been made for the three major resonance regions observed in terms of triad sequences indicated in the figure. The essential feature is that changes in overall composition of the copolymers lead to easily recognizable and qualitatively predictable changes in the carbonyl carbon resonance pattern.

Figures 2, 3 and 4 illustrate the differences between conventional mixed feed, staged and power-feed copolymers whose average composition is 50/50 ethyl acrylate/styrene. In the case of the staged polymerization, all the ethyl acrylate was fed first followed by the styrene. The power-feed copolymer was prepared with ethyl acrylate linearly increasing as a function of time, i.e., $x = 1.0$, $C^{o}_{EA} = 0$.

The differences in the three polymerization processes are immediately apparent in the spectra. For the conventional copolymer, Figure 2, there is a predominance of BAB triads (EA = A). The power-feed copolymer appears to have relatively higher concentrations of mixed (AAB = BAA) and homo (AAA) triads. In the case of the stage polymerization, Figure 3, the carbonyl resonance is essentially identical to that for poly(ethyl acrylate), that is, AAA triads.

One of the factors which can affect the sequence distribution of copolymers of this type is the value of the apparent reactivity ratios, r_1 and r_2. The fact that these ratios are apparent is important since the conversions obtained in these polymerizations at any point in time are considerably higher than those at which conventional reactivity ratio calculations should be applicable (16). Nonetheless, under otherwise similar conditions, changes in the apparent reactivity ratios may provide some information about the different systems.

Harwood (17) has described a technique using run number theory to calculate reactivity ratios based on sequence distribution. The run number, R, which characterizes a particular copolymer may be calculated as follows:

$$R = 2(\text{mole \% A})(P_{BAB})^{1/2} \qquad (5)$$

where:

$$P_{BAB} = \text{probability of BAB triad} = (P_{AB})^2$$
$$= \text{fraction of BAB triad present in the polymer.}$$

An equation can be derived which relates run number, and feed composition to reactivity ratios:

$$2\left(\frac{100}{R} - 1\right) \cdot \frac{A_f}{B_f} = \left(\frac{A_f}{B_f}\right)^2 \cdot r_1 + r_2 \,, \qquad (6)$$

where for the case of starved-feed: $A_f = \%\ A$, $B_f = \%\ B$ as mole percent.

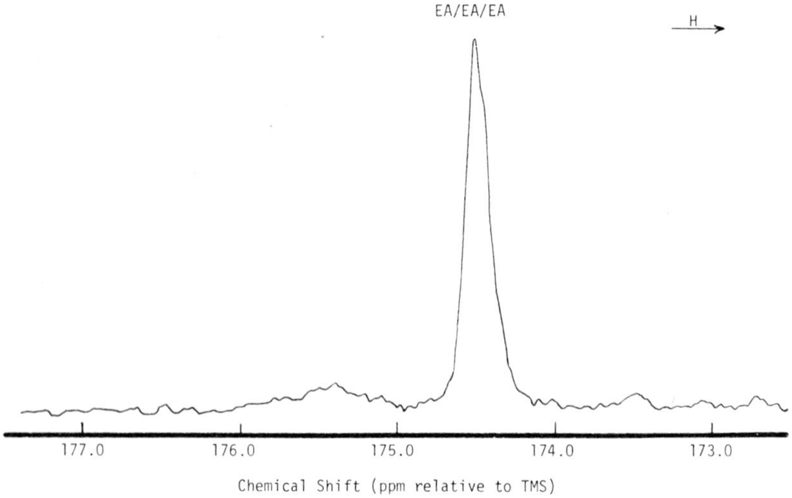

Figure 3. The ^{13}C NMR spectrum of a 50/50—styrene/ethyl acrylate copolymer prepared using a staged feed process in which the ethyl acrylate was fed first followed by the styrene

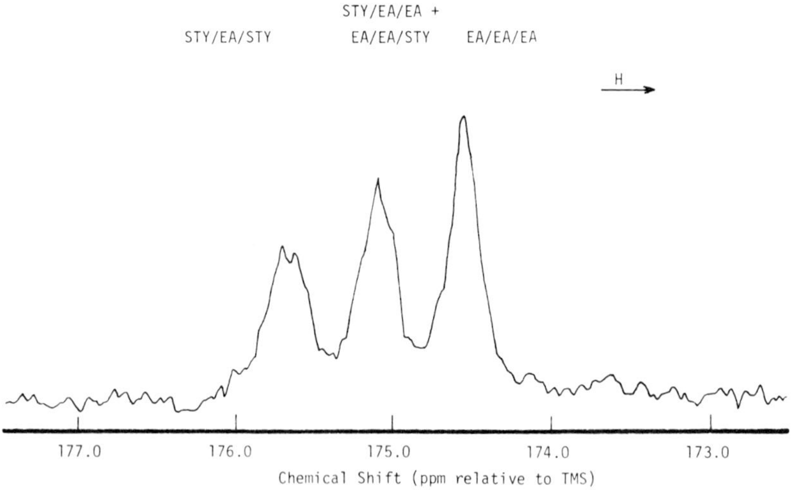

Figure 4. The ^{13}C NMR spectrum of a 50/50—styrene/ethyl acrylate copolymer prepared using a linear power feed profile with ethyl acrylate increasing $0 \rightarrow 1.0$, and styrene decreasing $1.0 \rightarrow 0$ with time

By using Equation 5 to determine run number values for each of our conventional semi-batch copolymers, and then regressing the data as

$$2\left(\frac{100}{R} - 1\right) \cdot \frac{A_f}{B_f} \text{ versus } \left(\frac{A_f}{B_f}\right)^2$$

the intercept is r_2 and the slope is r_1. When monomer 1 is ethyl acrylate and monomer 2 is styrene, these data are given in Table I. The expected values for r_1 and r_2 are 0.16 ± .04 and 1.01 ± .14 respectively (18). The agreement is quite good between the apparent reactivity ratios calculated from run number theory and those reported in the literature.

TABLE I

Run Number and Reactivity Ratios for Conventional Copolymers

Sample 42BRD-	%EA(A)	%STY(B)	P_{ABA}	R	$2\left(\frac{100}{R} - 1\right) \frac{A_f}{B_f}$	$\left(\frac{A_f}{B_f}\right)^2$
9	9.6	90.4	0.115	61.3	11.82	87.66
22	24.3	75.7	0.213	69.9	2.69	9.70
23	49.0	51.0	0.525	73.9	0.74	1.08
20	89.6	10.4	1.00	20.8	0.88	0.01

$r_1 = 0.12 \pm 0.04$

$r_2 = 0.97 \pm 0.15$

In order to predict sequence distributions based on run number theory, some model for the polymerization is necessary. Our model for the polymerization consists of two stages. The initial reaction stage, in which the rate of polymerization is slower than the rate of feed and some residual monomer accumulates in the reactor, is followed by the second reaction stage during which the rate of polymerization is essentially equal to the rate of feed. During this stage, the amount of residual monomer remains constant or decreases slightly. A simple linear power-feed system will illustrate the features of the model. The initial concentration of ethyl acrylate is 100% in the tank feeding the reactor and 0.0% in the second tank. This results in the feed profile shown in Figure 5. Only the first ten percent of this profile will be used in describing the feed during the first stage. Once the reaction has reached its second stage at $\alpha = 0.1$ the monomer feed rate equals the polymerization rate. Now the residual monomer pool takes on the function of a third tank. Thus, from this point on the equation for a three tank power-feed system must be used to describe the feed to the polymerization.

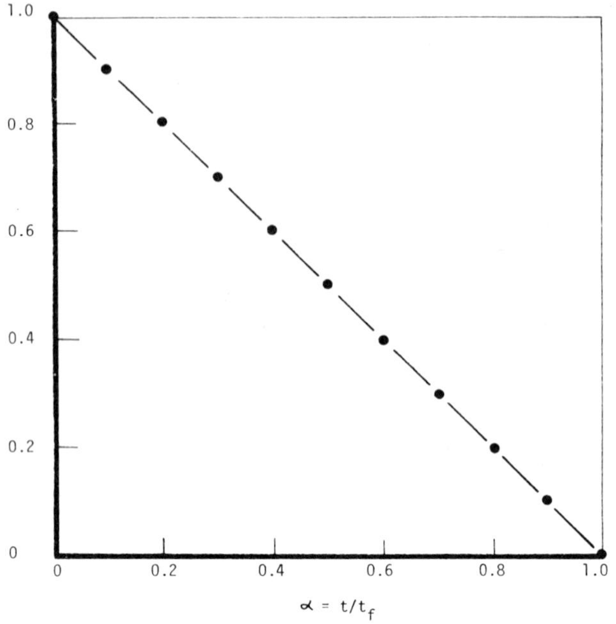

Figure 5. Linear power feed profile for a 50/50—styrene/ethyl acrylate copolymer in which ethyl acrylate decreases 1.0 → 0 and styrene increases 0 → 1.0 with time

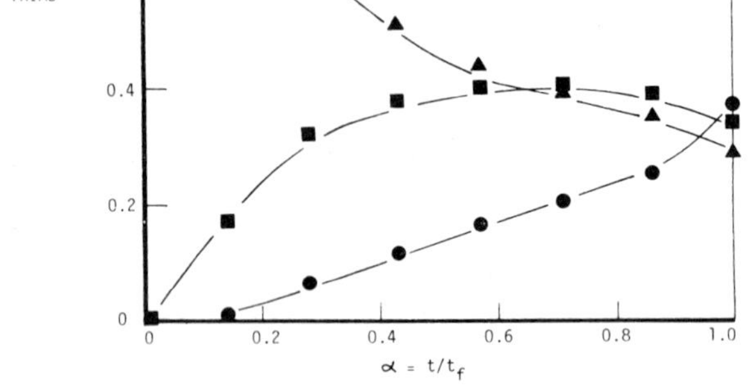

Figure 6. Calculated ethyl acrylate–centered triad probabilities as a function of conversion for the model copolymer described in Figure 5 ((▲) EA–EA–EA; (■) EA–EA–STY =STY–EA–EA; (●) STY–EA–STY)

The weight fraction conversion during the initial stage, 0.25 has been selected to create a residual monomer pool approximately the same size as that observed in the experimental systems. During the second stage, however, the amount of monomer being converted is identical to that being fed, hence the weight fraction conversion during this stage is 1.0.

A series of computer programs was written to test models for predicting sequence distribution for the power-feed copolymers. The programs consist of:

- a feed generator - calculates the input feed compositions as a function of reaction time.
- a copolymer generator - calculates the instantaneous copolymer and feed compositions at weight fraction conversions from the feed generator.
- a sequence distribution generator - calculates the instantaneous and cumulative triad fractions based on the changing feed composition from the copolymer generator.

The feed generator calculations are based on the two-tank and three-tank power-feed equations. The three tank equation, described in the previous paper (7), has the form:

$$C_1 = C_3^o + \frac{x}{y-x}(C_2^o - C_3^o)\left[(1-\alpha)^x - (1-\alpha)^y\right] + (C_1^o - C_3^o)(1-\alpha)^x. \quad (7)$$

where x = the first power exponent $= \frac{W_3 + W_2}{W_1}$

y = the second power exponent $= \frac{W_3}{W_2}$

The copolymer generator calculates a series of "mini-batch" copolymerizations and has been described in more detail by Molau (19) and Meyer and Lowry (20). The sequence distribution generator uses Harwood's (14) run number approach to calculate triad functions.

Table II shows the results of the final triad fractions calculated using the above model with reactivity ratio combinations corresponding to the range represented by the experimentally determined values.

Example number 4, $r_1 = 0.16$, $r_2 = 0.82$, shows the closest agreement with the experimentally observed values. A more rigorous test of the model is the prediction of the intermediate triad fractions. Figure 6 illustrates the change in triad probabilities as a function of conversion

TABLE II

Final Triad Fractions as a Function of Reactivity Ratio

Example No.	r_1	r_2	AAA[1]	AAB= BAA	BAB
1	0.08	0.82	.25	.36	.29
2	0.08	1.12	.30	.43	.27
3	0.12	0.97	.30	.40	.30
4	0.16	0.82	.30	.34	.36
5	0.16	1.12	.35	.41	.24
Experimental	--	--	.29	.34	.37

1) A = ethyl acrylate, B = styrene

as predicted by the model. Figure 7 shows the corresponding experimental data and Figure 8 shows the relationship between the calculated and measured triad fractions for all intermediate samples. The correspondence between the calculated and measured triad fractions is within the error of the NMR measurements. The same model has been applied to a linear profile in which the ethyl acrylate content is rising during the feed. Figure 9 shows the relationship between calculated and measured triad fractions for this case.

The development of the triad fractions as observed by NMR during the course of a polymerization of a 75/25 ethyl acrylate styrene copolymer is shown in Figure 10. Over the course of the feed, the ethyl acrylate content dropped from 100% to 50% linearly. The increase in styrene containing triads is evident from the NMR spectra and is predicted by the model. Figure 11 shows the correspondence between calculated and measured triad fractions for this system.

CONCLUSIONS

^{13}C nuclear magnetic resonance spectroscopy can be employed to study changes in copolymer sequence distribution brought about by differences in monomer feed profiles. Sequence distributions characteristic of conventional, staged, and power-feed copolymers are easily distinguishable in a model system of the type described here.

A method for calculating apparent reactivity ratios based on run number theory has been applied to "starved-feed" styrene/ethyl acrylate systems. The reactivity ratios found are in agreement with those determined from solution polymerization data. The further confirmation of the observed agreement between reactivity ratios determined at low conversions and those determined by run number theory in "starved-feed" high conversion copolymerization requires the analysis of other comonomer pairs.

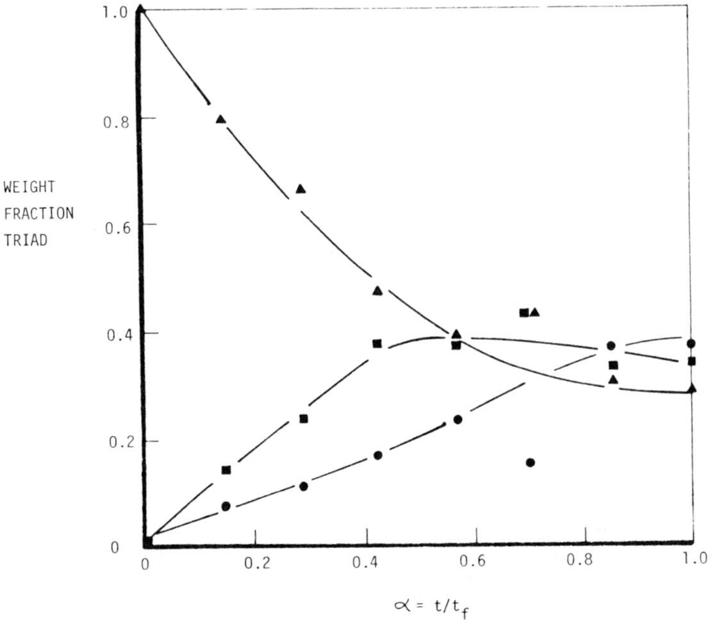

Figure 7. Experimental ethyl acrylate–centered triad probabilities as a function of conversion for the model copolymer described in Figure 5 ((▲) EA–EA–EA; (■) EA–EA–STY/STY–EA–EA; (●) STY–EA–STY)

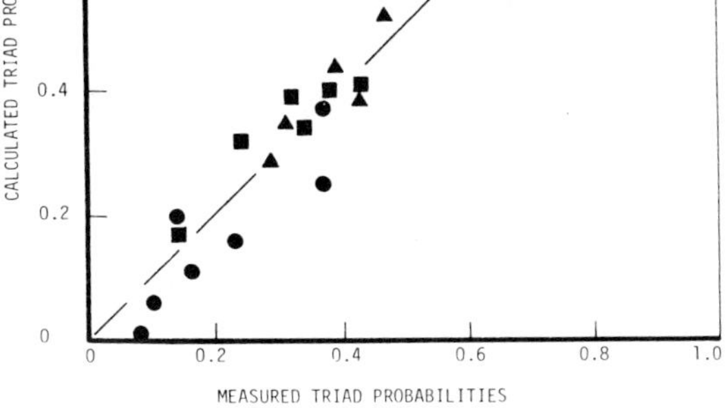

Figure 8. Correlation of calculated and measured triad fractions from Figures 6 and 7 (r_1 = 0.16; r_2 = 0.82; (▲) EA–EA–EA; (■) EA–EA–STY = STY–EA–EA; (●) STY–EA–STY)

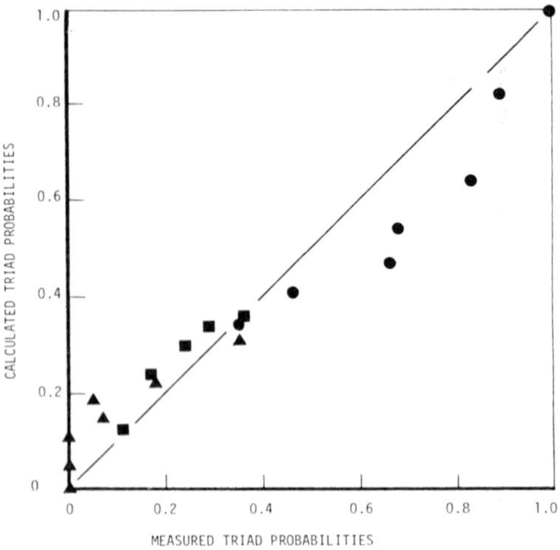

Figure 9. Correlation of calculated and measured triad fractions for a model 50/50—styrene/ethyl acrylate copolymer prepared with a linear power feed profile in which ethyl acrylate increases 0 → 1.0 and styrene decreases 1.0 → 0 with time (r_1 = 0.16; r_2 = 0.82; (▲) EA–EA–EA; (■) EA–EA–STY = STY–EA–EA; (●) STY–EA–STY)

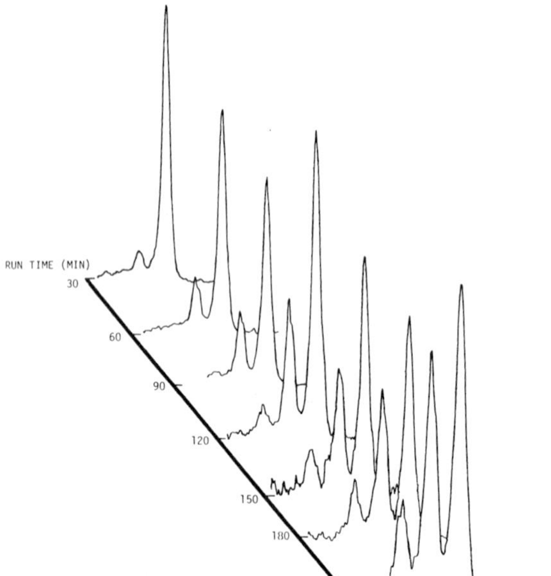

Figure 10. Development of triad fractions as observed by NMR for a model 25/75—styrene/ethyl acrylate copolymer prepared with a linear power feed profile in which ethyl acrylate decreased 1.0 → 0.50 and styrene increased 0 → 0.50 with time

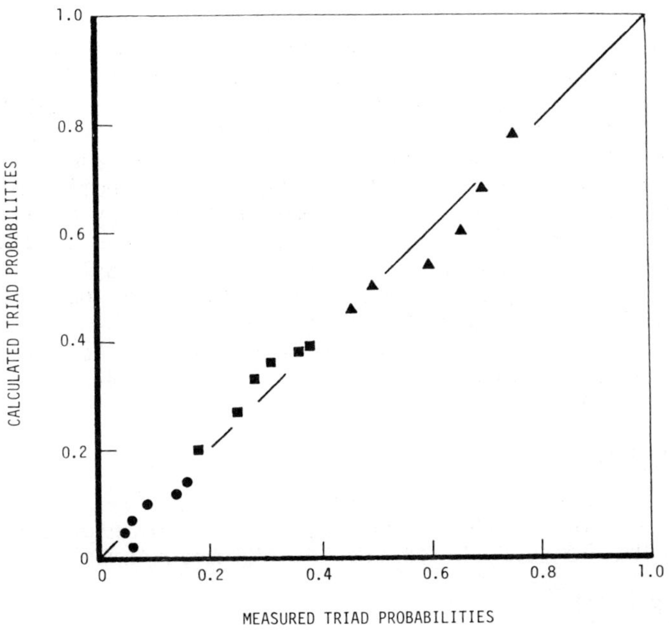

Figure 11. Correlation of calculated and measured triad fractions for the model copolymer described in Figure 10 (r_1 = 0.16; r_2 = 0.82; (▲) EA–EA–EA; (■) EA–EA–STY = STY–EA–EA; (●) STY–EA–STY)

A model has been developed in which the polymerization is described by two stages; a short first stage in which feed rate exceeds polymerization rate, and a second stage in which feed and polymerization rates are equal. The model requires the use of an additional "monomer tank" to simulate the residual monomer pool during this second stage. In addition, equivalent weights of monomer are entering and polymerizing during each portion of this stage; hence, the conversion is 100%. This model predicts the intermediate and final sequence distributions for a number of two-tank power-feed systems.

ABSTRACT

A "model" latex copolymerization system consisting of styrene and ethyl acrylate has been studied using ^{13}C nuclear magnetic resonance spectroscopy. Conventional semi-batch copolymers at five monomer ratios, a staged equimolar copolymer, and sequential, or "power-feed", copolymers of equimolar average composition have been prepared. Sequence distribution measurements for these samples using ^{13}C NMR spectroscopy clearly show differences arising from the use of different process designs. Triad sequence assignments have been made for ethyl acrylate-centered triads. Apparent reactivity ratios have been calculated for the semi-batch copolymers using run number theory. A model has been developed to describe the power-feed systems and predict the triad distributions in the incremental and final copolymer using the experimentally determined r_1 and r_2 values.

ACKNOWLEDGMENTS

The authors wish to thank Drs. K. L. Hoy, E. J. Derderian, and J. M. Whelan for their contributions to the progress of this work, and Mr. R. L. Pack for the ^{13}C NMR measurements.

LITERATURE CITED

1. Mirono, A.; Uchida, Y.; Bull. Chem. Soc. Jap. (1970), 43, 3259.
2. Schaefer, J.; Macromolecules (1971), 4, 98.
3. Inoru, 4.; Chiyo, R.; Nishioka, A.; Polymer Jour.(1973), 4 244.
4. Bailey, D. B.; Hendricks, P.M.; J. Poly.Sci.,A, (1978), 16 3185.
5. Kricheldorf, H. R.; Hull, W. E.; J. Poly. Sci., A, (1978), 16, 2553.
6. Spevacek, J.; Polymer (1978), 19, 1149.
7. Bassett, D. R.; Hoy, K. L.; This volume, preceeding paper.
8. Bassett, D. R.; Hoy, K. L., U. S. Patent 3,804,881.
9. Bassett, D. R.; Hoy, K. L.; U. S. Patent 4,039,500.
10. Hirai, H.; Koinuma, H.; Tanabe, T.; Takenchi, K.; J. Poly. Sci.; A, (1979), 17, 7339.

11. Stejskal, E. O.; Schaefer, J.; Macromolecules (1974), 7, 14.
12. Schaefer, J.; Macromolecules (1971), 4, 107.
13. Wehrle, F. W.; Wirthlen, T.; "Interpretation of Carbon-13 NMR Spectra", Heyden and Son, Ltd. London (1976).
14. Levi, G. C.; Nelson, G. L.; "Carbon-13 Nuclear Magnetic Resonance for Organic Chemists", Wiley-Interscience, New York (1972).
15. Randall, J. C.; "Polymer Sequence Determination, Carbon-13 NMR Method", Academic Press, New York (1977).
16. Flory, P. J.; "Principles of Polymer Chemistry", Cornell University Press, Ithaca, New York (1969).
17. Harwood, H. J.; Ritchey, W. M.; Polymer Letters (1964), 2, 601.
18. Luskin, L. S.; Myers, R. J.; "Enclyclopedia of Polymer Science and Technology", Interscience (1964), 1, 246.
19. Molan, G. E.; J. Poly. Sci., A, (1967), 5, 401.
20. Meyer, V. E.; Lowery, G. G.;, J. Poly. Sci., A, (1965), 3, 2843.

RECEIVED April 6, 1981.

25

Morphology of Two-Stage Latex Particles

Polystyrene and Styrene–Butadiene Copolymer Pair Systems

D. I. LEE

Designed Latexes & Resins, Michigan Division, Dow Chemical Company, Midland, MI 48640

Staged emulsion polymerization produces heterogeneous structured latexes which exhibit a wide variety of properties depending on their particle morphology. It is thus extremely important for the design of structured latex products to understand the factors controlling the particle morphology. In addition to this technological importance, a better understanding on the morphological aspect of latex particles may shed some light on the mechanism of emulsion polymerization as well as polymer mixing on a microscopic scale.

Williams and his co-workers (1-9) proposed a core-shell morphology for two-stage emulsion polymerization, based on a monomer-rich shell model. Gardon (10, 11) argued against a monomer-rich shell model, and suggested the possibility of surface polymerization due to the fast diffusion of monomer molecules. Napper (12) also argued against a monomer-rich shell on the basis of diffusion theory, but explained the core-shell formation by means of the surface-active oligomeric radicals in the absence of chain transfer activity. They all agreed on the possibility of surface polymerization, but the question remained whether surface polymerization alone would be sufficient for a core-shell morphology in two-stage emulsion polymerization over a wide range of variations in polymerization conditions, polymer compatibility, etc.

In this study, the morphology of two-stage (styrene//styrene-butadiene) latex particles was investigated as a function of polymer phase ratio (stage ratio), molecular weights (chain transfer agent level), polymer compatibility (butadiene level in S-B copolymers), polymerization sequence (polystyrene first and S-B copolymer second or vice versa), and polymerization conditions (flooded or starved, polymerization temperature, etc.). This paper describes the preparation of two-stage latex samples and the morphological characterization of their particles by electron microscopy.

0097-6156/81/0165-0405$05.00/0

Experimental

Materials. Styrene and butadiene monomers were polymerization grade, available from Dow Chemical Company. Acrylic acid was a technical grade monomer from Dow Badische. The polymerization surfactant was sodium dodecyldiphenyl oxide sulfonate available from Dow. The polymerization initiator was sodium persulfate, and bromoform and carbon tetrachloride the chain transfer agents.

Preparation of Latex Samples. Two-stage latex samples were prepared by emulsion polymerization of the second-stage monomer mix in the presence of the first-stage polymer latex. The first-stage latexes were either in-situ or separately made using an externally prepared polystyrene latex seed. The mode of polymerization was a semi-continuous process for both stages.

A series of latex samples with varying stage ratios (LS-1 to LS-4) was prepared at 90°C by emulsion polymerizing styrene and butadiene monomer mixes (S/B/AA/$CHBr_3$: 59/40/1/0.5) in the presence of polystyrene latex (S/AA: 99/1). Acrylic acid (AA) was used for latex stability. The stage ratios varied from 40/60 to 10/90. Both stage monomer mixes were continuously added over 4 hours, respectively, based on 100 parts monomer mix. One-hour cook-down was provided between the stages, regardless of the stage ratio variation. The aqueous mix containing sodium persulfate (0.7 parts), sodium dodecyldiphenyl oxide sulfonate (1 part), and sodium hydroxide (0.05 part) was continuously added over 6 1/2 hours without any interruption between the stages. The final particle sizes were 1500 Å and the final polymer solids were about 47%.

Two latex samples (LS-5 and LS-6) were prepared in the same manner as described above except that 5 parts of carbon tetrachloride were added in both the first and second stage monomer mixes, based on 100 parts monomer mix. The stage ratios were 50/50 and 20/80, respectively.

A series of latex samples with varying S/B ratios in the second stage (LS-7, LS-8 and LS-9) was prepared at 90°C, using an externally prepared polystyrene latex of 1350 Å. The second-stage S/B ratios were 70/30, 90/10, and 95/5, respectively. The stage ratio was 20/80. Neither vinyl acid nor chain transfer agent was used.

Two latex samples (LS-10 and LS-11) were prepared by switching the order of monomer addition between a styrene mix (S/AA: 98/2) and a high butadiene mix (S/B/AA: 28/70/2) at the stage ratio of 50/50. The polymerization temperature was 80°C during the continuous addition of monomer mixes and then raised to 100°C for one-hour cook-down.

All the latex samples are listed in Table I.

Table I.

The List of Latex Samples

Latex Samples	Compositions 1st Stage	2nd Stage	Stage Ratios 1st Stage/2nd Stage
LS-1	99S/1AA	59S/40B/1AA/0.5 $CHBr_3$	40/60
LS-2	99S/1AA	59S/40B/1AA/0.5 $CHBr_3$	30/70
LS-3	99S/1AA	59S/40B/1AA/0.5 $CHBr_3$	20/80
LS-4	99S/1AA	59S/40B/1AA/0.5 $CHBr_3$	10/90
LS-5	100S/5 CCl_4	60S/40B/5 CCl_4	50/50
LS-6	100S/5 CCl_4	60S/40B/5 CCl_4	20/80
LS-7	100S	70S/30B	20/80
LS-8	100S	90S/10B	20/80
LS-9	100S	95S/5B	20/80
LS-10	98S/2AA	28S/70B/2AA	50/50
LS-11	28S/70B/2AA	98S/2AA	50/50

S = Styrene, B = Butadiene, AA = Acrylic Acid

Electron Microscopy. For morphological characterization electron microscopy was extensively used in conjunction with the osmium tetraoxide staining method.

Results

Figures 1A and 1A' are the transmission electron micrographs at OsO_4-stained two-stage (styrene//styrene-butadiene) latex particles at the stage ratio of 40/60 (LS-1) and their ultra-thin cross-sections, respectively. It is clear from the micrographs that when styrene and butadiene monomers were polymerized in the presence of polystyrene latex particles, the second-stage S-B copolymer phase-separated as microdomains within the first-stage polystyrene phase. Figures 1B and 1B' show OsO_4-stained two-stage latex particles at the stage ratio of 30/70 (LS-2) and their ultra-thin cross-sections, respectively. At this stage ratio, the second-stage S-B copolymers are still dispersed in the polystyrene phase, however, some particles appear to be made up of two continuous polystyrene and S-B copolymer phases. Figure 1C shows OsO_4-stained two-stage latex particles at the stage ratio of 20/80 (LS-3). At this stage ratio, the second-stage S-B copolymer became a continuous phase. This suggests that phase inversion took place between the stage ratios of 30/70 and 20/80. It is interesting to note that this phase inversion may have been a result of the close packing of S-B copolymer microdomains. Figure 1D is the transmission electron micrograph of OsO_4-stained two-stage latex particles at the stage ratio of 10/90 (LS-4). This micrograph clearly shows an off-centered polystyrene phase encapsulated with the S-B copolymer.

Figure 2A shows the ultra-thin cross-sections of OsO_4-stained two-stage (styrene//styrene-butadiene) latex particles at the stage ratio of 50/50 (LS-5). Figure 2B is the ultra-thin cross-section of OsO_4-stained latex film made from LS-6 (stage ratio = 20/80). Unlike the micro-phase separation structure shown in Figures 1A and 1A', these two-stage latex particles prepared with high levels of carbon tetrachloride are composed of two distinct polymer phases, suggesting a complete phase separation due to the higher mobilities of polymer molecules involved. Figure 2A shows a hemispherical structure at the equal stage ratio. Figure 2B suggests that as the amount of the second-stage S-B copolymer increased above the equal stage ratio, the S-B copolymer began to encapsulate the first-stage polystyrene, resulting in an off-centered polystyrene phase in the continuous S-B copolymer phase.

Figures 3A, 3B, and 3C show the ultra-thin cross-sections of OsO_4-stained two-stage (styrene//styrene-butadiene) latex particles at the stage ratio of 20/80, whose S/B ratios in the second stage are 70/30 (LS-7), 90/10 (LS-8), and 95/5 (LS-9), respectively. It can be seen from the micrographs that the size of polystyrene phase domains decreases with decreasing butadiene level in the second-stage S-B copolymers and becomes so small at the S/B ratio

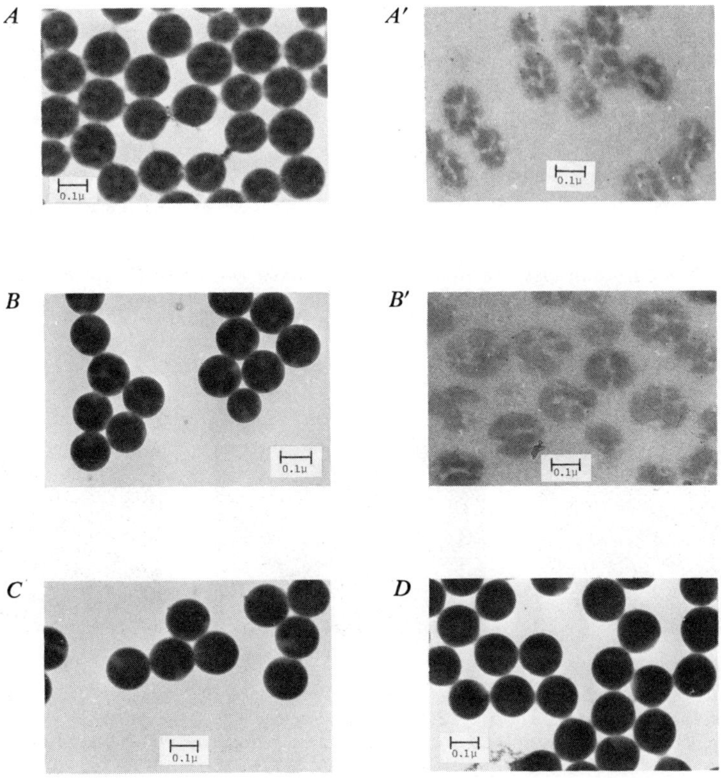

Figure 1. Transmission electron micrographs of OsO_4-stained two-stage (S//S–B) latex particles (A, B, C, and D) and their ultrathin cross sections (A′ and B′) showing the effect of stage ratio on morphology ((A and A′ 40/60 (LS-1); (B and B′) 30/70 (LS-2); (C) 20/80 (LS-3; and (D) 10/90 (LS-4))

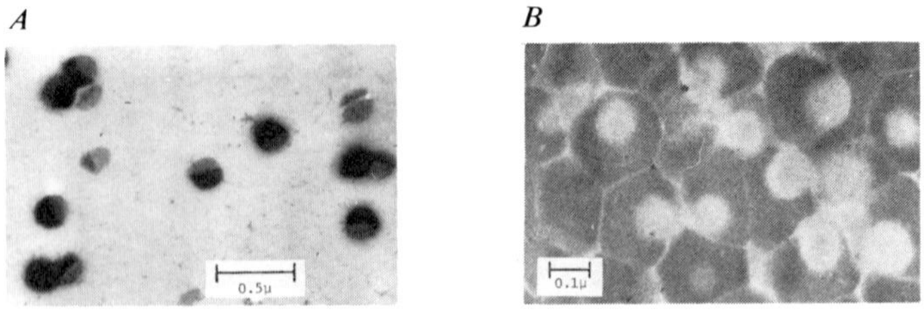

Figure 2. Ultrathin cross sections of OsO_4-stained two-stage (S//S–B) latex particles at the stage ratios of 50/50 and 20/80, respectively, showing the effect of molecular weight ((A) LS-5, (B) LS-6)

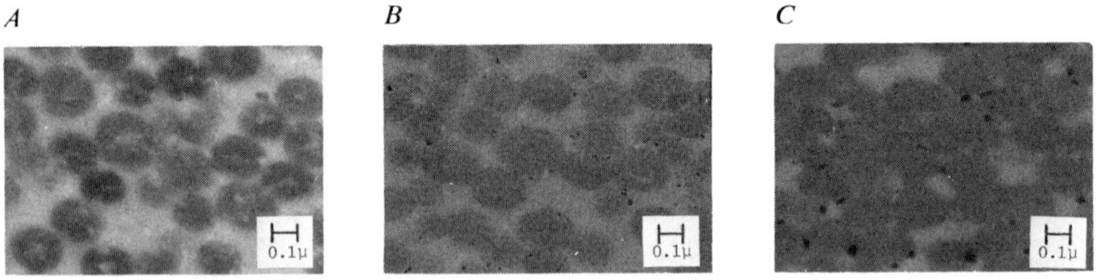

Figure 3. Ultrathin cross sections of OsO_4-stained two-stage (S//S–B) latex particles at the stage ratio of 20/80 showing the effect of polymer compatibility ((A) 30 parts butadiene (LS-7); (B) 10 parts butadiene (LS-8); and (C) 5 parts butadiene (LS-9))

of 95/5 that phase separation is no longer discernable by the present electron microscopy method.

Figures 4A and 4B are the ultra-thin cross-sections of OsO_4-stained two-stage (styrene//styrene-butadiene) and (styrene-butadiene//styrene) latex particles at the stage ratio of 50/50 (LS-10 and LS-11), respectively. Latex samples were mixed with a polymerizable monomer mix of butyl and methyl methacrylates, cured, and microtomed for examination. Figure 4A shows particle cross-sections much smaller than the actual particle size of LS-10. It appears that since the embedding monomer solution was a solvent for polystyrene, the continuous polystyrene phase was dissolved and small S/B copolymer microdomains were left behind. This is further evidence that the second-stage S-B copolymers phase-separated as microdomains within the first-stage polystyrene phase, as shown in Figures 1A and 1A'. Figure 4B shows somewhat swollen and deformed particle cross-sections, suggesting that the first-stage cross-linked S-B copolymers were a continuous phase. Indeed, the former (LS-10) behaved like a hard latex, but the latter (LS-11) behaved like a soft latex.

The morphology of two-stage latex particles was greatly affected by variations in polymerization conditions: e.g., batch vs. semi-continuous, flooded vs. starved, low vs. high polymerization temperature, etc. Two-phase structures of two-stage latex particles rearrange toward a more stable state (i.e., coarsening or intermixing), if the polymerizing particles are fluid. Such rearrangement was found to be accelerated by a high temperature, flooded polymerization. Conversely, a low temperature, starved polymerization immobilized two-phase structures during the polymerization process, leading to a core-shell morphology.

Discussion

A core-shell morphology was reported for two-stage combinations of moderately compatible polymers: PMA and PMMA (13), and PEA and PMMA (14). In addition, it was found that hydrophobic latex particles were easily encapsulated with hydrophilic polymers (14, 15). These observations, along with the kinetic evidence (16), support the fact that particle surface is a main locus of emulsion polymerization and this surface polymerization may lead to a core-shell structure for two-stage emulsion polymerization, if there is little or no interpenetration. On the other hand, our current study shows that in the case of incompatible polystyrene and S-B copolymers, the second-stage polymer forms a disperse phase within the first-stage polymer particle, despite surface polymerization. Matsumoto et.al. (17) also reported a microphase separation of the second-stage polystyrene in the first-stage polyethyl acrylate particle. Sperling et.al. (18, 19) studied the dynamic mechanical behaviors of two-stage latex interpenetrating networks and observed a shell-core phase

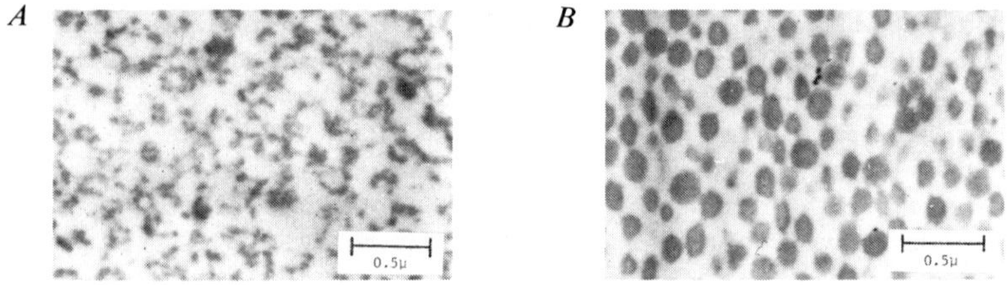

Figure 4. Ultrathin cross sections of OsO_4-stained two-stage (S//S–B) and (S–B//S) latex particles at the stage ratio of 50/50, respectively, embedded in a polymerizable mix of butyl and methyl methacrylates ((A) LS-10, (B) LS-11 (reverse of LS-10))

separation, in addition to a micro-phase separation of the second-stage polymer in the core. All these observations obviously suggest that the morphology of two-stage latex particles depends not only on surface polymerization, but also the kinetic and thermodynamic aspects of polymerizing latex systems. With these findings in mind, we can not only predict a priori the morphology of two-stage latex particles from the knowledge of polymerization systems, but also control their morphology.

Summary

The morphology of two-stage (styrene//styrene-butadiene) and (styrene-butadiene//styrene) latex particles was found to vary from a core-shell structure to a complete phase separation with various two-phase structures in between, depending on polymerization sequence, polymerization conditions, polymer compatibility, molecular weights, polymer phase ratio, etc.

Acknowledgement

The author would like to thank E. B. Bradford, R. A. Withers, and J. Cotter for their contributions in electron microscopy, R. A. Willency for his contribution of Micrographs 4A and 4B and E. F. Stevens for his assistance in the experimental work.

Abstract

The morphology of two-stage (styrene//styrene-butadiene) latex particles was studied with respect to variations in stage ratio, molecular weights, styrene-butadiene (S-B) copolymer composition, polymerization sequence, and polymerization conditions. The morphological features observed were mainly structures resulting from phase separation rather than the core-shell morphology expected from surface polymerization with no interpenetration. At intermediate chain transfer agent (CTA) levels in the second-stage polymer phase-separated as micro-domains within the first-stage polymer particles, then underwent phase inversion to become a continuous phase as the amount of the second-stage polymer increased. At high CTA levels, polymers were completely separated, resulting in a hemispherical morphology at the equal stage ratio. Upon further increase in the amount of the second-stage polymer, asymmetric encapsulation occurred. Phase domain sizes decreased with decreasing butadiene in the S-B copolymers, as expected from better compatibility. This study strongly suggests that the morphology of two-stage latex particles greatly depends on the thermodynamic nature of polymerizing latex systems, polymerization conditions, and the order of polymerization.

Literature Cited

1. Grancio, M. R.; Williams, D. J. J. Polymer Sci A-1 1970, 8, 2617.
2. Grancio, M. R.; Williams, D. J. J. Polymer Sci. A-1 1970, 8, 2733.
3. Williams, D. J. J. Elastroplast, 1973, 3, 187.
4. Williams, D. J. J. Elastoplast, 1973, 5, 6.
5. Keusch, P.; Williams, D. J. J. Polymer Sci., Polymer Chem. Ed. 1973, 11, 143.
6. Williams, D. J. J. Polymer Sci., Polymer Chem. Ed. 1973, 11, 301.
7. Keusch, P.; Prince, J.; Williams, D. J. J. Macromol. Sci-Chem. 1973, A7, 623.
8. Williams, D. J. J. Polymer Sci., Polymer Chem. Ed. 1974, 12, 2123.
9. Keusch, P.; Graff, R. A.; Williams, D. J. Macromolecules 1974, 7, 304.
10. Gardon, J. L. J. Polymer Sci., Polymer Chem. Ed. 1973, 11, 241.
11. Gardon, J. L. J. Polymer Sci., Polymer Chem. Ed. 1974, 12, 2133.
12. Napper, D. H. J. Polymer Sci. A-1 1971, 9, 2089.
13. Hughes, L. J.; Brown, G. L. J. Appl. Polym. Sci 1961, 5, 580
14. Matsumoto, T.; Okubo, M.; Imai, T. Kobunshi Ronbunshu, Eng. Ed 1974, 3, 1814.
15. Lee, D. I. unpublished data.
16. Wessling, R. A.; Gibbs, D. S. J. Macromol. Sci-Chem. 1973, A7, 647.
17. Matsumoto, T.; Okubo, M.; Shibao, S. Kobunshi Ronbunshu Eng. Ed. 1976, 5, 784.
18. Sperling, L. H.; Chin, Tai-Woo; Hartman, C. P.; Thomas, D. A. Intern. J. Polymeric Mater. 1972, 1, 331.
19. Sperling, L. H.; Chin, Tai-Woo; Thomas, D. A. J. Appl. Polym. Sci. 1973, 17, 2443.

RECEIVED April 6, 1981.

26

New Design for Producing Constant-Composition Copolymers in Emulsion Polymerization

Comparison with Other Processes

A. GUYOT, J. GUILLOT, C. PICHOT, and L. RIOS GUERRERO[1]

CNRS—Laboratoire des Matériaux Organiques,
BP 24 – 69390 VERNAISON, France

Except in very special cases (azeotropic copolymerizations), copolymerization via radical mechanism shows a drift in the composition of the copolymers produced through the polymerization process. Emulsion copolymerization obeys this rule too, although the special features of its mechanism can change the drift process. The most common way to obviate that composition drift is to use the semi-continuous process where, after polymerization has been initiated with a small percent of the total charge (say 10 to 20 %) like in the batch process, most of the charge is added continuously at a much smaller rate (Ra) than the rate (Rp) at the end of the batch period, so that the added charge is polymerized quite instantaneously (1, 2). Then, the composition drift is limited to the initial period and most of the product does possess actually a constant composition.

In this paper we would like to describe a new design, based on gas chromatographic analysis of the monomer mixture, for production of constant composition copolymers and its application to emulsion copolymerization. This design was already shortly described and applied to solution copolymerization (3) of methylmethacrylate and vinylidene chloride. Since then, the apparatus was made more simple, more reliable and more accurate. It is actually monitored by an analogic computering system which keeps the ratio of the monomers constant by controlling the addition of one of them. The process based on it can be called "corrected batch process" because the initial value of this ratio is kept up to the end.

[1] Current address : Universidad Autonoma Nacional de México MEXICO 20 D.F.

0097-6156/81/0165-0415$05.50/0

We will describe its use for controlling the styrene-acrylonitrile emulsion copolymerization system. Results concerning copolymer compositions, molecular characteristics and particle sizes will be compared to the corresponding ones from batch or semi-continuous processes.

Description of the apparatus

Basically, samples of the reaction medium are injected at regular time intervals in a GC apparatus ; signals from two (or more) comonomers are compared and a differential signal is used to monitor a metering system for introducing the monomer which is consumed more rapidly.

A scheme is presented in Figure 1.

From the reactor, the reaction medium is continuously circulated in a small tubing using a pump. In the circuit, an injection valve is inserted, which is worked by compressed air via an electrovalve governed by a timer, so that a given amount of the reaction medium, stored in a loop of the injection valve is injected into the GC apparatus. Because of flocculation problems, due to mechanical stress, occuring in the pump, injection valve or tubing, the solid contents of the reaction medium must be limited to about 30 %. The signal from the GC apparatus (with flame ionization detector) is treated by an electronic integrator with microprocessor based calculator (LTT-ICAP 10) which gives a differential signal from a reference one. A special electronic interface has been designed, which allows this differential signal to govern the rate of addition of one of the monomers to the reactor by a metering burette (Tacussel - Electroburex EBX 2). The variation of the addition rate tends to make null the differential signal. The time elapsed between two successive orders is dependent on the time necessary for GC analysis (a few minutes).

Styrene-acrylonitrile copolymerization - Results

Previous kinetic study of emulsion copolymerization of styrene (S) and acrylonitrile (AN) leads us to determine (4) the reactivity ratios as :

$$r_S = 0.44 \qquad r_{AN} = 0.10$$

using the Kelen-Tüdos method (5).

These values show that there will be a large composition drift in a wide range of composition of the monomer feed, where the styrene consumption will be higher than that of acrylonitrile. Experiments have been carried out to obtain a copolymer with 46 mole % of styrene units. Using the electronic computering system designed to obtain constant composition copolymers, just described above, the copolymers were found to be richer in styrene (Table 1). The effect seems not to be related to the stirring rate (if it is high enough) and then it is not a diffusion effect. The results at final conversion, obtained from gas chromatographic analysis of the residual monomer mixture, show that the styrene enrichment increases with conversion.

Table 1 - Acrylonitrile(AN)-Styrene(S) copolymerization with the "corrected batch process" - Effect of stirring rate

Stirring rate rpm	Final conversion %	Average Calculated[a)] AN	S	Composition Experimental AN	S
200	45	54	46	52	48
275	38	54	46	47	53
400	57	54	46	44	56
500	65	54	46	42	58
850	62	54	46	46	54

initial charge : water : 600 ml ; AN : 0.8 moles ; S : 0.2 moles
$K_2S_2O_8$: 10 mg ; $C_{12}H_{25}SH$: 445 mg ;
sodium laurylsulfate : 2 g ; temperature : 50° C

a) calculated values from the reactivity ratios

Explanation may be related to the heterogeneity of the reaction medium ; polymerization location is mostly limited to within the polymer particles, so the copolymer composition is dependent on the composition of the monomer mixture inside the particles, which may be different from that of the whole reactor, monitored by the apparatus. So, compositions of various phases, monomer droplets, polymer particles, aqueous phase, were measured using gas chromatographic (GC) analysis, after they were separated by ultracentrifugation (33,000 rpm).

The following experiments were carried out at 50° C, 250 rpm, keeping constant the initial AN/S ratio of 4, and the same initiator system ($K_2S_2O_8$: 10 mg ; $C_{12}H_{25}SH$: 445 mg in 600 ml H_2O).

A first set of experiments was carried out in batch, varying either the amount of emulsifier, sodium lauryl sulfate (SDS) from 2 to 8 g, or the amount of monomers from 1 to 2.7 moles (X = AN/S being kept constant at 4).

Typical conversion curves are shown in Figure 2. Their shapes are as expected. As expected too,polymerization rate increases with increasing emulsifier concentration (although the exponent of the concentration law is 0.74), but the limiting conversion surprisingly decreases with increasing emulsifier concentration. On the other hand, as shown in table II, higher final conversion may be reached if the initial monomer amount is larger. At the same time, it can be seen that the final amount of acrylonitrile in the water phase remains quite large at comparable values. In these experiments too, the initial copolymer composition is not very different from the calculated one using the above mentioned reactivity ratios.

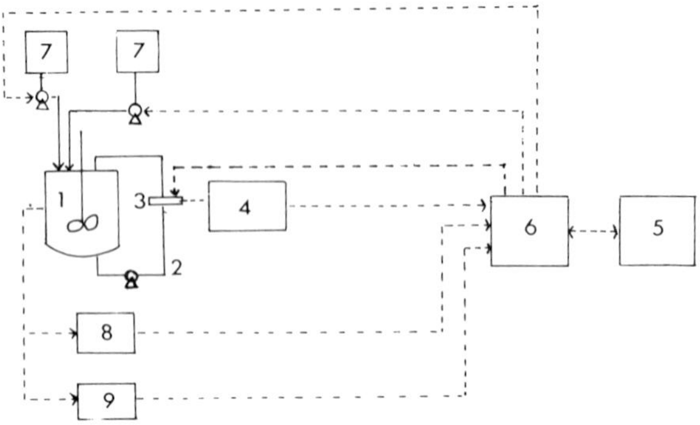

Figure 1. General scheme of the apparatus: (1) magnetically stirred reactor; (2) circulation path for sampling of the reaction mixture; (3) automatic injection device (1–2 μL); (4) gas chromatograph; (5) recorder; (6) electronic integrator + interface + timer; (7) monitor reagent addition system; (8, 9) additional sensors

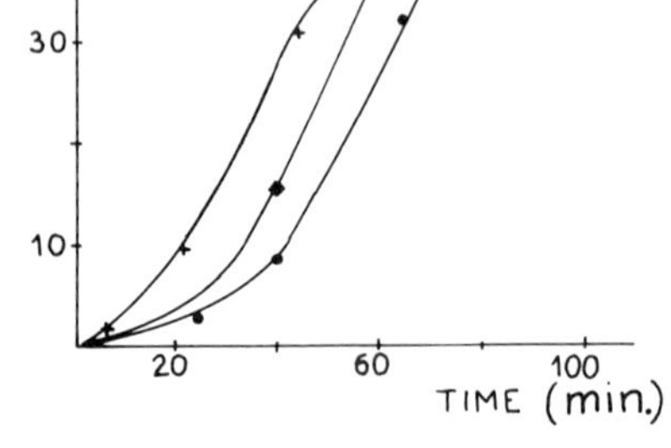

Figure 2. Styrene–acrylonitrile copolymerization in batch: conversion curve for varying sodium laury sulfate contents ((●) 2 g; (♦) 6 g; (×) 8 g))

Table II - Styrene-Acrylonitrile copolymerization in batch - Effect of monomer initial concentration

Run	Monomer/Water	Xp	Copolymer initial composition (mole %)				Final conversion (mole %)	Final amount of AN(g) in water	Mw 10^3	Mn 10^3
			Calculated		Experimental					
			AN	S	AN	S				
P_2	0.1	2.55	54	46	52	48	58	20	61	34
R_1	0.2	2.70	54.5	45.5	54	46	70	22	161	82
R_2	0.3	3.0	55.5	44.5	55	45	82	24	254	124

After separation of the phases through ultracentrifugation, the amount of each monomer in the various phases can be obtained. Typical results are given in figure 3 for run P_2 with 4 g SDS.

In these experiments, monomer droplets disappear after about 25 % conversion. Monomer amounts in the particles increase to a maximum around the point where droplets disappear, and then decrease.

But, as shown in figure 4, composition of the monomer mixture in the particles does not change very much when droplets are still present. After that point, however, the composition drift is very large, although styrene remains preferentially in the particles.

A large amount of acrylonitrile always remains in the water phase. The styrene content in the water phase is very much lower, it decreases continuously, and finally vanishes. At that point, the conversion practically stops.

The initial copolymer composition corresponds well to reactivity ratios measured (6, 7) from bulk or solution copolymerization (r_{AN} = 0.13 ; r_S = 0.34) taking into account not the whole monomer feed, but its composition within particles. So, the initial copolymer composition is practically kept constant as long as droplets remain. After their disappearance, the polymerization rate remains constant up to about 50 % conversion.

As shown in figure 5, the average compositions, using "solution" reactivity ratios (0.13 and 0.34) and taking into account monomer contents within particles (curves 1 and 2), correspond well to experiments, and are quite different from the ones calculated from "emulsion reactivity ratios" (0.1 and 0.44)(curves 5 and 6). As shown by the dotted lines (3 and 4) instantaneous compositions begin to change drastically after the droplet disappearance.

The monomer volume fraction in the particles (ϕ) (Figure 6) remains relatively constant if droplets are present and decreases to zero at the limiting conversion, so therefore it can be concluded that styrene is preferentially absorbed within the particles and allows a part of acrylonitrile to be absorbed too. When styrene has been totally consumed, no more acrylonitrile can be absorbed and polymerized. At the limiting conversion, the amount of acrylonitrile in the water phase is not very different from the initial one. The monomer volume fraction in the particles, decreasing after droplets have disappeared, does not seem to be consistent with the rate, which tends to remain constant. So the number and the size of the particles were measured using light scattering dissymetry method (8). Results shown in figure 7 are quite surprising : the average particle size tends to decrease initially and then although the particle number increases throughout, the polymerization rate remains constant and then levels off. Obviously, new particles are created at least up to very high conversion far after the disappearance of the droptlets, that is at variance for the most widely accepted theories.

The same behaviour is observed in other experiments where either the amount of emulsifier or the amount of monomers has been varied.

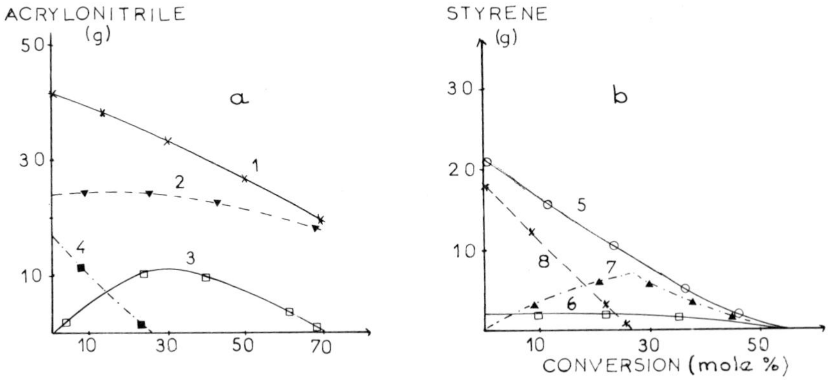

Figure 3. Monomer contents in the various phases. (a) AN: (1) whole emulsion; (2) water; (3) particles; (4) droplets; (b) S: (5) whole emulsion; (6) water; (7) particles; (8) droplets.

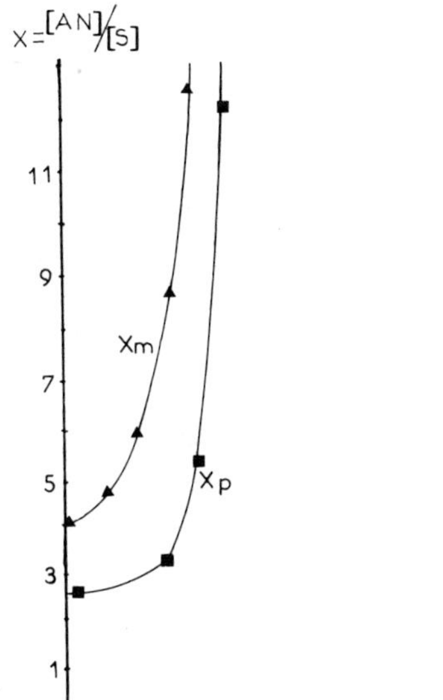

Figure 4. Experimental relative composition of monomer mixture in whole emulsion (Xm) and in particles (Xp)

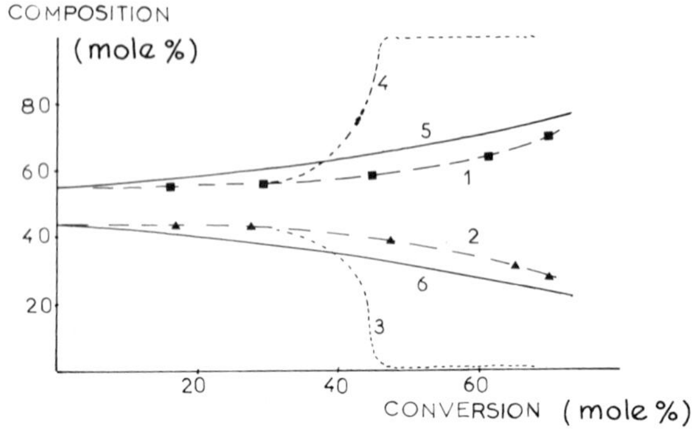

Figure 5. Average and instantaneous copolymer composition vs. conversion with Xp, $r_{AN\text{-}S} = 0.13$ and $r_{S\text{-}AN} = 0.34$. Overall values: (1) AN; (2) S; instantaneous values: (3) AN; (4) S, with Xm, $r_{AN\text{-}S} = 0.09$ and $r_{S\text{-}AN} = 0.44$; average values: (5) AN; (6) S.

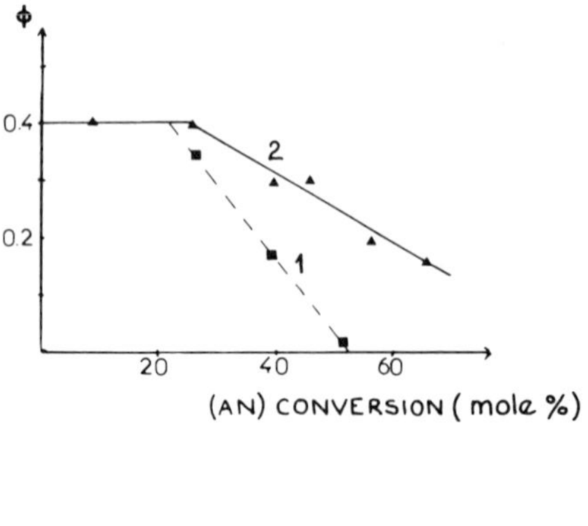

Figure 6. Monomer volume fraction (φ) vs. conversion: (1) batch; (2) corrected batch

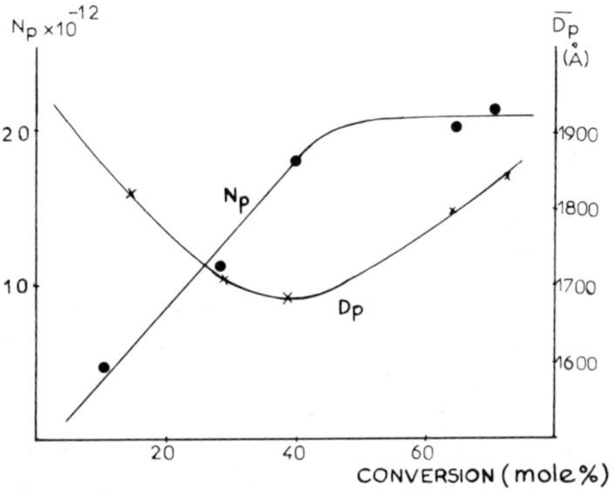

Figure 7. Number average particle size (Dp) vs. conversion (run P_2)

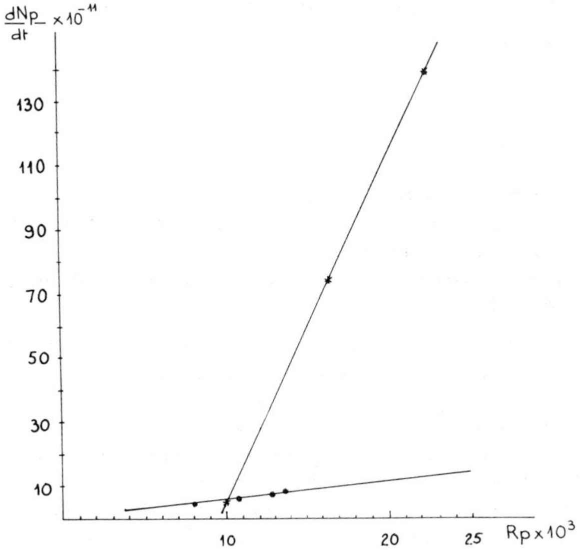

Figure 8. Rate of particle formation (dNp/dt) vs. polymerization rate at 20% conversion for various amounts of SDS (–●–) or monomer/water ratio (–×–)

It is quite interesting to see (figure 8) that in one of these series (varying emulsifier) a linear relationship is obeyed between the polymerization rate and the particle formation rate.

$$Rp = 1.46\ 10^{-14}\ \frac{dNp}{dt}$$

The same kind of law is valid for the second series but with a different coefficient

$$Rp = 0.9\ 10^{-15}\ \frac{dNp}{dt}$$

By increasing the initial amounts of monomer, particle size decreases and particle number, as well as the formation rate, increases drastically (Table III).

The final increase in particle size, shown in figure 7, is probably caused by limited flocculation,since particle coverage by emulsifier is very limited. Conductimetric titration of emulsifier shows that only a part of it is used for stabilizing particles.Typical results are shown in figure 9.

Curve 1 in figure 9 shows that SDS amount used for particle coverage is less than half the added emulsifier amount (4 g). Curve 2 shows that the actual particle coverage decreases as new particles are created. Data for curve 2 were calculated from Gardon (9) :

$$\Gamma = \frac{N_A(A_s)(SDS)}{NpDp_s^2Vem}$$

where $N_A.A_s = 1.25\ 10^5 Å^2$ per SDS g in 100 g water

Vem = volume of emulsifier

Dp_s = surface average diameter of particles

N_A = Avogadros number

A_s = area covered by one molecule SDS (value measured for polystyrene latex).

The flocculation process is more clearly observed in experiments with high monomer contents (2.7 moles) and rather moderate SDS concentration (4 g). As shown in figure 10, most of the emulsifier is used to stabilize the particles, when droplets are still present, and at the same time particle average size tends to decrease, while their number rapidly increases. At a point where practically all the emulsifier has been used, which possibly incidentally corresponds to disappearance of droplets, particle formation rate decreases to a lower value, particle size tends to increase, while a part of the emulsifier is desorbed.

Molecular weight remains rather limited. During an experiment, it does not change very much. It seems to increase a little with increasing emulsifier concentration (Mw from 36000 to 55000 when SDS varies from 2 to 8 g). As shown in table 2 it increases with amount of charged monomer (it seems to be linear). Polydispersity is found to be limited (Mw/Mn) to between 1.8 and 2. Molecular

Table III - Polymerization rate and particle formation rate

Run	Initial charge		20 % conversion				Final conversion		
	AN (moles)	S	$Np \times 10^{-14}$	$dNp/dt \times 10^{-12}$	$Rp \times 10^{3}$	Dp(Å)	$Np \times 10^{-14}$	Dp	Conversion(mole %)
P_2	0.8	0.2	0.082	0.62	10.6	1750	0.21	1830	58
R_1	1.6	0.4	4.8	7.5	16	720	9.1	650	70
R_2	2.14	0.53	3.6	14.0	22	750	6.6	760	82

Np : particles / cm^3 ; t : min. ; Rp : moles/min (whole reactor)

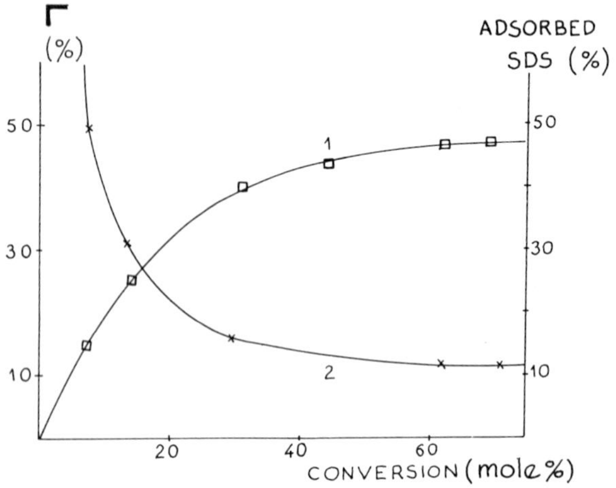

Figure 9. Amount of emulsifier adsorbed (1) on the particles and actual coverage (2) vs. conversion (run P_2)

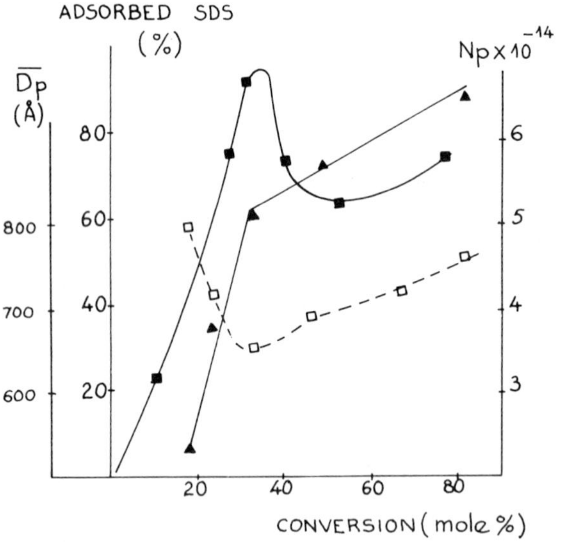

Figure 10. The % SDS adsorbed (–■–), average size (–□–), and number of particles (–▲–) at increasing conversion. Arrow shows the point where droplets disappear.

weight value is more probably governed by a transfer process onto mercaptan, so that the actual value should be dependent on mercaptan partition within the different phases.

Most of the above results are valid for the "corrected batch process" where styrene is added to the reactor in order to keep the AN/S ratio constant. Monomer distribution in the various phases, shown in figure 11, for an experiment with initial charge : 4 g SDS, AN : 0.8 mole, S : 0.2 mole and a total amount of 0.42 mole S added, is not drastically changed, except for styrene : its amount, which is kept constant in the water phase, decreases slightly in the particles after droplet disappearance, so that the AN/S ratio in the particles (Xp) tends to decrease : consequently, styrene content of the arising copolymer keeps increasing (Table IV).

Assuming copolymerization takes place chiefly in the particles, there is a good agreement between experimental average composition and calculated one using "solution" reactivity ratios (0.13 and 0.34).

As shown in figure 6, the monomer volume fraction in the particles follows a law similar to the case of the batch process, but due to the constant styrene addition, larger amounts of monomer (both acrylonitrile and styrene) are still being absorbed into the particles. Despite a large acrylonitrile amount remaining in water phase, higher conversion may be reached. Continuous formation of particles takes place again and a close parallelism between polymerization rate and nucleation rate is again observed. However, average particle size (figure 12), as well as molecular weight (Table IV), is kept constant.

Another difference lies in the value of average particle size (Table V). At low monomer contents (run P_1 and M_1) the size is smaller in the corrected batch process, but the reverse is true at high monomer contents (run R_1 and M_2) where the size also remains constant in the corrected batch process (or tends to decrease slightly from 2050 to 1950 Å). Although emulsifier amount has not been changed, no evidence for flocculation can be observed: the SDS amount used for coverage continuously increases, but coverage itself (Γ value) remains at values not higher than values shown in figure 9.

A final remark concerns molecular weight, which increases with amount of monomer charged, and is smaller in the case of corrected batch process than in batch process (table V).

For comparison, a few experiments were carried out using the semi-continuous process. In a typical experiment (N_2) corresponding to run P_2, a fourth of the monomer feed (AN : 0.19 mole, S : 0.05 mole) was initially charged with the total amount of water, emulsifier (4 g) and initiator system. After a constant rate (Rp) period has been reached (droplet disappearance), then the monomer mixture was added at the following rate : Ra = 0.75 Rp. Monomer distribution in various phases is shown in figure 13, while data concerning particle size and number is illustrated in figure 14 and data for coverage by emulsifier are given in figure 15. Final molecular weight is 57000 but polydispersity is higher (Mw/Mn= 4.7).

Table IV - Instantaneous and average copolymer composition versus conversion (run M_1)

$Xp = \frac{AN}{Sp}$	Instantaneous composition (mole %)		Average composition (mole %)				AN conversion (mole %)	Mw 10^3
			Calculated[(+)]		Experimental[(o)]			
	AN	S	AN	S	AN	S		
3.00	55	45	55	45	54	46	0	
2.58	54	44	54.5	44.5	54	46	30	110
1.74	50.6	49.4	53.0	47.0	53	47	53	110
1.40	48.7	51.3	51.0	49.0	52	48	66	110

(+) calculated with $r_{AN-S} = 0.13$ and $r_{S-AN} = 0.34$

(o) from the kinetic study

Table V - Comparison of batch and corrected batch process

Run	Type	SDS(g)	Monomer/Water	Average particle size (Å)		Particle number x 10^{13}		$\frac{Mw}{10^3}$
				20 % conversion	Final conversion	20 %	Final	
P_1	Batch	2	0.1	1800	1680	0.9	2.2	60
M_1	Corrected	2	0.1	1090	1090	1.1	3	110
R_1	Batch	4	0.2	720	650	48	91	161
M_2	Corrected	4	0.2	2100	2000	1.1	4	130

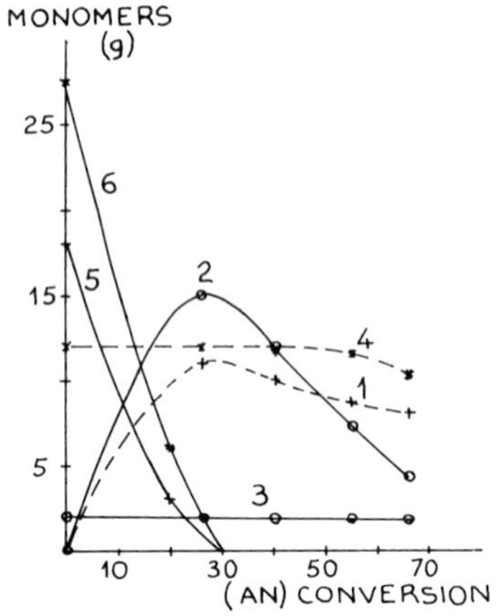

Figure 11. Corrected batch process—quantities of monomers in the different phases (in g): in particles: S_p (1); AN_p (2); in water: S_W (3); AN_W (4); in droplets: S_D (5); AN_D (6)

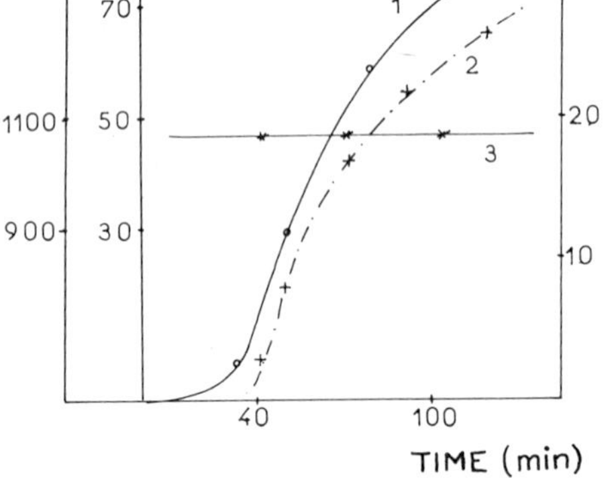

Figure 12. Corrected batch process—total particle number (1); AN conversion (2); particle size (3) vs. time

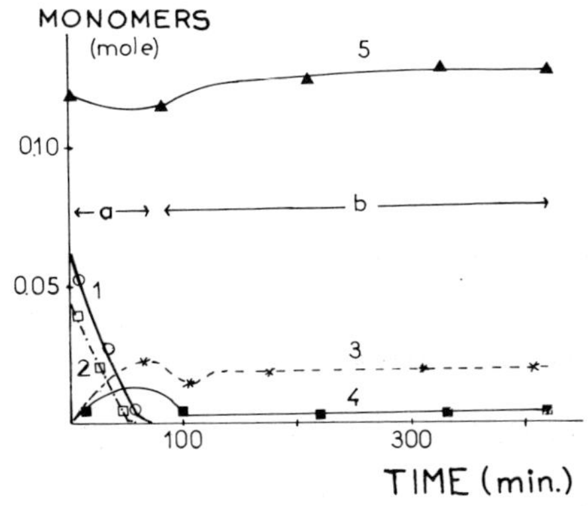

Figure 13. Unreacted monomers vs. time (run N_2)

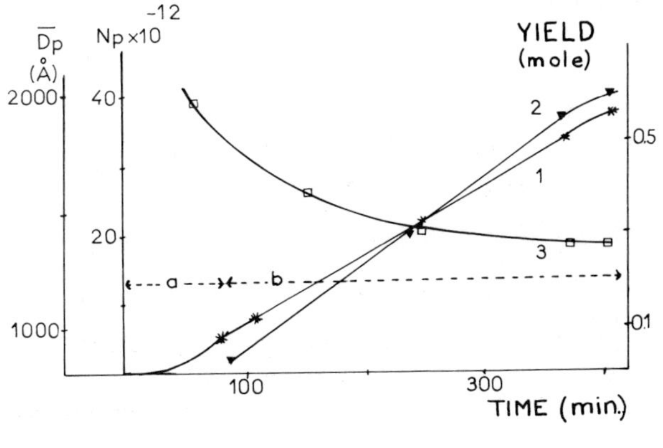

Figure 14. Conversion (1), total number (2), and average particle size (3) vs. time

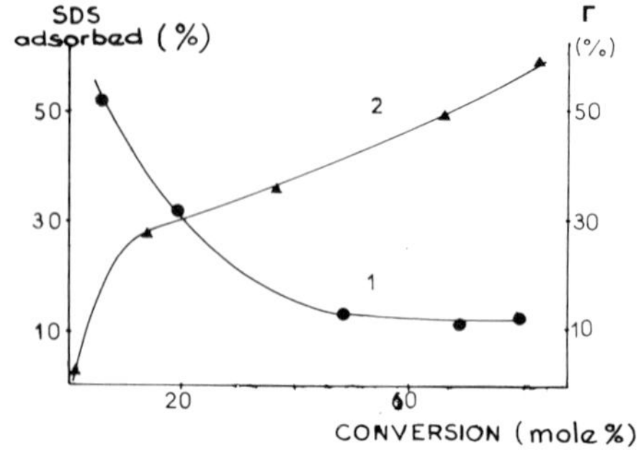

Figure 15. Coverage (1) and amount of adsorbed emulsifier on the particles (2) vs. conversion (run N_2)

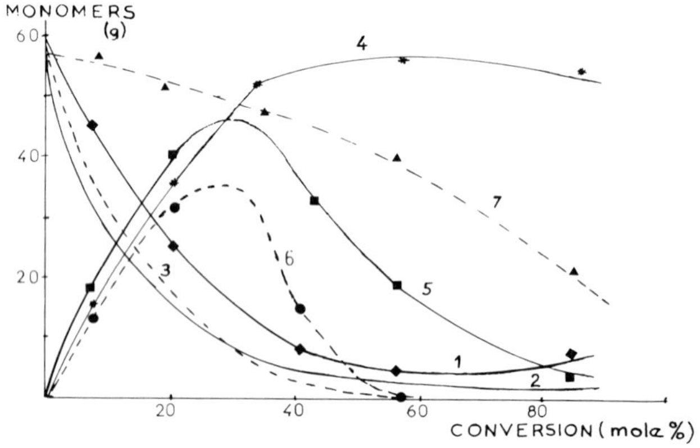

Figure 16. Monomer distribution in the different phases vs. conversion: droplets: (1) T; (2) AN; (3) S, water: (7) AN; particles: (4) T; (5) AN; (6) S, run Q_2

Constant composition copolymer may be produced in this process, but particle size changes during the process. Particle number increases continuously, again following polymer yield. Although emulsifier tends to be more completely used for covering particle, coverage itself remains weak.

Discussion

The study of monomer distribution among various phases, and the analysis of copolymer composition clearly shows that copolymerization takes place mainly within particles. Taking this fact into account, then it is easy to understand why the "corrected batch process" leads to a continuous drift of copolymer composition (with continuous enrichment in styrene) instead of expected constant copolymer composition. The drift is obviously caused by a large AN amount which remains dissolved in water phase and cannot be polymerized. Consequently that drift is less important if the total amount of monomer is higher because although a larger AN amount is then dissolved in the water phase its relative importance is smaller. So, in that case, the semi-continuous process remains the best way to get constant copolymer composition. In the batch process, polymerization inside the particles stops when most of the styrene has been consumed and it seems further that AN solubility inside particles is increased by the presence of styrene ; so, when styrene is continuously added in the "corrected batch process", a higher AN conversion may be reached. This point was checked through another set of experiments in batch, where 60 ml toluene was added to styrene. Toluene is distributed among the three phases, its solubility is close to that of styrene, but it is not consumed. Results of one experiment with high monomer contents (monomer/water ratio : 0.3) are shown in figure 16 for monomer and toluene distribution, and in figure 17 for conversion, number and size of particles. In such experiments droplets never disappear, high AN conversion may be reached even after all styrene has been consumed, and only at the end of the process, when acrylonitrile homopolymer is formed, some toluene is rejected from particles to droplets. The latter trend is more clearly visible in experiments with low monomer contents. Due to the presence of acrylonitrile homopolymer, the product is no longer totally soluble in THF, so that molecular weight was not studied.

A thermodynamic approach was put forward by one of us (10), based on the Flory-Huggins lattice theory of a polymer solution ; the chemical potentials of each monomer must be equal in each phase ; copolymerization increment causes a little change in the chemical potential in the particles : diffusion of monomers from the water phase will reequilibrate the system and in turn diffusion from droplets to water phase takes place. For instance, expression from monomer 1 in the particles is :

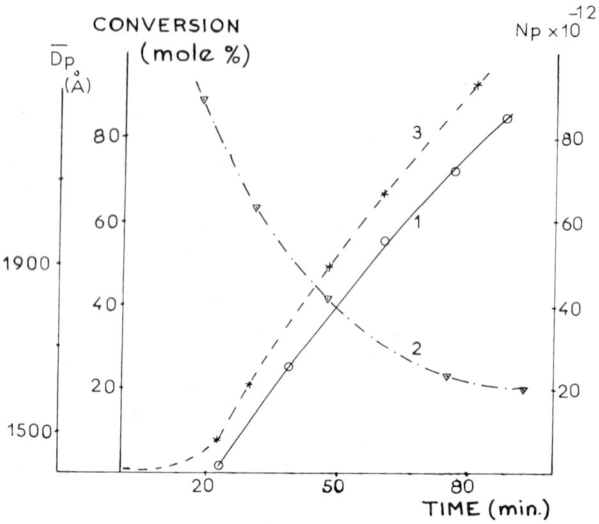

Figure 17. Total particle number (1), average particle size (2), and conversion vs. time (Q_2 experiment)

$$\mu_1^p = \mu_{01}^p + RT\left[\ln\phi_1 + (1-\phi_1) - \phi_2\frac{m_1}{m_2} - \phi_p\frac{m_1}{m_p} + (\chi_{12}\phi_2 + \chi_{1p}\phi_p)(\phi_2 + \phi_p) - \chi_{2p}\frac{m_1}{m_2}\phi_2\phi_p\right] + 2\frac{V_m\gamma}{r_o}\phi_p^{1/3}$$

where ϕ_1, ϕ_2 and ϕ_p are the volume fraction of monomer 1 and 2 and polymer inside particles

m_1 and m_2 the monomer weights, m_p the polymer weight

χ_{12}, χ_{1p}, χ_{2p} the Huggins interaction parameters

γ the interfacial energy between water and particles

r_o the radius of unswollen particles

V_m^o the molar volume of liquid inside particles

Similar but simpler expressions are valid for droplets or water phase.

Useful parameters are taken from the literature or calculated from the Hildebrand solubility parameters.

Polymerization causes a change in ϕ values, from which variation may be calculated. Differential equations describing monomer diffusion from one phase to the other can be integrated numerically through a computer program. Curves calculated in that way for monomer concentration in each phase are in good agreement with experiments. The calculation confirms that the presence of styrene favors AN solubility in the particles, so that at the end of the polymerization, some AN is rejected in water phase. The fact that copolymer composition does not change very much whenever droplets are present is also explained on this basis. However, it must be noted that the chemical potential of AN in the water phase is dependent on the emulsifier concentration : experimental determination of AN concentration in water allows to make the corresponding correction of the chemical potential. Anyway, the thermodynamic approach just described will be a useful tool for predicting how the copolymerization proceeds. The computer program allows an excellent simulation of the process and might be further used for monitoring the production of an actual constant composition copolymer.

Results concerning particle number and size are more puzzling. One fact is very clear : the Harkins-Smith-Ewart theory of emulsion polymerization is not obeyed by that system, since new particles are continuously created all over the process. Further, the polymerization rate follows very precisely particle formation rate. So, it seems that most of the particles are inactive at a given time. Polymerization probably takes place in a small number of active particles that very rapidly grow up to the limiting size and then stops. At the end of the process, some growth in the average size is observed in some cases, mostly in the batch process. Most probably, that growth process is the result of a limited floculation. It is interesting to note that similar results were observed for acrylonitrile homopolymerization by Morris and Parts (11): continuous formation of particles of rather constant size, with a trend to growing at the end of polymerization process.

Considering that polymerization takes place in particles, as shown by results concerning copolymer composition, these results seem surprising. The possibility of continuous nucleation is quite normal and explained through the Fitch theory (12) but questionable points are :

a) why does particle growth stop ?
b) which parameters are controlling the final particle size ?

It must be admitted that the monomer present in the inactive particles must leave them and again diffuse through the water phase to allow active particle growth : the inactive particles will act as a reservoir exactly as droplets do, such a fact might account for the rate not decreasing soon after droplet disappearance.

It must also be noted that, in these systems with high acrylonitrile contents, SDS is not efficient at all for covering particles. A large part of emulsifier remains in the water solution, where it tends to increase monomer solubility so emulsifier probably does not play a major role in particle stabilization. Its major role seems to make nucleation of new particles easier. Tentatively, it might be suggested that strong acid groups coming from the initiator, together with high polar nitrile groups, are able to inhibit not only the flocculation process, which remains limited, but also the capture of new radicals (primary or oligomeric) born in water phase.

A final remark is to underline the efficiency of the "corrected batch process" to regulate not only the copolymer composition, but also the molecular weight and more importantly particle size.

LITERATURE CITED

1. Wessling R.A. J. Applied Polym. Sci., 1968, 12, 309
2. Snuparek J. and Krska F., J. Applied Polym. Sci., 1976, 20, 1753
3. Chiang T.C., Graillat C., Guillot J., Pham Q.T. and Guyot A., J. Polymer Sci.(Polymer Chem. Ed.), 1977, 15, 2961
4. Rios L. Pichot C. and Guillot J., Makromol. Chem., 1980, 181, 677
5. Kelen T. and Tüdos F., J. Macromol. Sci.(Chem.), 1975, A9, 1
6. Hatate Y., Hano T., Miyata T, Nakashio F. ans Sakaki W., Kagaku Kogaku, 1971, 35 309 ; Chem. Abst., 1972, 76 15019f
7. Pichot C. Zaganiaris E. and Guyot A., J. Polym. Sci., Polym. Symp., 1975, 52 55
8. Cadle R.D., "Particle Size : theory and industrial applications" Reinhold, New York, 1965
9. Gardon J.L. in "Polymerization process" High Polymer Series, vol. XXIX, J. Wiley, 1977, chap. 6, 143
10. Guillot,J. Makromol. Chem. Rapid Comm., 1980, 1 697
11. Morris C.E.M. and Parts A.G., Makromol. Chem., 1976, 177 1433
12. Fitch R.M., Brit. Polym. J., 1973, 5 467

RECEIVED April 6, 1981.

27

Theory of Compartmentalized Free-Radical Polymerization Reactions

D. C. BLACKLEY

National College of Rubber Technology, The Polytechnic of North London, Holloway, London N7 8DB, England

The purpose of this paper is to summarise some of the advances which we have made in recent years in developing the theory of compartmentalised free-radical polymerisation reactions. These developments either have been, or will be, published and discussed in detail elsewhere. By the term "compartmentalised free-radical polymerisation reaction" we mean a free-radical polymerisation which is taking place within a large number of separate reaction loci. These loci are dispersed in a contiguous external phase. The details of the reaction model will be given subsequently, but it is noted here that whilst, in principle, the free radicals which initiate the polymerisation may be generated either within the external phase or within the reation loci, in all the cases to be considered in this paper the new radicals are assumed to be generated exclusively with the external phase. The reaction loci then acquire the radicals by absorption from the external phase.

The theory of compartmentalised free-radical polymerisation reactions of the type considered in this paper is of interest primarily because it is believed that most of the polymer which is formed in the course of an emulsion polymerisation reaction is formed by way of reactions of this type. The objective of the theory is to calculate the relative proportions of the reaction loci which at any instant contain 0, 1, 2, ..., r, ... propagating radicals, and also such properties of the locus population distribution as the average number of propagating radicals per reaction locus, and the variance of the distribution of locus populations. It is then a straightforward matter to write down an expression for the overall rate of polymerisation in the reaction system, once an expression has been obtained for the average number of propagating radicals per reaction locus.

The problem of the distribution of locus populations in reaction systems which have reached a steady state was completely solved several years ago by Stockmayer (1) and O'Toole (2). Our concern has been with the behaviour of reaction systems in a non-steady state, as they approach the steady state. The objective is to derive expressions which show the way in which the distribution of locus populations changes as the reaction proceeds towards the

0097-6156/81/0165-0437$05.00/0

steady state. Not only is the theory intrinsically interesting in itself; it is also of considerable additional interest because measurements of conversion as a function of time during the non-steady state offer the possibility of access to certain of the fundamental properties of reaction systems which are not otherwise available.

Reaction model assumed

The reaction model assumed is one in which free-radical polymerisation is compartmentalised within a fixed number of reaction loci, all of which have similar volumes. As has been pointed out above, new radicals are generated in the external phase only. No nucleation of new reaction loci occurs as polymerisation proceeds, and the number of loci is not reduced by processes such as particle agglomeration. Radicals enter reaction loci from the external phase at a constant rate (which in certain cases may be zero), and thus the rate of acquisition of radicals by a single locus is kinetically of zero order with respect to the concentration of radicals within the locus. Once a radical enters a reaction locus, it initiates a chain polymerisation reaction which continues until the activity of the radical within the locus is lost. Polymerisation is assumed to occur almost exclusively within the reaction loci, because the solubility of the monomer in the external phase is assumed to be low. The volumes of the reaction loci are presumed not to increase greatly as a consequence of polymerisation. Two classes of mechanism are in general available whereby the activity of radicals can be lost from reaction loci:

(i) Processes which are kinetically of first order with respect to the concentration of radicals within the reaction locus. These processes include exit from the locus into the external phase, termination by reaction with monomer within the locus, termination by reaction with adventitious impurities in the locus and spontaneous deactivation.

(ii) Processes which are kinetically of second order with respect to the concentration of radicals within the reaction locus. The most important of these processes is bimolecular mutual termination between pairs of propagating radicals within the same reaction locus. Radicals which are lost to the external phase by exit from the reaction loci are assumed to be not available for re-entry into the loci, or for the reinitiation of propagation. Insofar as it is desired to make predictions of the overall rate of polymerisation in the reaction system, the concentration of monomer within the reaction loci is assumed to be constant throughout the reaction. This implies that monomer droplets are present as a separate phase throughout the reaction, and that the rate of transfer of monomer to the loci from the droplets is fast relative to the rate of consumption of monomer in the loci by polymerisation.

The time-dependent Smith-Ewart differential difference equations; methods available for their solution

The fundamental equations which govern the behaviour of the

reaction system in the non-steady state are an infinite set of linear differential difference equations in which the variables are the populations of the various types of reaction loci classified with respect to radical occupancy, i.e., with respect to the number of propagating radicals which are present in a particular reaction locus. This set of equations is obtained by modifying the recurrence relationship of Smith and Ewart (3) in such a way as to allow for the possibility that the populations for the different classes of reaction locus are not necessarily stationary. The result is the following set of equations for i = 0, 1, 2, ..., with $n_{-1} = 0$:

$$\frac{dn_i}{dt} = (n_{i-1} - n_i)\sigma + \{(i+1)n_{i+1} - in_i\}k + \{(i+2)(i+1)n_{i+2} - i(i-1)n_i\}\frac{k_t}{v} \quad (1)$$

In this set of equations, n_i is the number of reaction loci per arbitrary volume of reaction system which contain i propagating radicals, σ is the average rate of entry of radicals into a single reaction locus, v is the volume of the reaction locus, k_t is the rate coefficient for the mutual termination of radicals, and k is a composite constant which quantifies the rate at which radicals are lost from reaction loci by first-order processes. For convenience we put $k_t/v = \chi$.

The eqns (1) can be derived by considering the rates of gain and loss of loci of class i as a consequence of processes of radical acquisition and radical loss. In these considerations, it is necessary to take into account the rates of certain of the transitions between the classes of loci containing i-2, i-1, i, i+1, and i+2 radicals, as indicated in Figure 1. The first term on the right hand side of the typical equation of the set (1) then arises as the difference between the rate at which loci of class i are formed from loci of class i-1 by acquisition of a radical, and the rate at which loci of class i are lost by acquisition of a radical, thereby causing them to become loci of class i+1. Similarly, the second term on the right-hand side arises from the balance between the rates of gain and loss of loci of class i by loss from, and gain to, loci of classes i+1 and i respectively by first-order processes The third term is the net rate of formation of loci of class i attributable to the loss of radicals from reaction loci by bimolecular mutual termination within loci. An equation which is equivalent to the typical equation of the set (1) can also be derived by considering the rates of transition of locus populations across a notional barrier situated between two neighbouring states of radical occupancy, as indicated in Figure 2.

Two approaches have been used to solve the set of eqns (1) generally:

(i) Napper and his co-workers (4-7) have re-cast the equations in matrix form, and then have sought to obtain approximate solutions of the consequent matrix equations after suitable truncation.

(ii) We have found it more profitable to transform the entire set of these equations into a single equation by introducing a locus-population generating function, $\Psi(\xi, t)$, defined as

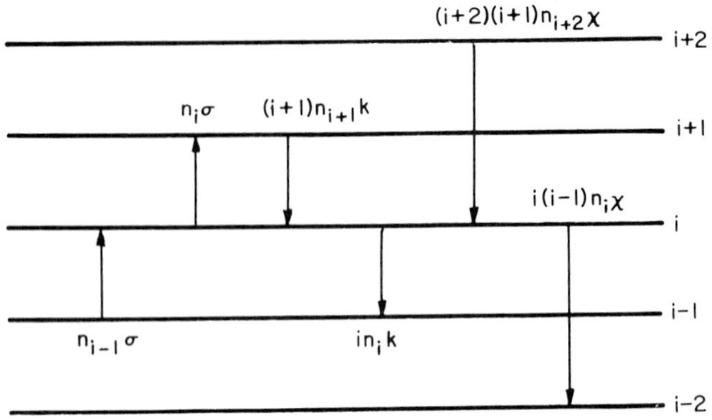

Figure 1. Transitions for derivation of time-dependent Smith–Ewart differential difference equations ($\chi = k_t/v$)

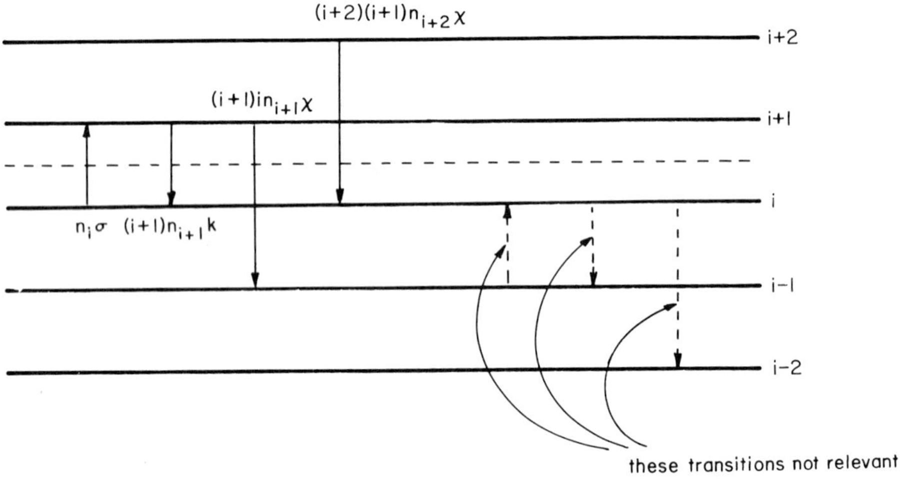

Figure 2. Transitions for derivation of equivalent equation

$$\Psi(\xi,t) = \sum_{i=0}^{\infty} n_i(t)\,\xi^i \qquad (2)$$

where ξ is an auxiliary variable. In this respect, we have generalised the approach of Stockmayer (1) and O'Toole (2), who used this approach in obtaining a complete solution to the steady-state (in which case, of course, the function Ψ does not contain t).

Attempts have also been made to provide approximate solutions by truncating the infinite set of differential difference equations to a small set which is then amenable to solution by standard methods for simultaneous linear differential equations. The most usual truncations which have been adopted are those which correspond to two- and three-state models in which each reaction locus can contain at most either one or two propagating radicals respectively.

In order to convert the set of differential difference equations (1) into a single differential equation with Ψ as the dependent variable, each equation for dn_i/dt is multiplied by ξ^i, and then all the equations so obtained are summed. If it is then noted that

$$\left.\begin{aligned} &\Sigma n_i \xi^i = \Psi && \Sigma i n_i \xi^i = \xi\,\partial\Psi/\partial\xi \\ &\Sigma n_{i-1}\xi^i = \xi\Psi && \Sigma(i+2)(i+1)n_{i+2}\xi^i = \partial^2\Psi/\partial\xi^2 \\ &\Sigma(i+1)n_{i+1}\xi^i = \partial\Psi/\partial\xi && \Sigma i(i-1)n_i\xi^i = \xi^2\partial^2\Psi/\partial\xi^2 \end{aligned}\right\} \qquad (3)$$

where in each case the summations are over all possible values of i, then the resultant single differential equation can be transformed into

$$\frac{\partial\Psi}{\partial t} = \sigma(\xi-1)\Psi + k(1-\xi)\frac{\partial\Psi}{\partial\xi} + \chi(1-\xi^2)\frac{\partial^2\Psi}{\partial\xi^2} \qquad (4)$$

It may be noted that Stockmayer's differential equation for his generating function follows immediately as a special case of eqn (4) by putting $\partial\Psi/\partial t = 0$ (corresponding, of course, to the steady state). The result is

$$\sigma\Psi - k\frac{d\Psi}{d\xi} - \chi(1+\xi)\frac{d^2\Psi}{d\xi^2} = 0 \qquad (5)$$

which is identical with eqn (4) of Stockmayer's paper, apart from differences of notation.

Having solved eqn (4) for Ψ, taking into account the relevant boundary conditions for the particular problem for which a solution is sought, the $n_r(t)$ can then readily be found from as follows:

$$n_r(t) = \frac{1}{r!}\left(\frac{\partial^r\Psi}{\partial\xi^r}\right)_{\xi=0} \qquad (6)$$

The average number of propagating radicals per reaction locus can be found from

$$\bar{\iota}(t) = \frac{(\partial\Psi/\partial\xi)_{\xi=1}}{\Psi(1,t)} \qquad (7)$$

Solution for case where radical loss is predominantly by first-order mechanism (8, 9, 10)

For this case we put $\chi = 0$. It is then possible to obtain an explicit solution to eqn (4). For the case where the boundary conditions are

$$\left.\begin{array}{l}\text{(i) } n_0(0) = N \\ \text{(ii) } n_1(0) = n_2(0) = \ldots = 0\end{array}\right\} \quad (8)$$

where N is the total number of reaction loci in the arbitrary volume of reaction system, the solution for Ψ is

$$\Psi(\xi,t) = N \exp\left\{\frac{\sigma}{k}(\xi-1)(1-e^{-kt})\right\} \quad (9)$$

This result shows that the distribution of locus populations with respect to radical occupancy is always Poissonian, i.e., the $n_r(t)$ at any instant t always form a Poisson distribution with respect to the r. The nature of the distribution is, however, time-dependent, in that the parameter of the distribution at any instant is equal to $(\sigma/k)(1-e^{-kt})$. The expressions which we obtain for $n_r(t)$ and $\bar{i}(t)$ are as follows:

$$n_r(t) = \frac{N}{r!}\left\{\frac{\sigma}{k}(1-e^{-kt})\right\}^r \exp\left\{-\frac{\sigma}{k}(1-e^{-kt})\right\} \quad (10)$$

$$\bar{i}(t) = \frac{\sigma}{k}(1-e^{-kt}) \quad (11)$$

Typical curves showing $n_r(t)/N$ as a function of t for r = 0,1,2 are given in Figure 3. The values chosen for the parameters σ and k are 1×10^{-5} sec^{-1} and 5×10^{-4} sec^{-1} respectively.

We have also found it possible to generalise this case to include reaction systems for which the parameters σ and k are time dependent. To emphasise the time-dependence of σ and k, they are written as $\sigma(t)$ and $k(t)$ respectively. We have found that the solution to eqn (4) always has the form

$$\Psi(\xi,t) = N e^{(\xi-1)\theta(t)} \quad (12)$$

where $\theta(t)$ is a function of time which satisfies the ordinary differential equation

$$\frac{d\theta(t)}{dt} = \sigma(t) - k(t).\theta(t) \quad (13)$$

and is subject to the initial condition $\theta(0) = 0$ for reaction systems for which the boundary conditions are as stated in eqns (8). This result for Ψ shows that the distribution of locus populations with respect to radical occupancy is at all times Poissonian, notwithstanding the time-dependence of σ and k, and that the parameter of the distribution is always equal to $\theta(t)$. It therefore follows that $\bar{i}(t)$ is always equal to $\theta(t)$.

Of the various particular cases of the variation of σ and k with t for which solutions have been obtained, probably the most important is that where σ decays exponentially with time and k is constant. This case is of considerable interest because the decomposition of most dissociative initiators is kinetically of first order with respect to initiator concentration, and therefore both the initiator concentration and the rate of decomposition of the initiator will decay exponentially with time. Putting $\sigma(t) = \sigma_0 e^{-\alpha t}$, we find that $\theta(t)$ is given by

$$\theta(t) = \frac{\sigma_0 e^{-\alpha t}}{k - \alpha}\left(1 - e^{-(k-\alpha)t}\right) \tag{14}$$

except for the unlikely special case where $\alpha = k$, in which case the result for $\theta(t)$ is

$$\theta(t) = \sigma_0 t e^{-\alpha t} \tag{15}$$

Examples of the variation of $\theta(t)$ with t for the case of exponential decay of σ with t, and k constant, are given in Figure 4. In constructing the curves of Figure 4, the values chosen for σ_0 and k were 1×10^{-5} sec^{-1} and 5×10^{-4} sec^{-1} respectively. Curves are given for several values of α ranging from 0 to 5×10^{-3} sec^{-1}. It appears that the exponential decay of σ has little effect upon the variation of $\theta(t)$ with t provided that $\alpha/k <$ ca.10^{-2}. But in all cases the effect of a non-zero α is to cause $\theta(t)$ to rise to a maximum and then eventually to decay to zero. This implies that eventually all the $n_r(t)$ for $r>0$ will decay to zero.

Other particular cases of the variation of σ and k with t for which solutions have been obtained are as follows:

(i) the case where σ decays linearly with t, and k is constant;
(ii) the case where σ decays as a consequence of second-order depletion of initiator, and k is constant;
(iii) the case where σ is constant but k varies because the loci grow at a constant rate; and
(iv) the case where σ varies because radicals lost to the external phase are available for re-initiation, and k is constant.

<u>Solution for case where generation of new radicals ceases</u> (11)

We have also been able to obtain an explicit analytic solution to eqn (4), and hence to the general time-dependent Smith-Ewart differential difference equations, for the case where the rate of formation of new radicals in the external phase is zero, i.e., $\sigma = 0$. Of course, if no radicals ever have been generated within the external phase of the reaction system, then the problem becomes trivial and admits of an obvious and simple solution, namely, that all loci are at all times devoid of propagating radicals, and the rate of polymerisation is always zero. This solution is clearly of no interest. The case which is of interest is that of a reaction system in which radicals have been generated within the external phase, so that a certain rate of polymerisa-

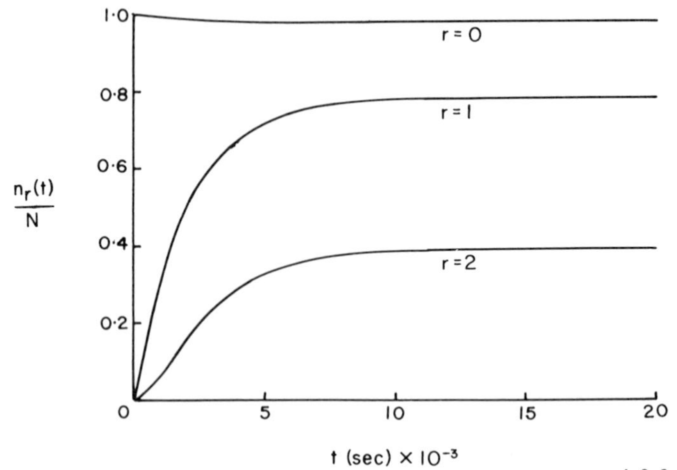

J. C. S. Faraday I

Figure 3. Fractional locus populations, $n_r(t)/N$, *as functions of time* t *for* $r = 0$, *1, and 2, for reaction system for which radical loss from reaction loci is exclusively by first-order processes (8). Values taken for* σ *and* k *are* $1 \times 10^{-5}\ s^{-1}$ *and* $5 \times 10^{-4}\ s^{-1}$*, respectively. The ordinates for* $r = 0$ *are* $n_0(t)/N$; *those for* $r = 1$ *are 40* $n_1(t)/N$; *those for* $r = 2$ *are 2000* $n_2(t)/N$.

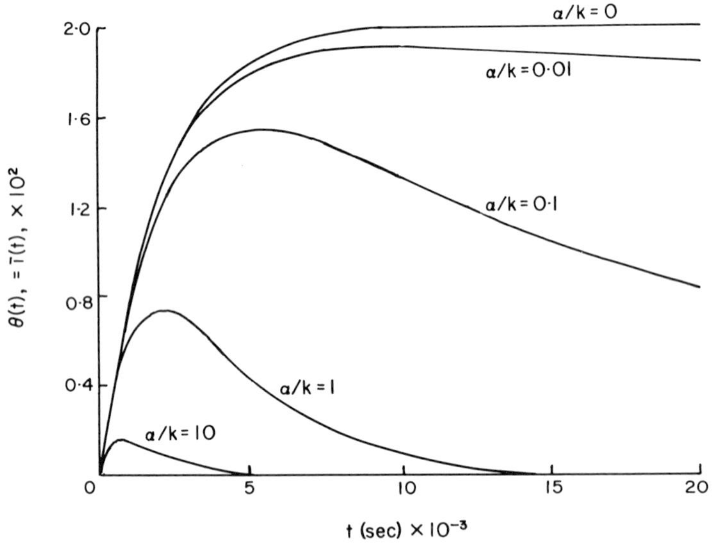

Figure 4. Locus–population distribution parameter $\theta(t)$ *and average number of radicals per locus* $\bar{i}(t)$ *as functions of time* t *for case where* $\sigma = \sigma_0 l^{-\alpha t}$ *and* k *is constant (10). Values taken for* σ_0 *and* k *are* $1 \times 10^{-5}\ s^{-1}$ *and* $5 \times 10^{-4}\ s^{-1}$*, respectively. Curves are given for various values of* α*, indicated by the* α/k *ratios appended to the curves.*

tion has developed; then the rate of generation of new radicals is suddenly reduced to zero, so that the reaction rate decays spontaneously. It is with the characteristics of the reaction during this period of decay, following the cessation of the generation of new radicals, that this aspect of the theory of compartmentalised free-radical polymerisation reactions is concerned. Possible ways in which the rate of generation of new radicals might be reduced to zero include (a) release of a radical scavenger into the external phase of the reaction system, and (b) reducing the intensity of radiation to zero in the case of a radiation-initiated reaction.

It is convenient to represent the ratio k/χ by the symbol m. It is then found necessary to consider two separate cases, namely, the general case for which $m \neq 0$, and the special case for which $m = 0$. The former corresponds to a reaction system in which radicals can be lost from reaction loci by first-order processes; the latter effectively corresponds to a reaction system in which radicals cannot be lost from reaction loci by first-order processes, although theoretically the case $m = 0$ also includes reaction systems for which χ is truly infinite (and therefore k_t is truly infinite) and k may or may not be zero. For the case $m \neq 0$, the behaviour of the reaction system during the decay period is characterised by the locus-population generating function

$$\Psi(\xi, t) = N \sum_{p=0}^{\infty} B_p J_p^{(-1, m-1)}(\xi)\, e^{-t/\tau_p} \tag{16}$$

where $$\tau_p = \{\chi p (p + m - 1)\}^{-1}, \qquad p = 0, 1, 2, \ldots \tag{17}$$

The J_p in eqn (16) are Jacobi polynomials, t is time measured from the instant at which the generation of new radicals ceased, and the B_p are coefficients whose values are determined by the requirement that $\Psi(\xi, t)$ at $t = 0$ must have a particular form $\Psi(\xi, 0)$. This particular form for $\Psi(\xi, 0)$ is in turn determined by the distribution of locus populations in the reaction system at the instant when the generation of new radicals cased. The general results obtained for $n_r(t)$ and $\bar{\iota}(t)$ are

$$n_r(t) = \frac{N}{2^r r!} \sum_{p=r}^{\infty} B_p (p + m - 1)_r\, J_{p-r}^{(r-1, m+r-1)}(0)\, e^{-t/\tau_p} \tag{18}$$

and

$$\bar{\iota}(t) = \frac{1}{2} \sum_{p=1}^{\infty} B_p (p + m - 1)\, e^{-t/\tau_p} \tag{19}$$

where a symbol of the type $(u)_\nu$ is a Pochhammer symbol denoting the function $\Gamma(u + \nu)/\Gamma(u)$.

Eqn (16) shows that the nature of the decay of the reaction following the cessation of the generation of new radicals is determined by, inter alia, a set of characteristic "relaxation times", τ_p , the number of which is in general infinite. Eqn (16) also shows that, whilst it is possible to obtain a perfectly general expression for the locus-population generating function throughout the decay period, an explicit solution for any particular reaction system can be obtained only if information is available concerning the nature of the distribution of locus populations which existed at the instant when the rate of generation of new radicals in the external phase was reduced to zero. More precisely, it is necessary to have this information in the form of the generating function for the distribution of locus populations which existed at that instant.

We have applied our general solution to obtain predictions for the decay of the reaction from the following three types of distribution of locus populations:

(i) a distribution of the Stockmayer-O'Toole type;
(ii) a Poisson distribution; and
(iii) a homogeneous distribution, i.e., a distribution in which all the reaction loci contain the same number of propagating radicals.

If the initial distribution of locus populations is of the Stockmayer-O'Toole type, then $\Psi(\xi,0)$ is given by

$$\Psi(\xi,0) = \frac{2^{(m-1)/2} N}{I_{m-1}(h)} (1+\xi)^{(1-m)/2} I_{m-1}\left(h\sqrt{\frac{1+\xi}{2}}\right) \quad (20)$$

where $h^2 = 8v\sigma/k_t$, σ being the average rate of entry of radicals into a single locus before the rate of generation of new radicals was reduced to zero, and I_η denotes the modified Bessel function of the first kind of order η. The result obtained for the coefficients B_p for this case is

$$B_p = \frac{2p+m-1}{p+m-1} \cdot \frac{I_{2p+m-1}(h)}{I_{m-1}(h)} \quad (21)$$

The case m = 0 requires separate treatment, because the argument which leads to the general result embodied in eqn (16) is invalid if m = 0. As has been pointed out above, m = 0 effectively implies that it is not possible for radicals to be lost from reaction loci by first-order processes. This in turn implies that the only processes by which the radical occupancy of a locus can change are second-order processes such as bimolecular mutural termination. Thus if the radical occupancy is to change, it must be by increments of -2. The mathematical complexities of this case appear to arise from these physical considerations. The general expression for $\Psi(\xi,t)$ obtained for this case is

$$\Psi(\xi,t) = N\sum_{p=0}^{\infty} K_p C_p^{(-1/2)}(\xi)\, e^{-t/\tau_p} \quad (22)$$

where $C_p^{(-1/2)}(\xi)$ is a Gegenbauer polynomial of order p and parameter $-1/2$, the τ_p are given by eqn (17) with m = 0, and the K_p are coefficients which are determined by the nature of $\Psi(\xi,0)$. The general results obtained for $n_r(t)$ and $\bar{\tau}(t)$ for this case are

$$n_r(t) = \frac{2^r N}{r!} \sum_{p=r}^{\infty} K_p (-1)^{(p-r)/2} \frac{\Gamma\left(\frac{p+r-1}{2}\right)}{\Gamma(r-\frac{1}{2})\,\Gamma\left(\frac{p-r}{2}+1\right)} e^{-t/\tau_p} \tag{23}$$

where the summation extends over all values of $p \geq r$ such that $p-r$ is even; and

$$\bar{\iota}(t) = -\sum_{p=1}^{\infty} K_p e^{-t/\tau_p} \tag{24}$$

For the case where the initial distribution of locus populations is of the Stockmayer-O'Toole type, $\Psi(\xi,0)$ being given by eqn (20), the coefficients K_p are found to be given by

$$K_p = -(p-\tfrac{1}{2}) \frac{I_{2p-1}(h)}{I_{-1}(h)} \tag{25}$$

Approximate "Poissonian" solution to the general case (12)

Although we have so far been unable to obtain a completely general explicit analytic solution to eqn (4)(and therefore to the set (1)), we have recently obtained an explicit analytic solution to a modified form of eqn (4), namely,

$$\frac{\partial \Psi}{\partial t} = \sigma(\xi-1)\Psi + k(1-\xi)\frac{\partial \Psi}{\partial \xi} + \chi(1-\xi)\frac{\partial^2 \Psi}{\partial \xi^2} \tag{26}$$

for reaction systems which are initially devoid of radicals, and in the external phase of which radicals suddenly begin to be generated at a constant rate. This equation can be obtained from a modified set of time-dependent Smith-Ewart differential difference equations in which the coefficient of k_t/v in the finalterm on the right-hand side of the set (1) is replaced by $\{(i+2)(i+1)n_{i+2} - (i+1)i\,n_{i+1}\}$. We believe the modified set of equations to be a reasonable approximation, provided that the rate of loss of radicals from reaction loci by second-order processes is not so great as to be in effect the dominant radical-loss mechanism.

Our analysis shows that, for reaction systems whose behaviour is governed by eqn (26), the distribution of locus populations is always Poissonian. Furthermore the parameter of the distribution, $\theta(t)$, is a function of time which satisfies the differential equation

$$\frac{d\theta(t)}{dt} = \sigma - k\theta(t) - \chi\{\theta(t)\}^2 \tag{27}$$

and is subject to the initial condition $\theta(0) = 0$ for reaction systems for which the boundary conditions are as stated in eqns (8). The explicit form which we obtain for $\theta(t)$ is

$$\theta(t) = 2\sigma \frac{\tanh\frac{at}{2}}{a + k\tanh\frac{at}{2}} \tag{28}$$

where $a^2 = 4\sigma\chi + k^2$. Examples of predictions of $\theta(t)$ (and hence of $\bar{\iota}(t)$) as functions of t given by this theory are shown in Figure 5 for various combinations of values of σ , k and χ.

Summary of cases for which explicit analytic solutions have so far been obtained

In reviewing the cases for which explicit analytic solutions have so far been obtained, it is helpful to recall that the Smith-Ewart differential difference equations are derived on the assumption that the state of radical occupancy of a reaction locus can change as a result of three distinct types of process:

(i) zero-order acquisition of radicals from the contiguous external phase;
(ii) first-order loss of radicals, e.g., by exit of radicals from the reaction loci into the contiguous external phase; and
(iii) second-order loss of radicals, e.g., by bimolecular mutual termination within reaction loci.

We have so far been able to obtain exact explicit analytic solutions for (a) the case where only processes (i) and (ii) are significant, and (b) the case where only processes (ii) and (iii) are significant. We have also obtained an approximate analytic solution for the case where all three processes (i), (ii) and (iii) occur, but where the loss of radicals occurs predominantly by process (ii) rather than by process (iii). As a generalisation of case (a), we have obtained a general solution which covers the case where the parameters which characterise the processes (i) and (ii) are themselves time-dependent. The general solution to case (b) requires modification if processes of type (ii) do not occur. Complete solutions have been obtained for three special cases of (b), namely, decay from a Stockmayer-O'Toole distribution of locus populations, decay from a Poisson distribution of locus populations, and decay from a homogeneous distribution of locus populations.

Numerical predictions (13)

We have also devised a numerical procedure for computing the solutions of the set of equations (1) in their most general form. Examples of the results of such calculations are given in Figures 6 and 7, and in Table I. All these calculations refer to reaction systems for which the boundary conditions are as stated in eqns (8), i.e., the reaction systems are initially devoid of radicals and at a certain instant of time, taken as t = 0, radicals suddenly begin to be generated in the external phase of the reaction system at a constant rate. Figure 6 illustrates the effect of increasing χ upon the variation of $\bar{\iota}(t)$ with t, σ and k being held constant. As expected, an increase in the value of χ leads to a reduction in the value of $\bar{\iota}$ at any instant. Figure 7 shows the effect of increasing χ and decreasing k upon $\bar{\iota}(t)$ as a function of t, σ being held constant. It is of particular interest

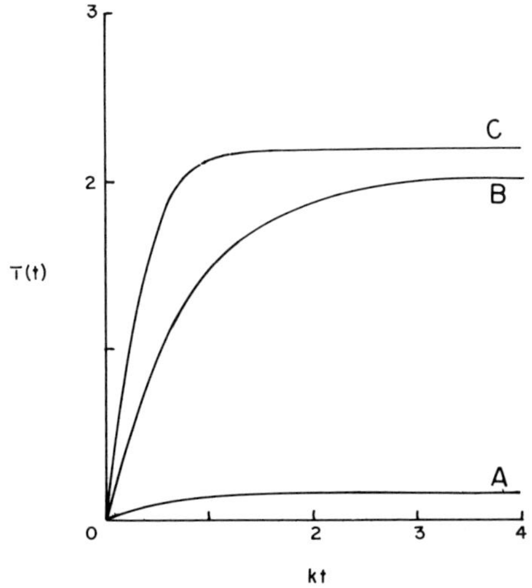

J. C. S. Faraday I

Figure 5. Predictions for i(t) *as function of* t *as given by Poissonian approximation for (A)* $\sigma = 1 \times 10^{-4}$, $k = 5 \times 10^{-4}$, $\chi = 5 \times 10^{-4}\ s^{-1}$; *(B)* $\sigma = 1 \times 10^{-3}$, $k = 4 \times 10^{-4}$, $\chi = 4 \times 10^{-5}\ s^{-1}$; *(C)* $\sigma = 1 \times 10^{-3}$, $k = 2 \times 10^{-4}$, $\chi = 1 \times 10^{-4}\ s^{-1}$ *(12)*

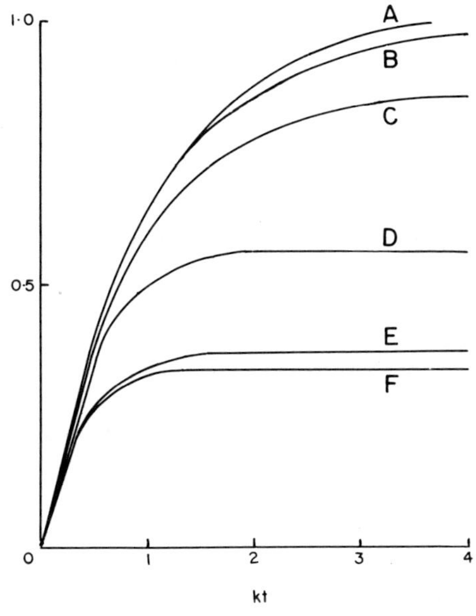

J. C. S. Faraday I

Figure 7. Variation of i(t) *with* t *for* $\sigma = 1 \times 10^{-5}\ s^{-1}$ *and various values of* k *and* χ *(in* s^{-1}*) as follows: (A)* $k = 1 \times 10^{-3}$, $\chi = 1 \times 10^{-8}$; *(B)* $k = 1 \times 10^{-4}$, $\chi = 1 \times 10^{-6}$; *(C)* $k = 1 \times 10^{-5}$, $\chi = 1 \times 10^{-4}$; *(D)* $k = 1 \times 10^{-6}$, $\chi = 1 \times 10^{-2}$; *(E)* $k = 1 \times 10^{-7}$, $\chi = 1 \times 10^{0}$ *(13)*

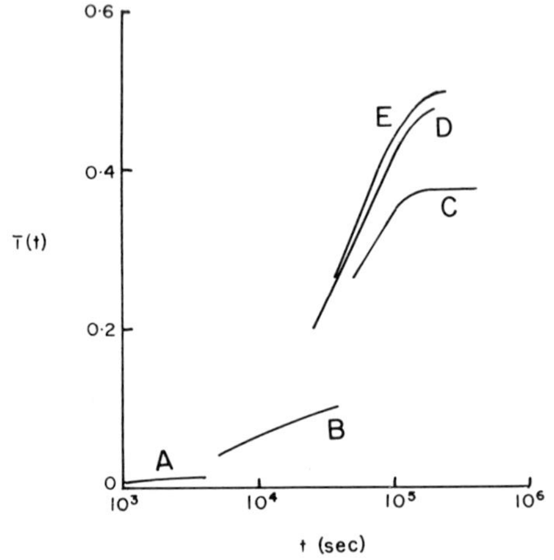

J. C. S. Faraday I

Figure 6. Variation of i(t) *with* t *for* $\sigma = 1 \times 10^{-3}$ s^{-1}, k $= 1 \times 10^{-3}$ s^{-1}, *and various values of* χ *(in* s^{-1}*) as follows: (A) 0; (B)* 1×10^{-5}*; (C)* 1×10^{-4}*; (D)* 1×10^{-3}*; (E)* 1×10^{-2}*; (F)* 1×10^{-1} (13)

Table I : Predictions for $n_r(t)/N$, $r = 0, 1, 2, 3$, as functions of t for reactions system for which $\sigma = 1x10^{-2}$ sec.$^{-1}$, $k = 5x10^{-4}$ sec.$^{-1}$, and $\chi = 1x10^{-5}$ sec.$^{-1}$ (13)

t (sec. x 10^{-3})	numerical predictions				Poissonian predictions			
	$\frac{n_0(t)}{N}$	$\frac{n_1(t)}{N}$	$\frac{n_2(t)}{N}$	$\frac{n_3(t)}{N}$	$\frac{n_0(t)}{N}$	$\frac{n_1(t)}{N}$	$\frac{n_2(t)}{N}$	$\frac{n_3(t)}{N}$
0.457	0.664	0.273	0.055	0.007	0.665	0.271	0.055	0.007
1.03	0.448	0.361	0.144	0.038	0.448	0.360	0.144	0.038
1.60	0.335	0.368	0.201	0.072	0.335	0.366	0.201	0.073
2.17	0.270	0.356	0.232	0.100	0.269	0.353	0.232	0.101
2.74	0.230	0.341	0.250	0.120	0.229	0.338	0.249	0.122
3.31	0.205	0.328	0.260	0.135	0.203	0.324	0.258	0.137
3.89	0.188	0.318	0.265	0.145	0.186	0.313	0.263	0.147
4.46	0.177	0.311	0.268	0.152	0.175	0.305	0.266	0.155
5.03	0.169	0.305	0.270	0.157	0.166	0.298	0.268	0.160
5.60	0.164	0.301	0.272	0.161	0.161	0.294	0.269	0.164
6.17	0.160	0.298	0.272	0.163	0.157	0.290	0.269	0.166
6.74	0.157	0.296	0.273	0.165	0.154	0.288	0.270	0.168
7.31	0.155	0.294	0.273	0.166	0.152	0.286	0.270	0.170
7.89	0.154	0.293	0.274	0.167	0.150	0.285	0.270	0.171
Stockmayer-O'Toole predictions	0.150	0.290	0.274	0.170	0.150	0.290	0.274	0.170

that, for large values of χ/k, the value of $\bar{\iota}(t)$ at long t approaches the Smith-Ewart "Case 2" value of 0.5. This is to be expected, since the requirements for a Smith-Ewart "Case 2" reaction system are k = 0 and $\chi = \infty$.

Table I shows the numerical predictions for $n_r(t)/N$, r = 0, 1, 2, 3, as a function of t for a combination of values of σ, k and χ such that the behaviour of the reaction system could be reasonably supposed to be represented by the "Poissonian" approximation described above. The corresponding predictions given by the Poissonian approximation are also shown. It is evident that these predictions do conform satisfactorily with the results obtained by numerical solution of the set of equations (1).

List of symbols

a = $\sqrt{4\sigma\chi + k^2}$

B_p = constant associated with term containing e^{-t/τ_p}

$C_p^{(-1/2)}(\xi)$ = Gegenbauer polynomial of order p and parameter $-\frac{1}{2}$

h = $\sqrt{8v\sigma/k_t}$

I_η denotes modified Bessel function of order η

i = number of propagating radicals per single reaction locus

$\bar{\iota}$ = average number of propagating radicals in single reaction locus.

$J_p^{(f,g)}(\xi)$ = Jacobi polynomial of order p

K_p = constant associated with term containing e^{-t/τ_p}

k = coefficient characterising rate of exit of radicals from reaction loci

k_t = rate coefficient for bimolecular mutual termination

m = kv/k_t = k/χ

N = number of reaction loci in arbitrary volume of reaction system

n_i = number of loci in arbitrary volume of reaction system which contain i propagating radicals

r = number of propagating radicals in single reaction locus

t = time

$(u)_\nu$ is a Pochhammer symbol which denotes $\Gamma(u+\nu)/\Gamma(u)$

v = volume of reaction locus

Γ denotes gamma function

$\theta(t)$ = time-dependent parameter of Poisson distribution

ξ = auxiliary variable of locus-population generating function

$\tau_p = \{\chi p(p+m-1)\}^{-1}$

$\chi = k_t/v$

Ψ = locus-population generating function

Acknowledgement

The substantial contribution which my colleague, Dr. D.T. Birtwistle, has made to the work described in this paper is gratefully acknowledged.

Literature cited

1. Stockmayer, W.H., J. Polymer Sci., 1957, 24, 314

2. O'Toole, J.T., J. Appl. Polymer Sci., 1965, 9, 1291

3. Smith, W.V.; Ewart, R.H.; J. Chem. Phys., 1948, 16, 592

4. Gilbert, R.G.; Napper, D.H., J.C.S. Faraday I, 1974, 70, 391

5. Hawkett, B.S.; Napper, D.H.; Gilbert, R.G., J.C.S. Faraday I, 1975, 71, 2288

6. Hawkett, B.S.; Napper, D.H.; Gilbert, R.G., J.C.S. Faraday I, 1977, 73, 690

7. Lichti, G.; Gilbert, R.G.; Napper, D.H., J. Polymer Sci.: Polymer Chem. Edn., 1977, 15, 1957

8. Birtwistle, D.T.; Blackley, D.C., J.C.S. Faraday I, 1977, 73 1998

9. Weiss, G.H.; Dishon, M., J.C.S. Faraday I, 1976, 72, 1342

10. Birtwistle, D.T.; Blackley, D.C., J.C.S. Faraday I, 1978, 74 2051

11. Birtwistle, D.T.; Blackley, D.C.; Jeffers, E.F., J.C.S. Faraday I, 1979, 75, 2332

12. Birtwistle, D.T.; Blackley, D.C., J.C.S. Faraday I, in press

13. Birtwistle, D.T.; Blackley, D.C., J.C.S. Faraday I, in press

RECEIVED April 6, 1981.

28

A Reinvestigation of Vinyl Acetate Emulsion Polymerization: Isotope Effect

M. H. LITT and K. H. S. CHANG

Department of Macromolecular Science, Case Western Reserve University, Cleveland, OH 44106

One of the most important factors in vinyl acetate polymerization is chain transfer of growing radical to the monomer (1-5). This was concluded because there is no dependence of molecular weight on initiator concentration. While chain transfer to polymer was found, it is unimportant below 30% conversion; there is no molecular weight increase in the polymer until higher conversions. Hydrolysis and reacetylation of low conversion polymer shows no molecular weight change, indicating that there is no back-biting on polymer acetyl groups.

An interesting and to us an important question is where is the main chain transfer site on vinyl acetate. All studies concluded that the main chain transfer site was the hydrogen atoms of the acetyl group (6-9). Some investigators found vinyl groups in the polymer and proposed some chain transfer on the vinyl hydrogens (8,9). On the other hand, we concluded on the basis of a kinetic analysis that chain transfer is mainly on the vinyl hydrogens of vinyl acetate (10).

One way to distinguish between the two possibilities is to study the isotope effect on the kinetics of vinyl acetate polymerization and on the polymer molecular weight. The deuterium isotope effect has been ascribed to the difference in the zero point energies of the stretching vibrations of the C-H and C-D bond (11). The rate of a reaction in which deuterium is transferred is slower than that of the corresponding reaction for hydrogen, since the C-D bond has a lower zero point energy. The magnitude of the isotope effect is temperature dependent. For example, the reaction at a C-H bond has a maxium isotope effect of 7 at 25° C and 2.1 at 500° C (12).

Urry (13) studied hydrogen abstraction from toluene α-d_1 versus toluene by chlorine, bromine and $CH_3COO\cdot$ radicals and found that the attack on toluene α-d_1 is slower than that on toluene. The isotope effect was found to be in the range of 2.2 to 9.9.

Bartlett and Tate (14) studied the polymerization of allyl acetate and allyl-1-d_2 acetate and found that the rate of poly-

0097-6156/81/0165-0455$05.00/0

merization for the deuterated allyl acetate was about 2.4 times as great as that of the undeuterated allyl acetate. An explanation was advanced that when deuterated allyl acetate was used there was less chain transfer on the monomer because of the stronger C-D bond. Bothe the polymerization rate and molecular weight increased by the same ratio when deuterated allyl acetate was used because in this system each radical produces one polymer molecule.

Thus, a relatively large isotope effect can be expected if deuterated vinyl acetate is used for polymerization. Since the degree of polymerization for vinyl acetate equals k_2/k_3, the ratio of the rate constant for propagation to that for chain transfer, the degree of polymerization should express the full isotope effect. It known that in emulsion polymerization, $Rp \propto k_3^{-\frac{1}{2}}$ (10), and thus the relative change in rate should be proportional to the square root of the isotope effect. We felt that synthesis and polymerization of two partly deuterated vinyl acetates, trideuterovinyl acetate ($D_2 = CD - OAc$) and vinyl trideuteroacetate ($CH_2=CH-O-\overset{O}{\overset{\|}{C}}-CD_3$) could settle the question of which end of the monomer is the major chain transfer site. If it is the acetyl group, when $CD_3COOCH=CH_2$ is used there will be an increase of the molecular weight compared to undeuterated vinyl acetate. On the other hand, if the polymerization of trideuterovinyl acetate shows the molecular weight increase, then the vinyl hydrogens are implicated. Also, it is expected that the emulsion polymerization rate would rise when trideuterovinyl acetate is used, if chain transfer on this site is important as we believe (10).

Experimental

Materials. Emulsifier: A very pure grade of sodium lauryl sulfate (Stepanol WA-100) was obtained from Stepan Chemical Company and was used directly from the bottle.

Initiator: Potassium persulfate was Fisher certified reagent grade and was used directly from the bottle in emulsion polymerization.

Seed Latex: The seed latex was prepared as described before (10). It had a solids content of 17.0% and its number average diameter was about 0.045 μ as measured by quasielastic laser light scattering (15).

Vinyl Acetate: Vinyl acetate was obtained from Eastman Kodak Company and purified by distillation through a Perkin Elmer spinning band column with 45 theoretical plates. The first and last 25% of the distillate was discarded for all the distillations. It was then stored at 5° C.

Vinyl Trideuteroacetate ($CD_3-\overset{O}{\overset{\|}{C}}-O-CH=CH_2$): The method of preparation was based on the interchange reaction of vinyl ester with acetic acid (16,17). 150 g (1.5 mole) of vinyl propionate

(Polyscience Inc.), 48 g (0.75 mole) of tetradeuteroacetic acid (Aldrich Chemical Co., 99.5 atom % D), 0.9 g of mercuric acetate, (Aldrich Chemical Co.), 0.155 g of concentrated sulfuric acid (Fisher Scientific, ACS reagent grade) and 0.001 g of hydroquinone (MC/B reagent grade) were put into a three neck flask which was connected to a Perkin Elmer spinning band column with 45 theoretical plates. The mixture was heated to 90°C for two hours; 2 g of sodium acetate (Fisher Scientific, ACS reagent grade) was added and the resultant vinyl trideutero acetate was distilled at atmosphere pressure. The fraction boiling from 71 to 75°C was collected. The crude vinyl trideuteroacetate was then redistilled and 58 grams (0.65 moles) of pure vinyl acetate, bp 74°C, was obtained. The yield was 87% based on the tetradeuteroacetic acid used. Proton N.M.R. spectrum of vinyl trideuteroacetate showed that the ratio of the proton peak intensities of methyl hydrogens to vinyl hydrogens was 0.018; the isotopic purity of $CD_3-\overset{O}{\overset{\|}{C}}-O-CH=CH_2$ was 98.2%.

Trideuterovinyl acetate ($CH_3-\overset{O}{\overset{\|}{C}}-O-CD=CD_2$): The reaction of acetylene with acetic acid (18,19) was adopted for this synthesis. To a three neck flask was added 30 g (1.50 moles) of deuterium oxide (Aldrich Chemical Co.), 162 g of acetic anhydride (1.59 moles) (Aldrich Chemical Co.) and 0.1 g conc. HCl (Fisher Scientific Co.). It was then stirred and heated to 100°C. After about 20 minutes, the reaction was complete; 180 g of monodeuteroacetic acid was formed. 7.0 g of mercuric acetate (Aldrich Chemical Co.) and 10 g of fuming sulfuric acid (Fisher Scientific Co.) were then added to the flask. Dideuteroacetylene was generated in another three neck flask by the slow addition of 120 g of calcium carbide (Fisher Scientific, 80% pure) (1.15 moles) to 400 g of deuterium oxide and passed continuously for five hours at 20°C into the flask containing monodeuteroacetic acid and mercuric acetate. After five hours 15 g of sodium acetate were added to the reaction mixture and the whole was distilled through a spinning band column at atmosphere pressure. Crude trideuterovinyl acetate was collected between 69 to 75°C, was obtained. The yield was 55% based on the dideuteroacetylene used. Proton N.M.R. spectrum of trideuterovinyl acetate showed that the ratio of proton peak intensities of vinyl to methyl hydrogens was 0.034; the isotopic purity of $CH_3-\overset{O}{\overset{\|}{C}}-O-CD=CD_2$ was 96.6%.

Polymerizations. Bulk Polymerization: 3.5 g of freshly distilled monomer was put into a 12 ml polymerization tube which was then degassed under a vacuum of 10^{-3} mm Hg four or five times and sealed. Three sealed samples of each monomer were immersed in 700 g of molten $NaH_2PO_4 \cdot 2H_2O$, m.p. = 60.0°C (made by adding 80.78 g of H_2O to 619.22 g of $NaH_2PO_4 \cdot H_2O$) at 60°C in a well insulated beaker. The tubes were then irradiated at 0.12×10^6

rad/hr using a ^{60}Co γ-ray source and samples were withdrawn at selected intervals. Conversion was measured by evaporating the monomer at 1 mm Hg and 40°C for 8 hours, raising the temperature to 65°C and weighing the residue at intervals until it reached constant weight.

Emulsion Polymerization: The polymerization apparatus was a dilatometer which was similar to though smaller than the one used before.(10) A graduated 10 ml column (adapted from a 10 ml pipette) was connected to a 50 ml volumetric flask with three indentations in it set at 120°. Latex volume changes were read by taking the emulsion to a reference point in the column and reading the height on a precision bore side tube.

The seeded polymerization procedure was the same as described earlier (10) except that only .175 of the recipe was used. Polymerizations of all monomers were run at 60.0C using four different initial initiator concentrations ($I_o = 2 \times 10^{-4}$ M to 10×10^{-4} M $K_2S_2O_8$) at a constant volume of organic phase per unit volume of aqueous phase ($V_{org}/V_{aq} = 0.33$). Rates were calculated from the slopes of the linear portion of the polymer/time curves at 30-85% conversion.

Analysis

NMR Measurement: 0.3 g of monomer was dissolved in 1 g of $CDCl_3$ using TMS as internal standard. NMR spectra were run using a Varian A-60A proton NMR instrument at 23°C.

Viscometry: The specific viscosity of each polymer from the bulk polymerization was measured in acetone at 30°C using an Ubbelohde dilution viscometer. Five concentrations in the range of 1.120 to 0.242 g/dℓ poly(vinyl acetate) and poly-(vinyl trideuteroacetate) and 0.385 to 0.084 g/dℓ (poly(trideuterovinyl acetate)) were run. Intrinsic viscosity was calculated by extrapolation of the η_{sp}/c versus c plot to zero concentration. Number average molecular weights were calculated using the equation(20) $[\eta] = 1.0 \times 10^{-4}[M_n]^{0.72}$ which is in the mid range of the equations listed.

Results and Discussion

Polymerization Rate.

Bulk Polymerization: The conversion versus time plot for the bulk polymerizations of vinyl acetate and its deuterated analogues is shown in Figure 1. Vinyl tridueteroacetate has a conversion rate of 9.9×10^{-3}/min which is identical with that of vinyl acetate (9.5×10^{-3}/min) within the experimental error. However, trideuterovinyl acetate has a much higher conversion rate (1.69×10^{-2}/min). The ratio of the rate of polymerization of tridueterovinyl acetate to the average of the other two monomers is 1.74 ± .03.

Vinyl acetate and vinyl trideuteroacetate were run together, while trideuterovinyl acetate was run later. The lag shown in the first two is almost certainly not due to residual oxygen, as this

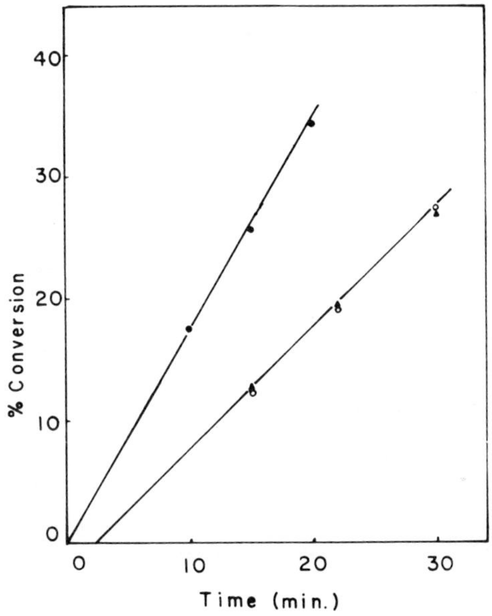

Figure 1. Plot of % conversion vs. time in bulk radiation polymerization for three monomers at 60°C ((●) $CD_2{=}CD{-}O{-}C({=}O){-}CH_3$; (▲) $CH_3{-}C({=}O){-}O{-}CH{=}CH_2$; (○) $CD_3{-}C({=}O){-}O{-}CH{=}CH_2$)

would probably be different for each sample. It is probably an experimental artifact due to putting the samples in the 60°C bath immediately before lowering them into the source.

Emulsion Polymerization: Table 1 shows the comparison of rates of emulsion polymerization of vinyl acetate and its deuterated analogues as the initial initiator concentration, I_o, was varied. Again, the rate of polymerization for vinyl trideuteroacetate is identical with that of vinyl acetate, while the rate of polymerization of trideuterovinyl acetate averages 1.76 times higher. This indicates that vinyl hydrogen is the major chain transfer site in vinyl acetate. Also, Table 1 shows that the chain transfer on acetyl hydrogens (shown later to be 6%) is kinetically insigificant. The accuracy of measurement of the rate is about 1% and the ratio of the rates of polymerization of vinyl acetate and vinyl trideuteroacetate average 1.00 to within ½%. If the chain transfer step is kinetically significant, the rate should have been increased by 2%, which would be easily seen. Our previous work[10,22] showed this step should have no kinetic effect.

Molecular Weights: The molecular weights of the three polymers at three different conversions are compared in Table 2. As can be seen, the molecular weight of poly(trideuterovinyl acetate) averages 2.59 times higher than that of poly(vinyl acetate), indicating again that the major chain transfer site is the vinyl hydrogens. Also there is some chain transfer on the acetyl hydrogens of vinyl acetate since the molecular weight of poly(vinyl trideuteroacetate) is about 1.04 times that of poly-(vinyl acetate).

The Fraction of Chain Transfer on Vinyl Hydrogens and the Magnitude of the Isotope Effect: If chain transfer is on both vinyl and acetyl hydrogen, the degree of polymerization of the polymers can be expressed as in equation (1)

$$\bar{X}_n = \frac{k_2}{k_{3A} + k_{3V}} = \frac{1}{C_{MA} + C_{MV}} \tag{1}$$

where C_{MA} and C_{MV} are the chain transfer constants on acetyl and vinyl hydrogens of vinyl acetate respectively. The kinetic model is given in the next section.

$$C_{MA} = \frac{k_{3A}}{k_2} \tag{2}$$

$$C_{MV} = \frac{k_{3V}}{k_2} \tag{3}$$

We know from Table 2 that the molecular weight of poly(vinyl trideuteroacetate) is 1.04 times that of poly(vinyl acetate).

Thus, taking into account isotopic purity,

TABLE I

The comparison of seeded emulsion polymerization, Rp, of three monomers (vinyl acetate, vinyl trideuteroacetate, and trideuterovinyl acetate) at various initiator concentrations. The polymerization was run at 60°C with $K_2S_2O_8$ as initiator.

Monomer/100 Rp (M/min)	$10^4 I_o$ (m/ℓ)			
	2.0	3.0	4.0	10.0
$CH_3-C(=O)-O-CH=CH_2$	2.63	3.28	4.44	6.89
$CH_3-C(=O)-O-CH=CH_2$	2.59	3.26	4.54	6.79
$CH_3-C(=O)-O-CD=CD_2$	4.57	5.85	7.77	12.27
Rp_3/Rp_1 [(a)]	1.74	1.78	1.75	1.78

(a) Rp_3 and Rp_1 are the polymerization rates of $CH_3-C(=O)-O-CD=CD_2$ and $CH_3-C(=O)-O-CH=CH_2$ respectively.

TABLE II

The intrinsic viscosities and molecular weights of poly(vinyl acetate), poly(vinyl trideuteroacetate), and poly(trideutero-vinyl acetate) at three conversions.

Polymer	Conversion	[η] (dℓ/g)	Molecular Weight[a]	M_w/M_{w_1} [b]
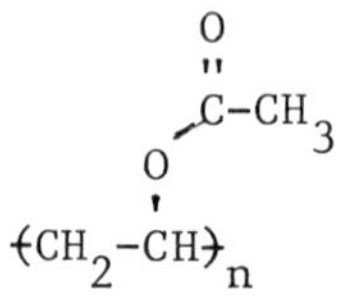	0.127	1.15	4.37×10^5	-
	0.195	1.17	4.48×10^5	-
	0.270	1.16	4.42×10^5	-
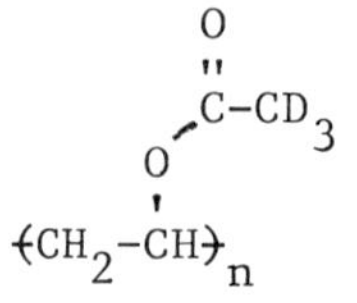	0.125	1.18	4.57×10^5	1.05
	0.193	1.19	4.61×10^5	1.03
	0.274	1.19	4.60×10^5	1.04
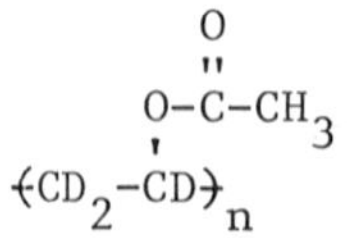	0.175	2.28	11.31×10^5	2.59
	0.257	2.32	11.58×10^5	2.58
	0.344	2.31	11.51×10^5	2.60

(a) Molecular weight was calculated by $[\eta] = KM^a$, where $K = 1.0 \times 10^{-4}$ (dℓ/g), a = 0.72 (20).

(b) M_w/M_{w_1} is the ratio of molecular weight of deuterated vinyl acetate to that of vinyl acetate.

$$\frac{\overline{X}_{nDA}}{\overline{X}_n} = 1.04 = \frac{C_{MA} + C_{MV}}{.982C_{MA}/x + .018C_{MA} + C_{MV}} \quad (4)$$

where $\overline{X}_n$ and $\overline{X}_{nDA}$ are the degrees of polymerization of vinyl acetate and vinyl trideuteroacetate respectively and x is the isotope effect.

Also, the molecular weight of poly(vinyl trideutero-acetate) is 2.59 times that of poly(vinyl acetate). Taking into account the remaining hydrogen, equation (5) results:

$$\frac{\overline{X}_{nDV}}{\overline{X}_n} = 2.59 = \frac{C_{MA} + C_{MV}}{C_{MA} + .966C_{MV}/x + .034C_{MV}} \quad (5)$$

where $\overline{X}_{nDV}$ is the degree of polymerization of trideutero-vinyl acetate.

Solving (4) and (5) for the unknowns, we obtain

$$\frac{C_{MA}}{C_{MV}} = 0.062 \quad (6)$$

$$x = 3.04 \quad (7)$$

Thus, only 6% of the chain transfer is on the acetyl hydrogens and 94% on the vinyl hydrogens of vinyl acetate. The overall isotope effect is 3.04.

The rate of polymerization in emulsion polymerization is proportional to $k_3^{-\frac{1}{2}}$, where k_3 is the fhain transfer step on the vinyl group (10). Substituting trideuterovinyl acetate for vinyl acetate raised the rate by a factor of 1.76. When the calculation for the isotope effect on rate is done accurately, taking into account the 3% H on the trideuterovinyl, we find that if the effect is purely on k_3, the rate should rise by a factor of 1.69 as compared to 1.76 ± .02. This is almost within the experimental error. There may be a very slight secondary isotope effect (23,24) on the propagation and re-initation rate constants k_2 and k_4, but it cannot be decided from these data.

A question has been raised by the reviewers about the exact nature of the transfer step. While no doubt exists that a vinyl hydrogen is transferred, the exact mechanism is disputed. This will be discussed later.

Model and Nature of Termination Step for Bulk Polymerization of Vinyl Acetate: The following is a reasonable kinetic model for the bulk polymerization of vinyl acetate.

1. $I \xrightarrow{k_1} R\cdot$

2. $R^{\cdot}_{n} + M \xrightarrow{k_2} R^{\cdot}_{n+1}$ $\qquad R\cdot \equiv$ Growing radical

3. $R\cdot + M \xrightarrow{k_3} RH + M\cdot$ $\qquad M\cdot \equiv$ Monomer radical from chain transfer

4. $M\cdot + M \longrightarrow R\cdot$

5. $R\cdot + R\cdot \longrightarrow x$ (termination)

6. $R\cdot + M\cdot \longrightarrow x$ (termination)

If k_4 is relatively large, e.g. 10^3, then M· would be present in such low concentration that step 5 would be the major termination step. If k_4 is <u>very</u> small (little or no reinitiation) then step 6 plus M· + M· would be the major termination steps. Such a case leads to allyl acetate type of kinetics with both $\overline{X}$ and Rp proportional to 1/k (<u>14</u>), which is not found. Thus M· + M· cannot be considered as a major termination step. We shall solve for extremes of the remaining termination steps which imply a reasonable amount of reinitiation, steps 5 and 6, assuming that each separately occurs almost 100%.

A: The expressions for the rate and degree of polymerization when the major termination step is step 5 are given in (8) and (9) respectively. The solution is based on the steady state assumption, $k_3(R\cdot)(M) = k_4(M\cdot)(M)$, which implies that the rates of Steps 3 and 4 are at least five times faster than termination.

$$Rp = k_2\left(\frac{k_1}{2k_5}\right)^{\frac{1}{2}} I^{\frac{1}{2}} M \qquad (8)$$

$$\overline{x}_n \approx \frac{k_2}{k_3} \qquad (9)$$

In this case, deuteration has little effect on the polymerization rate since k_3 (chain transfer constant) does not appear in equation (8). But there will be an increase in molecular weight compared to vinyl acetate since the chain transfer constant for trideuterovinyl acetate is smaller than that of vinyl acetate.

B: The major termination is step (6); the rate of polymerization is now given by (10)

$$Rp = k_2\left(\frac{k_1 k_4}{2k_3 k_6}\right)^{\frac{1}{2}} I^{\frac{1}{2}} M \qquad (10)$$

The equation for degree of polymerization is unchanged, (9). Thus an increase of polymerization rates by $(k_3/k_{3D})^{\frac{1}{2}}$ and molecular weight by k_3/k_{3D} should be observed when trideuterovinyl acetate is used (k_{3D} is the chain transfer rate constant for trideuterovinyl acetate).

We have seen in Tables 1 and 2 that both the polymerization rate and molecular weight of trideuterovinyl acetate increased compared to that of vinyl acetate. Since the rate increase is identical for bulk and emulsion polymerization and we know that almost all termination in emulsion polymerization is between monomer and growing radical (10), the implication is that the same is true in bulk polymerization. While this conclusion is initially surprising, it supports our argument that the vinyl radical is reasonably stable (10). Otherwise the radical would reinitiate rapidly and its concentration would be too low for termination with it to be important.

Further consideration shows that one should expect almost all termination to be by step 6. The steady state assumption can be written as

$$(R\cdot)/(M\cdot) = k_4/k_3 \tag{11}$$

k_4 was found (10) to be 29 while k_3 is known from the literature to be 2 to 2.5 at 60° (when $k_2 \approx 10^4$). Thus, $(R\cdot)/(M\cdot) \approx 13 \pm 2$. Since R· is a macromolecule with average degree of polymerization of four to five thousand while M· is a monomer radical, M· has a diffusion constant at least six to 100 times greater than (R·). Since termination rate constants are diffusion controlled, k_5 will be much smaller than k_6. Even though R· is the majority radical present, cross-termination is favored; $k_5(R\cdot)^2/k_6(M\cdot)(R\cdot) \leq 1/5$. On the other hand, the termination of 2 M·'s is disfavored even though this termination rate constant should be twice as high as k_6, since both radicals are small. Here we find $k_6(M\cdot)(R\cdot)/2k_6(M\cdot)^2 = (R\cdot)/2(M\cdot) \approx 6$. Thus by chance, we are in a region where even with no cross-termination enhancement (which usually occurs), the major termination must be by step 6, considering the rate constants found in our previous paper to be correct.

Nature of the Chain Transfer Step

Our kinetic work (10) showed that the small molecule radical produced by chain transfer with monomer had to be a stable radical. This was confirmed in the present paper by analysis of the isotope effect on the bulk polymerization rates. The isotope effect on molecular weights and rates unequivocally showed that almost 100% of the chain transfer involved the vinyl hydrogen. There is some evidence in the literature to support the idea of a stable vinyl radical. Phenyl acetylene acts as a retarder when copolymerized with styrene or methyl methacrylate (25). Thus the phenyl vinyl radical is very stable compared to the growing styryl or methacrylyl radical.

However, knowledge of bond strengths of various C-H bonds leads and has led to other conclusions, which is why the idea of vinyl chain transfer was not considered seriously in the past (6-9). There is unanimous agreement that the C-H bond strength where the carbon has sp_2 hybridization is higher than the C-H bond strength where the carbon has sp_3 hybridization (26,27,28). The question is how much higher and how is it affected by the adjacent oxygen. According to Sanderson (26), the intrinsic bond strengths are equal, but the vinyl carbon undergoes reorganization during cleavage to sp hybridization, which adds 11 kcal/mole to the energy needed to break the bond. Joshi (28) estimates the C-H vinyl bond energy about 2.6 kcal/mole higher than the C-H alkyl bond energy, due to the smaller C-H bond distance in the vinyl compounds. C-H sp_3 bonds adjacent to another oxygen are weaker by about 4 kcal/mole than those adjacent to carbon (26,28). Little is known about the bond strength of C-H groups adjacent to ester groups; however, for sp_3 hybridization there seems to be little effect compared to an adjacent carbon.

Thus, it may be possible that in the vinyl ester case, the C-H bond adjacent to the oxygen could effectively be no stronger than the acetyl methyl C-H bond. [$H-CH_2C=O$ was shown to have the same strength as $H-CH_2-CH_2-$ (27) against t-butoxy radicals, which are active.] For an active radical, and the growing radical from vinyl acetate is very active, chain transfer may involve little or no reorganization of the fragment before the bond is broken. Based on recent bond strength data and reactivity concepts, the vinyl C-H may then be as reactive as an aliphatic C-H.

Our problem is that the vinyl C-H in this case is much more reactive than the alkyl C-H, implying that the radical generated is more stable. Our kinetic data support this view, since the radical generated adds slowly compared to the growing radical. Are there other possibilities? A reviewer has suggested that perhaps chain transfer occurs by dismutation of a growing chain, particularly after a head-to-head addition. Such a step has been postulated for vinyl chloride (29), where the chlorine atom is lost from the chain and starts a new chain.

$$-\underset{\mathrm{Cl}}{\mathrm{CH}}-\underset{\mathrm{Cl}}{\mathrm{CH}}-\mathrm{CH_2}\cdot \longrightarrow -\mathrm{CH_2}-\underset{\mathrm{Cl}}{\mathrm{CH}}-\mathrm{CH}=\mathrm{CH_2}+\mathrm{Cl}\cdot$$

The loss of chlorine is favored both energetically and sterically as the C-Cl bond is weaker than the C-H bond and steric strain is removed.

The analogue for vinyl acetate, which could show the isotope effect, is loss of hydrogen

$$-\mathrm{CH_2}-\underset{\mathrm{OAc}}{\mathrm{CH}}-\underset{\mathrm{OAc}}{\mathrm{CH}}-\mathrm{CH_2}\cdot \longrightarrow \mathrm{CH_2}-\underset{\mathrm{OAc}}{\mathrm{CH}}-\underset{\mathrm{OAc}}{\mathrm{C}}=\mathrm{CH_2}+\mathrm{H}\cdot$$

This step is unlikely. There is no relief of steric strain as the acetoxy group is still present and the C-H bond strength is higher than the O=CO-C bond strength by about seven kcal/mole (26). In addition the acyloxy radical is unstable and decomposes exothermically (26). If this happens during dismutation, loss of acetoxy is favored even more, up to thirty kcal/mole.

$$-CH_2-\underset{OAc}{CH}-\underset{OAc}{CH}-CH_2\cdot \longrightarrow -CH_2-\underset{OAc}{CH}-CH{=}CH_2 + CH_3\cdot + CO_2$$

The resulting radical, $CH_3\cdot$, is very active and cannot be the stable radical that the kinetics require.

One further comment. Molecular weights are determined by the ratio of the propagation rate to chain termination rate. If dismutation, which terminates a given chain, is the main source of chain transfer, we would find:

$$\bar{X}_n = \frac{k_p R\cdot M}{k_D R\cdot} = \frac{k_p}{k_D} M$$

(The dismutation rate constant includes in it the fraction of H-H additions.) Here $\bar{X}_n$ depends on monomer concentration which is contrary to fact. There are about 1-2% H-H additions along the PVAc backbone (31), so it does not dismute as soon as one monomer adds backward.

Thus dismutation does not seem to be a possible explanation of our results. There should be no isotope effect on the preferred pathway for dismutation, and the resulting radical is very active. The pathway is also inadequate to explain the kinetics. We are left therefore where we started. The major chain transfer step is probably hydrogen abstraction from the vinyl group with the production of a stable vinyl radical.

The double contradiction of our results with the thermodynamics of bond strengths could be weakly rationalized as follows (we intend to study this). The vinyl radical stability may be kinetic rather than thermodynamic. Since the vinyl radical carbon rearranges to sp hybridization, the acetyl group is free to rotate. Low energy conformers for the radical (which after the reorganization to sp could be 11 kcal higher in energy than the growing radical) may have the acetyl group in such a position that it cannot easily move to the normal sp_2 hybrid orientation if the radical attacks a double bond. The difficulty of attack would then have partly energetic and partly steric origin.

The large proportion of attack on the vinyl group could be due to a phenomenon proposed many years ago but never proved (30): precomplexing of the radical with the monomer (solvation) before final collapse to the sigma bond state.

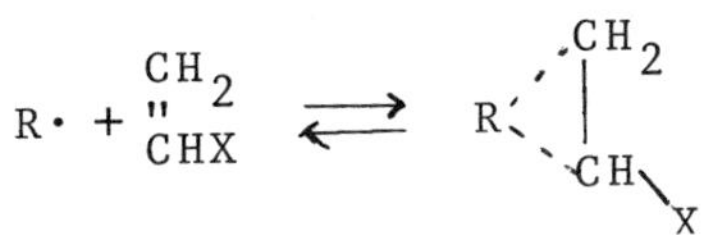

(Two 1.5 electron sigma bonds or a π complex)

If an alternative collapse could involve H abstraction, the same intermediate would be involved in propagation and vinyl chain transfer. The heat of formation of the complex then should be subtracted from the vinyl C-H bond energy as the radical now could spend most of its time associated with the double bond.

SUMMARY AND CONCLUSIONS

The bulk polymerization rate of trideuterovinyl acetate was found to be 1.78 times that of vinyl acetate. In emulsion polymerization the same result was found. The molecular weight of poly(trideuterovinyl acetate) was 2.59 times that of poly(vinyl acetate). The overall isotope effect on chain transfer to monomer was calculated to be 3.04. Chain transfer was shown to be 94% on the vinyl hydrogens and 6% on the acetyl hydrogens. The measurement of polymerization rates in emulsion polymerization showed that the chain transfer on acetyl hydrogens is kinetically insignificant.

The results definitely prove our hypotheses in the kinetic model for vinyl acetate emulsion polymerization (10), that vinyl radical, $CH_2{=}\dot{C}{-}OAc$, is the major monomer radical formed and is a stable radical which reinitiates relatively slowly compared to the propagation step.

The idea that 94% of the chain transfer at 60C is on vinyl hydrogens is completely different from the conclusions of all previous workers. However, they had to use indirect methods to determine the chain transfer site. Here deuterated vinyl acetates tested the isotope effect on the potential chain transfer sites directly and gave an unequivocal answer.

The question arose whether the production of new chains is due to dismutation of a head-to-head radical end or due to ordinary chain transfer. Analysis of dismutation energetics made that step seem very improbable in this case. We are still left with the problem that a bond which has higher strength in the ground state is preferentially attacked and the resulting radical is relatively stable. This is not understood, but nonetheless the vinyl chain transfer has been established by this work.

The termination step in bulk polymerization, which was previously assumed to be growing radical reacting with growing radical, has been shown here to be almost completely monomer radical reacting with growing radical.

ACKNOWLEDGMENT

The authors thank the Polymer Section of the Materials Science Division for partial support of this work on Grant No. DMR76-04562. We particularly want to thank Dr. W. E. Daniels of Air Products Corporation for suggesting the use of deuterated monomers.

LITERATURE CITED

1. Wheeler, O. L., Ernst, S. L., and Crozier, R. N., J. Polymer Sci., 1952, Vol. VIII, No. 4, 409-423.
2. Matsumoto, M., Imai, K., Maeda, M., Ohyanagi, Y., and Saito, T., Kobunshi Kagaku, 1955, 12, 398.
3. Imai, K., J. Chem. Soc. Japan, Ind. Chem. Sec., 1959, 62, 1127.
4. Matsumoto, M., and Ohyanagi, Y., J. Polymer Sci., 1960, Vol. XLVI, Issue 148.
5. (a) Friis, N., and Nyhagen, L. J., J. Appl. Polymer Sci., 1973, 17, 2311. (b) Friis, N., Thesis, Technical University, Copenhagen, Denmark (1973).
6. Lazar, M., Paulinec, J., and Manasek, Z., Collection Czech. Chem. Commun., 1961, 26, 1380.
7. Imoto, S., Ukida, J., and Kominami, T., Kobunshi Kagaku, 1957, 14, 101.
8. Wheeler, O. L., Lavin, E., and Crozier, R. N., J. Polymer Sci., 1952, 9, 157.
9. Howard, R. O., Ph.D. Thesis, Massachusetts Institute of Technology, Cambridge, Mass., 1952.
10. Chang, K., Litt, M., and Nomura, M., "A Reinvestigation of Vinyl Acetate Emulsion Polymerization (I) - Polymerization Rate".
11. Jenks, W. P., "Catalysis in Chemistry and Enzymology", McGraw-Hill Book Co., New York (1969).
12. Wiberg, K. B., Chem. Rev., 1955, 55, 713.
13. Urry, W. H., Abst. 12th Natl. Org. Symp., 1951, Denver, Colo., 1951.
14. Barlett, P. D., and Tate, F. A., J. Amer. Chem. Soc., 1953, 75, 91.
15. Chang, K., Litt, M., and Jamieson, A. M., "The Measurement of Latex Particle Size by Quasielastic Laser Light Scattering", J. Poly. Sci.-Phys., to be published.
16. Toussaint, W., and MacDowell, G., U.S. Patent 2,299,862.
17. Adelman, R., J. Org. Chem., 1949, 14, 1057-77.
18. Dykstra, H. B., U.S. Patent 1,849,616.
19. Skirrow, F. W., and Morrison, G. O., U.S. Patent 1,855,366.
20. Brandrup, J., and Immergut, E. H., eds., Polymer Handbook, Interscience, New York, 1975, p. IV--15.

21. Jenks, W. J., "Catalysis in Chemistry and Enzymology", McGraw-Hill Book Co., New York, 1969, p. 253.

22. Litt, M. H., Patsiga, R., and Stannett, V. T., J. Polymer Sci., A1, 1970, 8, 3607.

23. Thornton, E. R., Ann. Rev. Phys. Chem., 1966, 17, 349.

24. Streitwieser, A. Jr., Jagow, R. H., Fahey, R. C., and Suzuki, S., J. Amer. Chem. Soc., 1958, 80, 2326.

25. Higashiura, K., and Oiwa, M., J. Polymer Sci. A1, 1968, 6, 1857.

26. See, for example, Sanderson, R. T., "Chemical Bonds and Bond Energy", Academic Press, New York, 1976.

27. Korcek, S., Chenier, J.H.B., Howard, J. A., and Ingold, K. U., Can. J. Chem., 1972, 50, 2285.

28. Joshi, R. M., J. Macromolecular Sci., 1971, A5, 687.

29. Caraculacu, A., Buriana, E. C., and Robila, G., J. Polymer Sci. Chem., 1978, 16, 2741.

30. Walling, C., Briggs, E. R., Wolfstein, K. B., and Mayo, F. R., J. Amer. Chem. Soc., 1948, 70, 1537.

31. Flory, P. J., and Leutner, F. S., J. Polymer Sci., 1948, 3,880; 1950, 5, 267.

RECEIVED April 24, 1981.

29

Catalysis of Thermal Initiation of Styrene Emulsion Polymerization by Emulsifiers

ZAID F. M. SAID, SABAH A. HASSAN, and ALEXANDER S. DUNN

Chemistry Department, University of Manchester Institute of Science and Technology, Manchester M60 1QD, England

Monomers which have appreciable thermal initiation rates polymerize at much higher rates in emulsion than in bulk in the presence of some emulsifiers. Breitenbach (1) found that styrene emulsified with sodium oleate polymerized at 11.5 % hr^{-1} at 20 °C, 10,000 times faster than in bulk. Matsumoto *et al.* (2) found that thermal polymerization of 66% suspensions of methyl methacrylate, ethyl acrylate, styrene, and vinyl acetate but not of acrylonitrile was accelerated by a factor of about four in the presence of polymethacrylic acid. In the case of methyl methacrylate, the overall energy of activation was the same whether polymethacrylic acid was present or not. This effect is most probably attributable to the stabilization of latex particles nucleated in the aqueous phase. In a series of papers published in the years 1970-75 from the Institute of Industrial Science of the University of Tokyo, Asahara, Arita, Seno, and Shiraishi showed (3) that the thermal polymerization of methyl methacrylate was greatly accelerated by emulsification with sodium oleate, sodium dodecyl sulphate, sodium tetrapropylene benzene sulphonate, and especially sodium 2-dodecylbenzene sulphonate but not with cetyl pyridinium chloride and two non-ionic emulsifiers. The polymerization was inhibited by hydroquinone. The rate of polymerization was found (4) to be proportional to the cube root of the monomer concentration and the cube root of the sodium tetrapropylene benzene sulphonate concentration. The same rate of radical generation was observed through the rate of consumption of diphenyl picryl hydrazyl when the monomer was replaced by ethyl acetate. The overall energy of activation was 12 kJ mol^{-1}, very much lower than the value (71 kJ mol^{-1}) found by Matsumoto *et al.* (2) and, indeed lower than the energy of activation for the propagation reaction of methyl methacrylate (about 21 kJ mol^{-1} (5)) which might indicate that the rate of the initiation reaction decreases with increase of temperature. Acrylonitrile did not polymerize but styrene did polymerize with an overall activation energy of 58 kJ mol^{-1} when sodium tetrapropylene benzene sulphonate was used (6): the reaction was inhibited by hydroquinone and the energy of activation for initiation was calculated as 53 kJ mol^{-1}, much lower than 115 kJ mol^{-1} found for the thermal polymerization in bulk (7) although this calculation involves the assumption that the degree of polymerization of the polystyrene obtained was equal to the kinetic chain length i.e. was not limited by transfer to monomer, styrene dimer, or emulsifier contrary to the conclusion reached by Breitenbach (1). The higher alkyl carboxylates were also found to be effective above their critical micelle concentrations (c.m.c.) but with styrene the rate was independent of alkyl chain length (8). By contrast, the rate did increase with alkyl chain length for methyl methacrylate with sodium alkyl benzene sulphonate surfactants (9). Studies of the composition of copolymers formed confirmed that the reaction involved free radical intermediates (10). Originally (3) peroxides in the emulsifiers were suspected as the source of free radicals and, indeed, the earlier results (4) in which an approximately 2/5th order in emulsifier was found can be explained in this way. Later it was found that the rates were first order in emulsifier concentration (11) but 2/5th order when deliberately peroxidized emulsifier was added (12) and the energy of activation for methyl methacrylate was now (11) found to be 33 kJ mol^{-1}. It was also found (11) that the results could be fitted by the Michaelis-Menten equation derived for enzymatically catalysed reactions which involve the formation of an intermediate enzyme-substrate complex. These results do suggest that solubilization of monomers in emulsifier micelles reduces the activation energy of some step in the thermal initiation mechanisms of methyl methacrylate (and also ethyl, butyl, and hexyl methacrylates (11)) and styrene and it is relevant in this connection that surfactants with a cis double bond were found to be particularly effective (8) whilst those with a trans double bond were ineffective. Further confirmation of the catalysis of the thermal polymerization of styrene by an emulsifier is contained in two very recent publications from the University of Sydney (13, 14). The background thermal rate in seeded emulsion polymerizations of styrene using sodium dodecyl sulfate as emulsifier with either persulfate (13) or γ-ray initiation (14) was found to be much higher (by a factor of 225) than would have been expected for bulk thermal polymerization. No emulsifier micelles were present in these experiments. In a batch *ab initio* polymerization under similar conditions a rate 50 times faster than the bulk thermal rate was observed. The energy of activation for thermal initiation was found to be (14) 87 ± 10 kJ mol^{-1} significantly lower than the bulk value.

In recent years micellar emulsifiers have been found to affect the rate of many reactions (15,16). This phenomenon of *micellar catalysis* originally attracted attention as a model for enzymatically catalysed reactions although the analogy is rather strained since a large excess of the micellar catalyst is usually required severely limiting the potential of such methods for practical preparations whereas small amounts of enzyme can achieve conversion of large amounts of substrate. So far, however, only ionic reactions have been found to be subject to micellar catalysis or inhibition. In the case of bimolecular reactions the effect is often largely explicable by the increase in the concentration of the reagents on solubilization in the micelles, but catalysis of unimolecular reactions has also been observed. Interaction of surfactant and substrate has been postulated to explain the effect in these cases and the Michaelis-Menten equation has been found applicable to such systems.

The presence of micelles is not a necessary condition for emulsion polymerization: the essential characteristic is the isolation of polymerizing radicals in separate loci preventing mutual termination and producing high molecular weights which, at low initiation rates, may be limited by transfer. Nevertheless micelles are normally present during Interval I of an emulsion polymerization in which latex particles are nucleated. Micellar nucleation of latex particles is dominant for monomers which have only a low solubility in water (e.g. styrene). For such a monomer any effect of micellar catalysis is likely to be revealed by an increase in the number of latex particles formed which would also result in an increased rate of polymerization. The thermal emulsion polymerization cited above seem to be a *prima facie* case of micellar catalysis. The thermal emulsion polymerization of styrene is investigated further here.

Experimental

Materials. Styrene (BDH Chemicals Ltd.) was stabilized with 0.002 % t-butyl catechol. The stabilizer was removed by washing successively with 10 % potassium hydroxide solution and water, drying over calcium chloride for 24 hr, and vacuum distilling. The monomer was kept in a refrigerator until required. Effective removal of inhibitor was checked by gas chromatography and by dilatometric measurement of the rate of bulk thermal polymerization at 60 °C: this was 0.080 % hr^{-1} which compares with a literature value (17) of 0.089 % hr^{-1}.

Potassium octadecanoate was prepared *in situ* by neutralization of stearic acid (Hopkin & Williams Ltd.) using a slight excess (2 cm^3 of 0.1 M) potassium hydroxide solution to prevent hydrolysis. A concentration of 8.2 x 10^{-2} mol dm^{-3} was used in all experiments. Although labelled 'Pure', this sample of stearic acid was stated to contain a maximum of 6 % palmitic acid and 3 % oleic acid. Another sample (BDH 'Specially Pure') was stated to be 99 % by gas-liquid chromatography. A concentration of 8.23 x 10^{-3} mol dm^{-3} was used with 4.0 x 10^{-3} mol dm^{-3} excess potassium hydroxide to prevent hydrolysis.

Sodium dodecyl benzene sulfonate (BDH Chemicals Ltd.) was 'suitable for gas-liquid chromatography' i.e. was free from commercial detergent additives but contained 60 % sodium sulfate. It was used at a concentration of 4.6 x 10^{-3} mol dm^{-3} which is four times the c.m.c. (1.15 x 10^{-3} mol dm^{-3} (18)) in the absence of an additional electrolyte. The concentration of sodium sulfate was 1.7 x 10^{-2} mol dm^{-3}. The same concentration of excess potassium hydroxide was used (4.0 x 10^{-3} mol dm^{-3}) as in the experiments with potassium octadecanoate.

Polymerization procedure. Polymerizations were carried out in a five-necked flask immersed in a thermostat. The flask was fitted with a reflux condenser and stirrer. Styrene and water were separately freed from oxygen by passage of 'oxygen-free' nitrogen from which the residual oxygen was removed by passage through alkaline pyrogallol solution. A slow stream of nitrogen was maintained above the emulsion during polymerization. In an experiment in which the effluent gas was passed through a liquid nitrogen trap it was found that at 80 °C no more than 1.3 % of the monomer was lost to the gas stream.

Determination of particle size was primarily by electron microscopy because the particle size distributions were broad which meant that the average particle volume could not be calculated satisfactorily from the average diameters, $\bar{d}_T$ and $\bar{d}_{LS}$, obtainable by the light-scattering procedure (19) used in previous work. However these average diameters can be calculated from the particle size distributions obtained by measuring electron micrographs and provide a useful check on the results since, when the particle size distribution is broad the smallest particles may be out of focus and therefore easily overlooked on the electron micrographs. A drop of diluted latex was placed on a carbon-coated collodion covered grid and allowed to dry. The grid was dipped into hot water to remove emulsifier and photographed directly in the electron microscope at a magnification of x 15 000. Magnification was calibrated by use of a carbon replica grating. Prints enlarged x 4 were made on Kodagraph P84 lightweight projection paper for use with the Carl Zeiss TGZ-3 Particle Size Analyser which enables the number of particles in each of 48 size ranges to be counted. 800 particles were counted for each sample. To calculate the number of particles in a given volume of the water phase, the root mean cube diameter, $\bar{d}_{rmc}$, is required (20): this is the diameter of a particle with the number-average volume. The number-average diameter, $\bar{d}_n$, has often been used in the past but the number of particles per unit volume calculated on this basis will be too large unless the particle size distribution is nearly monodisperse. Conversely use of the higher averages derived from light scattering measurements $\bar{d}_\tau$ (originally written as $\bar{d}_w$) and $\bar{d}_{LS}$ (originally $\bar{d}_s$) give particle numbers which are too small.

$$\bar{d}_n = \frac{\Sigma n_i d_i}{\Sigma n_i}, \quad \bar{d}_{rmc} = \sqrt[3]{\frac{\Sigma n_i d_i^3}{\Sigma n_i}}, \quad \bar{d}_\tau = \sqrt[3]{\frac{\Sigma n_i d_i^6}{\Sigma n_i d_i^3}}, \quad \bar{d}_{LS} = \sqrt{\frac{\Sigma n_i d_i^8}{\Sigma n_i d_i^6}}$$

Results

TABLE I

Variation with Temperature of the Number of Latex Particles formed in the Thermal Polymerization of Styrene emulsified with 9.15×10^{-2} mol dm^{-3} Potassium Octadecanoate

Temperature/^{o}C	60	65	70	75	80
$\bar{d}_{rmc}$/nm	98.4	92.0	90.0	79.1	75.3
10^{-15} N/cm^3 water	1.03	1.27	1.35	1.99	2.31
$\bar{d}_n$/nm	90.6	75.9	70.4	67.0	61.8
$\bar{d}_\tau$/nm	107.2	129.7	142.0	110.8	129.8

A graph of $\log_{10}N$ against the reciprocal of the absolute temperature (Fig. 1) was obtained using a Least Mean Squares Program on the Hewlett-Packard 9810A Calculator: this has a gradient of -2.09×10^3 K with a correlation coefficient of -0.977. Hence the energy of activation for the nucleation of latex particles, E_N = 40.0 kJ mol^{-1}.

TABLE II

Variation with Temperature of the Number of Latex Particles formed in the Thermal Polymerization of Styrene emulsified with 4.6×10^{-3} mol dm^{-3} Sodium dodecyl benzene sulfonate in 1.7×10^{-2} mol dm^{-3} Sodium sulfate

Temperature/^{o}C	55	60	65	70
$\bar{d}_{rmc}$/nm	156.9	140.5	128.4	116.4
10^{-15}N/cm^3 water	0.200	0.346	0.454	0.600

The graph of $\log_{10}N$ against the reciprocal of the absolute temperature (Fig. 1) has a gradient of -3.47×10^3 K with a correlation coefficient of 0.986, whence E_N = 66 kJ mol^{-1}.

Discussion

Measurement of the energy of activation for the nucleation of latex particles, E_N, permits the energy of activation for

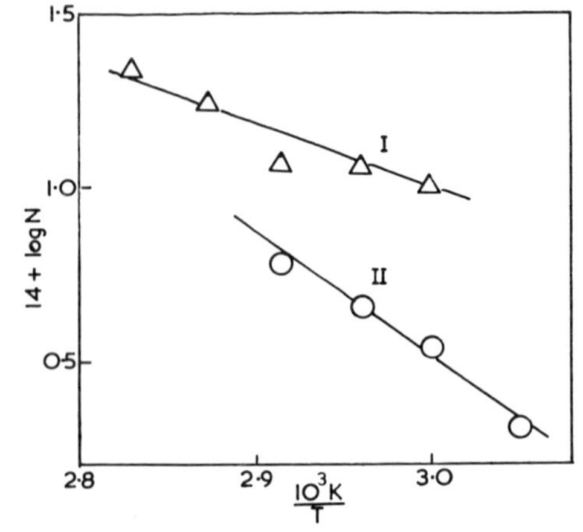

Figure 1. Arrhenius plots of dependence of number of particles formed per cm³ water on temperature in the thermal polymerization of styrene emulsified with (I) potassium octadecanoate and (II) sodium dodecyl benzene sulfonate

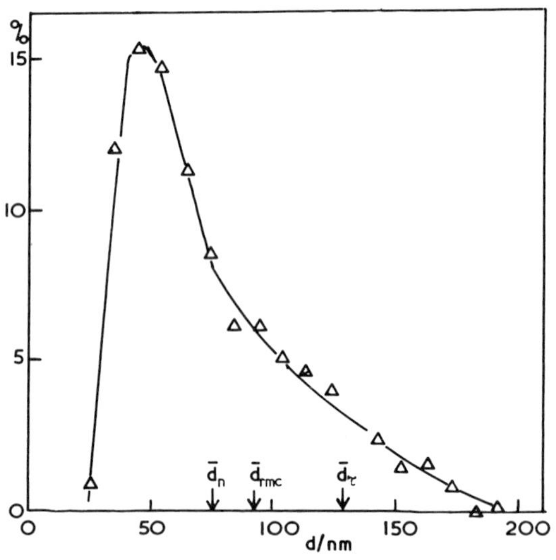

Figure 2. Particle size distribution for latex particles formed by thermal polymerization of styrene emulsified with potassium octadecanoate at 65°C

initiation of emulsion polymerization, E_i, to be calculated for a reaction which conforms to Smith and Ewart's Case 2 (21) throughout the temperature range provided, a_S, the area occupied by a surfactant molecule at the polymer/water interface is independent of temperature. This would imply that the enthalpy of adsorption of the surfactant is zero which seems unlikely.

Bartholmé *et al.* (22) found for styrene with persulfate initiation and a sodium alkyl benzene sulfonate emulsifier that there was a discrepancy between their measured value of E_N (21.7 kJ mol^{-1}) and that calculated (on the assumption that $\Delta H_s = 0$) from the expression $E_N = {}^2/_5(E_i - E_p)$. However their value of E_p (which was derived from measurements of the rate of seeded emulsion polymerization experiments in which N was the same at all temperatures) now seems to be too high probably because the average number of radicals per particle, $\bar{n} < 0.5$ at the lower temperatures: taking E_p = 32.5 kJ mol^{-1} as the best estimate, ΔH_s can be calculated from $\Delta H_s = {}^5/_2(E_N(\text{exp}) - E_N(\text{calc}))$ = -57.2 kJ mol^{-1}. The recent work of Piirma and Chen (23) clearly shows a significant increase in a_S for sodium dodecyl sulfate on polystyrene with temperature.

However, the Smith-Ewart equation for the dependence of the number of latex particles formed, N, on a_S, surfactant and initiator concentration, initiation and propagation rate constants does not apply to cases in which initiation is in the oil phase. Square root dependence on the rate of initiation and linear dependence on surfactant concentration have been observed experimentally (24) but the observations do not extend to the effect of variation of a_S or of the particle growth rate (involving the propagation rate constant, k_p). Consequently, it is not possible to calculate E_i from observations of E_N in this case. However the fact that different results are obtained with different emulsifiers (Fig. 1) indicates an effect of the emulsifier on the thermal initiation mechanism for styrene unless it can be shown that the effect can be accounted for completely by differences in the temperature dependence of a_S for the emulsifiers (i.e. differences in the enthalpies of adsorption, ΔH_s).

In the present experiments greatly enhanced rates of thermal emulsion polymerization were observed when potassium octadecanoate or sodium dodecyl sulfate (at 0.12 mol dm^{-3}) whereas sodium dodecyl benzene sulfonate and 'Triton' X-100 (Rohm & Haas, a non-ionic emulsifier octylphenoxypoly(ethyleneoxy)-ethanol) did not enhance the rate. The conversion after 12 hr at 60 °C with potassium octadecanoate was 69 % whereas with sodium dodecyl benzene sulphonate it was only 29 % (Fig. 2).

Increase of ionic strength reduces the stability of electrostatically stabilized latex particles and causes them to coalesce at sufficiently high values. However the difference in the effects of potassium octadecanoate and sodium dodecyl benzene sulfonate cannot be attributed to presence of sodium sulfate in the latter. Addition of 0.1 mol dm^{-3} sodium sulfate to potassium octadecanoate was sufficient to produce only a slight increase in particle size (Table III). Neither does the presence of minor amount of potassium oleate and potassium hexadecanoate in the sample of potassium octadecanoate used in most experiments have any significant effect although this would probably not have been the case if the impurities had had a lower c.m.c. than the bulk of the material.

TABLE III

Mean particle sizes of lattices obtained by thermal polymerization of styrene at 60 °C

Emulsifier	Conc./mol dm^{-3}	Na_2SO_4/mol dm^{-3}	$\bar{d}_{rmc}$/nm
Sodium dodecyl benzene sulfonate	0.0046	0.017	140
Potassium octadecanoate 93%	0.00915	-	100, 103
" " "	0.00915	0.1	110
" " 99%	0.00915	-	102
Sodium dodecyl sulfate	0.12	-	80

The analysis of their results by Asahara et al.(6) appears to be in error. Their use of the viscosity-average degree of polymerization, $\bar{P}_{\eta}$, where the number-average degree of polymerization, $\bar{P}_n$, is really required need not lead to any error in the energy of activation of the degree of polymerization, $E_{\bar{P}}$, so long as the molecular weight distribution does not change with temperature. However their assumption that termination of polymer molecules by all transfer processes may be neglected so that the degree of polymerization may be equated to the kinetic chain length seems very dubious since it is generally agreed that the molecular weight of polystyrene is limited at low rates of polymerization by transfer to monomer or styrene dimer which has a labile hydrogen atom and a much higher transfer coefficient than the monomer. Moreover with oil-phase initiation transfer to monomer (or, more probably to emulsifier to give a more water soluble radical) may be essential to allow one of a pair of radicals to escape to the water phase so as to leave an isolated radical in the polymer phase which is essential to permit emulsion polymerization to proceed at all. But their principal error is to have confused the rate of polymerization and the rate of propagation. It is of course true that in a polymerization reaction the overall rate of polymerization is equal to the rate of the propagation reaction. But the expression for the rate of the propagation reaction $R_{prop} = k_p((M)(M\cdot)$ contains the polymer radical concentration $(M\cdot)$ which in an emulsion polymerization conforming to Smith and Ewart's Case 2 may be equated to

$\frac{1}{2}$N/L (L = Avogadro's number). In an <u>ab initio</u> emulsion polymerisation (as distinct from a seeded reaction), N is not independent of temperature. It is correct to write (assuming transfer can be neglected) $\bar{P}_n = R_{overall}/R_i$ since the overall rate of polymerization depends on N and the overall energy of activation includes E_N. They give the overall energy of activation as 58 kJ mol^{-1} and find $E_{\bar{p}}$ = -21.7 kJ mol^{-1} for sodium tetrapropylene benzene sulfonate. Since $E_i = E_{overall} - E_{\bar{p}}$ (and not $E_{prop} - E_{\bar{p}}$ as they state), E_i = 80 kJ mol^{-1} (and not 53 kJ mol^{-1} which they derived from the erroneous expression). This is close to Russell and Tobolsky's bulk value of E_i 86 kJ mol^{-1} (<u>25</u>) which they quote. But in fact, Russell and Tobolsky's value seems to be too low and the amended value of 80 kJ mol^{-1} is still significantly lower than the median bulk value of E_i = 115 kJ mol^{-1}.

Duerksen and Hamielec (<u>26</u>) collated much of the literature data on the energy of activation for the thermal initiation of stryene in bulk or solution which was available in 1968 and concluded that the value of 115 kJ mol^{-1} derived from experiments in a Continuous Stirred Tank Reactor was concordant with the existing values which they quoted. Recent work by Bengough <u>et al.</u> (<u>27</u>, <u>28</u>) gives values of 121 and 118 kJ mol^{-1} supporting the median value of 115 kJ mol^{-1}.

It thus seems that sodium dodecyl sulfate, sodium tetrapropylene benzene sulfonate, and potassium octadecanoate do accelerate the thermal initiation reaction of styrene but that 'Triton X-100' and sodium dodecyl benzene sulfonate do not although the latter is effective in accelerating thermal initiation of alkyl methacrylates (<u>3</u>).

The mechanism by which emulsifiers could influence the rate of the thermal initiation reaction is obscure. Most probably the emulsifiers increase the efficiency with which one of the radicals produced in the thermal initiation process escapes into the aqueous phase so that emulsion polymerization may begin. If so those emulsifiers for which exchange between the micelle or the adsorbed layer on a latex particle and true solution in the aqueous phase is most rapid should be most effective in promoting the thermal polymerization. Recently the kinetics of micellization has attracted much attention (<u>29</u>) but the data which is available is inadequate to show whether such a trend exists. It would be necessary to determine the activation energy for thermal initiation in emulsion with two emulsifiers of similar structure selected to have exchange rates which differed as much as possible.

<u>Summary</u>

Evidence in the literature (<u>3</u>, <u>4</u>, <u>6</u>, <u>8-12</u>) showing that some emulsifiers can accelerate the thermal initiation reaction of monomers is confirmed although it is shown that there has been some confusion in calculating a value for the energy of activation for the thermal initiation of styrene emulsified with sodium tetrapropylene benzene sulfonate (6) which means that the value is not so low as was suggested. Nevertheless some emulsifiers do increase the rate of thermal initiation of styrene whereas others do not.

Literature Cited

1. Breitenbach, J.W. Anz. Akad. Wiss. Wien Math-naturw. Kl. 1946, 83, 9 cf. Chem. Abstr. 1949, 43, 8736b.
2. Matsumoto, T., Mune, I., Izutsu, H. Kobunshi Kagaku 1969, 26, 234.
3. Asahara, T., Seno, M., Shiraishi, S., Arita, Y. Bull. Chem. Soc. Japan 1970, 43, 3895.
4. Asahara, T., Seno, M., Shiraishi, S., Arita, Y. Bull. Chem. Soc. Japan 1972, 45, 2862.
5. Bamford, C.H., Barb, W.G., Jenkins, A.D., and Onyon, P.F. "The kinetics of vinyl polymerization by radical mechanisms" Butterworths, London, 1958; p. 82.
6. Asahara, T., Seno, M., Shiraishi, S., Arita, Y. Bull. Chem. Soc. Japan 1973.
7. Hui, A.W., Hamielec, A.E. J. Appl. Polym. Sci. 1972, 16, 749
8. Arita, Y., Shiraishi, S., Seno, M., Asahara, T. Nippon Kagaku Kaishi 1973, (5), 1042, cr. Chem. Abstr., 1973, 79, 53830h.
9. Arita, Y., Shiraishi, S., Seno, M., Asahara, T. Nippon Kagaku Kaishi 1972, (12), 2412, cf. Chem. Abstr. 1973, 78, 86295v.
10. Arita, Y., Shiraishi, S., Seno, M., Asahara, T. Nippon Kagaku Kaishi 1975, (7), 362, cf. Chem. Abstr. 1975, 83, 115062c.
11. Arita, Y., Shiraishi, S., Seno, M., Asahara, T. Nippon Kagaku Kaishi 1975, (2), 374, cf. Chem. Abstr. 1975, 82, 156808y.
12. Arita, Y., Shiraishi, S., Seno, M., Asahara, T. Nippon Kagaku Kaishi 1975, (2), 379, cf. Chem. Abstr. 1975, 82, 156809z.
13. Hawkett, B.S., Napper, D.H., Gilbert, R.G. J.C.S. Faraday Trans. I. 1980, 76, 1323.
14. Lansdowne, S.W., Gilbert, R.G., Napper, D.H., Sangster, D.F. J.C.S. Faraday Trans. I 1980, 76, 1344.
15. Fendler, J.H., and Fendler, E.J. "Catalysis in Micellar and Macromolecular Systems" Academic: New York, 1975.
16. Bunton, C.A. in Mittal, K.L. ed. "Solution Chemistry of Surfactants" Vol. 2, Plenum: New York, 1979; p.519.
17. Mayo, F.R., J. Amer. Chem. Soc. 1968, 90, 1289.
18. Gerrens, H. and Hirsch, G. in Brandrup, J. and Immergut, E.H. ed. "Polymer Handbook" Wiley: New York, 2nd Edn. 1975, II-485.

19. Burnett, G.M., Lehrle, R.S., Ovenall, D.W., Peaker, F.W. J. Polym. Sci. 1958, 29, 417.
20. Gardon, J.L. J. Polym. Sci. A-1 1968, 6, 623.
21. Smith, W.V., Ewart, R.H. J. Chem. Phys. 1948, 16, 592.
22. Bartholomé, E., Gerrens, H., Herbeck, H., Weitz, R.M. Z. Elektrochem. 1956, 60, 334.
23. Piirma, I., Chen, S.-R. J. Coll. Interface Sci. 1980, 74, 90.
24. Berezhnoi, G.D., Khomikovskii, P.M., Medvedev, S.S. Vysokomol. Soed. 1960, 2, 141.
25. Russell, K.E., Tobolsky, A.V. J. Amer. Chem. Soc., 1953, 75, 5052.
26. Duerksen, J.H., Hamielec, A.E., Polym. Preprints 1968, 9 (1), 757.
27. Barr, N.J., Bengough, W.I., Beveridge, G., Park, G.B. Eur. Polym. J. 1978, 14, 245.
28. Bengough, W.I., Park, G.B. Eur. Polym. J. 1978, 14, 889.
29. Muller, N. in Mittal, K.L. ed. 'Solution Chemistry of Surfactants' Vol. 1, Plenum: New York, 1978, p.267.

RECEIVED April 6, 1981.

30

Latex Seed Particle Growth at High Surfactant Surface Coverage

J. R. ERICKSON and R. J. SEIDEWAND

Glidden Coatings and Resins, Division of SCM Corporation,
16651 Sprague Road, Strongsville, OH 44136

The preparation of heterogeneous latex particles by two or more stage processes is a subject of considerable interest and importance (1,2,3). First stage (seed) particles are produced, and in the next stage(s) more monomer(s) of a different composition is added and polymerized. The additional monomer is intended to continue the growth of the seed particles to produce composite, heterogeneous particles. These hetero-particles are useful in plastics, adhesives, and coatings. Often they are characterized as being layered (4) or core//shell particles (5).

The journal (6,7) and patent (8,9,10) literature indicate that the amount of surfactant in the seed latex must be below the critical micelle concentration, cmc, before and during the next stage monomer addition and polymerization, if complete association (seed growth) is to occur. This is equivalent to stating that the seed particle surfactant surface coverage, S, must be below 100%, since complete coverage with a condensed and incompressible surfactant monolayer is equated with the attainment of the cmc in the surrounding aqueous phase. It is inconsequential whether the emulsifiers are anionic, nonionic, cationic, or mixtures, or the second stage monomer is batch charged or monomer fed, except that in the latter case, the instantaneous seed must have S significantly under 100%. Specifically, references 8 and 9 require the growing particles to have S<60-70%.

The present paper demonstrates that for polystyrene seed latexes and styrene-acrylic or all acrylic second stage monomers, complete association can take place when S>100%, if certain mixtures of anionic and nonionic surfactant are used. The morphology of some of the two-stage latexes is described.

Experimental

Several types of equipment were used to prepare the latexes. A five gallon glass reactor, five and two liter Morton flasks equipped with condensers, stainless steel stirrers, thermoregulators, and nitrogen inlets were used to prepare the seed latexes

0097-6156/81/0165-0483$05.50/0

in required quantities. The five gallon reactor contained internal thermoregulated cooling and heating coils while Morton flasks sat in thermoregulated water baths. Second stage polymerizations were generally carried out in twelve ounce beverage bottles. These were capped, placed in a thermostated tank, and tumbled end over end. The copolymer and acrylic components of the blend systems corresponding in composition to the two-stage latexes were also prepared in beverage bottles.

Reagent grade potassium persulfate, azobisisobutyronitrile (AIBN), .1N sodium hydroxide, dodecyl mercaptan, and sodium bicarbonate were used as received. Commercial grades of styrene (St), n-butyl acrylate (BA), methyl methacrylate (MMA), and methacrylic acid (MAA) were used without additional purification. Triton X-100 (polyoxyethylene isooctylphenyl ether from Rohm and Haas Co.) and Siponate DS-10 (a relatively clean commercial grade of sodium dodecylbenzene sulfonate from Alcolac Chemical Co.) were used as the nonionic and anionic surfactants in preparing the polystyrene seed and heterogeneous two-stage latexes. Two lots of Triton X-100 and three lots of Siponate DS-10 were used. The Triton X-100 contained ca. 1.6% moisture while the three lots of Siponate DS-10 varied in moisture content. The amount of active surfactant in each lot of Siponate DS-10 was determined by subtracting 2% (Siponate DS-10 nominally contains 2% Na_2SO_4 impurity) from the experimentally measured non-volatile contents. Aerosol MA (American Cyanamid Co.) and Triton X-114 (Rohm and Haas Co.) were the surfactants used for the preparation of the copolymer and acrylic components of the blend latexes. Deionized water was used in all of the experiments.

Surfactant titrations were carried out using a Cenco Tensiometer (Model No. 70545) equipped with a Platinum-Iridium ring.

Latex particle sizes and polydispersity were measured by use of an ICI Joyce Loebl disc centrifuge (JLDC) photosedimentometer (11,12).

For morphological studies, transmission electron microscopy (TEM) samples were prepared by adding 2-3 drops of latex to 20 ml of water and then placing one drop of the diluted latex solution on a carbon deposited copper grid and allowing the water to evaporate before placing in a Hitachi 100 kv electron microscope. Scanning electron microscopy (SEM) samples were prepared by making drawdowns on glass or teflon coated panels with a Gardner Ultra Applicator (Ser. No. 236) to give ca. a 4 mil wet film and ca. a 2 mil dry film. The fracture sections were prepared by immersion of the sample film in liquid N_2 and fracturing while immersed. The samples were then mounted and coated with an Au/Pd alloy and placed in a Cambridge Stereoscan (S4-10), equipped with a LaB_6 electron emitter.

Glass transition temperatures were measured by differential scanning calorimetry (DSC) using a DuPont 900 Differential Thermal Analyzer. The samples were cooled to -100°C in a closed pan and then scanned to 150°C at a rate of 15°/minute.

The minimum film temperatures (MFT) were determined using a MFT tester which consists of a graduated grooved metal temperature bar (13). Crack points were the highest temperatures at which visual discontinuities were present. The knife point MFT was the minimum temperature at which the film first exhibited continuity while being scrapped from low to high temperature.

Seed Latexes

Seed Preparation. Polystyrene seed latexes were produced by either the batch charge or the monomer feed technique. For batch charging, all the ingredients (Table I) were loaded into the reactor and purged with N_2 during upheating to 65°C. The initiator was added and the polymerizations were carried out at 65° under a N_2 blanket. For the monomer fed batches, the DDM was mixed into the styrene, and 10% of the mixture was added during the initial charging. When 65°C was reached, initiator was added. Approximately fifteen minutes later, the monomer feed was started and then added uniformly over three hours. Seeds 1-4 were used as ingredients in the two-stage polymerizations. Seed 9 was used for the determination of surfactant surface area.

Table I - Seed Latex Compositions

Ingredients	Seed 1	2 & 3	4	9
Siponate DS-10				
active	.127	.094	.156	.093
inert	.006	.006	.015	.006
Triton X-100				
active	3.40	1.31	3.37	1.31
inert	.06	.02	.05	.02
.1N NaOH	11.51	-	7.36	-
$NaHCO_3$	-	.090	-	.090
H_2O	133.33	132.67	111.58	132.88
Styrene	100.00	100.00	100.00	100.00
Dodecyl mercaptan	-	.200	-	.200
$K_2S_2O_8$	.180	.180	.101	.180
10% Siponate DS-10	-	4.94	-	-
Reactor	5 gal.	5 litre	5 gal.	2 litre
Type polymerization	batch	fed	batch	fed

Seed Characterization. A monodisperse seed latex of known initial diameter is required if continued seed growth by second stage polymers is to be unambiguously measured by particle size analysis. The seed recipes selected were known to give uniform sized particles (14,15). All five of the latexes were monodisperse as measured by JLDC (Table II) and exhibited a characteristic violet pink color (16). The monodispersity was further confirmed by TEM (Figure 1).

Table II - Seed Results

	Seed 1	2	3	4	9
Volume surface average diameter (μ)	.2464**	.2489**	.2530*	.2461*	.2408*
Polydispersity (Dw/Dn)	1.01	1.01	1.03	1.02	1.01
Particle density (g/cm^3)	1.057	1.057	1.057	1.056	1.058
% Conversion	99.1	99.0	99.3	99.3	98.4
Experimental % Polymer + residual monomer	40.2	41.8	41.8	44.3	42.5

*Duplicate analysis. **Triplicate analysis.

The batch charged recipes required more total surfactant, and a greater fraction of that surfactant as nonionic, than the monomer fed recipes employed to obtain the same particle size.

Surfactant Surface Area

The surface area per gram occupied by Siponate DS-10 and Triton X-100 were measured to allow the calculation of surfactant surface coverage on the seed latexes. The experimental procedure and analysis were based on Maron's technique (17,18). The procedure utilized is explained in detail below.

A surfactant solution (titrant) is used to titrate L grams of a diluted latex (mixture) to produce the typical plot shown in Figure 2, in which v ml of titrant are required to reach the cmc. The mass of surfactant per gram of mixture at the cmc is

$$\alpha = \frac{D + T + v(C_D + C_T)}{L + \rho v} \quad (1)$$

where D and T are the grams of active Siponate DS-10 and Triton X-100, respectively, in the amount of seed latex used; C_D and C_T are the concentrations (g/ml) of the active surfactants in the titrant, and ρ is the titrant density. At cmc, the mixed surfactant is distributed between two locations, as shown.

$$\alpha = \beta + \gamma M \quad (2)$$

Here β is the grams of surfactant in the water phase per gram of mixture, and γM is the grams of surfactant on the particle surface per gram of mixture. γ is the grams of surfactant at the particle surface per gram of polymer, while M is the grams of polymer (including residual monomer) per gram of mixture.

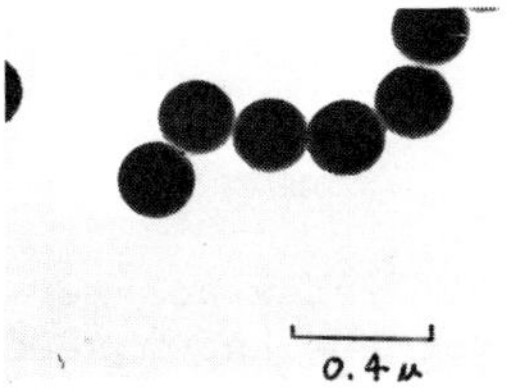

Figure 1. TEM photomicrograph of Seed 9

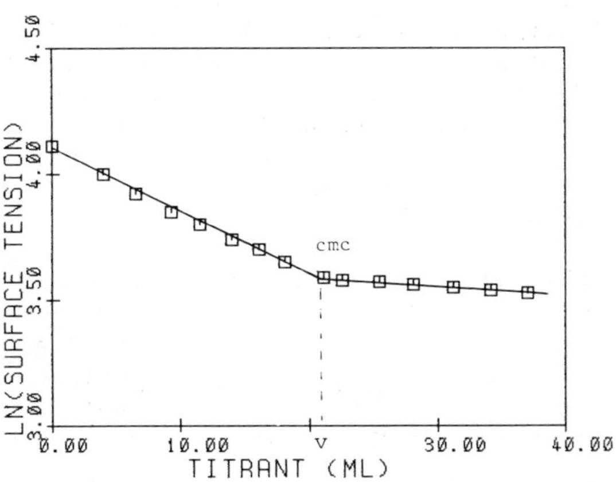

Figure 2. Titration of 12 g of Seed 9 polymer containing .0112 g of Siponate DS-10 and .1868 g of Triton X-100 with 1% Siponate DS-10 solution

Performing a series of Figure 2 type titrations, starting with different amounts of the same seed latex in the mixture and ending with the same fractional amount of nonionic surfactant, N (Equation 3) provides a set of α vs. M values. These can be plotted to provide estimates for β and γ.

$$N = \frac{T + C_T v}{D + T + v(C_D + C_T)} \tag{3}$$

The polystyrene surface area occupied per gram of mixed surfactant of composition N is given as

$$A_N = \frac{a}{\gamma_N} \tag{4}$$

Here a is the surface area (cm^2) per gram of latex polymer. The area/g of the latex polymer is obtained from the known seed diameter and density.

Starting with mixtures of Seed 9 and water, four sets of surface tension titrations were performed at 25°C. First, the mixtures were titrated with 0.01 g/ml Siponate DS-10. For the second and third sets, the surfactant composition of the starting mixtures were adjusted by adding known quantities of Triton X-100 solution. Titrations were then performed using the Siponate DS-10 titrant. Finally, unadjusted starting mixtures were titrated using 0.01 g/ml Triton X-100. The data collected from these titrations is given in Table III.

Table III - Titration Data for Seed 9

g polymer	D	T	$C_D v$	$C_T v$	Total g Surfactant	N	$L + \rho v$
7.01	.0065	.0918	.1718	0	.2701	.340	116.5
12.01	.0112	.1573	.2440	0	.4125	.381	127.2
17.00	.0159	.2227	.3352	0	.5738	.388	132.1
7.01	.0065	.1093	.1486	0	.2644	.413	116.2
12.00	.0112	.1868	.2196	0	.4176	.447	124.0
17.01	.0159	.2650	.2740	0	.5549	.478	130.5
7.01	.0065	.2197	.0324	0	.2586	.850	116.3
12.00	.0112	.3599	.0552	0	.4263	.844	126.0
17.01	.0159	.4973	.0783	0	.5915	.841	136.3
7.00	.0065	.0918	0	.1791	.2774	.977	118.3
12.00[(a)]	.0112	.1572	0	.2824	.4508	.975	134.7
17.00	.0159	.2227	0	.4171	.6557	.976	142.4

(a) Linear Figure 2 type plot used.

Using data from Table III, the experimental values of α, M, and average N were calculated. These are listed in Table IV.

Table IV - γ, M, and N Values for Figure 3 Type Plots

Average N	α	M
	.002318	.06017
.370	.003243	.09442
	.004344	.1287
	.002275	.06033
.446	.003368	.09677
	.004252	.1303
	.002224	.06028
.845	.003383	.09524
	.004340	.1248
	.002345	.05917
.976	.003347	.08909
	.004605	.1194

Figure 3 type plots were constructed using the data in Table IV. Referring to Figure 3,

$$\gamma = p + I \quad (5)$$

$$\beta = I(1-M) \quad (6)$$

where p is the least squares fit slope and I is the intercept, i.e. the cmc of the mixed surfactant in pure water. The γ, I, and computed A values obtained from the least squares fit of Figure 3 and the similar plots done for the remaining three N values are listed in Table V.

As seen in Figure 4, a linear fit of A to N is called for and should provide reasonably accurate estimates for A when $.3 < N < 1$.

Table V - Estimates of γ_N, I_N, and A_N

N	$\gamma \times 10^2$ (a)	$I \times 10^2$ (b)	$A \times 10^{-7}$
.370	3.007	.051	.785
.446	2.887	.059	.817
.845	3.306	.025	.714
.976	3.761	.008	.624

(a) Standard deviation of $\gamma = .148\times10^{-2}$ with 4 degrees of freedom.
(b) Standard deviation of $I = .014\times10^{-2}$.

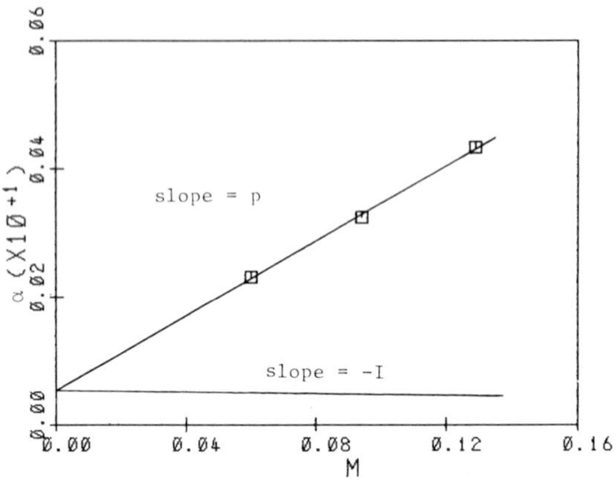

Figure 3. Maron-type plot for average N = *.370*

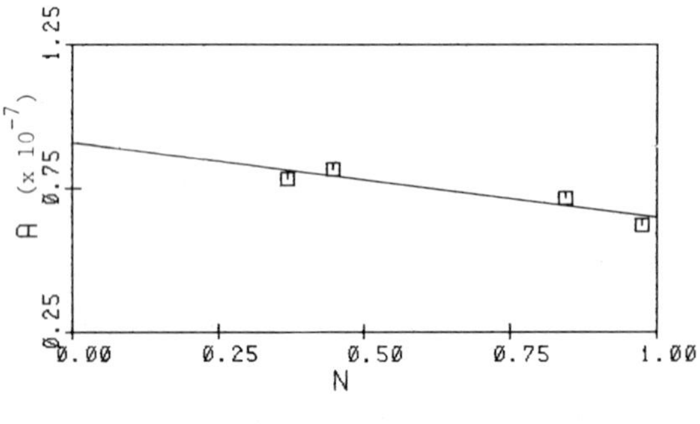

Figure 4. Surface area per gram of surfactant mixture (A) as a function of the nonionic surfactant fraction (N)

The linear equation is

$$A = (.913 - .268N) \times 10^7 \ cm^2/g \qquad (7)$$

It has a correlation coefficient of .94 and a standard error of estimate of $.036 \times 10^7$ with two degrees of freedom. When the estimated error for seed particle size determination and % polymer are included, the standard error of estimate is $.052 \times 10^7$. Thus, a fixed area per gram could be assigned to each surfactant and no confirmation of the anionic surfactant's area changing with the relative presence of nonionic (19) is indicated.

The fixed area/g for active Siponate DS-10 and Triton X-100 over polystyrene are shown below.

$$A_{N=0} = .913 \times 10^7$$

$$A_{N=1} = .645 \times 10^7 \ cm^2/g$$

These A_N values correspond to 53 and 67 $\mathring{A}^2$/molecule, respectively.

The initial surface coverages and N values for Seeds 1-4 and 9 have been computed using the determined area/gram information for the surfactants (Table VI).

Table VI - % Surface Coverages of Seeds as Synthesized

	Seed 1	Seed 2	Seed 3	Seed 4	Seed 9
S (%)	100	61	62	100	39
N	.96	.69	.69	.96	.93

Second Stage Polymerizations

Styrene-Acrylic Second Stage. Two series of polymerizations were done at a 35//65 first//second stage polymer ratio using a 70/28/2 parts by weight mixture of BA/St/MAA as the second stage monomer. A spread of S and N values were obtained on the seed latexes by adjusting them with additional anionic and nonionic surfactant. Table VII provides the basic compositions for the polymerizations. For bottle runs the ingredients were charged in the order shown (Table VII). Before adding the initiator, the bottles were purged with N_2, then the bottles were capped, shaken and placed in the thermostated bath at 65°C where they tumbled for ca. 16 hrs. When the polymerizations were done in 2 litre flasks, the seed, surfactants and water were added and mixed. The monomers were then added. During the ca. 45 minute heat up period to reaction temperature, the ingredients were purged with N_2. The initiator was added and the polymerization was carried to completion at constant temperature under a N_2 blanket.

Table VII - 35//65 - Two Stage Latexes Using Styrene/acrylate Second Stage Monomer

Ingredients	Parts by Weight	
Polystyrene Seed 1	87.03	-
Polystyrene Seed 2 or 3	-	83.68
Siponate DS-10*	Variable	Variable
Triton X-100*	Variable	Variable
H_2O	Variable	Variable
BA/St/MAA Monomer 70/28/2	65.00	65.00
$K_2S_2O_8$	.12	.23
Total	222.24	250.00
Reactor	12 oz. bottle	2 litre flasks

*Added as water solutions.

The weight fraction of the second stage polymer, y, which associated with the seed was obtained using Equation 6.

$$y = \frac{D_f^3 - D_i^3}{D_p^3 - D_i^3} \qquad (6)$$

Here D_f is the final, composite particle diameter, D_i is the initial (seed) diameter, and D_p is the projected diameter. D_f is obtained by using the previously reported (20) interative procedure. The method of calculating D_p can be found in the Appendix of that previous report. The independent estimate of error on y, determined from the present and related studies is .16 (22 degrees of freedom). Because of random error, y > 1.00 should often be observed when complete association occurs.

The results for the first series of polymerizations (using Seed 1) are shown in Table VIII. The fact that complete association occurs at S > 100% is seen. Further, with the y values ordered in the manner shown, the y decreases with increasing S and decreasing N.

Table VIII - Effect of S and N on y

Composition	S (%)	N	y
St//BA/St/MAA	314	.95	1.33
35//65 (70/28/2)	326	.93	1.06
	211	.83	.94
	127	.83	.93

Table VIII - (Continued)

Composition	S (%)	N	y
	418	.88	.36
	302	.77	.33
	543	.93	.32
	619	.88	.03

The N values in Table VIII are highly constrained because of the use of Seed 1. As made, Seed 1 had 100% surfactant surface coverage and N = .96. By adding additional surfactant, S could be varied upward, but relatively low values for N could not be achieved without producing extraordinarily high S values. To allow a wider range of N values, the monomer fed seed composition was developed and replicate Seeds 2 and 3 were prepared and used. These seeds had low S values and did allow relatively low N values to be obtained in the second stage polymerizations. Four S, N positions were checked. To determine whether the apparent y behavior with S and N values is independent of polymerization temperature, polymerizations were carried out at both 65° and 75°C at each of the four S, N points. Further, the polymerizations were done in 2 litre flasks so that % non-volatile monitoring could be done to determine if any correlation exists between the observed y values and the rate of polymerization. The results are given in Table IX.

Table IX - Effect of S, N, and Temperature on y and Rate

Composition	S (%)	N	T (°C)	Rate (%/hr.)	y
St//BA/St/MAA	251	.71	65	9*	.87*
35//65 (70/28/2)			75	22*	.87*
	203	.31	65	35	.29
			75	133*	.45*
	424	.71	65	48	.17
			75	100*	.35*
	336	.31	65	29	.26
			75	175*	.12*

*Average of two runs.

The rate of polymerization increased with increasing S, and decreasing N as might be expected if more second generation particles are being formed which is consistent with the determined y values. Increasing temperature also increased the rate of polymerization, but the temperature had no effect on the observed y values.

When the data from Table VIII & IX are combined (considering similar runs at different temperatures as replicates) as shown in Table X, there is little doubt that S and N control y in the manner previously indicated over a very wide range of S and N values. It should also be apparent that not only is it possible to find conditions above S = 100% which give complete second stage association but it is also possible to pick conditions which provide relatively exact amounts of second generation polymer to give predetermined in situ hetero-particle copolymer blends.

Table X - Complete Set of Styrene-acrylic Second Stage Results

Composition	S	N	y
	314*	.95	1.33
St//BA/St/MAA	326	.93	1.06
35//65 (70/28/2)	211	.83	.94
	127	.83	.93
	251***	.71	.87
	203**	.31	.39
	418	.88	.36
	302	.77	.33
	543	.93	.32
	424**	.71	.29
	336**	.31	.17
	619	.88	.03

*Average of two runs, **three runs, ***four runs.

<u>All Acrylic Second Stage</u>. Butyl acrylate and mixtures of butyl acrylate with methyl methacrylate, and methacrylic acid were used as second stage monomers. Polymerizations were done using both 65//35 and 35//65 first//second stage polymer ratios. Again, a spread of S and N values were obtained using a batch charge seed latex and adjusting it with additional anionic and nonionic surfactant. The initiator was $K_2S_2O_8$ except where noted. Table XI provides the basic compositions for the second stage polymerizations. All ingredients except for the initiator were charged into the bottle, purged with N_2, capped and tumbled for two hours in the 65°C bath. The initiator was then added and the bottles recapped and tumbled for 16 hours at 65°C.

Table XI - Two-Stage Composition with All Acrylic Second Stage

Ingredients	65//35[a]	35//65
Seed 4	146.89	79.08
Siponate DS-10	Variable	Variable
Triton X-100	Variable	Variable

Table XI - (Continued)

Ingredients	65//35[a]	35//65
H_2O	Variable	Variable
Monomer	35.00	65.00
Initiator	.15	.15
Total	250.00 g.	250.00 g.

(a) First stage//second stage polymer ratio.

The effect of S and N on y is shown in Table XII.

Table XII - Effect of S and N on y

Composition	S (%)	N	y
St//BA	194	.93	.93
65//35	128	.90	.89
	141	.83	.86*
	177	.72	.53
	211	.83	.46
	255	.90	.44
	316	.83	.26
	383	.90	.18
St//BA/MAA	127	.90	.99
65//35 (98/2)	194	.93	.90
	141	.83	.87*
	382	.90	.46
	211	.83	.45
	177	.72	.42
	316	.83	.33
	255	.90	.32
St//BA/MMA/MAA	193	.93	1.58
65//35 (64/34/2)	127	.90	1.57
	140	.83	1.25
	210	.83	1.02
	176	.72	.88*
	253	.90	.87*
	314	.83	.76*
	380	.90	.74
St//BA/MMA/MAA	170	.79	1.10
35//65 (64/34/2)	192	.67	1.02
	267	.86	.73
	340	.79	.46

Table XII - (Continued)

Composition	S (%)	N	y
	247	.51	.30
	292	.66	.24
	387	.66	.24
	512	.78	.21
St//BA 35//65	226	.92	1.31
St//BA 65//35 (AIBN)	153	.93	1.22

*Average of two runs.

Thus, regardless of the second stage composition or its relative amount compared to seed, y = 1 continues to occur when S > 100% and y continues to decrease with increasing S and decreasing N.

Latex Particle Morphology

Elucidation of the morphology of the two stage latex particles, which had complete second stage monomer association, was carried out by a comparison with the corresponding copolymer and mechanical blend systems using electron microscopy and thermal analysis techniques.

The copolymer and acrylic homopolymer and copolymer latexes required for the blend systems to correspond compositionally to the two-stage latexes were prepared by a batch charged process as outlined in Table XIII.

Table XIII - Copolymer Latex Synthesis

Ingredient	Parts by Weight
Aerosol MA-80	Variable (1.0-1.4)
Triton X-114	Variable (3.6-4.0)
H_2O	93.09
Monomer Mixture	100.00
5% $K_2S_2O_8$	7.57

All of the ingredients except the initiator were charged and purged with N_2. The initiator was then added and the reaction was carried out at 65°C (12-16 hr.).

The particle size information for the latexes selected for structural characterization is shown in Table XIV. The homopolymer polystyrene seed latex was used to prepare the two-stage latexes and was also blended with the BA homopolymer latex and the BA/MMA/MAA terpolymer latex to provide the blend systems corresponding to the two-stage latexes.

Table XIV - Latex Particle Size Statistics

Latex	Composition	Ratio	DVS(μ)[(a)]	D_w/D_n[(b)]
Homopolymer (Seed 4)	St	-	.2461	1.02
Homopolymer	BA	-	.2023	1.05
Two-Stage	St//BA	65//35	.2813	1.02
Two-Stage	St//BA	35//65	.3711	1.02
Two-Stage	St//BA/MMA/MAA	35//65	.3472	1.01
Copolymer	St/BA	65/35	.1695	1.02
Copolymer	St/BA	35/65	.2566	1.02
Copolymer	BA/MMA/MAA	64/34/2	.2034	1.02
Copolymer	St/BA/MMA/MAA	35/41.6/22.1/1.3	.2025	1.09

(a) DVS = volume surface average diameter.
(b) D_w/D_n = polydispersity, D_w = weight average diameter, D_n = number average diameter.

The transmission electron microscope was used to aid in the determination of latex particle structure. Figure 5 is a photograph of a 65/35 (St/BA) copolymer latex which shows spherical relatively non-deformed latex particles. The 65//35 (St//BA) latex shown in Figure 6 reveals the deformed, slightly coalesced, exterior portions of the latex particles. Comparison of Figures 5 and 6 readily reveal the softer latex particle surface in the two-stage latex, presumably due to the greater segregation of the soft p-BA second stage near the surface of the particle. The corresponding 65 + 35 (St + BA) blend system of Figure 7 shows the aggregation of the dark polystyrene spheres. It is difficult to determine if the polystyrene particles are actually dispersed in a p-BA matrix since the readily coalescable p-BA particles are relatively transparent under our TEM conditions. The 35//65 (St//BA) latex of Figure 8 reveals an apparent film formed from the coalescence of these softer two-stage particles. The relatively non-deformed darker spherical domains appear to be the first stage polystyrene seed particles. The better state of dispersion of the polystyrene spheres compared to the blend latex of Figure 7 can easily be seen.

The transmission electron microscopy results are consistent with a segregated latex particle consisting of a polystyrene rich core and a soft poly-n-butyl acrylate rich shell.

Scanning electron microscopy was used to probe the film morphology of the two-stage latex particles. Figure 9A is a surface photograph of the 65//35 (St//BA) latex showing a uniform packing

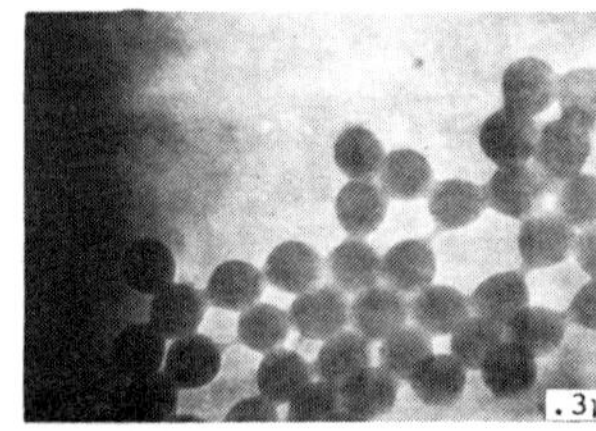

Figure 5. TEM photomicrograph of 65/35 (St/BA) copolymer latex

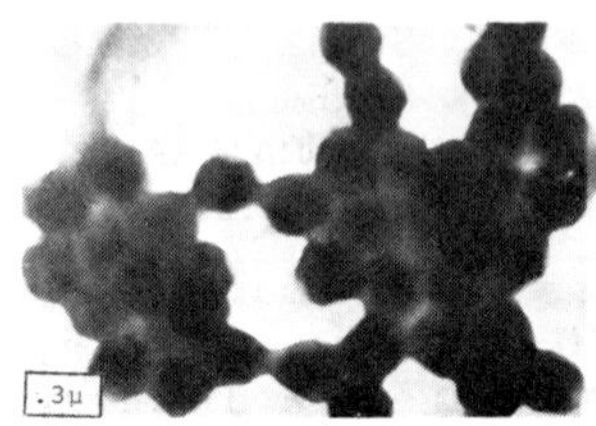

Figure 6. TEM photomicrograph of 65//35 (St//BA) two-stage latex

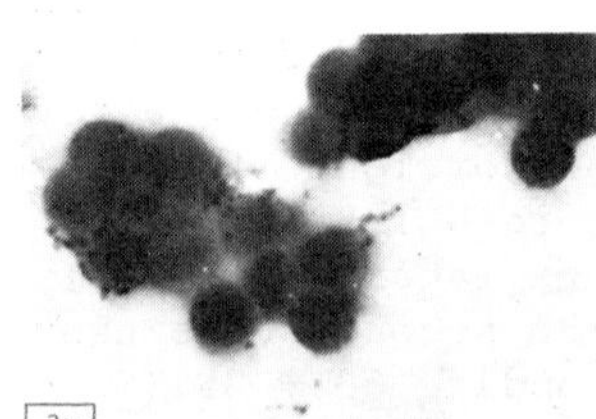

Figure 7. TEM photomicrograph of 65 + 35 (St + BA) blend latex

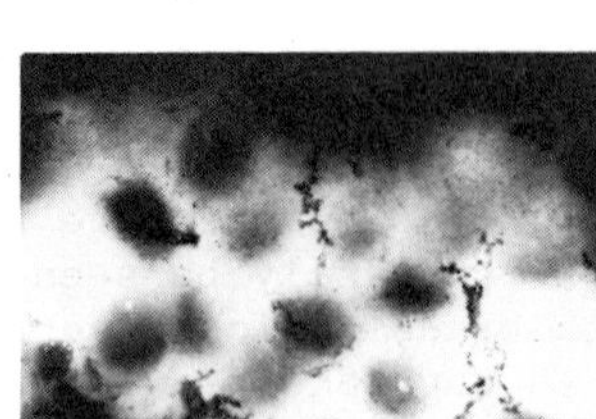

Figure 8. TEM photomicrograph of 35//65 (St//BA) two-stage latex

array for the slightly deformed latex particles. The freeze fracture section, shown in Figure 9B, likewise shows the slightly deformed latex particles. A 65//35 two-stage latex initiated in the second stage with AIBN provided similar SEM photographs. Figures 10A and 10B are the surface and freeze fracture photographs of the 65 + 35 (St + BA) blend system. Both pictures show the aggregated dispersion of the relatively non-deformed polystyrene particles. The tiny white spots on top of some of the particles in the fracture section are believed to be an artifact of the fracture process. Figures 11A and 11B are the surface and fracture sections of the 35//65 (St//BA) staged latex. The most interesting photograph is Figure 11B which shows the intertwined rope-like structure of the 35//65 two-stage latex film.

The SEM results appear to be consistent with the structural conclusions drawn in the analysis by transmission electron microscopy.

Differential scanning calorimetry was used to provide thermal analysis data to investigate latex particle structure. A glass transition at ca. 100°C for p-St and one at ca. -54°C for p-BA would be the expected transitions for a core of p-St and a shell of p-BA. A single, possibly broadened thermal transition, between these two values would be the anticipated result of a particle containing a good degree of mixing of the first stage and second stage polymer chains. Table XV lists the thermal analytical data for the two-stage latexes and the corresponding blend and copolymer latexes.

Table XV - Differential Scanning Calorimetry

Latex	Composition	Ratio	Transitions (°C)
Homopolymer	St	-	94
Homopolymer	BA	-	-54
Two-Stage	St//BA	65//35	-52,36,83
Blend	St+BA	65+35	-54,105
Copolymer	St/BA	65/35	33
Two-Stage	St//BA	35//65	-55,105

The thermal transitions obtained for the p-St and p-BA homopolymers, and for the blend of these two homopolymers, are the anticipated transitions. The St/BA (65/35) copolymer exhibits a single transition at 33°C which is in the temperature region expected for that copolymer composition. The 65//35 (St//BA) two-stage latex exhibits the p-St and p-BA transitions in addition to a transition in the corresponding copolymer range. This intermediate transition was also observed when a mixture of 98 parts n-butyl acrylate and 2 parts methacrylic acid was used as the second stage composition or if AIBN was used as the second stage initiator. A tentative interpretation is that this transition is

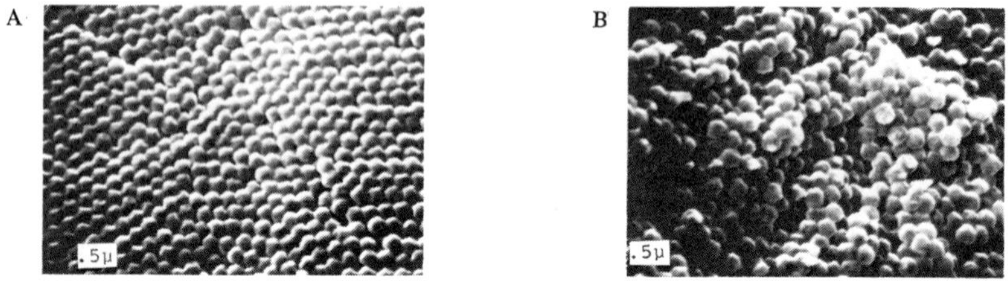

Figure 9. SEM photomicrograph of 65//35 (St//BA) two-stage latex film: (A) surface; (B) freeze fracture section

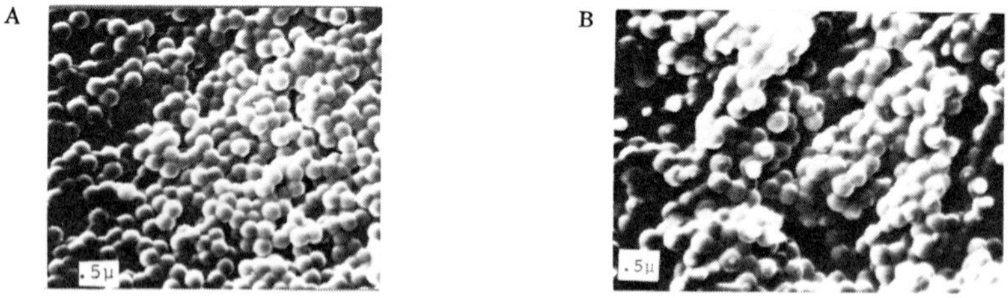

Figure 10. SEM photomicrograph of 65 + 35 (St + BA) blend latex film: (A) surface; (B) freeze fracture section

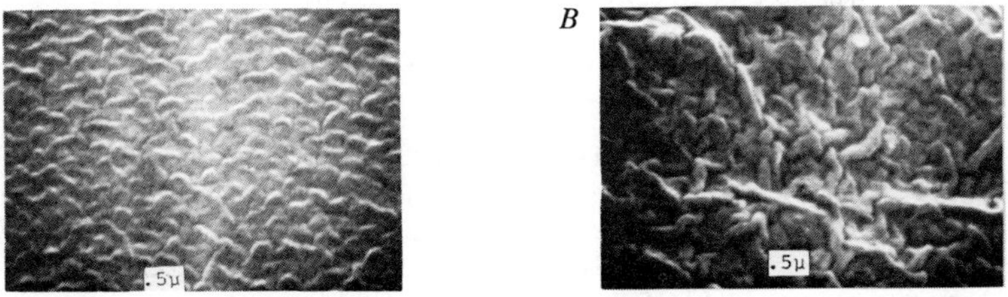

Figure 11. SEM photomicrograph of 35//65 (St//BA) two-stage latex film: (A) surface; (B) freeze fracture section

due to an interfacial region between a polystyrene rich core and a poly-n-butyl acrylate rich shell similar to an interpenetrating network. Reversal of the stage ratio to 35//65 (St//BA), unexplainably, does not show a mid-range transition, but does show the transitions for p-St and p-BA. With limited data it would appear that the stage ratio may exert some influence on whether one observes an interfacial glass transition.

The DSC results support a segregated two-stage latex particle which is rich in p-styrene and in p-n-butyl acrylate.

The surface characteristics of a two-stage polymer were compared against those of a corresponding blend and copolymer latex by minimum film temperature analysis (11) (Table XVI).

Table XVI - Minimum Film Temperature

Latex	Composition	Ratio	MFT (°C) Crack	MFT (°C) Knife
Two-Stage	St//BA/MMA/MAA	35//65(64/34/2)	12	18
Blend	St+BA/MMA/MAA	35+65(64/34/2)	9	11
Copolymer	St/BA/MMA/MAA	35/41.6/22.1/1.3	36	38

The blend system which consists basically of a dispersion of p-St in a BA/MMA soft film forming matrix has the lowest MFT values as would be anticipated. The copolymer latex has the highest MFT values in accord with the electron microscopy results. The two-stage latex polymer has MFT values much closer to the blend system which argues against a good degree of mixing of the first stage and second stage polymer chains.

The MFT results indicate the presence of the film forming second stage polymer at the surface of the latex particle and provides further evidence for a polystyrene rich core and polyacrylate rich shell for the two-stage latexes.

Conclusions

It has been generally accepted that second generation free heterogeneous particle latexes can be produced in a multiple stage latex polymerization process only if the seed particles have less than 100% surfactant surface coverage. The present work demonstrates that this view is not entirely correct. Using certain combinations of anionic and nonionic surfactants, the seed coverage may be greater than 100%. This is the case because the degree of second stage association increases with increasing fraction of the nonionic surfactant which compensates for the decreasing association caused by increasing the surfactant surface coverage.

The preparation of a controlled amount of second generation latex particles to give a relatively exact in situ blend of copolymer particles and hetero-particles can be affected by adjusting the nonionic/anionic ratio and the overall surfactant surface coverage.

Transmission and scanning electron microscopy, differential scanning calorimetry and minimum film temperature analysis supports a core/shell morphology for the two-stage latex polymers, consisting predominantly of a polystyrene rich core surrounded by a soft acrylic rich shell.

Acknowledgement

The authors appreciate the discussions with Prof. I. M. Krieger of Case Western Reserve University and his allocation of laboratory space and TEM facilities. The authors also thank Ms. L. Bender for synthesizing many of the latexes, Dr. K. J. Abbey for performing the soap titrations, and Mr. R. M. Holsworth for taking the SEM photomicrographs.

Literature Cited

1. Kato, K., Jap. Plast. Quarterly, 1968, 2, 6.
2. Dickie, R. A.; Cheung, M. F.; Newman, S., J. Appl. Poly. Sci., 1973, 17, 65.
3. Lee, D. I., U.S. Patent 4,156,669, 1979.
4. Pfluger, H. L.; Gebelein, C. G., U.S. Patent 3,291,768, 1966.
5. Dickie, R. A.; Newman, S., U.S. Patent 3,787,522, 1974.
6. Bradford, E. B.; Vanderhoff, J. W.; Alfrey, T., J. Colloid Sci., 1956, 11, 135.
7. Vandergaer, J. E., J. Appl. Poly. Sci., 1965, 9, 2929.
8. Powers, J. R., U.S. Patent 2,520,959, 1950.
9. Goodman, D.; Isgur, I. E.; Wacome, D. M., U.S. Patent 3,397,165, 1968.
10. Gallagher, R. E.; Hwa, J. C. H., U.S. Patent 3,657,172, 1972.
11. Fischer, N., Polym. Eng. Sci., 1974, 14, 332.
12. Provder, T.; Holsworth, R. M., Am. Chem. Soc. Coatings and Plastics Prepr., 1976, 36, 150.

13. Protzman, T. F.,; Brown, F. L., J. Appl. Polym. Sci., 1960, 4, 81.

14. Pierce, P. E.; Holsworth, R. M., U.S. Patent 3,423,351, 1969.

15. Woods, M. W.; Dodge, J. S.; Krieger, I. M., J. Paint Tech., 1968, 40, 541.

16. van den Hul, H. J.; Vanderhoff, J. W., In "Polymer Colloids", Fitch, R. M., Ed.; Plenum: New York, 1971.

17. Maron, S. H.; Elder, M. E.; Ulevitch, I. N., J. Colloid Sci., 1954, 9, 89.

18. Abbey, K. J.; Erickson, J. R.; Seidewand, R. J., J. Colloid and Inter. Sci., 1978, 66, 203.

19. Orr, R. T.; Breitman, L., Canadian J. Chem., 1969, 38, 668.

20. Seidewand, R. J.; Erickson, J. R., Poly. Eng. Sci., 1978, 18, 1182.

RECEIVED April 6, 1981.

31

On-Line Monitoring of Emulsion Polymerization Reactor Dynamics

F. J. SCHORK and W. H. RAY

Department of Chemical Engineering, University of Wisconsin, Madison, WI 53706

The available data from emulsion polymerization systems have been obtained almost exclusively through manual, off-line analysis of monomer conversion, emulsifier concentration, particle size, molecular weight, etc. For batch systems this results in a large expenditure of time in order to sample with sufficient frequency to accurately observe the system kinetics. In continuous systems a large number of samples are required to observe interesting system dynamics such as multiple steady states or limit cycles. In addition, feedback control of any process variable other than temperature or pressure is impossible without specialized on-line sensors. This note describes the initial stages of development of two such sensors, (one for the monitoring of reactor conversion and the other for the continuous measurement of surface tension), and their implementation as part of a computer data acquisition system for the emulsion polymerization of methyl methacrylate.

Emulsion Density and Monomer Conversion

Monomer conversion has traditionally been determined gravitimetrically by drying emulsion samples to constant weight. The procedure is slow, requiring several hours for analysis, and precludes automated data acquisition. A new method has been developed based on the DMA-series digital densitometers manufactured by Anton Paar of Austria, and marketed in the United States by Mettler Instrument Corporation. (Very recently Dr. Kirk Abbey made us aware of his parallel work in these directions and of some initial data reported from his laboratory [1,2]). This instrument is capable of immediate determination of the density of any test fluid, and, if equipped with a flow cell, can continuously monitor the density of a process stream. Results are displayed locally and can be transmitted digitally to a data acquisition computer.

0097-6156/81/0165-0505$05.00/0

Density measurement is accomplished by introducing a test fluid into a glass U-shaped sample tube which is rigidly supported at its open ends. The tube is electronically excited to vibrate at its natural frequency. The frequency of oscillation is continuously monitored electronically, and from the change of frequency caused by the test fluid within the tube, the density of the test fluid can be determined from the formula

$$(\rho - \rho_s) = k(T^2 - T_s^2) \tag{1}$$

where ρ and T are the density and period of oscillation for the test fluid and ρ_s and T_s are the density and period for a standard. Oscillation is continuous and the period of oscillation is updated every two seconds, making the instrument essentially continuous. The cavity surrounding the sample tube is filled with gas of high thermal conductivity; this in turn, is surrounded by thermostated liquid for accurate sample temperature control. An accessory flow adapter may be added to monitor the density of process streams. Output of either period of oscillation or actual density is possible. Models are available with four, five, or six place precision.

The instrument is normally calibrated against distilled water and air by rearranging Equation (1) as follows:

$$\rho = C(T^2 - D) \tag{2}$$

The values of C and D are determined from the known densities of water and air. Once calibrated, the instrument should not need recalibration unless the sample tube is replaced.

Since monomer conversion is a linear function of emulsion specific volume, the fractional conversion, x, may be calculated as

$$x = \frac{(\nu_e^0 - \nu_e)}{(\nu_e^0 - \nu_e^1)} = \frac{(1/\rho_e^0 - 1/\rho_e)}{(1/\rho_e^0 - 1/\rho_e^1)} \tag{3}$$

where the specific volumes at 0 and 100% conversion, ν_e^0 and ν_e^1, may be approximated as weighted averages of the component specific volumes:

$$\nu_e^0 = X_m\nu_m + X_A\nu_w \tag{4}$$

$$\nu_e^1 = X_m\nu_p + X_A\nu_w \tag{5}$$

where X_m and X_A are the initial weight fractions of monomer and aqueous phase respectively. For batch systems, and for a continuous system with start-up from an empty reactor, the weight fraction total monomer in the sample stream, X_m (including monomer incorporated in polymer chains), remains constant and equal to the initial or inlet value. If start-up is from a full continuous reactor (water or emulsion filled), X_m in the sample stream approaches the value of the feed as follows:

$$X_m = X_{m_{feed}} + (X_{m_i} - X_{m_{feed}}) \exp(-t/\theta) \tag{6}$$

where t is time and θ is the residence time of the reactor. In this case, at each sampling time, X_m must be calculated from Equation (6), ν_e^0 and ν_e^1 must be reevaluated based on the current value of X_m, and Equation (3) applied to evaluate x. In a system employing digital data acquisition, this computation is easily handled in real time.

For the purposes of conversion monitoring of emulsion polymerization, we have found the DMA40D with a precision of $\pm 1 \times 10^{-4} g/cm^3$ capable of resolving monomer conversion to $\pm 0.2\%$ in the absence of thermostating and sampling errors. On-line, in the presence of such possible errors, a resolution of at least $\pm 0.5\%$ can be expected. Care must be taken to ensure a representative sample and good temperature control of the sample stream before introducing it into the instrument. Some example results with this instrument are presented below.

Surface Tension and Free Emulsifier Concentration

The surface tension of the continuous phase of a polymer emulsion may be used as a measure of the free emulsifier concentration. The term "free emulsifier" is used here to denote surfactant which is dissolved in the continuous aqueous phase, rather than adsorbed onto polymer particles or monomer droplets, or aggregated into micelles. The distribution of surfactant within the emulsion has a critical effect on the dynamics of the polymerization system, and knowledge of this property is, perhaps, the key to understanding the abnormal system dynamics which are sometimes observed. Surface tension is a difficult property to measure, particularly for systems containing surfactants. Extreme cleanliness and good technique are required to obtain reproducible results. The most widely accepted technique is perhaps the Wilhelmy plate procedure which measures the vertical force necessary to balance the force exerted on the plate by the liquid surface as the plate is withdrawn from the liquid. This method can give good precision, but is slow and not at all suitable for automated on-line use. Gerrens [3] has published one set of data in which steady oscillations in the surface tension of a continuous emulsion polymerization system were monitored by counting the number of drops of emulsion per time coming from an orifice of known diameter. This method is also very slow and would not seem attractive for on-line applications.

A new method of surface tension determination has been developed which is continuous, automated, compatible with computer data acquisition systems, and capable of monitoring flowing process streams. The method is a variant of the well-known maximum bubble pressure technique. To illustrate the principles, we will describe the simplest initial configuration of the instrument here. Further details and a description of a refined version of the instrument will be reported later.

A test fluid is introduced into a sample cell (possibly as a continuous stream) as shown in Figure 1. Gas bubbles are formed below the surface of the liquid from two orifices of different diameters. The instantaneous difference in pressure between the two orifices is sensed continuously, and the output signal conditioned as shown in Figure 2 to yield a differential pressure ΔP_f. It may be shown [5] that the liquid surface tension is a linear function of ΔP_f.

In this simple configuration, the instrument was calibrated by measuring ΔP_f and surface tension for numerous test fluids. The independent measurements of surface tension were obtained by the tedious Wilhelmy plate method. Figure 3 illustrates such a calibration curve for one set of orifices and for five types of test fluids (methanol-water, ethanol-water, acetone-water, sodium lauryl sulfate in water saturated with methyl methacrylate, and polymethylmethacrylate latices). This is a "universal" calibration curve independent of the fluid being monitored. For the 63 data points shown in Figure 3, the least squares regression line is given by

$$\gamma = 0.060\Delta P_f + 2.76 \qquad (7)$$

with a standard deviation of 1.3 dynes/cm. The data were taken with an obsolete differential pressure transducer which has been discovered to have significant zero and span drift. Thus the capabilities of the instrument with more modern electronics have been found to be even better than noted here.

It is possible to obtain a theoretical calibration curve by precise measurements of the two orifice diameters. This theoretical curve is also shown in Figure 3, and is compared with the theoretical maximum bubble pressure curve to illustrate the differences between the two methods. In this simple configuration, the theoretical calibration calculation involves several approximations and still gives a remarkable a priori fit to the data.

The simple prototype unit discussed here is capable of continuously monitoring a process stream with rapid response to surface tension changes, and accuracy within 1-2%. Initial investigations indicate that a commercial unit based on this design would be capable of data acquisition, alarm monitoring, and/or closed-loop control of a process variable in a laboratory, pilot plant or production scale installation. A commercial instrument based on the work done in this laboratory is being developed and marketed.

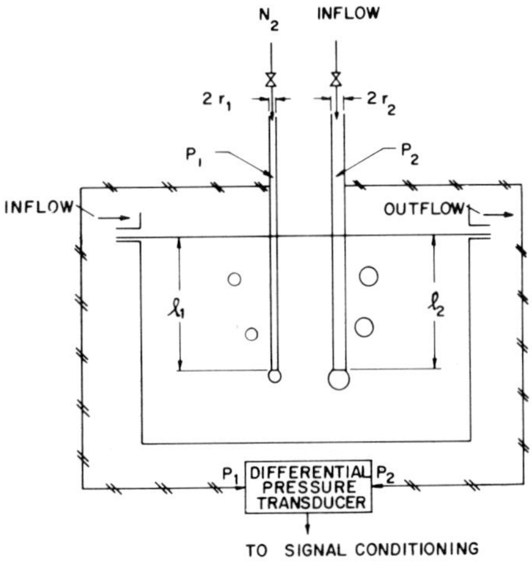

Figure 1. On-line surface tensiometer sample cell

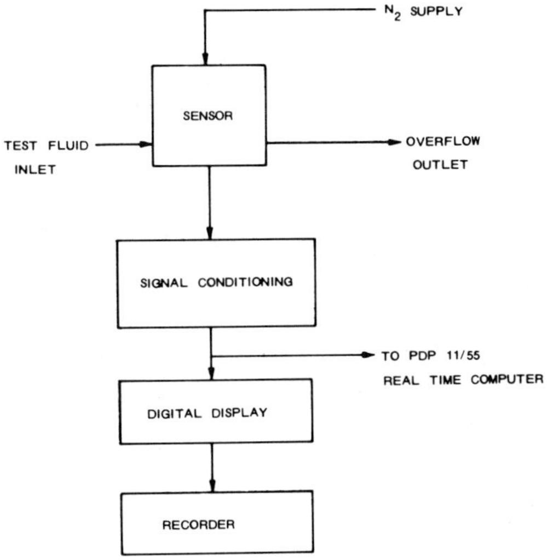

Figure 2. On-line surface tensiometer schematic

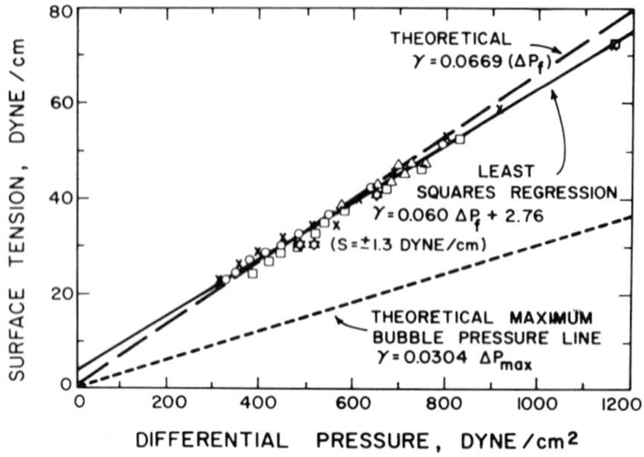

Figure 3. Example calibration curve for the surface tensiometer ((×) I1-MEOH-H_2O; (○) J2-ETOH-H_2O; (□) I2-acetone-H_2O; (△) T1-SLS-MMA-H_2O; (☼) R11 latex)

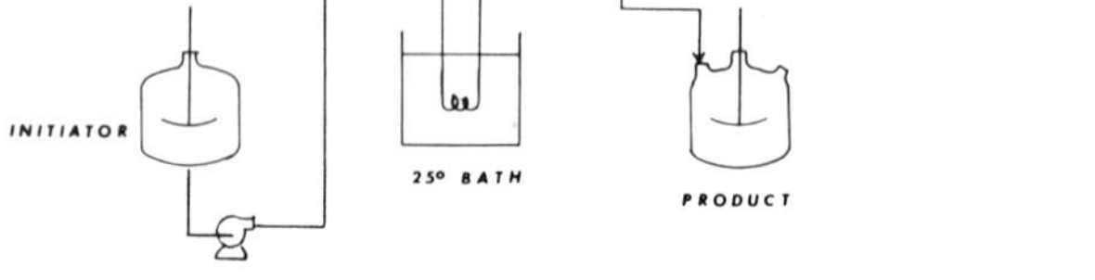

Figure 4. On-line monitoring of conversion and free emulsifier concentration in emulsion polymerization

Applications

The sensors described above have been used to monitor both the batch and continuous emulsion polymerization of methyl methacrylate as shown in Figure 4. In both cases, a small stream of emulsion is continuously pumped from the reactor during polymerization. To illustrate the performance of these monitoring devices, their application in a batch polymerization is shown in Figure 5. Also shown are conversion values determined off-line by the traditional gravimetric dry solids method. As may be seen, the off-line results agree quite well, and, in fact, the largest error can probably be attributed to inaccuracies in sampling and analysis for the off-line method. The application of these sensors to the monitoring of oscillatory behavior in continuous methyl methacrylate polymerization is shown in Figure 6.

Although we measure surface tension of the latex in Figures 5 & 6, it is really the free emulsifier concentration we wish to monitor. Recall that in emulsion polymerization, as the polymer particles grow with increasing conversion, the free emulsifier concentration falls due to additional surfactant being adsorbed onto the new polymer surface. As the free emulsifier concentration falls, the surface tension of the aqueous phase rises above the value at the critical micelle concentration. The free emulsifier concentration may be determined directly from the surface tension by reference to Figure 7. Plotted are surface tensions (as determined by the Wilhelmy plate method) of solutions of sodium lauryl sulfate (the polymerization emulsifier) in water. Curve A shows the results for pure water; B shows the results for water saturated (1.4 weight %) with methyl methacrylate. Since the methyl methacrylate concentration did not exceed its solubility in water, no monomer droplets were present. As may be seen from the graph, the presence of methyl methacrylate has little effect on the critical micelle concentration, or the surface tension at the critical micelle concentration. This is due to the fact that methyl methacrylate is only slightly surface active, and its effect is far overshadowed by that of the sodium lauryl sulfate. At low surfactant concentrations, however, the effect of the methyl methacrylate becomes significant, and the surface tension in the presence of methyl methacrylate is much lower than in its absence. Thus by making use of curves such as those shown in Figure 7, one may relate measured surface tension to free emulsifier concentration for any emulsion polymerization.

It would appear that these sensors are quite valuable in exploring the dynamics of emulsion polymerization systems both at the laboratory and pilot plant scales. In addition, it appears that these instruments, in more rugged design would have applications for monitoring, alarm, and control functions in industrial-scale installations. Further refinements and applications are under study at present.

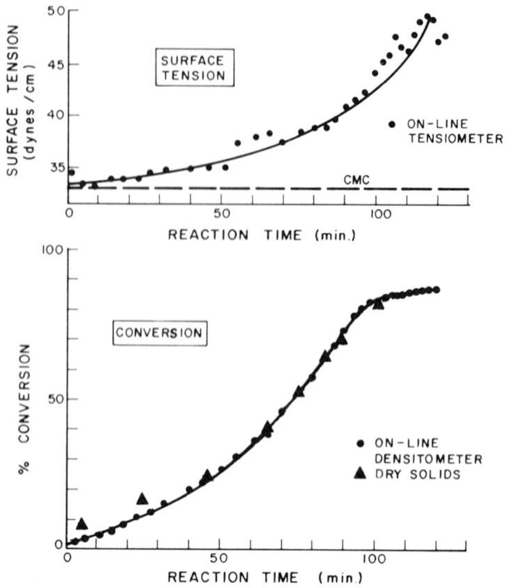

Figure 5. Example data acquisition for the batch emulsion polymerization of MMA at 40°C (initiator (ammonium persulphate) = 0.01 gmol/L H_2O; emulsifier (SLS) = 0.02 gmol/L H_2O; wt. ratio monomer/water = 0.43)

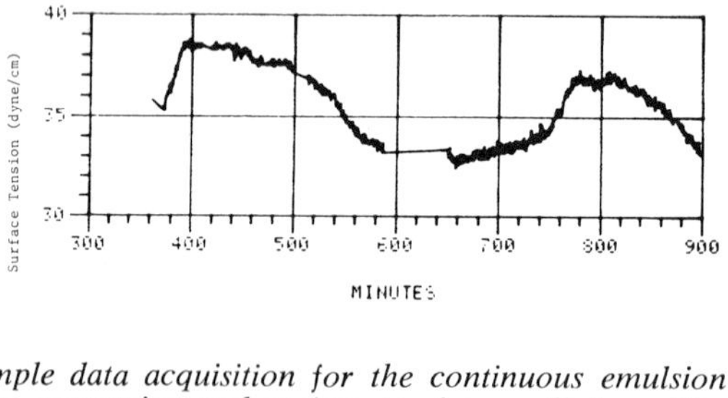

Figure 6. Example data acquisition for the continuous emulsion polymerization of MMA showing conversion and surface tension oscillations (Run 15, Recipe 8; T = 40°C; initiator (ammonium persulfate) = 0.01 gmol/L H_2O; emulsifier (SLS) = 0.02 gmol/L H_2O; wt. ratio monomer/water = 0.43)

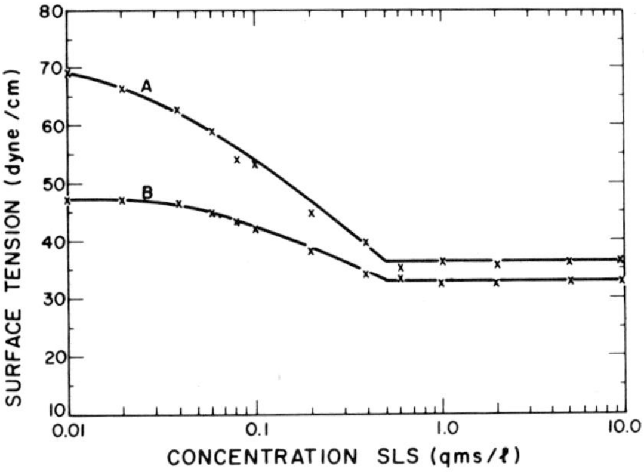

Figure 7. Wilhelmy plate measurements of surface tension for SLS solutions: (A) no MMA; (B) saturated with MMA

Acknowledgements

The authors are indebted to the National Science Foundation and the Mobil Foundation for research support and to the Rohm & Haas Company for contributing the MMA monomer.

Literature Cited

1. Abbey, K. J., personal communication, February (1980).
2. Abbey, K. J., paper presented at the 1979 Pittsburgh Conference on Analytical Chemistry and Applied Spectroscopy.
3. Ley, G., and H. Gerrens, Makromol. Chem. 1974, 175, 563.
4. Padday, J. F., in "Surface and Colloid Science 1", E. Matijevic, Ed., Wiley, N.Y., 1969.
5. Schork, F. J., Ph.D. Thesis, University of Wisconsin, (1981).

RECEIVED April 6, 1981.

32

Control of Particle Size Distribution Through Emulsifier Metering Based on Rate of Conversion

DANIEL L. GORDON and KARL R. WEIDNER

Diamond Shamrock Plastics Corporation, Painesville, OH 44077

In emulsion polymerization of PVC, water and vinyl chloride monomer are charged to the reactor. The reactor is heated to the desired reaction temperature. Then the pumps, to continuously meter initiator and emulsifier into the reactor, are activated. As will be explained later, the particle size distribution is a function of the amount of emulsifier present at all times during the polymerization. The extent of polymerization can be followed by measuring the conversion of monomer which can be followed by quantifying the heat liberated from the reactor. By combining the knowledge of the extent of conversion with the effect of emulsifier on particle size distribution (PSD), an algorithm can be generated to produce a given PSD.

Experimental Materials

The reactor system used for these experiments is a 190 liter, jacketed, stainless steel vessel equipped with initiator and emulsifier metering system. The reactor is monitored and controlled by a minicomputer. The computer monitors: the reactor temperature and pressure, the jacket water inlet and outlet temperatures and flow rate, and the initiator and emulsifier flow rates. The computer calculates the amount of heat transferred through the jacket from the process measurements and transmits signals to control the reactor temperature and metering pumps.

Determining Conversion

Determining the conversion of monomer can only be as accurate as the method of quantifying the heat liberated from the reaction. The usual method is to take the difference between the inlet and outlet jacket water temperatures multiplied by the specific heat and flow rate of the water. This steady-state energy balance equation is:

$$Q = wCp\ (T_{in} - T_{out})$$

0097-6156/81/0165-0515$05.00/0

This equation is easily implemented in an industrial plant using analog instrumentation. With computers becoming available in the plants, more accurate methods of obtaining Q (cal/s), can be employed.

The steady-state (SS) method is completely acceptable and correct if, and only if T_{in} and T_{out} are not changing. But if they are, the value calculated for Q will be in error. For example, with the reactor empty, a step change in the jacket inlet water temperature was made. The responses of the jacket inlet and outlet temperatures are shown in Figure 1. With negligible heat being transferred, Q should be zero, but the SS method gave the Q profile shown in Figure 2. The SS method has obvious limitations to its use.

With no heat being transferred through the jacket, an accurate prediction of the jacket water outlet temperature should be possible by knowing the history of the jacket inlet temperature and the mixing characteristics of the jacket. Therefore, the difference between the actual and predicted outlet temperatures multiplied by the flow rate and specific heat of the water should also equal Q.

UN-SS Energy Balance. To predict the outlet temperature, an unsteady-state (UN-SS), energy balance (1) must be written around the reactor jacket. That balance is:

$$\rho VCp \frac{dT_{out}}{dt} = wCp\,(T_{in}-T_{out}) + UA\,(T_{in}-T_{out}) - U'A'(T_{out}-T_E)$$

The rate of change of the outlet temperature times the heat capacity of the jacket equals the heat accumulated by the water flow plus the heat transferred from the reactor to the jacket minus the heat transferred from the jacket to the environment.

The UN-SS equation reduces to the SS equation if $dT_{out}/dt = 0$ and the heat transfer to the environment is negligible:

$$Q = -UA\,(T_R-T_{out}) = wCp\,(T_{in}-T_{out})$$

Dividing the UN-SS equation by wCp produces:

$$\frac{\rho VCp}{wCp}\,\frac{dT_{out}}{dt} = (T_{in}-T_{out}) + \frac{UA}{wCp}\,(T_R-T_{out}) - \frac{U'A'}{wCp}(T_{out}-T_E)$$

Let: $\tau_I = \frac{\rho V}{w}$

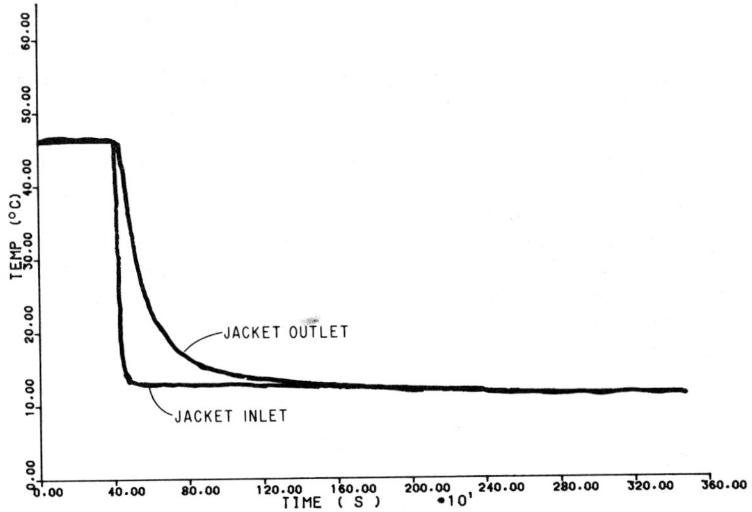

Figure 1. Jacket water inlet and outlet temperatures in response to a −33°C (Test II) change in the inlet. The response of the jacket outlet temperature typifies a first-order mixing model.

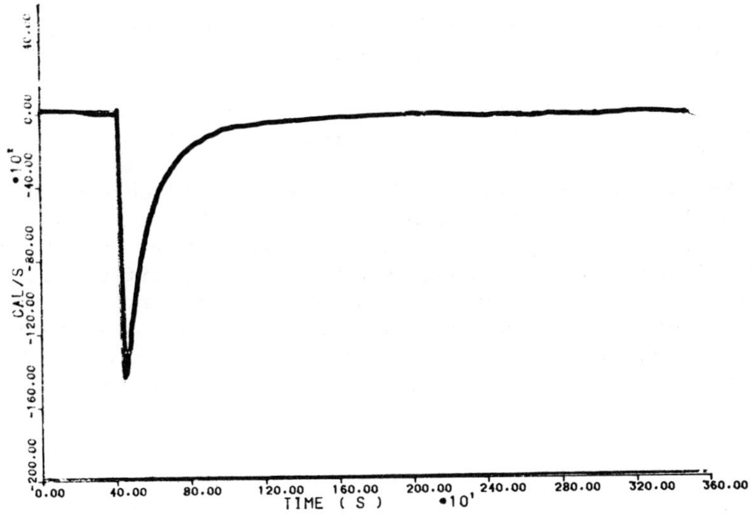

Figure 2. Plot of the heat transferred from the reactor to the jacket using the SS equation on the data from Figure 1

This equation is first order in T_{out} with respect to t. A first order mixing pattern has been assumed, and a first order pattern is exhibited by most "well-mixed" vessels that do not have baffles or flow directing nozzles. How closely this first order equation fits the actual process will be determined later.

Determining Model Parameters. To simplify fitting this first order equation to the actual data, the UN-SS equation must be reduced. Heat transfer to the environment, the U'A' term, may be assumed negligible (for a first approximation anyway). With no reaction occurring and the reactor empty, the heat transfer from the reactor to the jacket, and the heat retained in the jacket walls, the UA term, may be assumed to be zero. The remaining equation (2,3) is:

$$\tau \frac{dT_{out}}{dt} = (T_{in} - T_{out})$$

Given the derivative (dT_{out}/dt) and the temperature T_{in}, this equation will predict the value of the outlet temperature with all external heat transfer equal to zero.

The solution of this differential equation (assuming $dT_{in}/dt = 0$) is:

$$T_{out} = T_{out(i)} e^{-t/\tau} + T_{in}(1-e^{-t/\tau})$$

To account for lag or dead time as with plug flow, a delay, Θ, may be included in the equation.

$$T_{out} = T_{out(i)} e^{-(t-\Theta)/\tau} + T_{in}(1-e^{-(t-\Theta)/\tau})$$

The two parameters, Θ and τ, need to be determined (4). Both may be obtained from a plot of the jacket inlet and outlet temperature to a step change in the jacket inlet (see Figure 1). For the first approximations, Θ will equal the difference in time between the observed changes of the inlet and outlet temperatures. τ equals the time taken for the outlet temperature to achieve 63.2% of its final change. (The definition of a time constant or τ in this case is the time for a response to attain 63.2% of its final value.) Values for the two tests are shown below.

TABLE I

	TEST 1	TEST 2
Step Change	$-39.5^{o}C$	$-33.0^{o}C$
Flow Rate	$567.7 cm^3/s$	$525.7 cm^3/s$
θ	18.5 s	23.5 s
K_1	10500	12350
τ	130.2 s	143.4 s
τ_I	86.4 s	93.6 s
K_2	1.502	1.532

Earlier, τ_I was set equal to $\rho V/w$. The volume, density, and flow rate are known so that this "ideal" τ could have been calculated. Realistically, the ideal and actual, τ , will differ. K_2 is the compensating factor that relates the ideal to the actual, and hence:

$$\tau = \tau_I K_2$$

θ was assumed to be a funtion of w, and K_1 is used to relate θ to w as in:

$$\theta = wK_1$$

The first order curves were compared to the actual data in Figures 3 and 4. Average K values for θ and τ were used. The fits are relatively good and substantiate the assumptions of first order with a delay time.

Before the UN-SS equation can be implemented on the computer the temperature sampling rates and method of calculating dT_{out}/dt must be decided. The smaller the time constant, the smaller the sampling rate should be. All data in this report was obtained at 3 second intervals (about 2.2% of the time constant).

The derivative, dT_{out}/dt, may be calculated by performing a least squares fit around 'N' number of points. N should be chosen inversely proportional to the accuracy and stability of the temperature measurements. The higher the accuracy, the lower N can be. N was chosen to be 7 in these tests.

A comparison of the SS and UN-SS equations are shown in Figure 5 as applied to the earlier step change in Figure 1.

The UN-SS equation was rearranged so both sides equal the heat transferred through the reactor. The UN-SS equation plotted is:

$$\text{UN-SS:} \quad Q = -UA\,(T_R - T_{out})$$

$$= (T_{in@(t-\theta)} - \tau_I K_2 (dT_{out}/dt) - T_{out})\,wCp$$

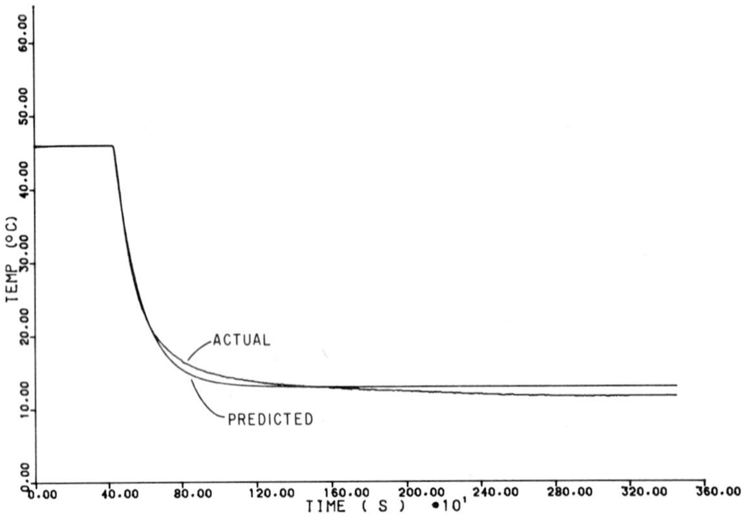

Figure 3. Comparison of the actual and predicted jacket water outlet temperatures. Average values of $\theta = 21.7$ and $\tau = 142.3$ were used for the predicted temperature.

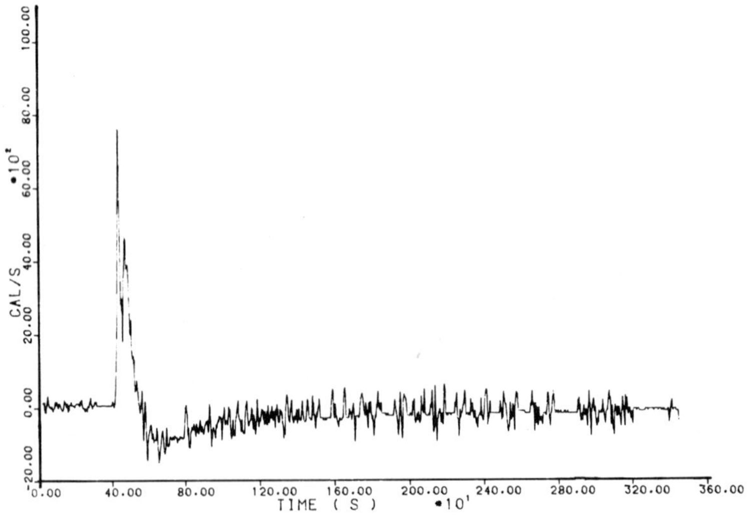

Figure 4. Plot of the heat transferred from the reactor to the jacket using the UN–SS equation on the data from Figure 1

The UN-SS equation is an improvement over the SS equation. The total area between the curve and 0.00 cal/s should be zero. The UN-SS equation does have some drawbacks. The dT_{out}/dt term is sensitive to noise and to the flucuations in the last bit of the digital to analog converter in the computer. This noise may be controlled by varying 'N' until an acceptable agreement is reached between noise and accuracy.

To further improve the UN-SS equation, the variables, θ and τ, may be varied to optimize the fit, or a higher order model may be necessary for a smoother response. U'A' may be determined by trial and error, or the UN-SS equation may be solved for a non-zero U'A' to improve the accuracy of the equation.

Experimental Polymerization. The only real test of the UN-SS equation is with actual data. Figure 6 illustrates the jacket inlet, jacket outlet, and reactor temperatures during an emulsion run. The control was intentionally underdamped to insure the temperatures oscillated, thereby imposing a fairly stringent test on the two methods. Figure 7 is the SS curve, and Figure 8 is the UN-SS. Both models account for the major changes in the jacket temperature, but the UN-SS equation handled the spike in the jacket inlet much better. Unlike the SS equation, at no appreciable time was the jacket "heating" the reactor (showing a positive cal/s) when it was thermodynamically impossible to do so. The UN-SS equation reduced the effect by reducing the area encompassed by the spike in the heat transfer curve.

With an accurate total and rate of heat evolved from the reactor, the total and rate of conversion is easily calculated. Now an algorithm relating emulsifier addition to conversion can be developed and properly implemented.

Applications

Algorithms, computer process control equations, have been developed to control emulsifier metering rates based on information obtained from rate of conversion. Using this technique of emulsifier metering, particle size growth can be modified during the polymerization. Particle size distribution is normally controlled for emulsion PVC through homogenization (5) or by using seeded polymerization (6,7). Seeded polymerizations normally produce a bimodal particle size (PSD). Figure 9 shows a typical PVC latex bimodal PSD (6,7). Note one fraction of particles is labeled "larges" and the other fraction is labeled "smalls". Particle size increases from right to left. Particle size was determined by an ICI-Joyce Loebl Disc Centrifuge MK III with photosedimentometer. This instrument measures particle size and particle size distribution by centrifugation using a photocell detection system (8,9).

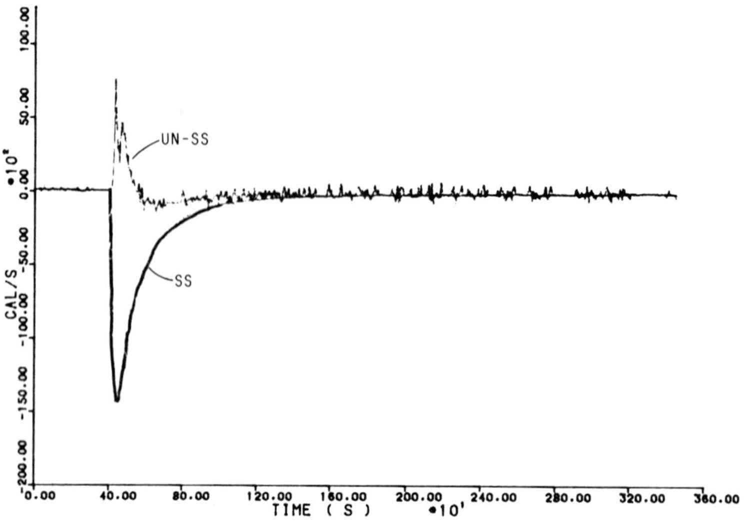

Figure 5. Superposition of Figures 2 and 4. The area between the curves and 0.0 cal/s represents the error in each calculation.

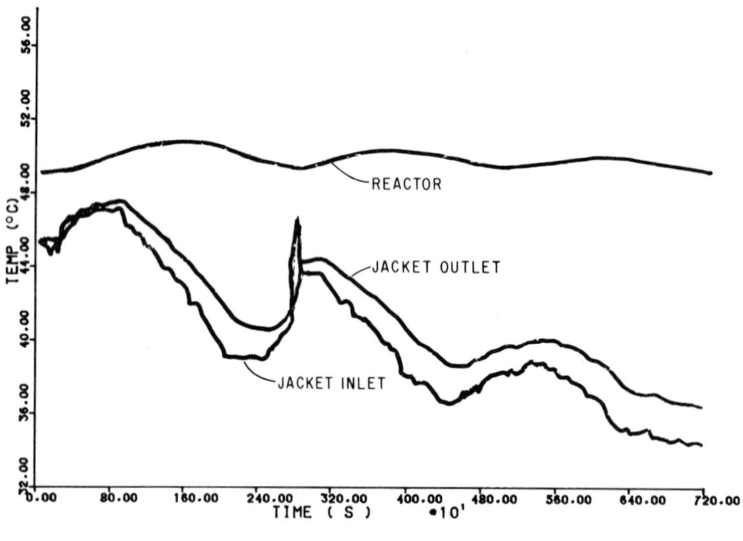

Figure 6. Temperature responses from an emulsion run using an underdamped controller to simulate adverse conditions

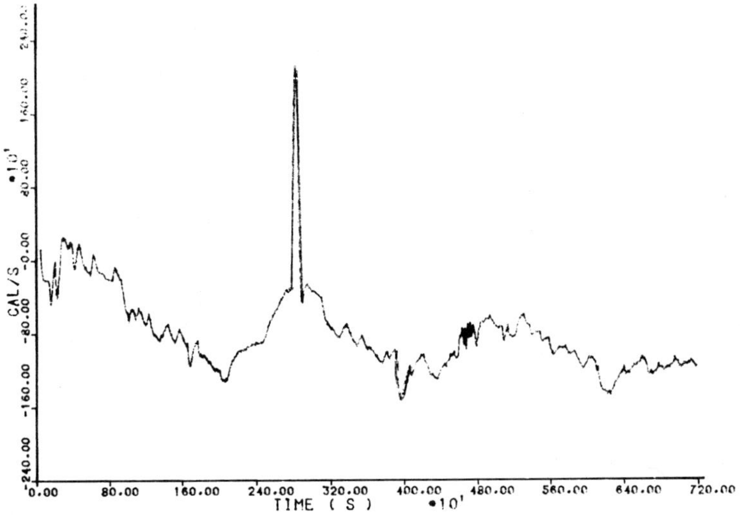

Figure 7. Plot of the heat transferred from the reactor to the jacket using the SS equation on the data from Figure 6

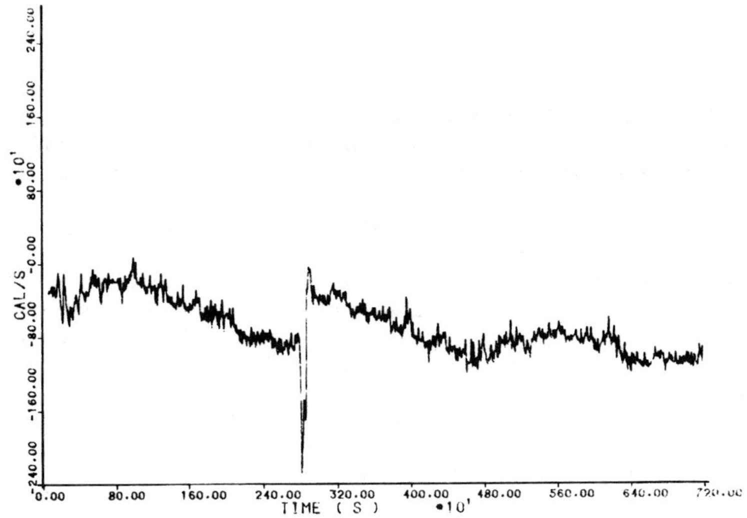

Figure 8. Plot of the heat transferred from the reactor to the jacket using the UN–SS equation on the data from Figure 6

The distribution of larges (Figure 9) has been found to be a primary function of the distribution of particles in a seed (Figure 10). This can be seen by comparing the distribution at the base of the large particle peak in Figure 9 to the distribution of the seed in Figure 10. The seed in Figure 10 was used to make the seeded latex in Figure 9. K. W. Min (10) has indicated that the distribution of smalls is a function of the excess emulsifier over that which is needed to cover the total growing particle surface area during polymerization. Figure 11 shows Min's work for a typical computer simulated run. In this figure, excess emulsifier over that needed to cover all surface area, occurs at approximately 180 minutes and continues until 420 minutes of run. During this period, micelle and subsequent particle formation can occur. For this particular run, new particle or smalls formation occurs over a long period of the polymerization creating a wide distribution for a smalls peak (10) (Figure 12).

By plotting the conversion curve for the same simulator run in Figure 13, a number of similarities occur between this curve and the particle surface area growth curve. During Interval I, the slopes of conversion and particle surface area curves initially are similar. This is the period of seed growth only. The seed will form the large particles shown in Figure 9. During Interval II, emulsifier metering begins. Both smalls and larges begin to compete for emulsifier. Surface area growth becomes a function of both smalls and larges growth. Finally in Interval III, the surface area growth of the smalls becomes predominant. From these observations, it becomes evident that smalls growth can be inferred from changes in slope of the conversion curves.

From a practical standpoint, the emulsifier metering should be able to be altered based on rates of change of conversion to give a desired particle size distribution of the small particles. To prove this, runs were made in a pilot plant to determine if a narrow distribution of small particles could be produced through emulsifier metering techniques. Run 11127-88 was used as a control. Emulsifier was metered at a constant rate in a 190 liter computer controlled reactor. This is identical to the type of emulsifier metering presented by Min in Figures 11 and 13. As shown for Run 11127-88 in Figure 14, shortly after emulsifier metering is started the conversion curve slope increased significantly. The increase in conversion is probably a result of new micelle and particle formation. The greater number of particles, now present, competing for monomer would increase the overall rate of conversion (11). The slope of the conversion curve remained constant until the emulsifier was turned off. From the particle size curve in Figure 15, the smalls distribution was as wide as expected indicating small particle growth over a long period of conversion.

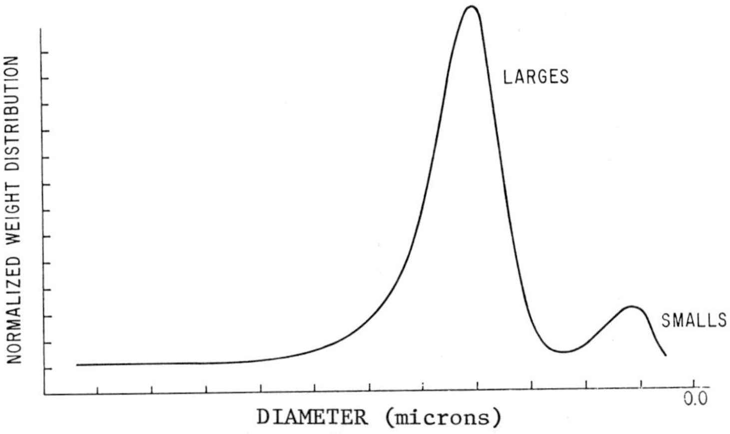

Figure 9. Particle size distribution of a seeded PVC latex (bimodal distribution)

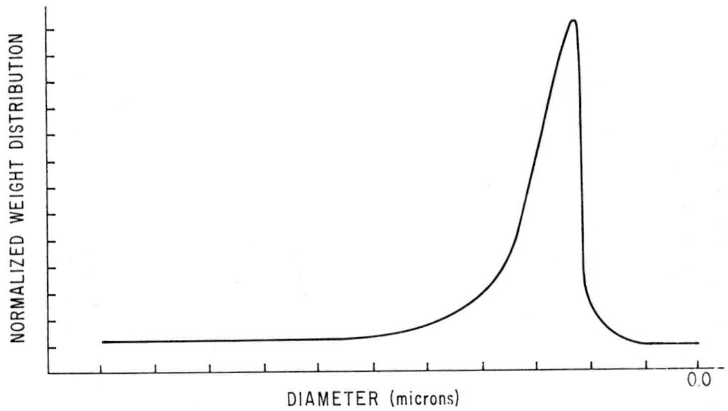

Figure 10. Particle size distribution of a typical seed

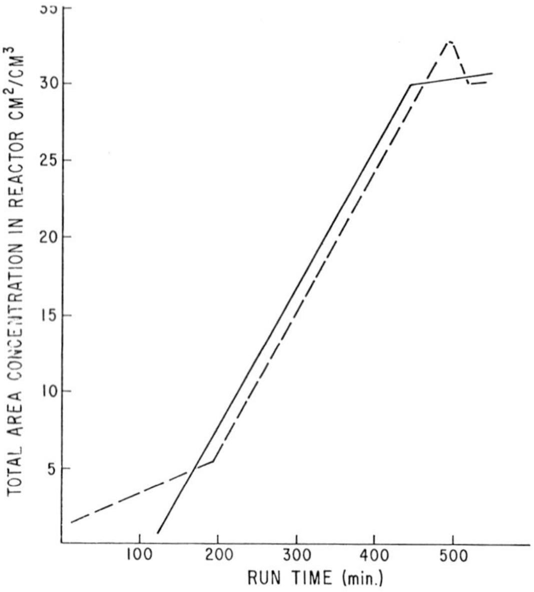

Figure 11. Particle surface area growth ((——) total emulsifier; (– – –) total particle surface area)

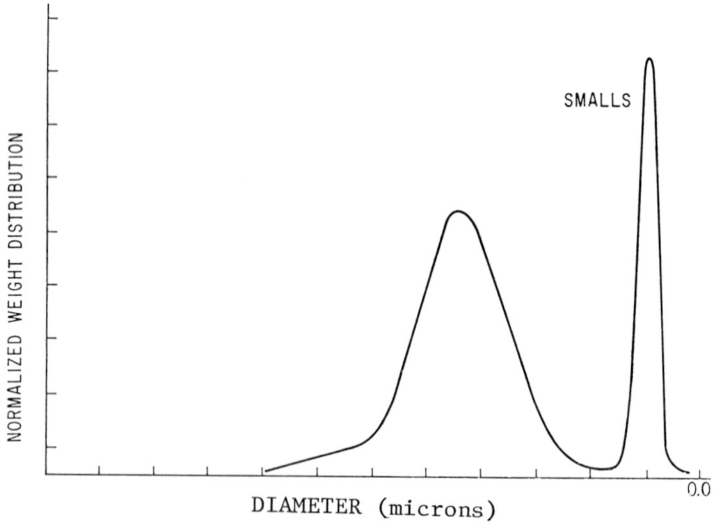

Figure 12. Simulator particle size distribution

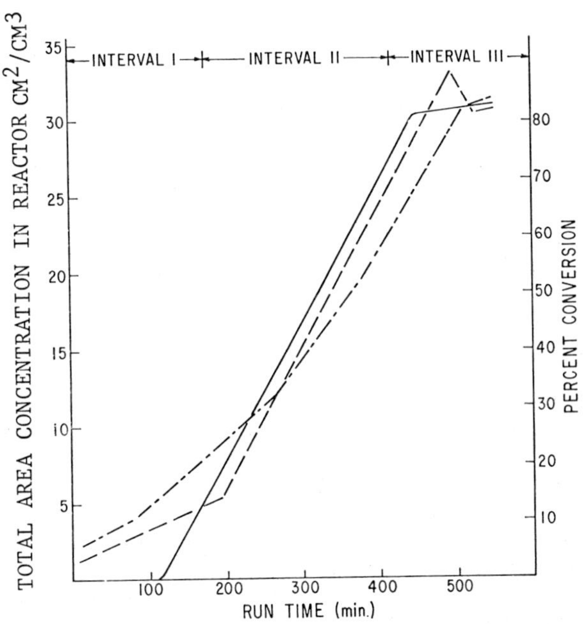

Figure 13. Particle surface area growth ((——) total emulsifier; (– – –) total particle surface area; (— – —) conversion)

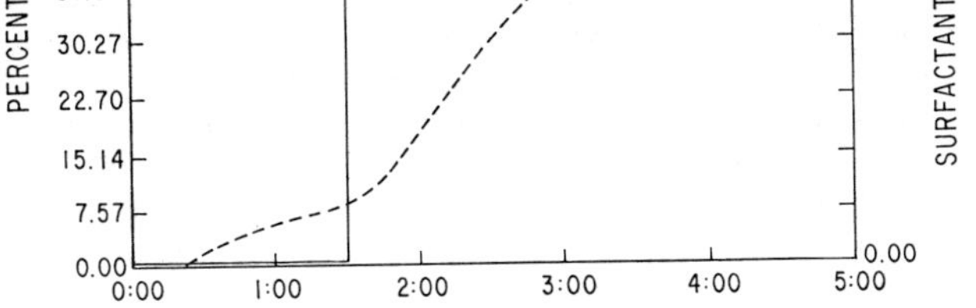

Figure 14. Rate of conversion and emulsifier metering vs. time (Run No. 11127-88; start date: 03-JUN-80; start time: 05:55:42)

For the next run to narrow the small particle distribution, it was envisioned that emulsifier metering should be carried out rapidly over a short period of time. The metering would then be stopped for a given time so that micelle formation would cease. The emulsifier would be restarted to give enough emulsifier to stabilize the growing particles, yet not exceed the emulsifier level needed to cover all the growing particle surface area (not produce new micelles). With the aid of a computer controlled 190 liter reactor, this was accomplished in Run 11127-90.

Emulsifier metering for Run 11127-90, Figure 16, was started at the same conversion as for Run 11127-88. The rate was kept very low until 20% conversion. The rate was then increased to the same rate as Run 11127-88. With this change in emulsifier metering, the conversion curve began to increase slightly. To ensure smalls growth, the metering rate was increased 2.5 times at 30% conversion. A definite increase in rate of conversion was noted. To stop new micelle formation, the emulsifier was turned off after one hour at the high rate. Emulsifier metering ceased for approximately 8% conversion. It was then restarted at a low rate to maintain latex mechanical stability as the particles grew. No new micelle formation should have occurred at the low emulsifier rate since no change in the conversion curve was noted compared to the zero emulsifier metering rate. The particle size distribution curve in Figure 17 shows that the above emulsifier metering technique worked to produce a very narrow distribution of smalls. Run 11127-88, the control, had a particle size distribution approximately three times that of the experimental Run 11127-90.

This work has shown that by monitoring conversion curves by a computer, emulsifier metering can be varied to produce a desired particle size distribution of smalls in a seeded PVC emulsion polymerization.

Abstract

The particle size distribution of emulsion poly (vinyl chloride) can be controlled by varying the rate of emulsifier addition during the polymerization. For reproducibility, the emulsifier should be metered according to a reaction dependent variable such as percent conversion. By monitoring the heat liberated by the reaction, the rate of monomer conversion, and therefore, the progress of the reaction can be followed. But this liberated heat must be accurately quantified if the conversion measurement is to be exact. The common, steady-state method of measuring the heat given off can lead to an appreciable error in the measurement. In some instances, an unsteady-state equation may be better suited to calculate the heat transferred. With conversion accurately measured, algorithms can be developed that

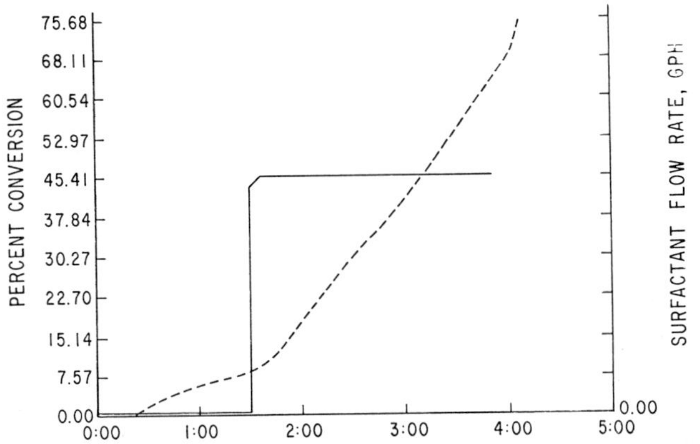

Figure 15. Particle size distribution of control (Run No. 11127-88; start date: 03-JUN-80; start time: 05:55:42)

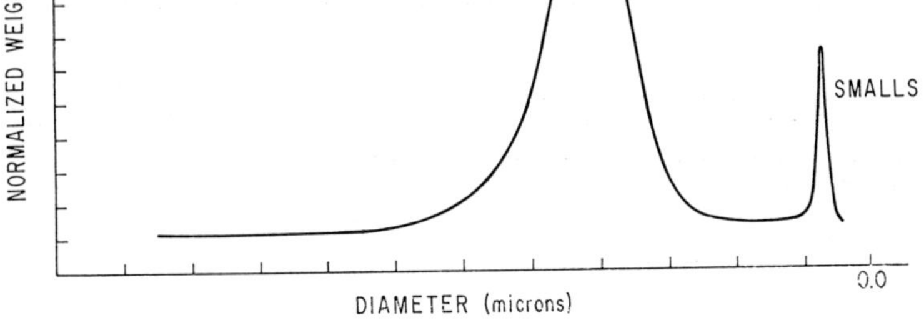

Figure 16. Percent conversion and emulsifier metering vs. run time (Run No. 11127-90)

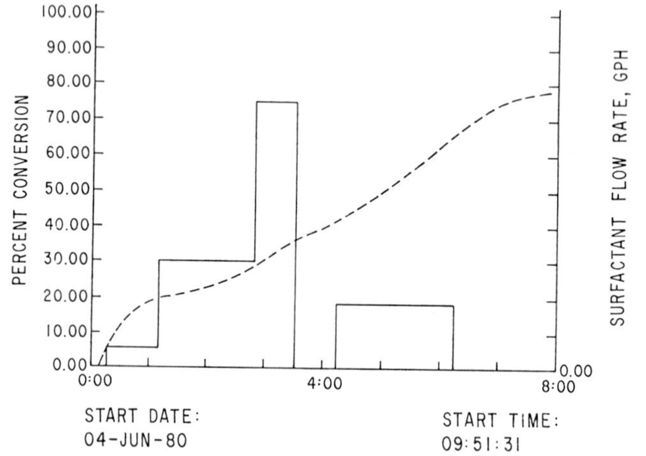

Figure 17. Particle size distribution (Run No. 11127-90)

will control the addition of emulsifier as a function of conversion and therefore, control the particle size distribution of emulsion PVC.

Legend of Symbols

Cal = calories
C_p = specific heat of water - cal/g °C
K_1 = compensating factor for θ
K_2 = compensating factor for τ_I
Q = heat transferred to/from reactor - cal/s
s = seconds
T_E = environmental temperature - °C
T_{in} = jacket water inlet temperature - °C
T_{out} = jacket water outlet temperature - °C
T_R = reactor temperature - °C
t = time
UA = heat transfer coefficient and area to reactor - cal/s °C
U'A' = heat transfer coefficient and area of environment - cal/s °C
V = volume of jacket - liters
w = mass flow rate of jacket water - g/s
θ = dead, delay or lag time - seconds
ρ = density of water - g/cm^3
τ = experimental time constant - seconds
τ_I = ideal time constant - seconds

Acknowledgement

Thanks are expressed to Dr. J. Robert Fowler, Jr. and Dr. Paul A. Husted for their assistance in developing and using the beforementioned equations.

Literature Cited

1. Himmelblau, David M.; "Basic Principles and Calculations in Chemical Engineering", Prentice-Hall, Englewood Cliffs, New Jersey, 1967.
2. Weber, Thomas W.; "An Introduction to Process Dynamics and Control", John Wiley & Sons, New York, New York, 1973.
3. Coughanowr, Donald R.: Koppel, Lowell B.; "Process Systems Analysis and Control", McGraw-Hill, New York, New York, 1965.
4. Smith, Cecil L.; "Digital Computer Process Control", International Textbook, 1972.
5. Bush, C. N.; Mikofalvy, K. M.; B. F. Goodrich Co. "Process for Producing Homopolymers or Copolymers of Vinyl or Vinylidene Halides by Emulsion Polymerization", U. S. Patent #4186259, January 29, 1980.

6. Uglestad, J.; Logstad, H. F.; Hansen, F. K.; Ellingsen, T.; J. Polymer Science, 1973, 12, 473-485.

7. Bovey, F. A.; Kolthoff; I. M., Medalia, A. I.; Mechan, E. J.; "Emulsion Polymerization", Interscience Publishers, New York, New York, 1955.

8. Mueller, J.; Stauffer, W.; Moser, K.; Chimia 27, 1973, 2, 82-86.

9. Provder, T.; Holsworth, R. M.; Coat. Plast. Prept. Pap. Meet. (Am. Chem. Soc., Div. Org. Coat. Plast. Chem.), 1976, Chem. Paper 36, 150-155.

10. Min, K. W.; Gostin, H. I.; Ind. Engr. Chem. Prod. Res. Dev., 1979, 18, 272-278.

11. Uglestad, J.; J. Polymer Science, Part C., 1976, 27, 49.

RECEIVED April 6, 1981.

33

A Simulation Study on the Use of a Dead-Time Compensation Algorithm for Closed-Loop Conversion Control of Continuous Emulsion Polymerization Reactors

KENNETH W. LEFFEW and PRADEEP B. DESHPANDE

Department of Chemical and Environmental Engineering, University of Louisville, Louisville, KY 40292

Although continuous emulsion polymerization is an area of considerable industrial interest and importance, it has received very little attention in the literature when compared to the vast amount of published work available for the batch process. Largely in response to the growing interest expressed by industry, the amount of work on continuous emulsion polymerization appearing in the technical literature has been increasing in recent years. The majority of this work has dealt with the steady-state operation of a single continuous stirred tank reactor (CSTR). Few workers have dealt with the problems associated with operation of a train of reactors, which is the type of reaction system most commonly found in industry. Even less attention has been given to control strategies for these multi-reactor trains, in which the system dynamics are of concern. In the commercial manufacture of polymers by continuous emulsion polymerization, perhaps the primary concerns are maintenance of uniform product quality and avoidance of production of poor quality material resulting from swings away from the steady-state levels of the process variables. Therefore, a control system which provides tight regulatory control of the process is a need of the industry. Design of such a system for many continuous emulsion polymerized monomers is complicated by the occurrence of a steady-state limit cycle in the number of polymer particles produced and in the monomer conversion achieved in the reaction system. In this paper, control strategies are suggested, and demonstrated by simulation of the vinyl acetate system, which are designed to provide the required regulatory control of a series of continuous emulsion polymerization reactors.

Background

The two prominent variables to be controlled in a continuous polymerization system are the reaction temperature and the monomer conversion achieved in the reaction system. Final polymer properties are directly influenced by changes in these process

0097-6156/81/0165-0533$08.25/0

variables. Several control systems for continuous emulsion polymerization have been suggested. Wismer and Brand (1) manipulated reaction temperature over a portion of a continuous reactor train to control the monomer conversion from the final reactor in a styrene-butadiene emulsion polymerization. Changes in temperature were based on a feed-forward control scheme developed from a very simple linear process model and the control algorithm was implemented on a digital computer. The applicability of this control strategy is limited to systems which obey the simple process model and to those where changes in reaction temperature can be made without affecting polymer properties. Francis and Sonntag (2) suggested an alternate control strategy in which the residence time in the last of a train of continuous stirred tank reactors was manipulated as required to control the conversion exit that reactor. This technique also has limited applicability due to the relatively slow response of the control loop maintaining the last reactor level and also because the polymerization rate of many systems has been found to be nonlinearly affected by reactor residence time (Poehlein and Dougherty (3)).

The most common continuous emulsion polymerization systems require isothermal reaction conditions and provide for conversion control through manipulation of initiator feed rates. Typically, as shown in Figure 1, flow rates of monomer, water, and emulsifier solutions into the first reactor of the series are controlled at levels prescribed by the particular recipe being made and reaction temperature is controlled by changing the temperature of the coolant in the reactor jacket. Manipulation of the initiator feed rate to the reactor is then used to control reaction rate and, subsequently, exit conversion. An aspect of this control strategy which has not been considered in the literature is the complication presented by the apparent dead-time which exists between the point of addition of initiator and the point where conversion is measured. In many systems this dead-time is of the order of several hours, presenting a problem which conventional control systems are incapable of solving. This apparent dead-time often encountered in initiation of polymerization.

Several control techniques have been developed to compensate for large dead-times in processes and have recently been reviewed by Gopalratnam, et al. (4). Among the most effective of these techniques and the one which appears to be most readily applicable to continuous emulsion polymerization is the analytical predictor method of dead-time compensation (DTC) originally proposed by Moore (5). The analytical predictor has been demonstrated by Doss and Moore (6) for a stirred tank heating system and by Meyer, et al. (7) for distillation column control in the only experimental applications presently in the literature. Implementation of the analytical predictor method to monomer conversion control in a train of continuous emulsion polymerization reactors is the subject of this paper.

The analytical predictor, as well as the other dead-time compensation techniques, requires a mathematical model of the process for implementation. The block diagram of the analytical predictor control strategy, applied to the problem of conversion control in an emulsion polymerization, is illustrated in Figure 2(a). In this application, the current measured values of monomer conversion and initiator feed rate are input into the mathematical model which then calculates the value of conversion T' units of time in the future assuming no changes in initiator flow or reactor conditions occur during this time. Here, T' is the sum of the process dead-time, Θ_d, and one-half of the sampling time, T_s,

$$T' = \Theta_d + \tfrac{1}{2} T_s \qquad (1)$$

(The dynamic effect of sampling is equivalent to that of a pure dead-time of one-half the sampling time.) The model-predicted future value of conversion is then compared with the conversion setpoint to generate an error signal which is used by the control algorithm to achieve the new initiator feed rate setpoint. Utilizing this strategy, the control scheme for the first reactor of a series appears as Figure 3, where temperature is maintained by manipulating the reactor jacket temperature, and exit conversion is controlled by means of the analytical predictor algorithm. By comparison, the block diagram depicting the standard feedback control strategy in which the initiator feed rate setpoint is manipulated by the primary conversion controller is shown in Figure 2(b) and a schematic of the implemented cascade control system is identical to Figure 3 with a standard discrete form of a PID controller as the control algorithm (i.e. no prediction).

A recent paper by Kiparissides, et al. (8) details a mathematical model for the continuous polymerization of vinyl acetate in a single CSTR. Operating conditions were shown to exist in which either steady-state operation or sustained conversion oscillations would occur for vinyl acetate. Experimental results for both cases were successfully simulated by their model. In addition, regulatory conversion control policies were considered in which both initiator feed rate and emulsifier feed rate were used as manipulated variables (Kiparissides (9)). The problem of conversion control in the operating region in which sustained conversion oscillations occur is one of significant commercial importance. Most commonly, however, a uniform concentration of emulsifier is required in the emulsion recipe and, hence, emulsifier flow rate cannot be used as a manipulated variable.

The objective of this paper is to illustrate, by simulation of the vinyl acetate system, the utility of the analytical predictor algorithm for dead-time compensation to regulatory control of continuous emulsion polymerization in a series of CSTR's utilizing initiator flow rate as the manipulated variable.

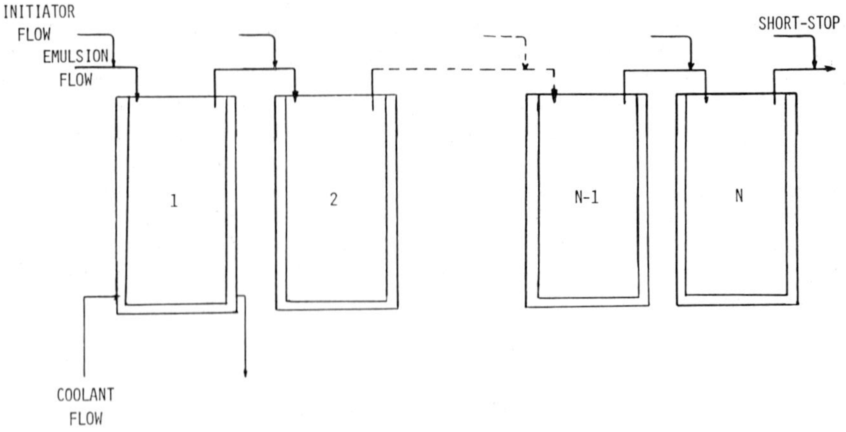

Figure 1. Typical continuous reactor train for emulsion polymerization

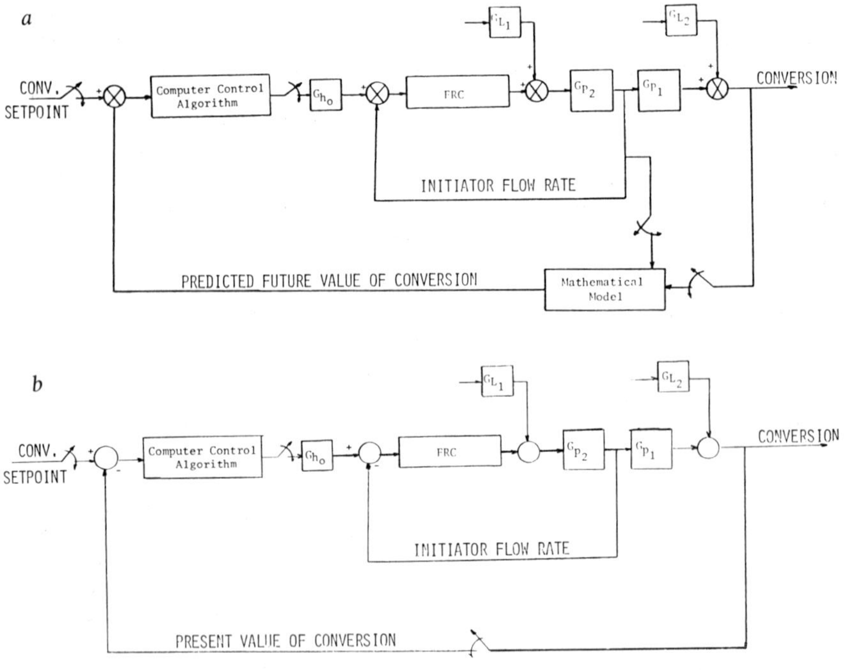

Figure 2. (a) Block diagram of conversion control loop utilizing the analytical predictor technique of dead-time compensation; (b) conventional feedback control loop ((G_{P_1}) reactor process transfer function; (G_{P_2}) initiator flow transfer function; (G_{L_1}) initiator flow load; (G_{L_2}) reactor load; (G_{H_0}) zero-order hold)

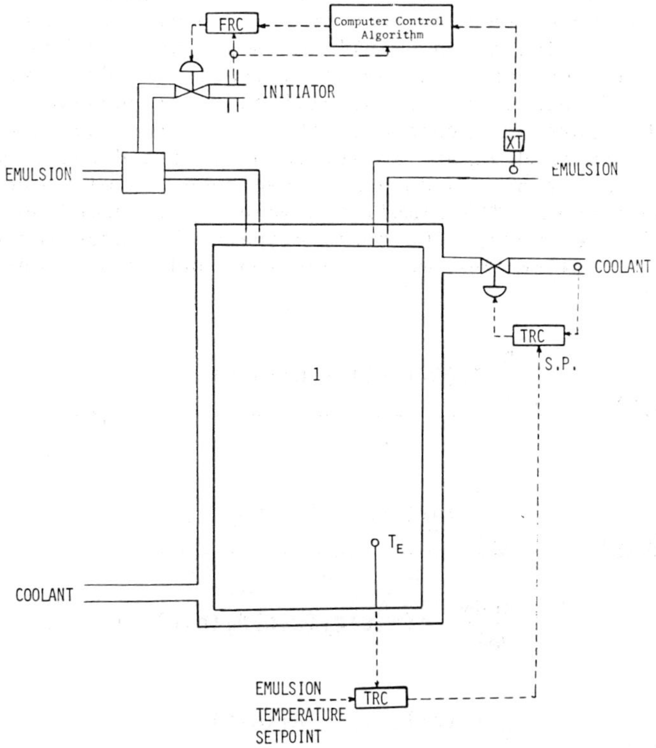

Figure 3. Schematic of a control system for the first reactor of a series using supervisory setpoint control of initiator feed rate to control reactor conversion

Closed-loop response to process disturbances and step changes in setpoint is simulated with the model of Kiparissides extended to predict the behavior of downstream reactors. Additionally, a self-optimizing control loop is simulated for conversion control of downstream reactors when the first reactor of the train is operating under closed-loop control with dead-time compensation.

Model Development

Kiparissides, et al. (8) developed mathematical models of two levels of sophistication for the vinyl acetate system: a comprehensive model that solved for the age distribution function of polymer particles and a simplified model which solved a series of differential equations assuming discrete periods of particle nucleation. In practice, the simplified model adequately describes the physical process in that particle generation generally occurs in discrete intervals of time and these generation periods are short in duration when compared with operation time of the system. The simplified model is expanded here for a series of m reactors. The total property balances for number of particles, polymer volume, conversion, and area of particles, are written as:

$$\frac{dN_i(n,t)}{dt} = \frac{\sum_{j=1}^{k}(N_j(n-1,t)) - N_i(n,t)}{\Theta_n} + f(n,t) \qquad (2)$$

$$\frac{dVp_i(n,t)}{dt} = \frac{\sum_{j=1}^{k}(Vp_j(n-1,t)) - Vp_i(n,t)}{\Theta_n} + \frac{kpdm}{N_a d_P}\,\phi(n,t)\bar{q}_i(n,t)N_i(n,t) \qquad (3)$$

$$\frac{dX_i(n,t)}{dt} = \frac{\sum_{j=1}^{k}(X_j(n-1,t)) - X_i(n,t)}{\Theta_n} + \frac{kpdm}{M_T M_W}\,\phi(n,t)\bar{q}_i(n,t)N_i(n,t) \qquad (4)$$

$$Ap_i(n,t) = (36\pi)^{1/3}(\bar{V}p_i(n,t))^{2/3}N_i(n,t) \qquad (5)$$

where $Ap_i(n,t)$, $\bar{V}p_i(n,t)$, $X_i(n,t)$, and $N_i(n,t)$ are the total particle area, total volume, total conversion and total number of particles in reactor n of the ith particle generation. Also,

$\sum_{j=1}^{k}(N_j(n-1,k))$ is the total number of particles coming into reactor n from the discharge of reactor n-1, summing the number of particles in all k generations that occurred in reactor n-1. If more than one particle generation occurs in a reactor, then the total properties at the exit of the reactor will be the sum of the properties of the individual generations:

$$X(n,t) = \sum_{i=1}^{k}(X_i(n,t), A_p(n,t) = \sum_{i=1}^{k} A_{p_i}(n,t), V_p(n,t)$$

$$= \sum_{i=1}^{k} V_{p_i}(n,t), N(n,t) = \sum_{i=1}^{k} N_i(n,t) \qquad (6)$$

The other variables in equations (2)-(6) are defined in the nomenclature. The series of differential equations above are solved simultaneously with material balances on the initiator and emulsifier concentrations in the reactors:

$$\frac{d(I_n)_w}{dt} = \frac{1}{\Theta_n}((I_{n-1})_w - (I_n)_w - k_d(I_n)_w \qquad (7)$$

$$\frac{dS_n}{dt} = \frac{1}{\Theta_n}(S_{n-1} - S_n) \qquad (8)$$

This particular model allows for particle nucleation to occur by either micellar or homogeneous nucleation mechanisms. The details of the mathematical development are available in the paper by Kiparissides (9). Solution of the set of differential equations (2)-(8) requires the additional iteration over the number of reactors in the train.

The controller in Figure 2 was chosen arbitrarily to be a proportional + integral + derivative type, the discrete form (for implementation on the digital computer) of which is:

$$m_n - m_{n-1} = \Delta m_n = K_c((e_n - e_{n-1}) + \frac{T_s}{\tau I}e_n$$

$$+ \frac{\tau D}{T_s}(e_n - 2e_{n-1} + e_{n-2})) \qquad (9)$$

where m_n = controller output at sample period n
K_c = controller gain
e_n = error signal (setpoint-feedback value) at sample period n
T_s = sampling period

τ_I = reset time
τ_D = derivative time

The feedback value used in the calculation of error is the measured conversion for the standard feedback loop, while it is the predicted value of future conversion for the analytical predictor algorithm, the predicted value being obtained from the model of Kiparissides.

Simulation Results

Kiparissides considered the continuous polymerization of vinyl acetate in a single CSTR at emulsifier concentrations which resulted in either steady-state operation (0.06 moles surfactant/liter of water) or sustained oscillations in number of particles and conversion (0.01 mol/l). The open-loop performance of a two equal-sized reactor train in terms of conversion (Figure 4), number polymer particles (Figure 5) and free soap area (Figure 6) for an initiator feed concentration of 0.005 mol/l for regions of sustained oscillations and steady-state operation indicates that the second reactor is quite similar in performance to the first. It should be observed, however, that in both regions of surfactant concentration, particle formation occurs in both reactors. Kiparissides chose to manipulate emulsifier feed rate and initiator feed rate to control reactor performance. In many commercial polymerizations the emulsifier concentration in the reaction system is fixed by technical or economic reasons and is, therefore, not an available manipulable variable. We, therefore, have considered conversion control utilizing a single variable, either initiator flow rate or reactor temperature, at both high and low, fixed levels of surfactant.

Conversion Control of the First Reactor. Initially, values for the tuning constants, K_c, τ_I, and τ_D for the controller were selected by a crude trial and error method and are given in Table I. Later, "optimum" values were determined from the IAE tuning relations for load disturbances which will be discussed below. A set of constants are shown in Table I for conversion control both by manipulation of initiator flow rate and by manipulation of reactor temperature. For the arbitrarily selected tuning constants, the simulated system response during start-up, while under closed-loop control, is shown in Figures 7 through 10. In these simulations, the initiator flow rate was held constant for three residence times of the reactor at which point the control action was initiated. Figure 7 illustrates the conversion profile in the first reactor at the low emulsifier level (0.01 mol/l) with the control system manipulating initiator feed rate. The calculated control action for both standard feedback and analytical predictor algorithms is shown in Figure 8. Despite prediction of the future occurrence of a particle generation, these control algorithms were not capable of preventing or

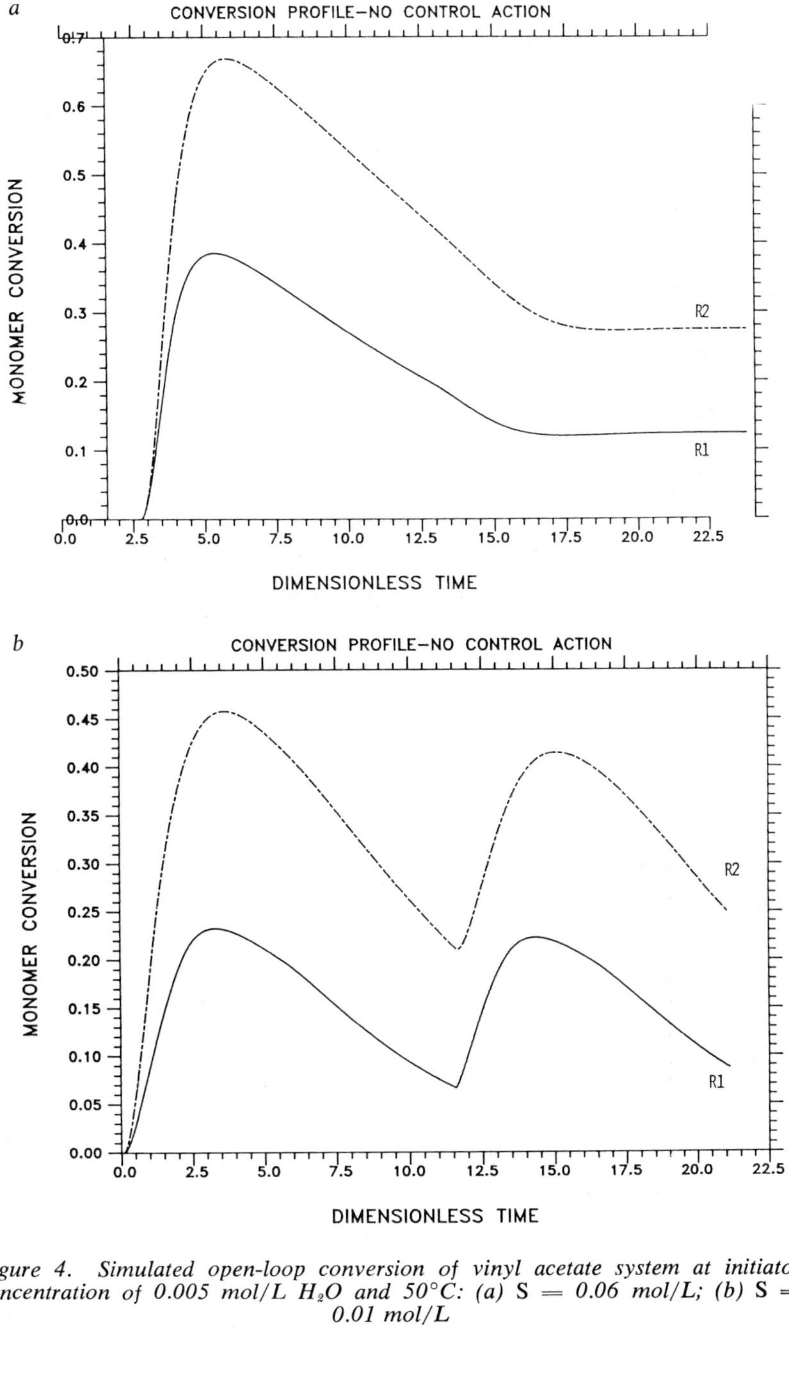

Figure 4. Simulated open-loop conversion of vinyl acetate system at initiator concentration of 0.005 mol/L H_2O and 50°C: (a) S = 0.06 mol/L; (b) S = 0.01 mol/L

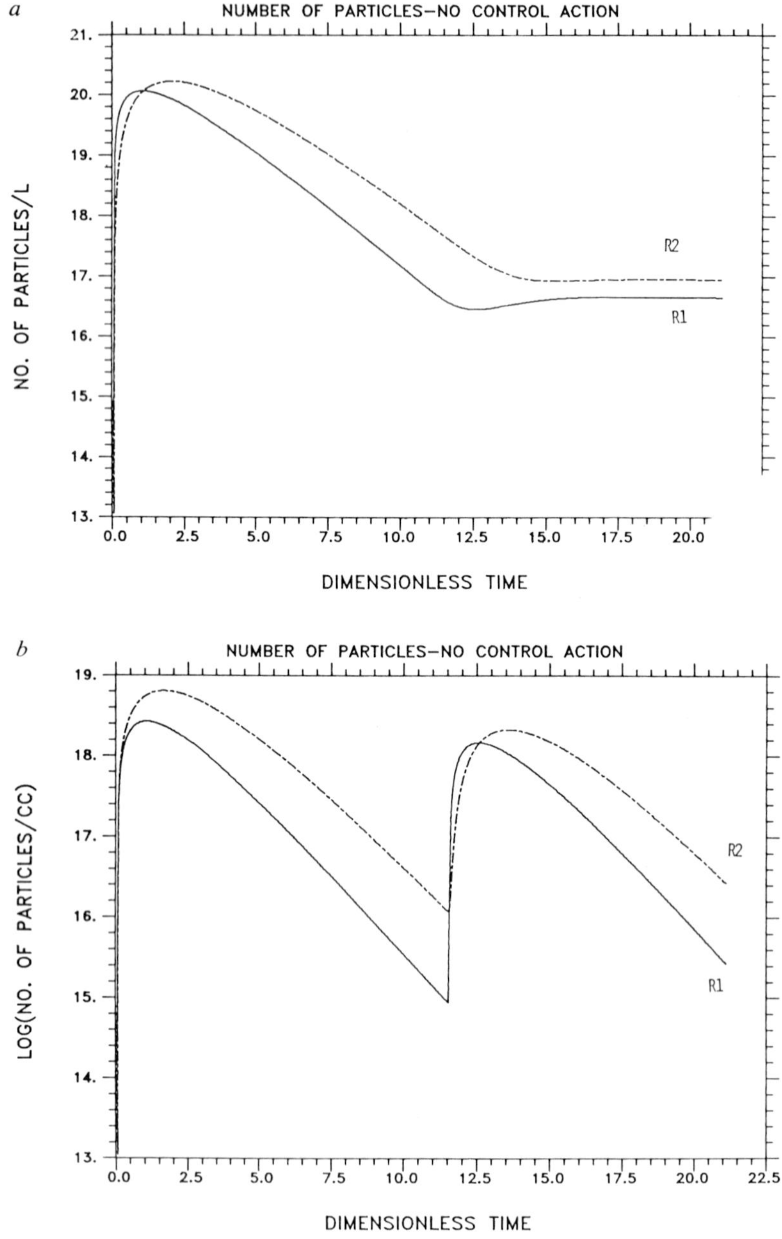

Figure 5. Number of particles under open-loop conditions at initiator concentration of 0.005 mol/L H_2O and 50°C: (a) S = *0.06 mol/L; (b)* S = *0.01 mol/L*

a

b

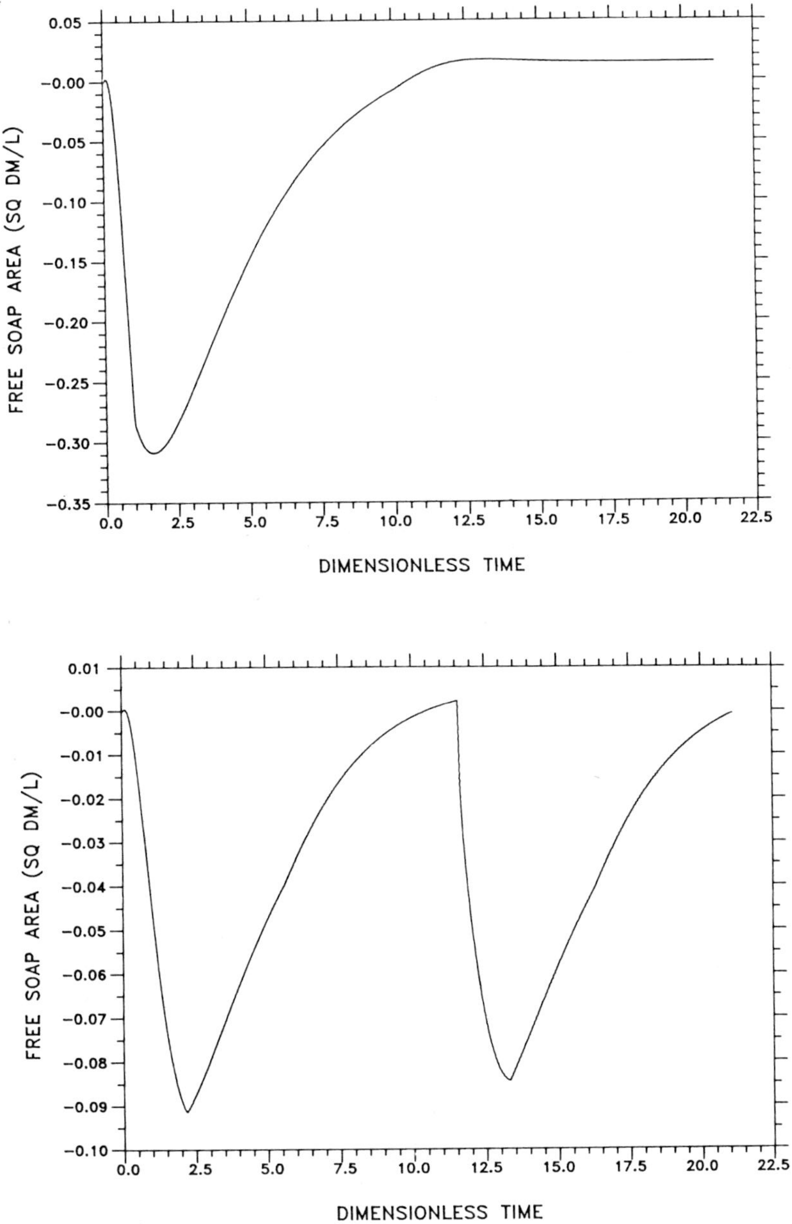

Figure 6. Free soap area in the first reactor with no control action at initiator concentration of 0.005 mol/L and 50°C: (a) S = *0.06 mol/L; (b)* S = *0.01 mol/L*

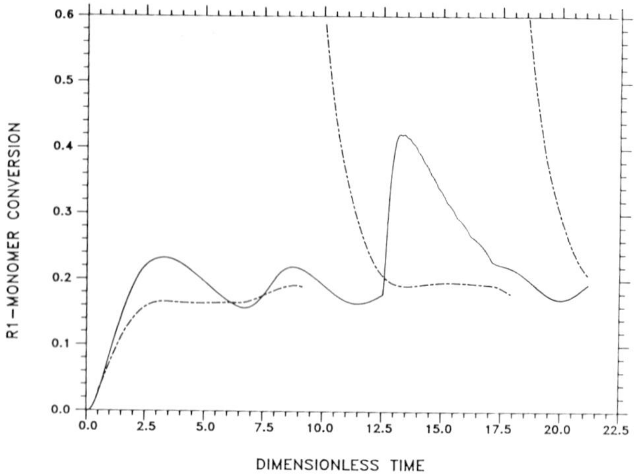

Figures 7. Simulated start-up of vinyl acetate polymerization at low emulsifier level (0.01 mol/L H_2O) under closed-loop control with arbitrarily selected controller tuning constants and manipulation of initiator flow rate at 50°C: conversion in R1—STD feedback (——) vs. DTC (– – –)

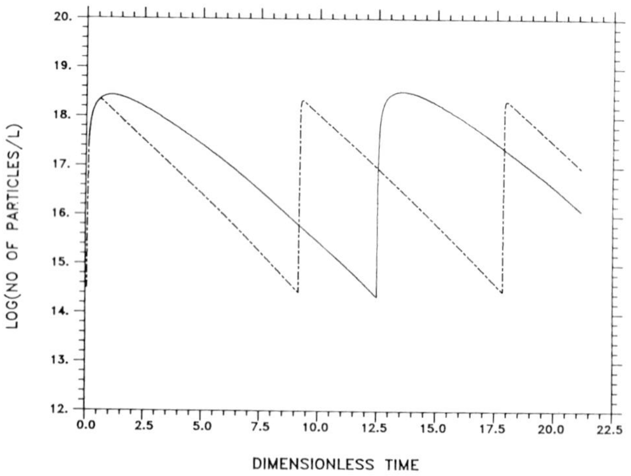

Figures 8. Simulated start-up of vinyl acetate polymerization at low emulsifier level (0.01 mol/L H_2O) under closed-loop control with arbitrarily selected controller tuning constants and manipulation of initiator flow rate at 50°C: R1—no. of particles—STD feedback (——) vs. DTC (– – –)

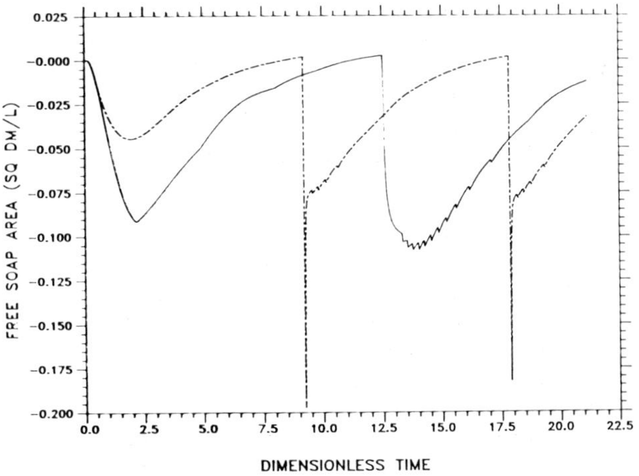

Figure 9. Simulated start-up of vinyl acetate polymerization at low emulsifier level (0.01 mol/L H_2O) under closed-loop control with arbitrarily selected controller tuning constants and manipulation of initiator flow rate at 50°C: free soap area—STD feedback (——) vs. DTC (– – –)

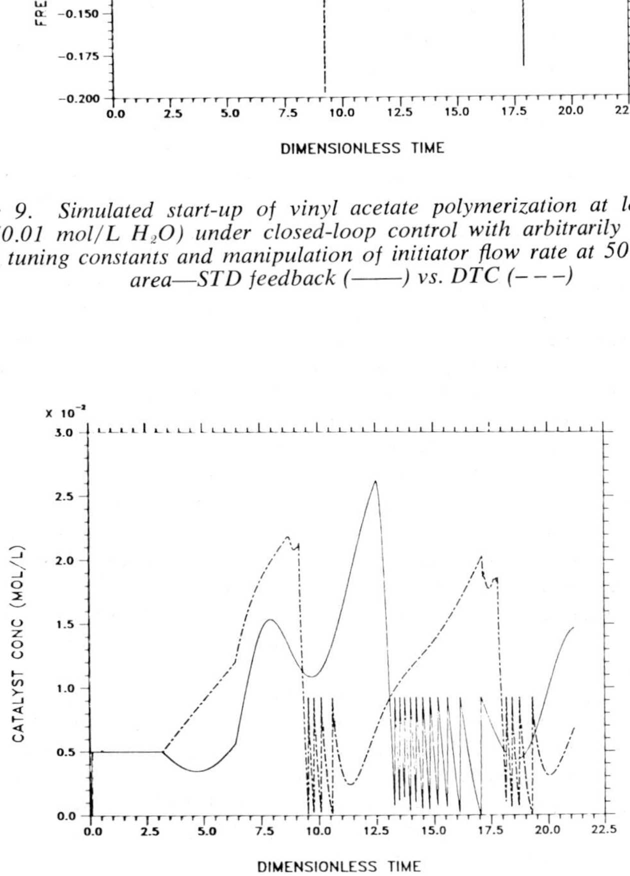

Figure 10. Simulated start-up of vinyl acetate polymerization at low emulsifier level (0.01 mol/L H_2O) under closed-loop control with arbitrarily selected controller tuning constants and manipulation of initiator flow rate at 50°C: catalyst feed concentration—STD feedback (——) vs. DTC (– – –)

even severely dampening particle nucleation at the low emulsifier concentration as shown in Figure 9. The same type of response was simulated when temperature was the manipulated variable (Figures 11-13). The wide excurisons in temperature that were simulated, of course, would not be possible in practice, but were allowed in the simulation to see if the sporadic nucleation could be precluded. The advantages of dead-time compensation under these conditions are faster return to setpoint after nucleations and tighter regulatory control between nucleations. It is apparent from these simulations, however, that reactor control in the regions of low emulsifier concentration is very poor, even with dead-time compensation. These results clearly demonstrate the difficulty of reactor control in the presence of sporadic particle nucleations. Prevention of formation of new particles appears to be impossible by means of a control strategy if only one variable is manipulated. The free soap area in the first reactor is shown as a function of time in Figure 10 for closed-loop control utilizaing initiator feed rate. The successful control strategy would provide a constant free soap area and, hence, constant particle generation rate in the reactor. Because this reactor is being run under a "soap-starved" condition, the free soap area is normally negative. It might be possible to start-up the reactor under conditions that resulted in a constant particle generation rate and then slowly decrease the feed soap concentration to the desired level without initiating the limit cycle performance. However, this would not be an acceptable operating condition because the introduction of any disturbance into the process would be likely to send the system into the limit cycle condition. For this operating region, therefore, elimination of the unsteady-state condition by means of control strategy appears unlikely and solution of the problem requires a design modification that would result in a constant particle generation rate in the system. Poehlein and Daugherty (3) suggested one possible means of achieving this goal; by feeding a "seeded" latex into the continuous reactor system where the seed is formed either in a batch reactor or in a tubular reactor placed in front of the CSTR train. The concentration of seed in the feed latex must be high enough to preclude further particle nucleation in the continuous reactors. This type of design modification appears to be necessary to achieve stable operation of a train continuous emulsion polymerization reactors when it is necessary to run at low emulsifier concentrations.

At high emulsifier concentration, the utility of the analytical predictor is much more apparent. Under these operating conditions, the first two reactors reach a steady-state level of polymer particle generation rate and monomer conversion as was shown in Figures 4-6. Again using the arbitratily selected tunging constants, the simulated system response during start-up is shown in Figurs 14 through 17. Figure 14 illustrates the conversion profile for both the standard feedback and DTC

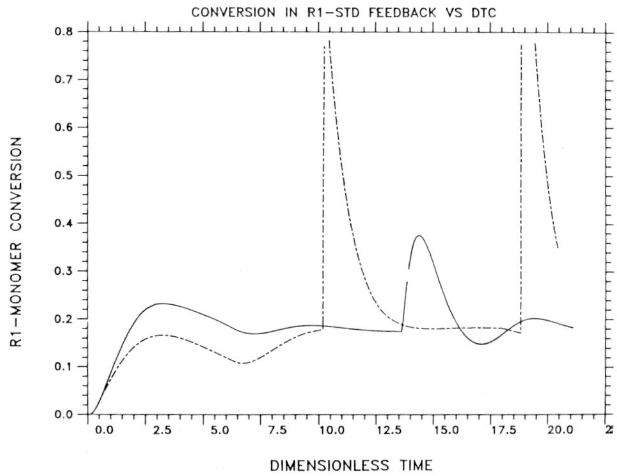

Figure 11. Simulated start-up of vinyl acetate polymerization at low emulsifier level (0.01 mol/L H_2O) under closed-loop control with arbitrarily selected controller tuning constants and manipulation of reactor temperature at initiator concentration of 0.005 mol/L: conversion in R1—STD feedback (——) vs. DTC (- - -)

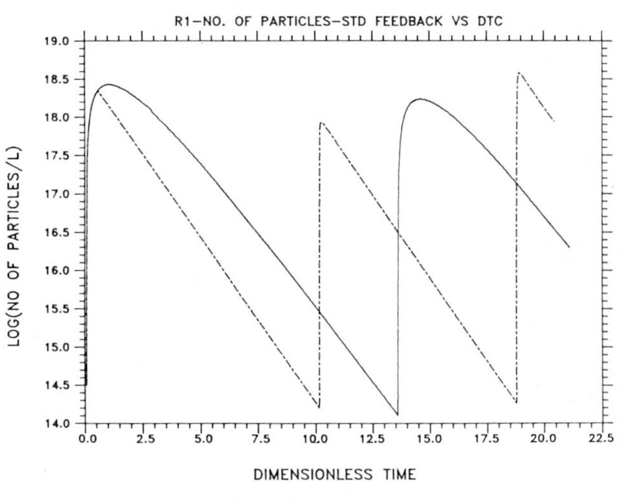

Figure 12. Simulated start-up of vinyl acetate polymerization at low emulsifier level (0.01 mol/L H_2O) under closed-loop control with arbitrarily selected controller tuning constants and manipulation of reactor temperature at initiator concentration of 0.005 mol/L: R1—No. of particles—STD feedback (——) vs. DTC (- - -)

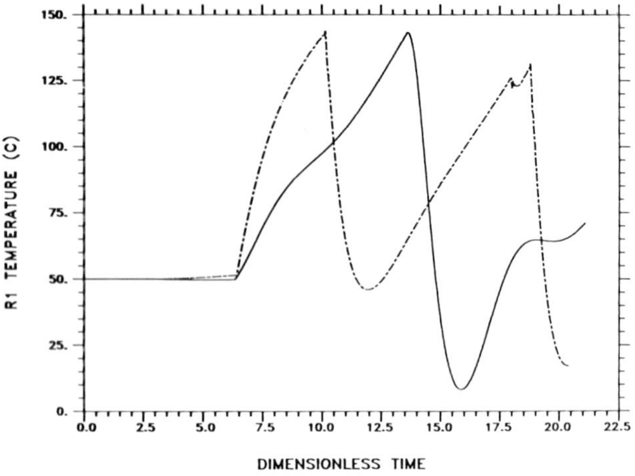

Figure 13. Simulated start-up of vinyl acetate polymerization at low emulsifier level (0.01 mol/L H_2O) under closed-loop control with arbitrarily selected controller tuning constants and manipulation of reactor temperature at initiator concentration of 0.005 mol/L: reactor temperature—STD feedback (——) vs. DTC (– – –)

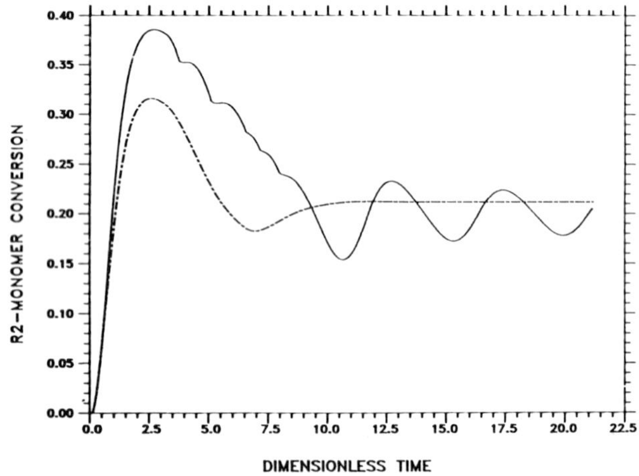

Figure 14. Simulated start-up of vinyl acetate polymerization at high emulsifier level (0.06 mol/L H_2O) under closed-loop control with arbitrarily selected controller tuning constants and manipulation of initiator flow rate at 50°C: conversion in R1—STD feedback (——) vs. DTC (– – –)

algorithms manipulating initiator flow rate. Figure 15 indicates the degree of conversion control achieved in the downstream reactors when the front reactor is under control. At the higher level of emulsifier, free soap area is available at steady-state resulting in a constant particles generation rate (Figure 16). The inherent advantages of dead-time compensation under this operating condition are faster approach to setpoint and tighter regulatory control. As seen in Figure 17 the DTC algorithm requires smaller and less frequent changes in the manipulated variable. The system response, utilizing temperature as manipulated variable, is similar (Figures 18 - 19). The "leading" tendency of the DTC algorithm is apparent in Figure 19 where the temperature trace is presented.

More typically, instead of setpoint changes, the regulatory problem of responding to a system disturbance is encountered in commercial reactors. For this reason, the optimum tuning constants for the PID controller were developed from the IAE ralations for load disturbances. First, however, it is necessary to obtain a process model of the system. Brantley (10) has developed a process identification technique which fits process data to the second order plus dead-time form:

$$G_p(s) = \frac{K_p e^{-\Theta_d s}}{(\tau_1 s+1)(\tau_2 s+1)} \tag{10}$$

For the continuous polymerization of vinyl acetate at 0.06 mol/l emulsifier concentration, the model prameters as determined by introducing step changes into the mechanistic model are shown in Table II. Given the form of equation (10), the "optimum" controller tuning constants for a PID controller were determined by the method of Gallier and Otto (11), which minimizes the integral of the absolute error (IAE) that results following a process disturbance. The effective time delay, T', used in calculating the optimum tuning constants as shown by Meyer, et al (7) is Θ_d+T_s for the standard feedback loop and T_s for the analytical predictor loop. The degree of control achieved by these algorithms is measured by the common integral criteria:

1. Integral of the square error (ISE):

$$ISE = \int_0^\infty (e(t))^2 dt$$

2. Integral of the absolute error (IAE):

$$ISE = \int_0^\infty (e(t))dt$$

3. Integral of time multiplied by the absolute value of the error (ITAE):

TABLE I: CONTROLLER PARAMETERS FOR SIMULATION STUDY

Manipulated Variable	Control Scheme	K_{c_1} $\frac{\text{lb/hr or °C}}{\text{fraction conversion}}$ By Trial	K_{c_1} IAE Optimum	τ_I, min By Trial	τ_I, min IAE Optimum	τ_D, min By Trial	τ_D, min IAE Optimum
Initiator Flow							
	Std. Feedback	8.0	1.68	1.0	16.83	2.0	4.29
	Analytical Predictor	8.0	37.21	1.0	3.79	2.0	0.11
Reactor Temperature							
	Std. Feedback	40.0	173.8	1.0	15.73	2.0	3.42
	Analytical Predictor	40.0	2750.	1.0	2.37	2.0	0.12

TABLE II: PROCESS MODEL PARAMETERS

$$\text{Model Form:} \quad G_p(s) = \frac{K_p e^{-\Theta_d s}}{(\tau_1 s+1)(\tau_2 s+1)}$$

Manipulated Variable	Θ_d, min	K_p $\frac{\text{lb/hr or °C}}{\text{fraction conversion}}$	τ_1, min	τ_2, min
Initiator Flow	30.0	0.2793	3.6826	0.0200
Temperature	30.0	0.002529	2.2540	0.0200

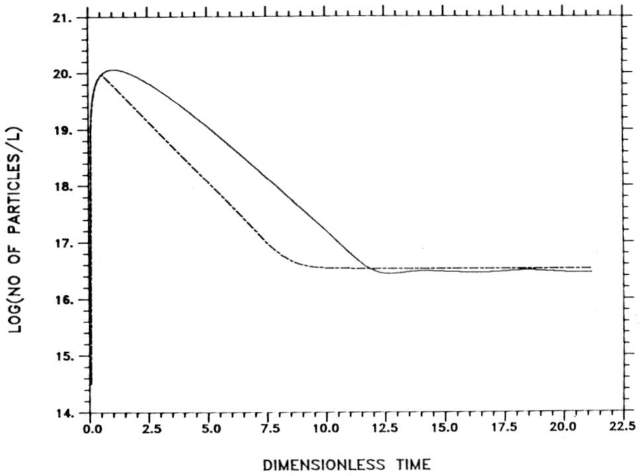

Figure 15. Simulated start-up of vinyl acetate polymerization at high emulsifier level (0.06 mol/L H_2O) under closed-loop control with arbitrarily selected controller tuning constants and manipulation of initiator flow rate at 50°C: R1—no. of particles—STD feedback (——) vs. DTC (– – –)

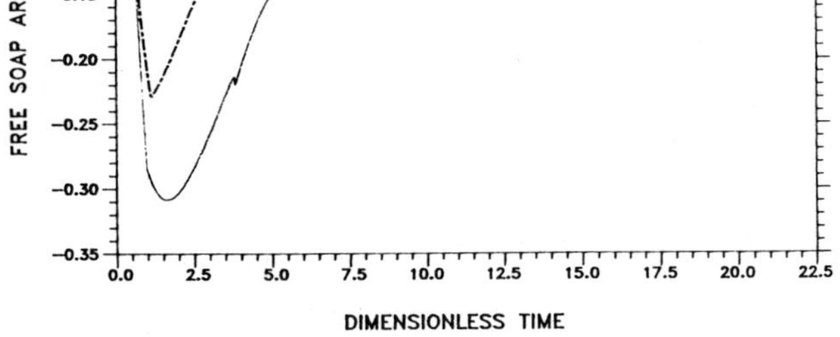

Figure 16. Simulated start-up of vinyl acetate polymerization at high emulsifier level (0.06 mol/L H_2O) under closed-loop control with arbitrarily selected controller tuning constants and manipulation of initiator flow rate at 50°C: free soap area—STD feedback (——) vs. DTC (– – –)

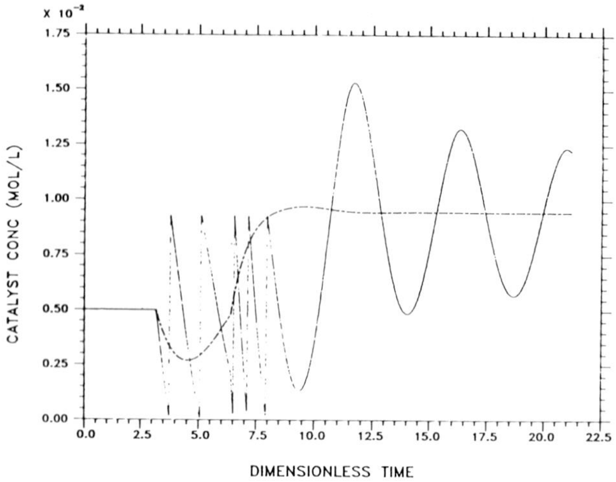

Figure 17. Simulated start-up of vinyl acetate polymerization at high emulsifier level (0.06 mol/L H_2O) under closed-loop control with arbitrarily selected controller tuning constants and manipulation of initiator flow rate at 50°C: catalyst feed concentration—STD feedback (——) vs. DTC (– – –)

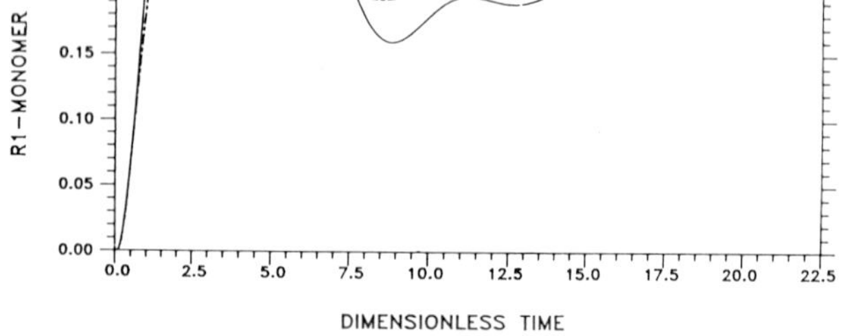

Figure 18. Simulated start-up of vinyl acetate polymerization at high emulsifier level (0.06 mol/L H_2O) under closed-loop control with arbitrarily selected controller tuning constants and manipulation of reactor temperature at initiator concentration of 0.005 mol/L: conversion in R1—STD feedback (——) vs. DTC (– – –)

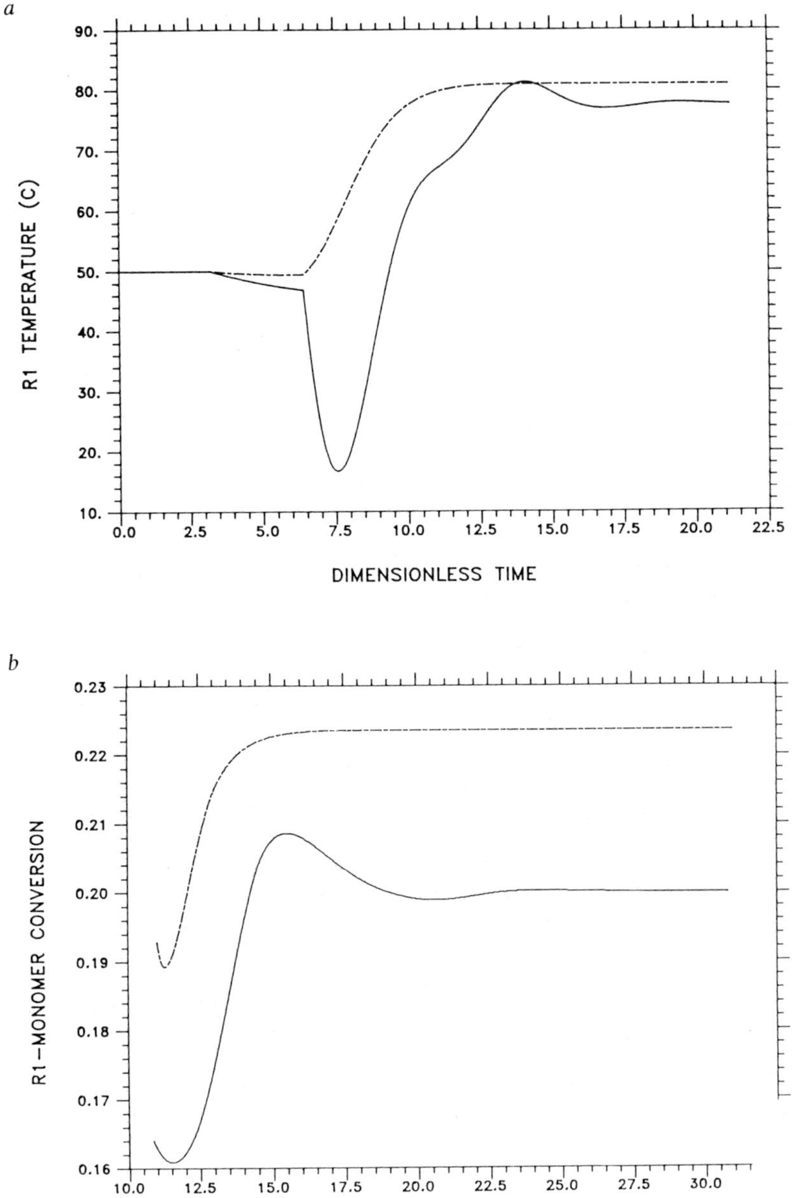

Figure 19. Simulated start-up of vinyl acetate polymerization at high emulsifier level (0.06 mol/L H_2O) under closed-loop control with arbitrarily selected controller tuning constants and manipulation of reactor temperature at initiator concentration of 0.005 mol/L: (a) R1 reactor temperature—STD feedback (——) vs. DTC (– – –); (b) conversion in R1—STD feedback (——) vs. DTC (– – –)

$$\text{ITAE} = \int_0^\infty t(e(t))dt$$

where e(t) is the difference between the setpoint value and the actual value of conversion.

These "controllability" criteria are used to judge the performance of the various control systems described above in Table III. As seen in Table III, dead-time compensation markedly improves system performance at the high emulsifier concentration. A system load disturbance was simulated by introducing a step change in the propagation rate constant, k_p, of -5% after the system was at steady-state under closed-loop control. This type of disturbance in which the system reaction rate changes suddenly, in most cases due to the introduction of an unknown disturbance, is common in commercial processes. The improved performance of the analytical predictor with "optimum" controller tuning is shown in Figure 20 for regulatory control. Notice that these controller settings result in "ringing" of the control valve in the analytical predictor algorithm that does not occur with the arbitrarily selected tuning constants.

These simulated results for the high emulsifier concentration operating condition demonstrate the utility of dead-time compensation to the control of conversion from the first reactor in a train. With implementation of this degree of control on the first reactor, control schemes for downstream reactors can be simplified as discussed in the next section.

Conversion Control of Downstream Reactors. With the first reactor of the train under closed-loop conversion control, response of downstream reactors to changes in initiator flow can be approximated very closely by the second order plus dead-time model of equation (10), provided that particle nucleation in the downstream reactor occurs at a constant rate or is totally precluded because of low free soap concentration in the feed to the reactor. A danger of operating these reactor trains at high emulsifier concentration to provide stability is that, although the front reactors do reach a steady-state level of particle nucleation rate, there may exist the condition in a downstream reactor that leads to sporadic nucleations and, hence, oscillations in conversion. The required condition for stability, then, would be that the feed to the last reactor would allow for continuous particle nucleation, i.e., that positive free soap area be present in the finished emulsion. For many systems this requirement forces a level of emulsifier in the feed that results in prohibitively high emulsion viscosities. The solution to this problem again requires a reactor design modification which provides separation of the regimes of particle nucleation and particle growth. The following development for control of downstream reactors is applicable, however, to those reactors in which a constant or negligible rate of particle nucleation occurs.

TABLE III: COMPARISON OF CONTROL SYSTEM PERFORMANCE - FIRST REACTOR

Manipulated Variable	Surfactant Conc. (mol/1)	Control Method[1]	K_c	τ_I	τ_D	ISE	IAE	ITAE
			Setpoint Change					
None	0.01	None[2]	-	-	-	2.123	26.03	7608
None	0.01	None[3]	-	-	-	3.518	35.46	9552
Initiator Flow	0.01	SFB	8.0	1.0	2.0	2.868	24.73	7703
" "	0.01	DTC	8.0	1.0	2.0	19.65	50.51	15520
" "	0.06	None[2]	-	-	-	3.774	38.99	9512
" "	0.06	None[3]	-	-	-	5.987	36.72	4595
" "	0.06	SFB	8.0	1.0	2.0	2.123	22.78	3309
" "	0.06	SFB	1.08	16.83	4.29	2.010	19.30	1912
" "	0.06	DTC	8.0	1.0	2.0	0.397	9.11	1725
" "	0.06	DTC	37.21	3.79	0.11	0.999	10.65	1080
Temperature	0.01	SFB	40.0	1.0	2.0	1.111	15.57	4523
"	0.01	DTC	40.0	1.0	2.0	23.18	56.72	18610
"	0.06	SFB	40.0	1.0	2.0	2.007	17.62	1442
"	0.06	SFB	173.8	15.73	3.42	1.955	16.25	1146
"	0.06	DTC	40.0	1.0	2.0	0.521	10.06	1600
"	0.06	DTC	2750.	2.37	0.12	2.057	24.75	4717
			System Disturbance					
Initiator Flow	0.06	SFB	1.68	16.83	4.29	0.0952	3.548	305.4
" "	0.06	DTC	37.21	3.79	0.11	0.0399	1.693	83.78

[1] None = open-loop process
SFB = standard feedback control
DTC = dead-time compensation
[2] Initiator Concentration = 0.005 mol/1
[3] Initiator Concentration = 0.01 mol/1

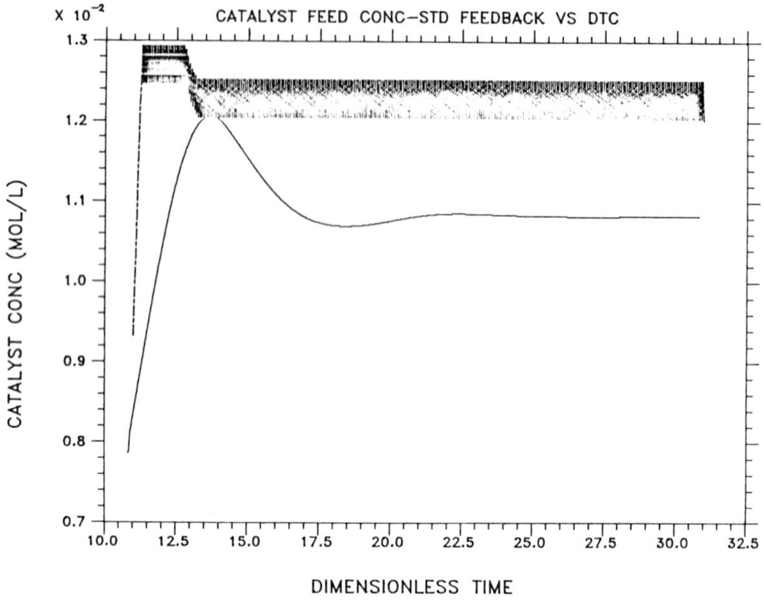

Figure 20. Simulated conversion response of continuous polymerization system to a load disturbance under closed-loop control with IAE optimum controller tuning constants and manipulation of initiator flow rate at 0.06 mol/L H_2O surfactant and 50°C: catalyst feed concentration—STD feedback (——) vs. DTC (– – –)

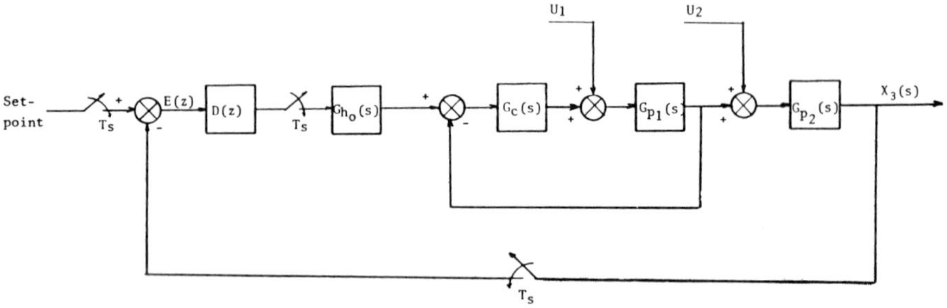

Figure 21. Block diagram of digital control loop for downstream reactors

For the vinyl acetate system at 0.06 mol/l of emulsifier in the feed, the process identification technique of Brantley (10) was used to find the conversion response of the third equal-sized reactor in the train to changes in feed rate of initiator to that reactor:

$$G_p(s) = \frac{X_3(s)}{\dot{m}_3(s)} = \frac{0.0525e^{-30.0s}}{(19.864s+1)(0.020s+1)} \quad (11)$$

Brantley's method uses a multivariable search technique to determine the model parameters of a second order plus dead-time model that minimizes the square of the error between the actual output and the model-predicted output for a given set of input values.

The empirical model of equation (11) predicted the response of the mechanistic model to a step change in initiator flow very closely (the average absolute deviation between the empirical model and mechanistic model was 0.8% of the response). Three algorithms have been considered for control of the downstream reactor modeled by equation (11).

1. Optimum PID Controller: This technique follows the work of Gallier and Otto (11) and is described in the previous section.
2. Z-Transform-Designed Algorithm: This technique involves specifying the desired response to a load or setpoint change and using that response to design the controller. The digital process control loop for downstream reactor control is shown in Figure 21. The controller transfer function, D(z), can be obtained from this loop as:

$$D(z) = \frac{1}{HG(z)} \cdot \frac{C(z)/R(z)}{1- C(z)/R(z)} \quad (12)$$

where HG(z) is the Z-transform of the zero-order holds times the process transfer function, or

$$HG(z) = \left(\frac{1-e^{-sT}}{s} \cdot G(s)\right) \quad (13)$$

Typically, one specifies the desired response, C(z)/R(z), which yields from equation (12) the required design of the controller, D(z). In practice, however, this design technique results in a controller which requires excessive valve movement, an undesirable situation. Consequently, Kalman (12) developed a Z-transform algorithm which specifies the desired output, C(z), and the desired valve travel, M(z) for a setpoint change. The desired response and valve travel for a unit step change in setpoint is shown in Figure 22. The system response, then, follows the expression:

$$C(z) = z^{-N}(C_1 z^{-1} + z^{-2} + z^{-3} + \cdots) \quad (14)$$

where N is the integer number of sampling periods in the process dead-time.
Likewise, the valve travel can be described as:

$$M(z) = M_o + M_1 z^{-1} + M_f z^{-2} + M_f z^{-3} + \cdots \quad (15)$$

The transfer functions relating response and valve travel to a unit change in setpoint ($R(z) = 1/(1-z^{-1})$) are then:

$$\frac{C(z)}{R(z)} = (1-z^{-1})z^{-N}(C_1 z^{-1} + z^{-2} + z^{-3} + \cdots)$$

$$= C_1 z^{-N-1}(1-C_1)z^{-N-2} = P_1 z^{-N-1} + P_2 z^{-N-2}$$

$$= P(z) \quad (16)$$

and,

$$\frac{M(z)}{R(z)} = (1-z^{-1})(M_o + M_1 z^{-1} + Mf z^{-2} + Mf z^{-3} + \cdots)$$

$$= M_o + (M_1 - M_o)z^{-1} + (M_f - M_1)z^{-2}$$

$$= q_o + q_1 z^{-1} + q_2 z^{-2} = Q(z) \quad (17)$$

It can be seen from Figure 21 that the ratio of C(z) to M(z) is the process pulse transfer functions, HG(z):

$$HG(z) = \frac{C(z)}{M(z)} = \frac{P(z)}{Q(z)} \quad (18)$$

It follows from the controller design equation (12) that:

$$D(z) = \frac{Q(z)}{1-P(z)} \quad (19)$$

Therefore, the Kalman designed algorithm is specified by obtaining the modified Z-transform of the process pulse transfer function ot get P(z) and Q(z) and then substituting these values into equation (19).

3. Self-Tuning Algorithm: This algorithm incorporates the method described just above with the added feature of

on-line tuning of the controller. Here, process data are compared each sampling period with the model prediction obtained from equation (11). When significant error develops, typically as a result of the introduction of a process disturbance, the process identification technique of Brantley (10) is used to calculate new parameters of the process model. These new parameters result in a new controller algorithm as obtained by the Kalman method described above. This control loop is shown schematically in Figure 23.

The performance of the algorithms described above in controlling downstream reactor conversion is illustrated for a step change in setpoint of the third reactor in the vinyl acetate polymerization model in Figure 24 and for the introduction of a disturbance (10% decrease in k_p) in Figure 25. The control criteria results for these algorithms are given in Table IV. Manipulation of initiator flow rate to the third reactor is shown for each controller for a step disturbance in Figure 26. The self-tuning algorithm shows no improvement in control over the Z-transform technique over the period of operation illustrated in Figure 25. However, the process model had been updated to account for the reduction in k_p by the process identification technique to the form:

$$G_p(s) = \frac{0.00745e^{-30s}}{(0.024s+1)(0.02s+1)} \qquad (20)$$

Consequently, the self-tuning algorithm would respond to the introduction of a setpoint change or another disturbance faster than the original model, illustrating the value of the self-tuning algorithm.

The free soap area of the third reactor of the vinyl acetate polymerization is shown in Figure 27. Notice that for the reactor conditions chosen for this simulation that particle formation occurs continuously and approaches a steady-state level in the third reactor. For the case of vinyl acetate, the homogeneous nucleation rate is sufficient at low free soap concentration to provide a constant rate of particle generation in the absence of micelles and, hence, the free-soap area is actually depleted in the third reactor but reaches a steady-state level so that micellar nucleation never occurs. As discussed above, this condition does not always exist and the potential for initiation of sustained oscillations exists for downstream reactors where the free soap area is negative and the homogeneous nucleation rate is not significant compared with the micellar nucleation rate. This situation must be considered in reactor design and in selection of operating conditions.

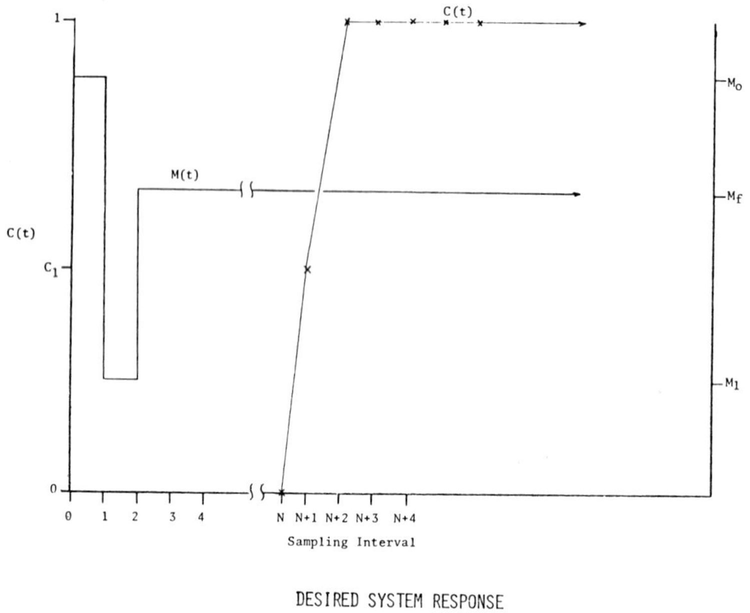

Figure 22. Desired system response C(z) and valve travel M(z) for a unit step change in setpoint according to Kalman's approach

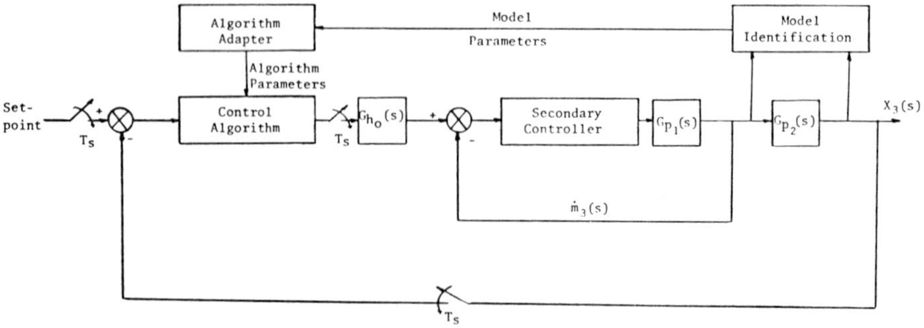

Figure 23. Block diagram of self-tuning control loop for downstream reactors

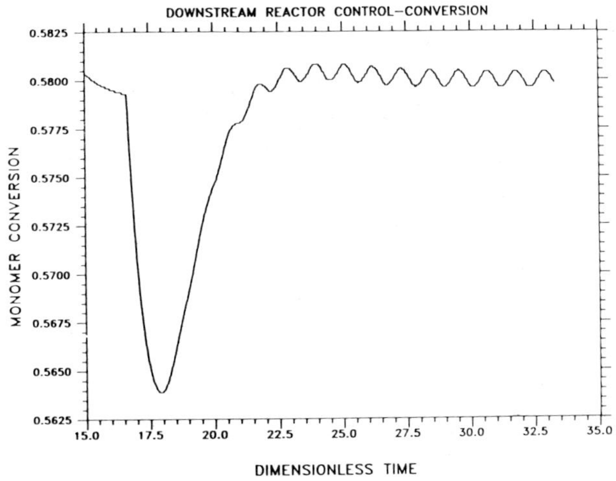

Figure 24. Simulated response of third reactor of a continuous vinyl acetate polymerization to a step change in setpoint at high emulsifier feed concentration (0.06 mol/L H_2O) and manipulation of initiator flow rate to the third reactor at 50°C ((——) optimum PID); (– – –) Z transform)

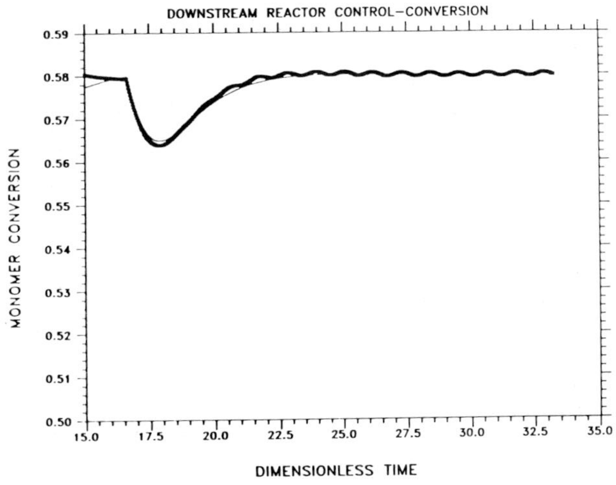

Figure 25. Simulated response of third reactor of a continuous vinyl acetate polymerization to a step disturbance at high emulsifier feed concentration (0.06 mol/L H_2O) and manipulation of initiator flow rate to the third reactor at 50°C ((——) optimum PID); (– – –) Z transform; (×××) self-tuning algorithm)

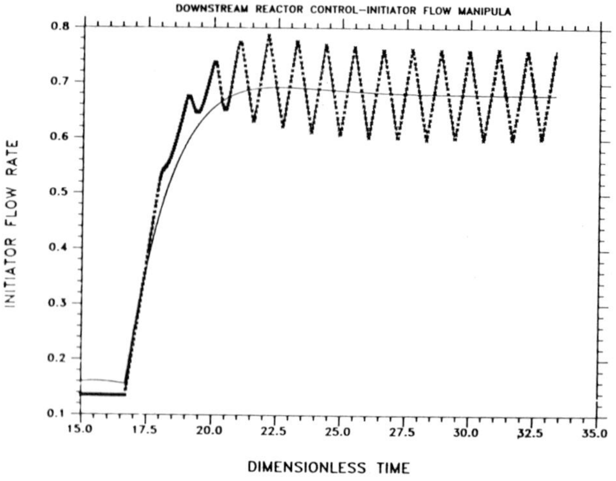

Figure 26. Simulated initiator flow rate manipulation for closed-loop control of the third reactor in response to a step disturbance at high emulsifier feed concentration (0.06 mol/L H_2O) and 50°C ((——) optimum PID; (– – –) Z transform; (×××) self-tuning algorithm)

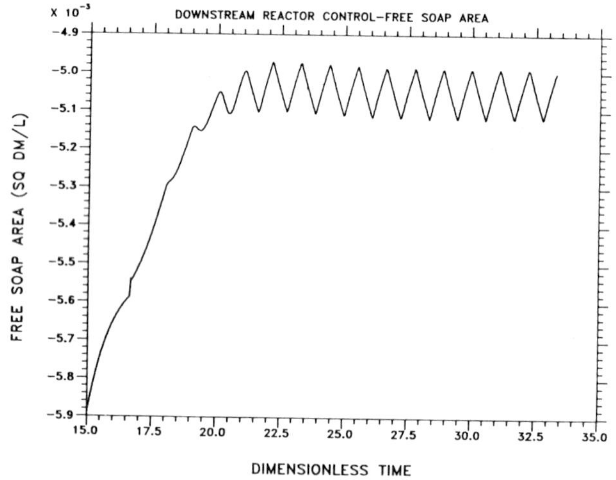

Figure 27. Free soap area of emulsion in third reactor of a continuous vinyl acetate polymerization under closed-loop control at high emulsifier feed concentration (0.010 mol/L H_2O) and 50°C ((——) optimum PID; (– – –) Z transform)

TABLE IV: CONTROL SYSTEM PERFORMANCE - DOWNSTREAM REACTOR

Controller Algorithm	ISE	IAE	ITAE
	Step Change in Setpoint		
1. Optimum PID	0.0611	2.234	94.17
2. Z-Transform	0.0612	2.219	94.43
	Step Change in Disturbance (10% decrease in k_p)		
1. Optimum PID	0.0140	1.339	89.29
2. Z-Transform	0.0147	1.370	104.0
3. Self-tuning	0.0147	1.370	104.0

Process Model: $$\frac{X_3(s)}{\dot{m}_3(s)} = \frac{0.05245e^{-30s}}{(19.864s+1)(0.020s+1)}$$

Conclusions

The utility of the analytical predictor method of dead-time compensation to control of conversion in a train of continuous emulsion polymerizers has been demonstrated by simulation of the vinyl acetate system. The simulated results clearly show the extreme difficulty of controlling the conversion in systems which are operated at "soap-starved" conditions. The analytical predictor was shown, however, to provide significantly improved control of conversion, in presence of either setpoint or load changes, as compared to standard feedback systems in operating regions that promote continuous particle formation. These simulations suggest the analytical predictor technique to be the preferred method of control when it is desired that only one variable (preferably initiator feed rate) be manipulated.

Three algorithms were also suggested for control of conversion from downstream reactors when the first reactor of the train is operating at steady-state under closed-loop control: a conventional PID controller with optimum IAE tuning constants, a Z-transform designed algorithm based on an approximate linear process model, and a self-regulating system which updates the controller algorithm based on changes in the process model achieved by on-line process identification. All three of these controllers were shown to provide excellent control of downstream reactors in which continuous polymer particle generation occurred, for both setpoint and load changes.

Nomenclature

A_p = total surface area of polymer particles
$C(z)$ = system output or response in z-domain
d_m = monomer density
d_p = polymer density
$D(z)$ = controller transfer function
e = system error signal
$f(n,t)$ = net rate of particle nucleation in reactor n at time t
G_p = system open-loop transfer function
$(I)_w$ = initiator concentration
$HG(z)$ = process pulse transfer function
M_T = total monomer concentration in the emulsion
M_W = molecular weight of monomer
$M(z),m$ = value of manipulated variable
$\dot{m}_3$ = initiator flow rate to third reactor in series
N = total number of particles
$\overline{q}_i(n,t)$ = average number of radicals per particle in reactor n at time t
$R(z)$ = setpoint value
S = emulsifier concentration
T_s = sampling period
T' = sampling period

V_p = total volume of polymer particles
X = total monomer conversion
X_3 = monomer conversion from third reactor in series

Subscripts

i = having to do with the ith generation of particles within a given reactor
n = reactor number or sample number

Greek Symbols

ϕ = monomer volume fraction in a particle
Θ = mean residence time of a reactor
Θ_d = process dead-time
τ = time constant

References

1. Wismer, D. A., Brand, W., Joint Automatic Control Conference, Stanford, California, (1964), 147-154.
2. Francis, D. H. and H. R. Sontag, Chemical Engineering Progress, 1949, 45, (6).
3. Poehlein, G. W. and D. J. Dougherty, Rubber Chemistry and Technology, 1977, 50, (3).
4. Gopalratnam, P. C., P. B. Deshpande, and R. H. Ash, paper presented at ISA National Conference, Chicago, Illinois, paper No. C.I. 79-619 (1979).
5. Moore, C. F., Selected Problems in the Design and Implementation of Direct Digital Control, Ph.D. Thesis, Louisiana State University (1969).
6. Doss, J. E. and C. F. Moore, 74th National AICHE Meeting, New Orleans, Louisiana (1973).
7. Meyer, C., D. E. Seborg and R. K. Wood, Ind. Eng. Chem. Process Des. Dev., 1978, 17, (1).
8. Kiparissides, C., J. F. McGregor and A. E. Hamielec, Journal of Applied Polymer Science, 1979, 23, 401-418.
9. Kiparissides, C., Continuous Latex Reactor Modelling and and Experimental Studies, Ph.D. Thesis, McMaster University (1978).
10. Brantley, R. O., M.S. Thesis, University of Louisville, Louisville, Kentucky (1981).
11. Gallier, P. W. and R. E. Otto, Instrumentation Technology, 1968, 15, (2), 65-70.
12. Kalman, R. E., Trans. AIEE, 1954, 236-247.

RECEIVED April 6, 1981.

Experimental Study of the Seeded Polymerization of Vinyl Acetate in a Tube

C. K. LEE[1] and T. H. FORSYTH[1]

The University of Akron, Akron, OH 44325

While vinyl acetate is normally polymerized in batch or continuous stirred tank reactors, continuous reactors offer the possibility of better heat transfer and more uniform quality. Tubular reactors have been used to produce polystyrene by a mass process (1, 2), and to produce emulsion polymers from styrene and styrene-butadiene (3-6). The use of mixed emulsifiers to produce mono-disperse latexes has been applied to polyvinyl toluene (5). Dunn and Taylor have proposed that nucleation in seeded vinyl acetate emulsion is prevented by entrapment of oligomeric radicals by the seed particles (6). Because of the solubility of vinyl acetate in water, Smith-Ewart kinetics (case 2) does not seem to apply, but the kinetic models developed by Ugelstad (7) and Friis (8) seem to be more appropriate.

The objective of this study was to investigate the feasibility of using a tubular reactor for the seeded emulsion polymerization of vinyl acetate, and to study the effect of process variables on conversion rate and latex properties.

EXPERIMENTAL

Figure 1 shows a schematic view of the tubular reactor. Seamless tubing, of 1/4-inch OD, type 316 stainless steel, was used for the preheater and the reactor. The reactor itself was 187 feet long, wound into a helix of 1 ft. diameter. The reactor and preheater were immersed in separate drums, which were filled with water, and maintained at constant temperature.

Two feed tanks were used, one to hold the vinyl acetate emulsion and the other to hold the initiator solution. Constant

[1] Current address: BFGoodrich R&D Center, Brecksville, Ohio 44141

0097-6156/81/0165-0567$05.00/0

feed flow rates were possible by using gear pumps, with the capacity of the emulsion pump twice as great as the initiator solution pump. The preheater was used to obtain a reaction temperature of 50°C, prior to combining the two streams to start the reaction. Flow rates varied between 0.5 and 1.2 cc/sec., and these were laminar conditions.

Upon start-up, and between runs, the reactor was full of water. The initial run required operation for a time equivalent to 2.1 residence times to displace the water, with 1.5 residence times between each run. The difference is because the preheater was not filled with water between runs. Samples were regularly and periodically collected after 1.5 or 2.1 residence times for further characterization. Molecular weights and distributions were measured by GPC.

Initial Batch Reactor Studies. An agitated 2000 ml thick-walled glass reactor was blanketed with nitrogen and operated at 50°C. Vinyl acetate containing about 15 ppm hydroquinone was used without purification. The ionic emulsifier was Sipex EST-30, advertised as a sodium tridecyl ether sulfate, and the nonionic surfactant was Siponic L-25, a lauryl alcohol ethoxylate. Table I shows the recipes and properties of the three seed latexes produced in the batch reactor. Essentially complete conversions were obtained in 30 to 45 minutes, but with a temperature rise of almost 50°C.

TABLE I

Recipe used to Produce Seed
(Prepared in a 1500 ml Batch)

	I	II	III
Water	100	100	100
Vinyl Acetate	52	52	52
Sipex EST-30	0.57	0.93	1.89
Siponic L-25	0.19	0.31	0.63
Sodium Carbonate	0.08	0.08	0.08
Potassium Persulfate	0.25	0.19	0.19
Anionic:	non-ionic	3:1	
Shelf Life	> 6 month	-	-
$M_W \cdot {}^{-5}$	8.64	8.75	7.58
M_W/M_N	3.6	3.1	3.5
Particle Size, Å	1690	1270	(smallest)

Tubular Reactor Studies. The first run in the tubular reactor was with the same recipe as for Seed I in Table I, but the conversion was very low, and there were two distinct phases. The residence time in the tube was equal to the batch reaction time. Apparently the more nearly constant temperature of the tubular reactor prevented rapid polymerization. In the next run, initiator and emulsifier levels were doubled, but still conversion was low, although phase separation was not so severe. With seed latex and still more emulsifier, Run I shown in Table II, monomer conversions of about 60% were obtained at 50 minutes average residence time in the reactor. No phase separation was evident, but later tests indicated that some phase separation was occurring.

TABLE II

Recipes for Tubular Reactor Studies

	Seed Latex Type								
	I			II			III		
Run	T04	T06	T08	T09	T10	T11	T12	T13	T14
Total Emulsifier (%)	2.2	2.2	2.2	2.2	2.2	2.2	2.2	4.4	6.9
$K_2S_2O_8$ Initiator (%)	0.5	1.0	1.5	1.0	1.0	1.0	1.0	1.0	1.0
Seed Polymer (%)	0.4	0.4	0.4	0.4	1.3	2.6	0.4	0.4	0.4

Percentages based on total water; 0.08 gm Na_2CO_3/100 gm water in each

Water: Vinyl Acetate 2:1 and ionic emulsifier: nonionic emulsifier 3:1

The monomer conversion in the tubular reactor was regularly measured during the run, but it did not always reach a steady state value, even after reaction times equivalent to four residence times. At low conversions, steady state was normally obtained. At exit conversions between approximately 30 and 60%, conversion increased to a maximum at two residence

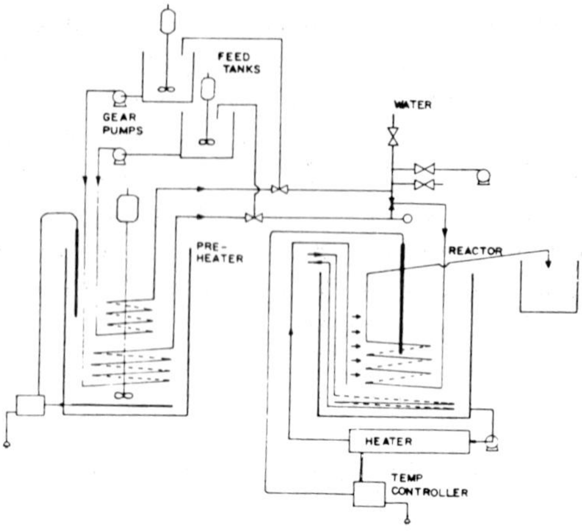

Figure 1. Tubular reactor assembly

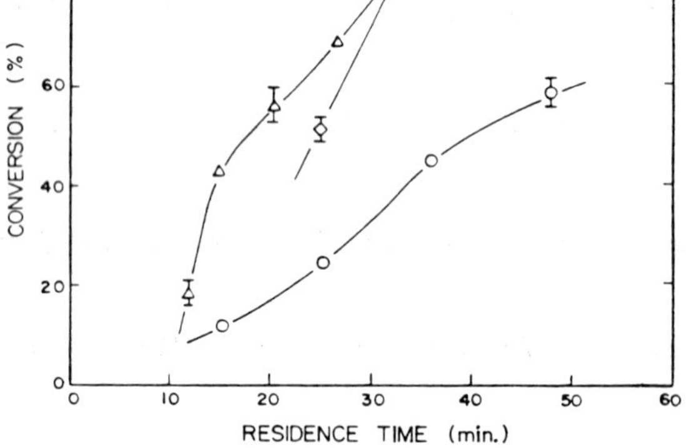

Figure 2. Effect of initiator concentration on conversion rate in the tubular reactor ((I): (○) 0.49%; (◇) 0.98%; (△) 1.47%)

times, then oscillated with a constant frequency and amplitude. This oscillation is due to competing functions of the surfactant. As conversion increases, the particle size increases, and more surfactant is required to cover the surface. As the surfactant is adsorbed on the surface, there are fewer total particles, and the reaction rate decreases. At long residence times, steady states were observed in several runs, with oscillations in other runs. The use of latex seeds and agitation in the emulsion tank reduced the oscillations.

At low conversions of vinyl acetate, much of the ionic and nonionic emulsifier will be within the pools of vinyl acetate. As these pools disappear at higher conversions, the emulsifier becomes available to stabilize the polymer particles.

Conversion Rate. Figures 2-5 show the conversion-time curves as a function of initiator concentration, seed concentration, mixed emulsifier concentration and size of seed latex. The values of conversion, as plotted in Figures 2 to 5, were selected at times equivalent to two residence times, since this was either a steady state value or a consistent maximum value. Where duplicate samples were collected at two residence times, the range of actual values is shown in the plot. It can be seen in Figures 2 and 3 that increased initiator and emulsifier levels increased the rate of reaction. While Figure 2 is consistent with theories of vinyl acetate polymerization, the increased amounts of nonionic emulsifiers in Figure 3 might be expected to retard polymerization, as explained by Netschey (9). Since the negatively charged oligomeric radical is unable to penetrate the micelle border because of the viscous nonionic surfactant and the anionic surfactant, the rate of polymerization is retarded. For the data presented in Figure 3, however, there is additional anionic surfactant (used to colloidally stabilize the emulsion) which is also available to create new micelles and increase the reaction rate.

Figure 4 shows that higher concentrations of seed latex decrease the reaction rate, at constant mixed emulsifier concentrations, while Figure 5 shows that the smallest latex particle gave the lowest reaction rate. This effect is explained by the decreased availability of emulsifier to create micelles, since the small seed latex particle or high seed latex concentrations adsorb more of the surfactant, thus removing it from the water phase.

Napper and Parts (10) have shown that the reaction rate is accelerated when the polymer separates from the water phase

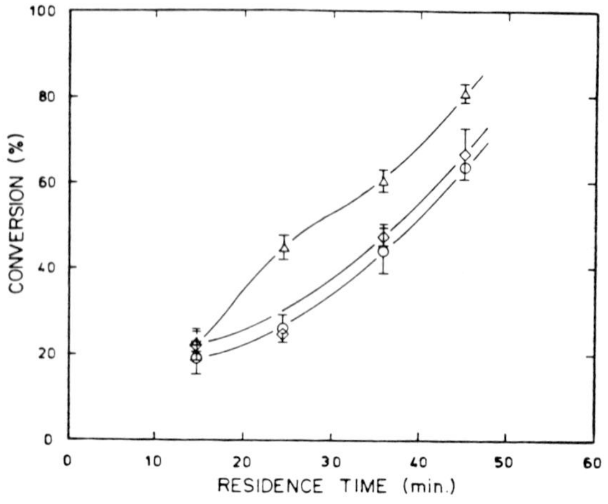

Figure 3. Effect of mixed emulsifier concentration on conversion in tubular reactor ((E): (○) 2.2%; (◇) 4.4%; (△) 6.9%)

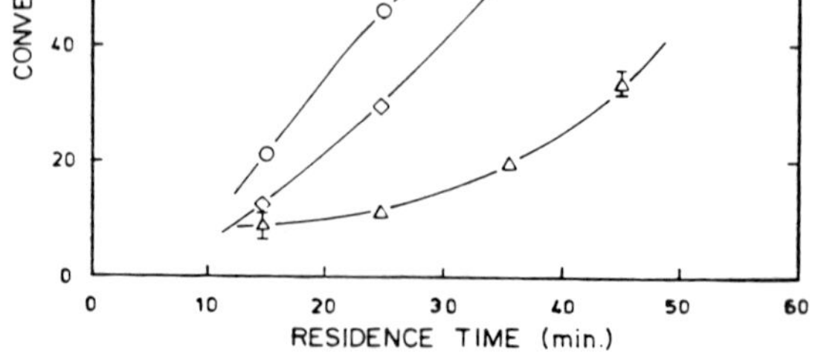

Figure 4. Effect of seed concentration on conversion in tubular reactor ((S): (○) 0.38%; (◇) 1.29%; (△) 2.61%)

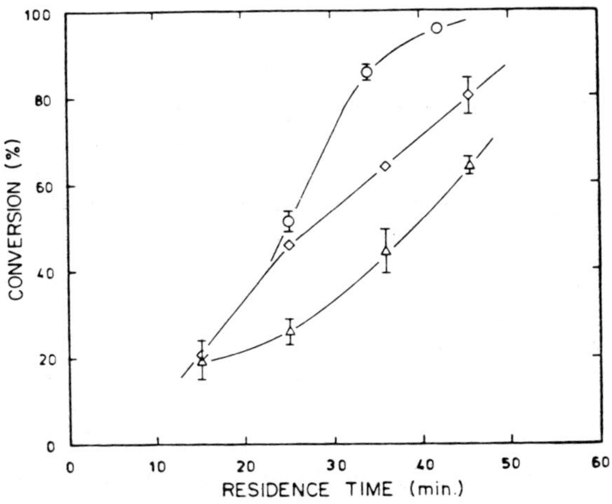

Figure 5. Effect of different seed latexes on conversion in the tubular reactor ((seed): (○) Latex I; (◇) Latex II; (△) Latex III)

to form a polymer particle. This indicates that the principle locus of polymerization is the monomer-swollen polymer particle. For all of the data in Figures 2-5, there are polymer particles (seed) available, so there is no slow retarded polymerization rate, during which the polymer particles are being produced.

Particle Size. Electron microscopy showed a wide, bimodal range of particle sizes. Many particles were about 1000 Å in diameter and were produced by secondary nucleation. There were particles of 10,000 to 30,000 Å, produced by growth of the seed latex particles. Further studies are planned to eliminate the secondary nucleation and produce a narrower range of particles.

Molecular Weight. Molecular weight values were obtained with a GPC unit. This unit was standardized for polystyrene, so the polyvinyl acetate values are not quantitative. The weight average molecular weights were between 500,000 and 900,000. No trends of molecular weight average or distribution could be deduced.

Latex Shelf Life. The shelf life of seed I was greater than 6 months, but none of the latexes produced in the tubular reactor had a shelf life of more than a week. Initially, a thin liquid layer formed at the top. After several weeks, a solid layer formed in the bottom of the container. Despite the high levels of emulsifier used, the large polyvinyl acetate particles separated quickly. No studies were made to increase the shelf life.

CONCLUSIONS

This study of the continuous, tubular, seeded emulsion polymerization of vinyl acetate has led to the following conclusions:

1. Complete monomer conversion is possible with high concentrations of initiator and mixed emulsifier.

2. The use of seed improves the latex stability, but phase separations still occur at low monomer conversions.

LITERATURE CITED

1. Ghosh, M.; Foster, D. W.; Lenczyk, J. P.; and Forsyth, T. H.; "Continuous Polymerization Reactors", edited by T. C. Bouton and D. C. Chappelear, A. I. Ch. E. Symposium, 160, 102 (1976).
2. Wallis, J. P. A.; Ritter, R. A.; and Andre, H.; A. I. Ch. E. J., 21, 686 (1975).
3. Ghosh, M. and Forsyth, T. H., in "Emulsion Polymerization", edited by I. Piirma and J. L. Gardon, ACS Symposium 24, p. 367 (1976).
4. Feldon, M.; McCann, R. F. and Laundrie, R. W., India Rubber World, 128, 51 (1953).
5. Rollin, A. L.; Patterson, I.; Huneault, R. and Rataille, P.; Can. J. Chem. Eng., 55, 565 (1977).
6. Idem, "Polymerization Reactors and Processes Symposium", edited by Henderson and Bouton, ACS Symposium Series, 104, 113 (1979).
7. Dodge, J. S.; Woods, M. E. and Krieger, I. M., J. Paint Technology, 42, 71 (1970).
8. Dunn, A. S. and Taylor, P. A., Die Makromol. Chem., 83, 207 (1965).
9. Ugelstad, J.; Mork, P. C.; Dahl, P. and Rangnes, P., J. Pol. Sci., C-27, 49 (1969.
10. Friis, N. and Nyhagen, L., J. App. Pol. Sci., 19, 97 (1975).
11. Netschey, A.; Napper, D. H. and Alexander, A. E.; J. Pol. Sci. B-7, 829 (1969).
12. Napper, D. H. and Parts, A. G., J. Pol. Sci., 61, 113 (1962).
13. Nomura, M. and Harada, M., in "Emulsion Polymerization", edited by I. Piirma and J. L. Gardon, ACS Symposium 24, p. 102 (1976).

RECEIVED May 14, 1981.

INDEX

INDEX

D

M

N

O

P

R

U

V

Jacket design by Joan Wolbier.
Production by Susan Moses and V. J. DeVeaux

Elements typeset by Service Composition Co., Baltimore, MD.
Printed and bound by The Maple Press Co., York, PA.